A Chorus of Praise for **Harold McGee's**
On Food and Cooking

"I don't think anyone who really loves cooking and eating could pick it up without being fascinated at once."—M. F. K. Fisher

"A unique book on the science of cooking, examining questions like: Are copper bowls really better for preparing egg whites? Terribly important to understanding the nature of fermentation and other invisible processes that go on in the kitchen."—Julia Child

"This is by all odds a minor masterpiece and a welcome addition to any cookbook library."—Mimi Sheraton

"It is a fascinating and indispensable guide."—*Bon Appétit*

"A must-have companion to your cookbook collection."—*Cuisine*

"Brilliant and unputdownable . . . An epoch-making book, a veritable score for cooking."—Egon Ronay, London *Sunday Times*

"McGee has chosen the grand way, and has produced a book which is to be a standard reference in the English-speaking world for decades."
—Alan Davidson

"This compendium of challenging revelation and gastronomic lore is, in sum, an obligatory reference book. . . . It is one of the best books of 1984—and I suspect many years to come."—*Travel & Leisure*

"Good books of food scholarship are rare; this one is to be treasured. . . . It is the kind of book that will bring the food lover back again and again—the way a paella does—in search of another morsel."
—Richard Flaste, *The New York Times Book Review*

"A thoughtful popular treatment of the scientific aspects of food . . . successful in filling one of the crying needs of the field . . . [with] dozens of marvelous quotations from old recipes and food authorities . . . really splendid."—*Kirkus Reviews*

"How refreshing it is to be treated to an inspiring blend of botany, zoology, physiology, anatomy, biochemistry, and, of course, gastronomy. In the chapter on sauces . . . the enthusiasm of the gastronome, the artistic skill of the cook and the critical approach of the scientist shine most brightly. This is a veritable saucier's vade mecum. With the publication of *On Food and Cooking*, gastronomes and scientists alike can rejoice."—*Nature*

"This book may not unlock the secrets of the universe, but it comes close if your heart lies in the kitchen. . . . It will provide considerable thought for food for quite some time."—*Science '85*

"[An] encyclopedia repository of culinary science and lore, a work of scholarship and, clearly, of love."—*The Sciences*

". . . more than a well-written companion to recipe books for curious cooks . . . [It has] great value as an educational device devoted to what many people consider an esoteric branch of human thought—natural science."—*Chemical and Engineering News*

"A fascinating and assiduously researched study . . . the book constitutes a unique opportunity to know what is knowable about food so that we can make informed decisions."—*San Francisco Chronicle*

"The best research on food chemistry and cooking that's been done."
—Los Angeles *Herald-Examiner*

"A superb reference book for the cook and food lover. The 680 pages contain a unique blend of culinary lore and the science behind cooking."
—JeanMarie Brownson, *Chicago Tribune*

"A well-researched volume that makes for fascinating and informative reading."—*Los Angeles Magazine*

"Not only straightforward and clear but [it] is also a pleasure to read. . . . When is the last time you read a book about food that refers, naturally and appropriately, to Aristotle and Augustine, Boswell and Dr. Johnson; or shows you what fun you've been missing by not knowing the home truths about oxidation and decomposition, or the chemistry of Scotch malt whisky?"—*Publishers Weekly*

On Food and Cooking

The Science and Lore of the Kitchen

HAROLD McGEE

SCRIBNER
New York   London   Toronto   Sydney   Singapore

SCRIBNER
1230 Avenue of the Americas
New York, NY 10020

First Scribner trade paperback edition 2003

SCRIBNER and design are trademarks of Macmillan Library
Reference USA, Inc., used under license by Simon & Schuster,
the publisher of this work.

For information about special discounts for bulk purchases,
please contact Simon & Schuster Special Sales at
1-800-456-6798 or business@simonandschuster.com

Manufactured in the United States of America

10   9   8

Library of Congress Cataloging-in-Publication Data
McGee, Harold.
   On food and cooking : the science and lore of the kitchen / Harold
McGee.
   p.   cm.
   Reprint. Originally published : New York : Scribner, 1984.
   Bibliography : p.
   Includes index.
   1. Cookery.   2. Food.   I. Title.
[TX651.M27   1988]
641.5—dc 19   88-19738   CIP

ISBN 0-684-84328-5

To my mother and father

Contents

ACKNOWLEDGMENTS

It is a pleasure to thank the many people who have helped me put this book together.

Professors P. R. Azari, Linda M. Bartoshuk, Doris H. Calloway, Robert E. Feeney, Donald F. Klein, Carl J. Pfeiffer, William D. Powrie, Paul Rozin, and Romeo T. Toledo kindly answered queries about their research. Lloyd R. Stolich of the California Artichoke Advisory Board, Kay Engelhardt of the American Egg Board, Mary P. McGinley of the National Livestock and Meat Association, Ginny Blair of the Popcorn Institute, and Sarah Setton of the Sugar Association, Inc. passed on their institutions' pertinent resources. Wesley Wong, librarian of the Oakes Ames Library of Economic Botany at Harvard, gave me many productive leads and kept the lights on past closing on many a winter evening.

Photographic studies of food structure and microstructure have been generously provided by Drs. John E. Bernardin, Miloslav Kalab, Kenneth Lorenzen, Alex W. MacGregor, and Henry B. Trimbo, and Professors B. L. Dronzek, Ann Hirsch, R. Carl Hoseney, Pauline C. Paul, William D. Powrie, and Alastair T. Pringle. I am especially grateful to Professor Paul for hunting down a dozen elusive photographs, and to Dr. Kalab and Professor Hirsch for producing micrographs expressly for this book. Bern Dibner of the Burndy Library and B. G. Murray of Cadbury Schweppes, Ltd. provided historical illustrations.

I have also profited from the avocational interests of several scientists. John Torrey led a tour of maple sugaring operations near Petersham, Mass. In Palo Alto, Peter Ray gave his illuminating course in "The Natural History of Wines." Craig and Renu Heller took me along to their 1982 harvest and crush of Zinfandel grapes in the Napa Valley. And by setting up and supervising a number of sessions on the spectrophotometer and computer, Winslow Briggs and Joe Berry made it possible to test a theory about the influence of copper on whipped egg whites.

Those good friends who have shared their various expertises, culinary, scientific, or otherwise, have passed on the odd reference or fact, and have provided steady encouragement over the past four years are Fred Ausubel and Stephanie Bird, Welcome Bender, Susan Cohen and Bryan Tyson, Marti Crouch, Richard Drake, Pearl Fles, Louise and Chuck Hammersmith, Jan Harris and Tim Peltason, Christie and Jean-François Le Coz, Harold and Florence Jean Long, Rich and Susan Long, Russ McDuff, Chuck McGee, Michael McGee, Jon and Andrea Miller, Rebecca Newman and Gary Drucker, Scott Poethig, Harvey Rishikof and Miri Abramis, Peter and Lois Smith, Ann and Ray Starr, Ian Sussex, Joan and Richard Thomas, Elissa and Terry Tyler, Rommert van den Bos, Virginia Walbot, Bob West and Karen Reinhard, and Jim Wong.

Individual chapters have profited from the close attention of Corey Goodman, Russ McDuff, and Ann Starr. Jan Harris, Ann McGee, and Peter Smith read better than half the manuscript, and Sharon Long reviewed the whole thing. Their suggestions have greatly improved it, as have the patient care and persistent questioning of Megan Schembre and Kathy Heintzelman at Scribners.

Finally, three special debts. Without the early enthusiasm of Charles Scribner, Jr., for an approach that blends science with history, this project would have been much narrower in scope. My illustrator and sister, Ann McGee, has been an ideal collaborator: prompt in the execution, endlessly tolerant of changes of mind, a rich source of graphic ideas and botanical knowledge. And my wife, Sharon Long, planted the seed of this book when she became curious about the chemistry of cooking and did some browsing in the literature. She has since been chief scientific consultant, house critic, and, as always, boon companion in bitter times and in sweet.

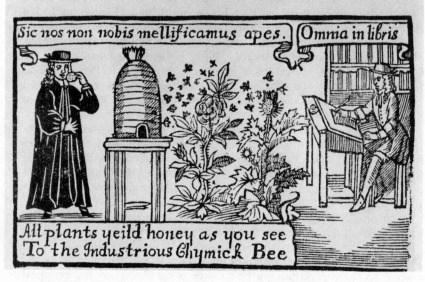

Seventeenth-century woodcut likening the scholar to the alchemist bee, which transmutes the nectar of all plants—prized rose or lowly thistle—into golden honey. The Latin captions read, "Thus we bees make honey, [but] not for ourselves," and "All things in books." The bee does indeed bring about a chemical transformation when it makes nectar into honey. (From the collection of the International Bee Research Association, Gerrards Cross, England.)

INTRODUCTION

Cooking and Science

The Professor to his cook: You are a little opinionated, and I
have had some trouble in making you understand that the
phenomena which take place in your laboratory are nothing
other than the execution of the eternal laws of nature, and that
certain things which you do without thinking, and only because
you have seen others do them, derive nonetheless from the
highest scientific principles.

—Brillat-Savarin, *The Physiology of Taste* (1825)

    Thanks to his witty masterpiece on the appreciation of food, Jean
Anthelme Brillat-Savarin is now one of the immortals in the pantheon of
gastronomy. And yet, more than 150 years after this mock-stern lecture to
his cook, professional opinion remains opposed to his view of the kitchen as
a laboratory. Most writers on food either ignore the scientific principles, high
or low, that underlie cooking, or else disparage the value of such information
on the grounds that art cannot be reduced to the test tube. Experience, intuition, and a discriminating palate, they imply, are what really count. Of
course, there is a great deal of truth in this attitude, as the traditional form
of writing about food attests. Cookbooks are, after all, sets of explicit directions developed by an expert, and are meant to help us prepare food as did
Brillat-Savarin's cook: without the distraction of having to think. On the
other hand, amateurs who have not yet logged years at the stove might well
profit from an explanation of the order that governs this sometimes disor-
derly pursuit. A general understanding of what is going on in the pan,
together with a little thinking, can compensate for a lack of familiarity with
particular ingredients or techniques. And many people, serious cooks or not,
may simply be curious about what foods actually are and how cooking works.

Those who do want to understand the principles of cooking are in good company, for in addition to the tasteful Brillat-Savarin, some eminent philosophical minds have suggested that cooking be looked into more deeply. In Plato's *Gorgias*, Socrates compares it unfavorably with medicine, which, having "investigated the nature of the subject it treats," is a true art. But cookery is a mere "routine" because, "exclusively devoted to cultivating pleasure," it does so

> without once having investigated the nature of pleasure or its cause; and without any pretense whatever and practically no attempt to classify, it preserves by mere experience and routine a memory of what usually happens.

Experience, routine, what *usually* happens: the phrase remains a good description of both the appeal and the limitations of cookbooks. Then two thousand years after Plato, in the course of a particularly inspiring meal, the less abstractly minded Samuel Johnson remarked (according to Boswell) that

> I could write a better book of cookery than has ever been written; it should be a book upon philosophical principles. Pharmacy is now made much more simple. Cookery may be made so too. A prescription which is now compounded of five ingredients, had formerly fifty in it. So in cookery, if the nature of the ingredients be well known, much fewer will do. . . . you shall see what a book of cookery I shall make!

Unfortunately for us all, Johnson did not live to fulfill his intention of demystifying culinary tradition and uncovering its true principles.

Though this is not a book of cookery—it offers no expert recipes—it is meant to supply what Dr. Johnson and Plato found wanting, and what Brillat-Savarin tried in part to write. It explains the nature of our foods, what they are made of and where they came from, how they are transformed by cooking, when and why particular culinary habits took hold. Chemistry and biology figure prominently in this approach, but science is by no means the whole story. History, anthropology, and etymology also contribute to our understanding of food and cooking.

The application of science to cooking was once quite popular. Just a couple of decades after *The Physiology of Taste* appeared, the great German chemist Justus von Liebig—along with Richard Wagner, he was one of the mad King Ludwig's tutors at Munich—wrote a short book which was quickly translated into English under the title *Researches on the Chemistry of Food*. Within ten years, he had revolutionized the way the English and Americans cooked their meat (see chapter 3 for the details). The idea of scientific cookery caught on, and every few years would bring a recipe book that claimed to include the latest advances. One of the last, and the only one still around in some form, was Fannie Merritt Farmer's *Boston Cooking School Cook Book* of 1896. Miss Farmer expressed the wish that her book "may not only be looked upon as a compilation of tried and tested recipes, but that it may awaken an interest through its condensed scientific knowledge which will lead to deeper thought and broader study of what to eat." She then began with 30 pages on the composition of our major foods. In editions that appeared after her death, the wish was preserved but the knowledge was not. By the second major revision, in 1936, the book opened with a chapter on menu making that encouraged "gaiety" and "self-expression" above all else. Biology and chemistry had become more complicated (Liebig's theory about meat turned out to be wrong), their application to cooking less obvious, and cooks more interested in other matters. The science of cooking faded from the scene.

In recent years, however, it has made something of a comeback. Marion Rombauer Becker has expanded the occasional half-page introductions in her mother Irma Rombauer's *Joy of Cooking* into better than 100 pages of lore about ingredients and methods. And some writers, notably Madeleine Kamman and Julia Child, will interject a chemical point or two to elucidate a procedure. But their books are still mainly recipe books, their primary purpose—for which we are all grateful—to present carefully tested, reliable routines. So it's not surprising that technical terms may be used loosely, that factual errors creep in, and we begin to feel that the details are beyond us. Nor is it just a matter of being told that cellulose is a protein, or salt a chemical antagonist to acid. While this book was being written, a well-known food columnist announced that it was quite all right to eat green potatoes. In fact, the green color is usually a sign that toxic substances are present in possibly dangerous amounts.

On Food and Cooking is meant not as an alternative to the recipe books, but as a companion to them. It offers few new techniques, but explains how it is that the traditional techniques work. The most obvious advantage of such knowledge is that it may free the cook from too great a dependence on following directions. No cookbook can foresee and forestall all the emergencies and mistaken turns that an ingenious amateur can invent, and someone who understands the basic nature of the foods and methods at hand is in the best position to recoup and come up with a nutritious and palatable dish.

Nor need a knowledge of protein coagulation diminish the pleasure we feel in contemplating and consuming a well-made roast or meringue. In fact, it may enhance our pleasure if it has helped us to snatch success from the jaws of failure. An understanding of what food is and how cooking works does no violence to the art of cuisine, destroys no delightful mystery. Instead, the mystery expands from matters of expertise and taste to encompass the hidden patterns and wonderful coincidences of nature. How remarkable it is, when you come to think in such terms, that heat has such fortunate effects on the flavor and digestibility of plant and animal tissues, that roast and meringue are two different outcomes of the same process, that wheat proteins have just the right balance of properties to make raised bread possible, that bread, cheese and yogurt, beer and wine are all the result of controlled spoilage! Science can enrich our culinary experience by deepening its significance, by disclosing its connections with the rest of the world.

Reestablishing these connections may be an especially appropriate aim today, when, instead of making our own, we get our flour and bread in sealed bags, our meat in small pieces surrounded by several kinds of plastic, our beer in seamless cans. Most of us have very little idea of what goes into these foods or how they are made. We may be up on Black Holes or the Big Bang or the Primordial Soup, but we can't explain how flour thickens soup or soap cleans the bowls. Our everyday, half-conscious routines may be less spectacular subjects than the heavens or the origins of life, but they can be just as intriguing and of more immediate interest. It is worth remembering that the "father" of the experimental method and so of all modern science, astrophysics and molecular biology alike, died in the service of what now goes under the humble name of food science. In March of 1626, while riding in a coach from his rooms in London to the suburb of Highgate, Sir Francis Bacon decided to take advantage of a fresh snowfall "to try an experiment or two, touching the conservation and induration of bodies," as he later wrote from his sickbed. He stopped at a cottage, bought a chicken, and stuffed it with snow to study the preservative effects of freezing. In the process he caught a chill, and died of bronchitis two weeks later.

Everyday life is worth studying for its own sake, but *our* own sake may also be involved. At the same time that the quality and extent of science education in our public schools is declining, the influence of science on our lives continues to grow, and to raise important issues of public policy. The use of additives in processed foods, the feeding of hormones and antibiotics to livestock, the dependence on genetically uniform crop plants, the intensive cultivation of the soil: these practices have made cheap and abundant food supplies possible, but they also exact certain costs and run certain risks. Similarly, there are both benefits and costs involved when we tap various power sources in our wall sockets, use water freely, or make plastics and other synthetics into eyeglass frames, nonstick pans, and silky lingerie. These and many other unspectacular choices are having or will have significant effects on our world and well-being. The more we are aware of our everyday life, of what we have and do, and why, the more likely we will be to deal productively with the consequences, rather than being content with finding scapegoats. The kitchen is only one arena of the everyday, and this book does not delve much into the wider implications of food technology. But perhaps it will encourage some curiosity about other, more consequential aspects of our material life.

Finally, a few words about the scientific approach to cooking and the organization of this book. Foods, like most everything else on earth, are mixtures of different chemical compounds, and the qualities that we aim to

influence in the kitchen—taste, texture, color, nutritiousness—are all manifestations of chemical properties. So it is to the world of molecules and their
reactions with each other that we must turn.

The prospect may seem a daunting one. There are a hundred-plus
chemical elements, thousands of different compounds in foods—and the
forces that influence them are described by complicated mathematics. Fortunately, only a handful of elements is significant to the behavior of food,
and they tend to be found in only a handful of arrangements: those substances we call water, proteins, carbohydrates, and fats. Their behavior can
be pretty well described with the most elementary principle of electricity—
like charges repel each other, opposite charges attract—and with the idea
that heat is a manifestation of molecular movement.

Part 3 is devoted to just such a description. Chapter 13 gives details
about the makeup, shapes, and typical behavior of the four major food molecules, and chapter 14 analyzes our cooking methods and the materials from
which we make our utensils. Systematic readers may want to begin with this
basic information. But because it is concentrated and somewhat abstract, you
may instead want to work up an appetite for the details by starting out with
the chapters on particular foods, in part 1, or the chapters on our bodies'
recognition and use of food, in part 2. Then you can browse in part 3 to get
better acquainted with molecules and methods, and return to those pages
when some clarification of pH or protein coagulation or fat oxidation or
browning would be helpful.

Most people today have at least a vague idea of what atoms, molecules,
and energy are, and a vague idea will be enough to follow most of the explanations in this book. For those readers who remember little from high school
chemistry beyond broken beakers and singed hair, or who would like to
sharpen their vague ideas of heat or hydrogen bonding, the Appendix will
serve as a quick refresher course in the basic vocabulary of science.

A NOTE ABOUT UNITS OF MEASUREMENT

Throughout the text of this book, temperatures are given in both
degrees Fahrenheit (°F), the standard units in the United States,
and degrees Centigrade (°C), the units used by most other countries and in science. The Fahrenheit temperatures shown in
graphs and charts can be converted to Centigrade by using the
formula °C = (°F − 32) × 0.56.

For many of the photographs taken through the microscope, the
scale is given in microns (μ). 1μ is one millionth of a meter, or
about 40 millionths of an inch.

Part 1

Foods

CHAPTER 1

Milk and Dairy Products

THE NATURE OF MILK

The History of Dairying

What better subject for the first chapter than the food with which we all begin our lives? By our very biological nature as mammals (from the Latin *mamma*, meaning "breast"), we humans take mother's milk as our first food, and have done so from the beginning of our species, a million or more years ago. Less easy to ascertain is when we first drank the milk of *other* mammals with any regularity. This probably came well after the domestication of animals, which were first of all a source of meat and skins. The archaeological evidence suggests that sheep and goats were domesticated in Eurasia around 11,000 and 9,500 years ago respectively, at least a millennium before the larger, much less manageable cattle, which came under human control about 8,500 years ago. The smaller animals would have been much easier to milk, and were probably the first dairy animals. Rock drawings from the Sahara show that dairying was known by 4000 B.C., and what appear to be the remains of cheese have been found in Egyptian tombs dating back to 2300 B.C. In any case, dairy products are a relatively recent addition to the human diet, roughly contemporaneous with such other innovations as bread, beer, and wine.

By the time that the Old Testament began to be set down, roughly 3,000 years ago, milk and its products had become familiar enough to serve as metaphors or analogies for less immediate, more abstract conditions. The Promised Land is described over and over again as a land "flowing with milk and honey": a durable image of plenty. And in a sentence that suggests both

3

creation and violence—the bringing forth of form and substance—Job asks God rhetorically, "Hast thou not poured me out as milk, and curdled me like cheese?" The process of curdling seems to have been especially intriguing to those who pondered the transition from chaos to order. Aristotle used it in *On the Generation of Animals* to explain human conception.

> The male provides the "form" and the "principle of the movement," the female provides the body, in other words the material. Compare the coagulation of milk. Here, milk is the body, and the fig-juice or rennet contains the principle which causes it to set.

A much more recent and wonderfully strange version of the cheese analogy of creation has been uncovered by the Italian scholar Carlo Ginzburg in the records of the Inquisition. According to his study, *The Cheese and the Worms,* a miller named Menocchio, from the town of Friuli, was put to death in 1599 for various heretical views, including this account of the making of earth *and* heaven:

> I have said that, in my opinion, all was chaos, that is, earth, air, water, and fire were mixed together; and out of that bulk a mass formed—just as cheese is made out of milk—and worms appeared in it, and these were angels. The most holy majesty declared that these should be God and the angels. . . .

Still other analogies live on unnoticed in contemporary English. From the Greek for milk, *gala,* came *galaxis,* "milky way," the origin of our "galaxy." And from the Latin *lac* came *lactuca,* ancestor of our "lettuce" (which exudes a milky sap when cut from its roots).

Dairy products were important foods all over early Europe, though preferences varied from region to region. Neither fresh milk nor butter was very popular in Greece or Rome, while cheese was. The reverse was true of northern Europe and Asia. Milk and butter would have spoiled quickly in the Mediterranean climate, which offered the olive as an alternative source of oil. The Greeks and Romans commonly referred to the barbarians as "milk-drinkers" (Greek: *galaktopotes*), so remarkable did this habit seem to them. In the 5th century B.C., the Greek historian Herodotus described the Massagetai, inhabitants of the Caucasus, in this way: "They sow no crops but live on livestock and fish, which they get in abundance from the river Araxes; moreover, they are drinkers of milk." And 500 years later, the Roman Pliny said that butter was considered to be "the most delicate of foods among barbarous nations, and one which distinguishes the wealthy from the multitude at large." He wondered at the fact that "the barbarous nations, who live on milk, do not know or disdain the value of cheese," but also noted that the cheeses most favored at Rome came from the provinces that are now parts of France and Switzerland. By Pliny's account, fresh milk was as much a cosmetic as a food, at least among the ruling classes.

> Milk is valued for giving a part of its whiteness to the skin of women. Poppea, wife of Domitius Nero, took 500 nursing asses everywhere in her travelling party, and soaked herself completely in a bath of this milk, in the belief that it would make her skin more supple.
>
> (Book 11)

Over a thousand years later, another writer from the rim of the Mediterranean reported on the strange uses to which northerners put milk. The Venetian Marco Polo traveled to, from, and in China between 1271 and 1295, and observed the nomadic Tartars as they prepared milk from their mares (horses being more mobile and versatile than cattle). The Tartar armies, wrote Polo,

> can march for ten days together without preparing meat, during which time they subsist upon the blood drawn from their horses, each man opening a vein and drinking from his own. They make provisions also of milk, thickened or dried to the state of a hard paste, which they prepare in the following manner. They boil the milk, and skimming off the rich or creamy part as it rises to the top, put it into a separate vessel as butter; for so long as that remains in the milk, it will not become hard. The milk is then exposed to the sun until it dries. [When it is to be used,] some is put into a bottle with as much water as is thought necessary. By their motion in riding, the contents are violently shaken, and a thin porridge is produced, upon which they make their dinner.

Besides this precursor of powdered milk, the Tartars also enjoyed a dairy product "with the qualities and flavor of white wine." Not living the settled life that makes it possible to brew beer from grain or wine from grapes, they coupled the alcoholic fermentation of yeasts with milk to produce *koumiss*. A similar beverage has been made in the Balkans for many centuries; there it is called *kefir*.

Changes in the handling of milk came very slowly from the Middle Ages through the eighteenth century. Milking, churning, and cheese making, all hard work, were done by hand, and, at least in England, done mostly by women. The word *dairy* was originally *dey-ery*, with *dey* meaning a woman servant in Middle English (in Old English, it meant "kneader," "maker of bread"; "lady" shares this root). In other European languages, the words equivalent to "dairy" have, appropriately, something to do with milk.

The making of cheese, yogurt, and other fermented products was largely uncontrolled, with microbes from the air or left over from the previous batch, whether desirable or not, colonizing the milk. Apparently none of these foods was known in North America until the arrival of the Europeans, and the first dairy herd was not established here until about 1625. While farmers may have enjoyed wholesome milk, city-dwellers generally saw only watered-down, adulterated, disease-carrying milk hauled in open containers through the streets (see Tobias Smollett's description on page 501). With the Industrial Revolution, of course, much changed. Railroads, steam power, and refrigeration made fresher milk available to a larger population. Milking machines and automatic churners appeared in the 1830s, specialized cheese factories in the 1850s, margarine in the 1870s. By the turn of the century, purified bacterial cultures were being used to control the quality of cheese more closely. Today, dairying has split up into several very big businesses, with nothing of the *dey* left about them.

Milk in the Diet

Milk is a white, opaque liquid secreted by female mammals to nourish their newly born offspring. In a sense, it takes on the duties performed by the mother's blood during gestation, and early students of human development and medicine often spoke of these two fluids as related to each other. Milk is an especially valuable food, then, because, like the contents of the chicken egg, it is meant to be the sole sustaining food of the animal during part of its early life. Though nearly all milks contain the same battery of substances—water, proteins, fat, milk sugar, or *lactose*, various vitamins and minerals—the relative proportions of these substances vary greatly from species to species. Generally, animals that grow rapidly are fed with milk high in protein and minerals, and those that need to develop a thick layer of insulating fat (in particular, seals, sea lions, and walruses) receive high-fat, low-sugar or no-sugar milks. A calf doubles its weight at birth in 50 days, a human infant in 100—and cow's milk contains about three times the protein and minerals of mother's milk.

Two other differences between human and bovine milks are worth mentioning. Human milk is more easily digested both because there is less protein in it, and because less of that protein curdles in stomach acid (molecules at the center of the curd are not as easily reached by digestive enzymes as are those that remain in solution). Homogenization, pasteurization, and cooking all cause milk proteins to form weaker, looser curds than they normally do, and so improve the digestibility of cow's milk. Second, human milk alone contains the so-called "bifidus factor," an as-yet-unidentified substance that promotes the growth of *Lactobacillus bifidus*. This harmless bacterium populates the infant's digestive tract and excretes as a waste product lactic acid, which helps to inhibit the growth of other, harmful microbes. This growth factor, together with antibodies from the mother

THE COMPOSITION OF HUMAN AND COW MILKS

	Percentage by Weight	
	Human	Cow
water	87	87
protein	1.1	3.5
fat	3.8	3.7
lactose	6.8	4.9
minerals	0.2	0.7
Of the Proteins:		
casein	40	82
whey	60	18

(Casein proteins are those that coagulate when acid is added to milk; whey proteins remain in solution. See page 11 for a discussion of these proteins.)

against such ingestible pathogens as polio, salmonella, and coliform bacteria, help the baby over the transition from a shielded development in the womb to life in an environment filled with germs.

Lactose Tolerance

In the animal world, humans are exceptional for drinking milk of any kind after they have started eating solid food. In fact, those people who drink milk after infancy are the exception within the human species. This is not simply a matter of tradition or choice. The milk sugar lactose, which accounts for about half of the calories in milk, is a disaccharide; each of its molecules consists of one glucose and one galactose unit joined together. It is found in no other food: in fact, the only sources of lactose currently known aside from the mammal's breast are forsythia flowers and a few tropical shrubs, which produce it only in very small quantities. Now, all multiunit sugars, including table sugar (sucrose), must be broken down into their components by digestive enzymes in the small intestine before they can be absorbed and used by the body. The lactose-breaking enzyme, lactase, reaches its maximum levels in the human intestine shortly after birth, and thereafter declines slowly, with a steady minimum coming between 1½ and 3½ years of age. The logic of this trend is obvious: it is a waste of resources for the body to produce an enzyme when it is no longer needed. Once most mammals are weaned, they never encounter lactose again, and this was also true of humans until the beginnings of animal husbandry. But if substantial amounts of lactose are consumed in the absence of lactase, then the sugar passes through the small intestine without being absorbed and reaches the colon intact. Some colonic bacteria ferment lactose, producing carbon dioxide gas in the process, and thereby cause real distress. The presence of sugar in the colon can also cause water retention, which may be manifested in a bloated feeling or diarrhea.

It was not until the late 1960s that Western medical science realized that adult lactase deficiency—the inability to digest milk sugar—was the rule rather than the exception, and was not really a deficiency at all but the natural state of affairs. It took this long because most Westerners, in particular those of northern European background, are capable of digesting lactose in adulthood. Their lactase levels, and those of a couple of nomadic African tribes, do not drop off as drastically as those of the rest of the world's population. The "ethnic chauvinism," as it has been called, of assuming that everyone could digest milk led to such policies as shipping surplus powdered milk to famine areas where it could do as much harm as good. This attitude was overcome only when researchers noticed that about 70% of American blacks are intolerant of lactose, while the figure for whites is closer to 10%. A few years of study showed that lactose-tolerant adults are a distinct minority on this planet.

This is not to say that only a minority can eat dairy products. Most lactose-intolerant adults can consume about a pint of milk per day, which provides valuable amounts of several nutrients, without severe symptoms. (This is *not* true of those people who are allergic to milk proteins. Lactose intolerance is not an allergy.) And cheese, yogurt, and other cultured foods are practically free of lactose because the fermenting bacteria use it as fuel. But the practice of dairying would have been unlikely to develop in populations with a low tolerance for milk itself. It has been suggested that the genetic trait of continuing lactase production arose in the people of northern Europe because it conferred the advantages of increased calcium intake and improved absorption of that mineral (one of the effects of lactose in the small intestine) on a group whose dark, cold environment developed little vitamin D in the skin. Of course, the ability to drink milk also simply widens the range of foods on which people can survive. In any case, lactose tolerance is probably a very recent adaptation. Mother's milk has been drunk on our branch of the phylogenetic bush for millions of years, but sheep, goats, and cows have been milked only for a few thousand.

The Milking Cycle

A cow must give birth before it will begin to give milk. The mammary glands are activated by hormones produced at the end of pregnancy, and are stimulated to continue secreting milk by regular milkings. It is possible to initiate the secretion of milk by administering hormones to the cow artificially, but this technique is not used; dairy herds do need to be perpetuated. The interval between insemination and birth is 282 days, and even with con-

tinuous milking a cow will dry up about 10 months after calving. The optimum sequence for milk production is to milk a cow for 10 months after it
calves, let it go dry for 2 months, have it calf again, and then repeat the
cycle. In order to meet this schedule, the cow is bred 90 days after calving.

The first fluid secreted by the mammary glands is not milk, but *colostrum*, a clear solution of concentrated protein, vitamins, and antibodies. As
soon as the colostrum flow has ceased and the milk is salable—a matter of a
few days—the calf is put on a diet of reconstituted and soy milks, and the
cow is milked by the farmer two or three times a day to keep the secretory
cells working at full capacity. Many factors can influence the particular balance of nutrients and the taste of the milk, including the breed and age of
the animal, its feed, the stage in its lactation period, the season of the year,
and even odors in the barn.

The Composition of Milk

Both physically and chemically, milk is a complex material. Globules of
milk fat, complexes of protein and salts, and dissolved sugar, vitamins, and
other salts and proteins all swim in the water that accounts for the bulk of
the fluid. The fat and proteins are by far the most important components,
and we will examine them in some detail. A few words only about the
remainder. A wide range of *salts* is found in milk, with sodium, potassium,
calcium, magnesium, chloride, phosphate, sulfate, citrate, and bicarbonate
ions among the more populous. Milk is also slightly acidic, with a pH
between 6.5 and 6.7. Both the salt concentrations and the acidity affect the
behavior of the proteins, as we shall see. All of the recognized *vitamins* are
present in milk, though some, such as vitamin C, come only in tiny quantities. Vitamin A and its chemical precursor carotene are carried in the fat
globules and give milk and butter a yellowish cast; riboflavin, which has a
greenish color, can sometimes be seen in skim milk or in the whey that separates during cheese making. Finally, milk contains *lactose*, a sugar that is
only one tenth as soluble in water as table sugar. As a result of this low solubility, lactose crystals form readily in such products as ice cream and condensed milk, and do not dissolve very quickly on the tongue: they can give
a persistently gritty, "sandy" texture to these foods.

Fat Globules
Milk fat is important nutritionally, aesthetically, and economically. It
carries the fat-soluble vitamins, essential fatty acids, and about half the calories; it contributes to milk's characteristic taste and texture; and the higher
the milk fat content, the more butter or cream can be made from the milk,
and so the higher the price it will bring. Most cows secrete more fat in winter,
more as a particular lactation period gets on, and more at the end of a particular milking than at the beginning. Specialized cells in the mammary

gland release it in the form of globules that are 1 to 5 microns, or about 0.0001 inch, in diameter. This is close to the size of the cells themselves, and in fact each globule brings with it a remnant of its origin: a thin membrane or outer sheath that was part of the cell's membrane as it squeezed the globule out (see page 50 for a view of the globule membrane through the scanning electron microscope). The membrane is composed largely of proteins and phospholipids (phosphate-fatty acid complexes) which stabilize the globules by preventing them from pooling together into one large mass of fat; it also carries most of the vitamin A and cholesterol.

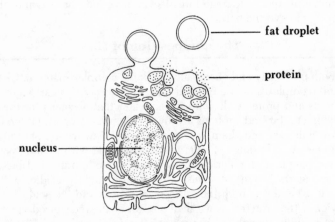

fat droplet

protein

nucleus

A cell in the cow's mammary gland which secretes milk fat and protein. The droplets of fat apparently carry part of the cell membrane along with them.

While the globule membranes may prevent the sheathed fat from coalescing into blobs, they do appear to cause the discrete globules to cluster together. When fresh milk is allowed to stand for a while, the fat globules tend to rise and form a fat-rich layer at the top of the container. This phenomenon is called *creaming*, and it was the means by which cream was obtained from milk until the last century, when centrifuges were developed to do the job much more rapidly and thoroughly. The globules rise, of course, because fat is lighter than water. Given the viscosity of the water phase and the size of the globules, it should be possible to calculate how fast creaming will occur in milk; but the actual rate greatly exceeds the theoretical prediction. It is thought that the membranes are responsible for this, by causing the initial aggregation of globules into larger masses that can rise more rapidly. Few of us see creaming any more because most milk is now homogenized precisely to prevent it from separating into layers.

Two Protein Classes: Curds and Whey

There are a great many different kinds of protein—perhaps dozens—floating around in milk, from those meant to nourish the calf, to the globulins that fight disease organisms, to various enzymes that break down and build up other molecules. One of these is the enzyme that synthesizes lactose from its constituent units. The presence of such biochemical machinery has led to the description of milk as an "unstructured tissue," since even after it leaves the animal's body it retains some of the characteristics of a living system. When it comes to the culinary behavior of milk, we can fortunately reduce the protein population to two basic groups: little Miss Muffet's curds and whey. The two are distinguished by their reactions to acid and rennin, an enzyme from the stomach of a calf that is used to make cheese. The curd protein, *casein*, coagulates and forms solid clumps, while the whey proteins, principal among them *lactoglobulin*, remain suspended in the liquid.

Casein Particles

Take casein first. It is not a single kind of molecule, but consists of several different components that are bound up, together with some calcium and phosphate ions, into bundles about ⅒ micron (a few millionths of an inch) across. These bundles are called *micelles*, and along with the larger fat globules, they give milk its milky, opalescent appearance by deflecting light rays as they pass through the liquid. The major component of casein itself has a subunit that somehow stabilizes that component and keeps it dispersed in separate micelles. When this protein subunit is removed from the casein

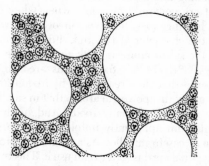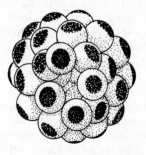

A model of the milk protein casein, which occurs in small bundles a fraction of the size of fat globules *(left)*. A single micelle of casein *(right)* is thought to consist of several kinds of subunits, one of which—the dark patches in this model—prevents the further accumulation of subunits, and so limits the size of the micelle. When this shielding component is disturbed, the casein micelles can clot together, and the milk curdles.

micelles—and this is just what the calf enzyme rennin does—then the micelles react with free calcium ions, which act as bridges between them. The micelles clot together and form the curd that is treated further to make cheese.

Casein proteins will also coagulate to form curds under other conditions, and the extraction of a chemical subunit need not be involved. If enough acid is added to milk—whether in the form of fruit or vegetable tissue, or the lactic acid produced by certain bacteria—curdling will occur. For all proteins there is a particular range of surrounding acidity that will cause them to aggregate, while outside this range the molecules have a net electrical charge that repels them from each other. The pH of milk is normally about 6.5, and if it is lowered to about 5.3, the casein micelles lose their negative charge and clump together. Added salt will have the same effect since it releases positive and negative ions into the solution and changes the electrical environment, but the threshold level for curdling is unbearably salty to the palate. Milk is normally resistant to heat curdling, but small amounts of acid or salt can sensitize the casein fraction and cause it to clump when cooked.

Apart from its function as a food, casein has been exploited for its adhesive properties in the manufacture of paper, plastics, and glue.

Whey Proteins

In contrast to casein, the whey proteins, and especially the antibodies, are quite resistant to denaturation and curdling, perhaps because it is important that they remain intact in the gut to do their protective work. It takes sustained temperatures near boiling and a pH of about 4.6 to coagulate the whey proteins, conditions that are generally met only when whey cheeses are being made. Less drastic heating does, however, involve the whey in important changes that affect the character of the milk. For one thing, the lactoglobulin contains most of the sulfur atoms in milk along its side chains. When it is heated to about 165°F (74°C), these atoms are sufficiently exposed to the surrounding liquid that they react with hydrogen ions and form hydrogen sulfide (H_2S), a gas with a powerful aroma that in small quantities contributes to the cooked aroma of many foods. And in the same temperature range, the lactoglobulin apparently unfolds into a form that interferes with the aggregation of casein micelles. As a result, the process of clotting is slowed down, and the final curd is softer than usual. This turns out to be one of the problems with using pasteurized milk in cheese making, but is an advantage in making yogurt, as we shall see.

The Flavor of Milk

The flavor of milk is normally rather bland—or subtle—and alterations in it are easily noticed and quickly off-putting. The major contributors to the flavor of fresh milk are the slightly sweet lactose, the salts, and the light odors of short-chain fatty acids and sulfur compounds. Undesirable off-flavors can be caused by several different conditions. They may be native to the milk because the cow breathed foul air or ate tainted feed. The growth of bacteria that excrete lactic acid will sour milk, while the by-products of other microbes can give it stale, bitter, or putrid flavors.

UNFERMENTED DAIRY PRODUCTS

Milks

Milk is sold in a variety of forms today, but very few of us ever see raw milk any more. Ordinary drinking milk is routinely subjected to pasteurization, homogenization, and vitamin fortification, and condensed products to even more drastic treatments.

Pasteurization

Most milk sold for direct human consumption has been pasteurized, or heated hot and long enough to destroy all disease-causing organisms and most others as well. Since it is by design a very nutritious substance, milk is quite hospitable to microbes, and is easily contaminated by contact with bovine or human skin and with milking equipment. Tainted milk has been known to transmit such serious diseases as tuberculosis and undulant fever to people. In the 1820s, long before the germ theory of disease had been proposed, some books on domestic economy advocated boiling all milk before using it. The great French scientist Louis Pasteur studied the spoilage of wine and beer in the 1860s and developed a heat treatment that preserved these fluids without greatly injuring their flavor. Pasteurization was not applied to milk in the United States until around the turn of the century, but by the 1940s individual states had begun to require it. Nowadays, with a system of distribution that includes handling large quantities at once, shipping fairly long distances, and stocking in supermarkets, pasteurization is a practical necessity. It extends the shelf life of milk not only by killing microbes, but also by inactivating enzymes native to milk, especially the fat-splitters, whose slow but steady activity can make it unpalatable.

There are many different combinations of temperature and time that pasteurize milk, but a few are by far the most commonly employed. One standard method is to heat the milk to 144°F (62°C) and hold it there for 30 minutes; another keeps it at 160°F (71°C) for 15 seconds. The first has the advantage of staying well below the temperature at which a cooked flavor develops, and while the second flirts with this limit, it is much faster. Pasteurization does not normally liberate the strongly aromatic gas hydrogen sulfide from the protein lactoglobulin, but it does change the flavor slightly by evaporating away some volatile molecules and creating new ones. Ultra-pasteurization, in which a temperature of 280°F (138°C) is held for one second, is a more severe treatment that does leave behind a cooked flavor. It is normally applied only to cream, which we tend to use more slowly and keep longer than we do milk.

Homogenization

This treatment, whose name comes from the Greek for "of the same kind," involves forcing the milk at high pressure through a very small nozzle onto a hard surface; it breaks the fat globules up into more uniform particles about a quarter of their original size. Homogenization was developed in France around 1900 to prevent creaming, because once the fat globules have aggregated and risen, it is difficult to redisperse them evenly in the milk. When broken down to about a micron in diameter, the individual globules are too small to rise alone, and because their surface area has multiplied beyond the covering capacity of their membranes, some of the other dissolved milk proteins fill in, and apparently interfere with globule aggregation. As a result, the fat remains evenly dispersed in the milk. Homogenized milk is whiter, blander, less stable to heat, and more sensitive to spoilage by light than unhomogenized milk, and the process can also have undesirable effects on the whipping of cream and on cheese making, as we shall see.

Fresh milk is never homogenized as is, because it will go rancid in a matter of minutes. When stripped of some of its protective membrane, the fat is exposed to the activity of fat-splitting enzymes in the milk, and these quickly produce unappetizing quantities of odiferous free fatty acids. The enzymes are inactivated by high temperatures, and accordingly all milk is pasteurized before or simultaneously with homogenization.

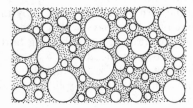 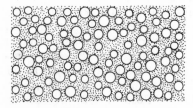

Homogenization decreases the size of the larger fat globules in milk, and thereby prevents them from rising to form a separate layer of cream at the top of the bottle.

Nutritional Alteration

In addition to being pasteurized and homogenized, most milk is fortified with the fat-soluble vitamins A and D; it is a good source for these vitamins, riboflavin, calcium and phosphorus, and high-quality protein. The diversity that we see in the dairy case comes with a number of further alterations in milk composition. Low-fat milks are made by centrifuging some of the globules off before homogenization. Whole milk is about 4% fat, low-calorie milks 1 or 2% ("99% fat-free" doesn't mean that 99% of the fat has been removed, but only that the percentage has been lowered from 4 to 1), and skim milk between 0.1% and 0.2%. Additional milk solids can be used to fortify low-fat milks with more protein and restore some of the body that is lost when the fat is removed. "Acidophilus" milk is designed for people with lactose intolerance; it has been cultured with *Lactobacillus acidophilus*, a bacterium that consumes the lactose and produces lactic acid in the process, and sometimes has a sour taste.

Storage and Cooking

Milk is a highly perishable food. Even Grade A pasteurized milk contains tens of millions of bacteria in every half-gallon, and will sour quickly unless refrigerated. A temperature just above freezing would be ideal, though few home refrigerators are set that low.

It is less well known that the flavor of milk can be altered by the electrical energy carried in ordinary daylight. In the chemical reaction called "autoxidation," light energizes an oxygen atom, which then invades the long, regular chains of carbon and hydrogen atoms in the fats and disrupts their structure (see page 603). Fragments of the hydrocarbon chain split off and react with each other to produce various small, volatile, and odorous molecules; oily, fishy, metallic, and other unpleasant overtones accumulate. Indirect daylight is intense enough to fuel this deterioration. Direct sunlight, with its far greater intensity, causes not only autoxidation, but "sunlight flavor" as well. This burnt or cabbagelike flavor is the result of a reaction between an amino acid, methionine, and the vitamin riboflavin, which is destroyed in the process. So for both nutritional and gustatory reasons, clear glass or plastic containers of milk should be kept in the dark as much as possible.

Whether fluid milk is used to make a soup or a sauce, scalloped potatoes or hot chocolate, the tendency of its proteins to coagulate can cause problems. The skin that forms on the surface of boiled milk or cream soups is a complex of casein and calcium, and results from evaporation of water at the surface and the subsequent concentration of protein there. If you skim off the skin, you remove significant amounts of valuable nutrients. Skin formation can be minimized by covering the pan or whipping up a little foam; both actions slow evaporation. Milk scorches easily because relatively dense complexes of casein micelles and whey proteins fall to the pan bottom, stick, and burn. A moderate flame or a double boiler is the best safeguard against scorching. Large foreign molecules—starch, vegetable or fruit cellulose and hemicelluloses, sugars, fats—can also cause some general coagulation, and so a curdled appearance, by providing sites at which the milk proteins can gather and begin to clump. And acid in the juices of all fruits and vegetables, and phenolic compounds in such things as potatoes and tea, make the milk proteins more sensitive to coagulation and curdling at cooking temperatures. Fresh milk and careful control of the burner are the best weapons against curdling.

Condensed Milks

Condensed milk products are useful because they supply milk fat and solids in concentrated form, and are treated to keep for at least several months. Ordinary condensed or evaporated milk is made by rapidly evaporating off about half its water, not by heating it, but by putting it under a vacuum. The resulting liquid is then sterilized so that it will keep indefinitely, and then homogenized. The high temperature reached in sterilization, together with the greater concentration of lactose, causes that sugar to undergo some browning, and this gives evaporated milk its characteristic note of caramel. In *sweetened condensed milk*, the caramel flavor is replaced by more intense sweetness. Sugar is added in the amount of about 2 pounds per 10 pounds of condensed milk, a level that makes the liquid uninhabitable for microbes by raising its osmotic pressure (see page 170). Accordingly, sweetened condensed milk need not be sterilized. *Powdered* or *dried milk*, of course, is the result of taking evaporation to the extreme. With all but a tiny fraction of the original water removed, it is impossible for microbes to grow on it. Most powdered milk is made from low-fat milk because the fat quickly goes rancid in contact with solid salts and plenty of atmospheric oxygen, and because it tends to coat the particles of protein and so makes subsequent mixing with water difficult. Powdered milk will keep for several months in dry, cool surroundings. It is primarily used in and best suited to batters and doughs for baked goods, where the lack of water is an advantage and the lack of flavor no great disadvantage. Without any fat or volatile aroma molecules left, dried milk is not very appealing when mixed with water and sampled straight.

Cream

The word *cream* comes from the Greek *chriein*, which means "to anoint," and which is also the root of *Christ* ("the anointed one"). The link between ancient ritual and rich food is oil, the substance used to anoint the chosen, and the defining element of cream. Cream is a form of milk in which the fat globules have become more concentrated than usual, whether by rising to the top in a bottle or spinning off from the heavier water phase in a centrifuge. There are three grades of cream marketed today: light cream is between 18 and 30% butterfat, light whipping cream between 30 and 36%, and heavy whipping cream between 36 and 40%. Whole milk, by contrast, is closer to 4% fat. "Half-and-half" is, as its name suggests, intermediate in composition between milk and cream; it must be at least 10.5% fat. Cream

A scene from Bartolomeo
Scappi's *Works* (1570), showing
the making of whipped cream
and butter in 16th-century Italy

is chiefly valued for its thick, smooth texture and rich taste, and in some ways
it is a handier cooking ingredient than milk. Because its proteins have been
greatly diluted by fat globules, it is less likely to form a skin when heated or
even boiled down for a sauce, and it is fairly immune to curdling in the
presence of acidic or salty foods. Perhaps most important, cream can be
whipped into a stable foam.

Whipped Cream

On the other hand, cream is infamous for its seeming fickleness when
whipped. Sometimes it doubles its original volume, rising in soft, light, long-
lived peaks, and other times it stays stubbornly liquid, even to the point of
turning into a mixture of buttermilk and butter. It has its reasons, and these
have to do with the nature of foam reinforcement in cream. Like whipped
egg whites, whipped cream is a foam of air and water that is stabilized by
the proteins contained in the liquid. When the beating action introduces air
bubbles into the liquid, some proteins are caught in the walls of the bubbles,
and because of the imbalance of forces there, the molecules are distorted
from their normal shape; they react with each other, and form a thin film of
coagulated molecules (for more detail, see the discussion of beating egg
whites in chapter 2). This film gives the liquid foam a solid if very delicate
reinforcement, and prevents it from collapsing under the force of gravity—
for a few seconds, or indefinitely, depending on the materials. Milk foams

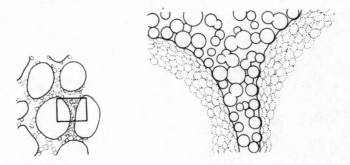

When cream is whipped, the fat globules are numerous enough to surround the air bubbles and give the foam solid reinforcement.

are very unstable and collapse almost immediately after they form, because the milk proteins simply don't unfold and coagulate very much, and because the liquid is so thin that it easily runs out of the bubble walls back into a large pool. Egg white, by contrast, is very viscous and slow-moving, and some of its proteins readily denature and coagulate when foamed.

The fact that cream can succeed where milk fails clearly has to do with the greater concentration of fat globules. For one thing, that concentration has a noticeable effect on the viscosity of the liquid: cream flows less rapidly than milk. More important is the globules' activity. They apparently cluster together in the bubble walls, where surface forces rupture some of their membranes. The exposed spheres of soft fat then stick to each other and form a rigid but delicate network that the milk proteins alone cannot provide. In milk, the globules are too few and far between to do the job, while in cream, beginning at about 20% fat content, their number is adequate to support the foam. Milk also has the disadvantage of being homogenized. Whey proteins leave the solution to fill in the gaps of the globules' suddenly larger surface area, and this leaves less to stabilize the interface between water and air. In addition, the initial clustering of globules is probably disturbed by the change in their covering membranes. For precisely these reasons, whipping cream is not homogenized. Pasteurization, which is generally required for creams as well as milk, has only a slightly detrimental effect on whipping.

Cookbooks commonly advise the cook to store bowl and beaters in the freezer before trying to whip cream, while egg whites are said to whip best at room temperature or even slightly above it. There is nothing mysterious about this divergence, and too high a temperature is frequently the cause of failure to whip cream into a good foam. We all know that the properties of milk fat change drastically within the range of ordinary kitchen temperatures. Butter is stiff, even brittle, right out of the refrigerator, but spreads easily at room temperature, and liquefies completely on hot summer afternoons. Milk fat actually melts at around 90°F (32°C), but it gets soft long before that point. Now, if the fat globules lining the bubble walls are too soft, they will be deformed by the weight of the foam, and the whole structure will be weakened. And if even a small amount of fat escapes as liquid from a globule (the equivalent of a drop of egg yolk in whipped whites), it will interfere with the ordered system of water, unfolded protein, and air, and a stable foam will not form in the first place. It also appears that globules cluster more readily at low temperatures. And the fluid as a whole is much more viscous when cold than when warm, so that it is slower to drain from the foam. All in all, then, the colder the cream and beating utensils, the better, especially considering the fact that both the activity of beating and the incorporation of room-temperature air will heat the cream up.

Cooling the cream in the freezer before whipping can be a help, especially in the summer, but be careful not to let it begin to freeze. The water leaves the solution to form ice, and this segregation of phases makes an even redispersion of fat, and so a good foam, difficult to achieve afterward. A temperature of 45°F (7°C) or lower is recommended; above about 70°F (21°C), even heavy cream is too thin and its globules too soft to make a stable foam. Light (30%) whipping cream is considered ideal for making foams; heavy cream more readily turns lumpy and buttery. It also appears that larger globules produce stiffer foams than small ones; if you happen to have a choice of cows, be advised that Jerseys and Guernseys give milk with the largest average globule size (the other common dairy breed, the Holstein, is the more copious producer). Vegetable gums or gelatin are sometimes added to cream to improve its foaming properties (they make it more viscous and help stabilize the bubble walls); if you feel that such aid would compromise your achievements, check carton labels carefully. Sugar will decrease the final volume and stability of whipped cream when added at the beginning of the process, probably by interfering with the clumping proteins on the globule membranes. These effects are diminished by sweetening the cream after the whipping is mostly done. As is the case with egg whites, the point at which a cream foam is most stable does not coincide with its greatest volume. If you want whipped cream that will last for a while, stop beating it when it seems the stiffest, before it begins to turn soft and glossy.

Butter and Margarine

A sure sign of failure in whipping cream is the appearance of small lumps in the liquid. These lumps are butter, and once they have formed there is no longer any chance of getting a good volume of stable foam, even if everything is put back into the refrigerator to cool down. In fact, butter has always been made by a process much like whipping, though generally it is less delicate, and *churning* is usually the word applied. Cream is removed from the whole milk, then agitated, and eventually butter granules form, grow larger, and coalesce. In the end, there are two phases left: a semisolid mass of butter, and the liquid left over, which is the *buttermilk*.

Exactly how churning works is still unknown. Current theory runs along these lines: just as happens in whipped cream, some air is incorporated into the liquid, bubbles form, and the fat globules collect in the bubble walls. But where whipping cream is kept cold, and the agitation stopped when a a stable, airy foam is produced, churned cream is warmed to the point that the globules soften and to some degree liquify. The ideal temperature range is said to be 55° to 65°F (12° to 18°C). Persistent agitation knocks the softened globules into each other enough to break through the protective membrane, and liquid fat cements the exposed droplets together. The foam structure is broken both by the free fat and the released membrane materials, which include emulsifiers like lecithin. These materials disrupt thin water layers and so burst bubble walls, and once enough of them have been freed in the process of whipping or churning cream, the foam will never be stable again. As churning continues, then, the foam gradually subsides, and the butter granules are worked together into larger and larger masses.

The process of butter making can be described as an inversion of the original cream emulsion. The system of fat droplets dispersed in water is converted into a continuous phase of fat that contains water droplets. The final product is about 80% milk fat, 18% water, and 2% milk solids, mainly proteins and salts carried in the water. The physical structure of butter is, however, a bit more complicated. The continuous, amorphous phase of solid fat surrounds not only the water droplets, but also air bubbles, intact fat globules, and highly ordered crystals of milk fat that have grown during the cooling process. The proportion of continuous or "free" fat can vary from 50% of the total to nearly 100%, and it has a direct influence on the behavior of butter. The more fat there is in discrete globules or crystals, the harder and more crumbly the butter, even to the point of brittleness. A preponderance

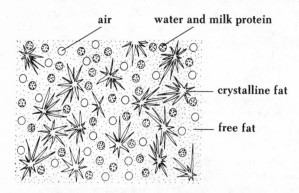

The structure of butter. Milk fat that solidifies into highly ordered crystals imparts a crumbly, stiff texture to butter, while the random arrangement of the solid but noncrystalline, "free" phase makes for smoothness and a tendency to melt readily.

of free fat, on the other hand, makes for a malleable butter that softens readily and may even weep some liquid fat in the process. The difference is a matter of both large-scale and molecular arrangements. In a mass where the free fat merely fills the small interstices between globules and crystals, the texture will be largely that of the separate particles. And it takes more energy to separate the molecules ordered in a crystal than it does to disrupt an already disordered phase of the same molecules. Mostly crystalline butter, then, will be relatively stiff and not as smooth as mostly amorphous butter. The ideal, of course, lies somewhere between the two extremes, and is attained by manipulating the cooling process (much as one controls the texture of candy).

Clarified Butter

Butter is frequently heated in a saucepan and then used either as one of several ingredients in a dish, or as a medium in which to fry other foods. When the temperature reaches the boiling point of water, the dispersed water droplets vaporize and bubble off the melted fat; this is what causes the butter to sizzle. And as the temperature continues to rise, the small amount of initially white sediment—the milk proteins and salts—will turn brown and eventually burn, thereby imparting a harsh flavor to the butter and to delicate foods that may be cooked in it. To avoid this, or simply to improve the appearance of a melted butter sauce, the cook "clarifies" butter by skimming off the froth, which contains whey proteins, and then carefully pouring the melted fat off the white sediment of casein and salts. In the Indian version of clarified butter, *ghee*, the solids are allowed to brown before being removed, and this imparts a nutty flavor to the fat.

Storage

These days, any deterioration in the flavor of butter is probably due not to bacterial spoilage, but to oxidation of the fats and the liberation of odorous short-chain fatty acids. In very small concentrations, these molecules give milk and butter characteristic notes of flavor, but in larger amounts, they are disgusting. This chemical process is slowed down at low temperatures, and butter will keep for months in the freezer, provided it is wrapped air-tight and cannot pick up odors from the appliance. Before the advent of refrigeration, bacteria were also a problem, and this is where the practice of salting butter originated: salt was added in sufficient amounts to act as a preservative. Today, it is used primarily as a flavoring.

Margarine: A French Pearl

Margarine, a butter substitute made originally from other animal fats, but nowadays exclusively from vegetable oils, is, like homogenization and pasteurization, a French innovation. It was developed in 1869 by a pharmacist and chemist, Hippolyte Mège-Mouriés, after Napoleon III offered a prize for the formulation of a synthetic edible fat. Western Europe was running low on fats and oils; petroleum hydrocarbons were as yet unexploited, and the growing industrial need for lubricants and the popular demand for soaps (caused by a rising standard of living and interest in hygiene) were cutting into vegetable sources.

The name *margarine* comes from a minor scientific error. Michel Chevreul, a chemist whose investigations of color influenced the painter Seurat, also worked on natural fats early in his career, isolating and naming many fatty acids and establishing a model of analytic research in the heretofore rather casual field of organic chemistry. In 1813, Chevreul isolated a substance from animal fat that formed pearly drops, and, thinking it to be a new fatty acid, he named it margaric acid, from the Greek for "pearl" (*margaron*, also the root of "Margaret"). As it later turned out, there was no such thing as margaric acid (a *synthetic* fatty acid has since been given that name), but Mège-Mouriés used an extract of animal fat that supposedly contained a great deal of it, and so gave his concoction the name *margarine*. He was looking for a butter substitute, and so of course had to use animal fats, which are semisolid at room temperature. Mège-Mouriés was not the first to give suet a buttery texture, but he was the first to make it palatable by flavoring it with a small amount of milk. It was not until 1905, after French and German chemists had developed the process of hydrogenation for hardening normally liquid vegetable oils (see page 604), that these oils could be made into a butter substitute.

Margarine caught on quickly in both Europe and the United States, where patents began pouring out in 1871, and large-scale production was under way by 1880. At the turn of the century, Mark Twain overheard a conversation between two businessmen aboard the Cincinnati riverboat, and recorded it in *Life on the Mississippi*.

> Why, we are turning out oleomargarine *now*, by the thousands
> of tons. And we can sell it so dirt-cheap that the whole country
> has *got* to take it—can't get around it, you see. Butter don't stand
> any show—there ain't any chance for competition.

Little did this enthusiast suspect what resistance margarine would meet from the dairy industry and from government. First it was defined as a "harmful drug" and its sale restricted. Then it was heavily taxed, stores had to be licensed to sell it, and, like alcohol and tobacco, it was bootlegged. The government refused to purchase it for use in the armed forces. And, in an attempt to hold it to its true colors, some states did not allow margarine to be dyed yellow (animal fats and vegetable oils are much paler than butter); the dye was sold separately and mixed in by the consumer. Two world wars, which brought butter rationing, probably did the most to establish margarine's respectability. But it was not until 1950 that the federal taxes on margarine were abolished, and not until 1967 that yellow margarine could be sold in Wisconsin. Today, we consume nearly three times as much margarine as we do butter. Both price and the current concern about cardiovascular disease are responsible for this differential. Margarine, once far cheaper than butter, is still marginally so, and contains none of the cholesterol and less of the saturated fats that have been implicated in heart disease. (A fat's hardness at a given temperature is an index of its saturation; the proportion of saturated fats in liquid oil, tub margarine, stick margarine, and butter increases in that order. See page 603 for an explanation of saturation.)

Like its model, margarine is about 80% fat, 20% water and solids. It is flavored, colored, and fortified with vitamin A and sometimes D to match butter's nutritional contribution. A single oil or a blend may be used. During World War I, coconut oil was favored; in the thirties, it was cottonseed, and in the fifties, soy. Today, soy and corn oils predominate. The raw oil is pressed from the seeds, purified, hydrogenated, and then fortified and colored, either with a synthetic carotene or with annatto, a pigment extracted from a tropical seed. The water phase is usually reconstituted or skim milk that is cultured with lactic bacteria to produce a stronger flavor, although pure diacetyl, the compound most responsible for the flavor of butter, is also used. Emulsifiers such as lecithin help disperse the water phase evenly throughout the oil, and salt and preservatives are also commonly added. The mixture of oil and water is then heated, blended, and cooled. The softer tub margarines are made with less hydrogenated, more liquid oils than go into stick margarines.

Ice Cream

Though ice cream seems to be one of those quintessentially American foods, it is at least a century older than this country. Exactly where and how it got its start is unclear. It may have begun as a way of preserving milk, but pleasure soon became the dominant motive. Some historians give credit to Catherine de Medici and her Florentine cooks, who arrived at the French court in the middle of the 16th century. Fruit ices were enjoyed there soon after, and Charles I, who reigned early in the 17th century, is the first on record to have served "cream ice." A hundred years later, ice cream was a standard item in middle-class English cookbooks. Here, for example, is the popular Hannah Glasse's recipe, from her *Compleat Confectioner*, about 1760. (She plays loose with pronouns.)

> *To make ice cream.* Take two pewter basons, one larger than the other; the inward one must have a close cover, into which you put your cream, and mix it with what you think proper, to give it a flavour and colour, as raspberries, etc. then sweeten it to your palate, cover it close, and set it in the larger bason; fill it with ice,

and a large handful of salt under and over and round about; let
it stand in the ice three quarters of an hour, uncover, and stir it
and the cream well together, then cover it again; let it stand half
an hour longer, and then turn it into your plate. . . .

In the American colonies at about the same time, newspaper ads gave notice
of ice cream for sale, and it was frequently served at dinner by prominent
citizens.

If America did not invent ice cream, it certainly did pioneer in its
refinement, and—no surprise—in its commercial development. Hannah
Glasse's ice cream, like modern "freezer" versions, would have been relatively dense, coarse, and crystalline because it did not involve the continuous
mixing of ingredients as they froze. Other early directions called for the
inner bowl to be rocked or shaken while in the brine, but this would have
been an awkward, messy business. The problem received its classic solution
in 1846, when one Nancy Johnson, an American about whom little else is
known, invented the hand-cranked freezer that is still used in many homes
today. This design, which was patented two years later by William G. Young,
employed a simple and steady mechanical action to keep the mix moving,
thereby cooling it evenly, preventing the growth of large ice crystals, and
incorporating some air. The second fateful advance came in 1851 when
Jacob Fussell, a Baltimore milk dealer, decided to use up his surplus milk by
freezing it into ice cream, and thereby became its first large-scale
manufacturer.

From this point on, the country's considerable creative forces went into
action. The following is just a partial honor roll of milestones in the progress
of ice cream. In 1874, it was substituted for cream in sodas, a concoction that
had been around since the thirties. In the nineties, Midwestern laws against
Sunday sales of ice-cream sodas incited the invention of the sodaless, legal
"sundae." At the St. Louis World Exposition of 1904, an ice cream vendor
ran out of dishes, substituted the wafers sold at a nearby stand, and thus
invented the World's Fair Cornucopia, or ice cream cone. And in the few
years between 1919 and 1924, three immortals were conceived: the Eskimo
Pie, ice cream in a chocolate shell; the Good Humor Bar, ice cream on a
stick; and the Popsicle, originally lemonade on a stick. Nothing developed
since has matched the popularity of these frozen novelties.

The privations of World War II seem to have whetted the American
appetite for ice cream. When asked what they would do first at war's end,
soldiers and civilians alike spoke of gorging themselves on ice cream. On
March 13, 1943, the *New York Times* reported that American fliers stationed
in Britain had come upon a most efficient way of making their own in the
line of duty. In a story titled "Flying Fortresses Double as Ice-Cream Freezers," it was disclosed that the airmen "place prepared ice-cream mixture in
a large can and anchor it to the rear gunner's compartment of a Flying For-
tress. It is well shaken up and nicely frozen by flying over enemy territory
at high altitudes."

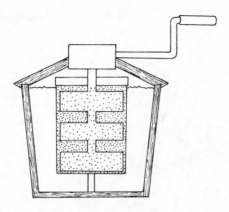

The classic hand-cranked ice cream freezer. The container of mix is surrounded by ice and brine. Continual stirring of the mix produces an even, smooth texture.

The all-time record for per capita ice cream consumption in this country, something over 20 quarts, was set in 1946. Since then, sherbets and ice milks have cut into ice cream's share, which now accounts for about 15 quarts out of 24 for all frozen desserts. The postwar years brought home refrigeration, carry-out and drive-in services, and soft ice cream. And with the increasing use of emulsifiers, stabilizers, and other additives that can compromise the quality of foods, a hierarchy developed. At the top was traditional but relatively expensive ice cream; at the bottom, a cheaper product made with inferior ingredients (dry milk solids, for example), filled with additives to cut production costs, and pumped up with air. With the 1960s came a craze for different, often bizarre flavors, and the 1970s saw a revival of interest in premium quality, no matter what the price.

The Composition of Ice Cream
Ice cream is a foam that is stabilized by freezing much of the liquid. When examined under a microscope, it reveals four phases. Even at freezer temperatures, there is some liquid left, containing dissolved salts, sugars, and suspended milk proteins. There are tiny ice crystals, composed of pure water, and solid globules of milk fat. Finally, there are air cells, which should be very small, perhaps double the size of the ice crystals, or 0.1 millimeter (a few thousandths of an inch across). Sometimes a fifth, undesirable phase is also present: crystals of lactose, which give a gritty texture to the whole.

Each phase makes its own contribution to the character of ice cream. The liquid (*kept* liquid primarily by the dissolved sugar, which lowers the freezing point) prevents it from being a solid block of ice, while the substances it carries bring flavor, nutritional value, and, in the case of the proteins, body and foaming aid. The solid ice crystals stabilize the foam by

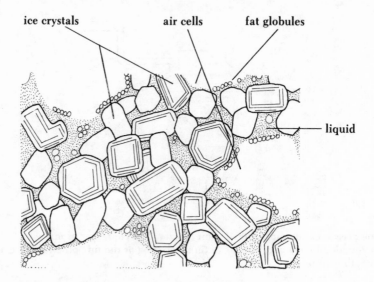

The structure of ice cream. It is a foam in which air bubbles are trapped by freezing much, but not all, of the liquid phase. Both sugar and milk solids are dissolved in the liquid.

trapping the other phases in their interstices, and the fat provides richness, smoothness, and body. The air cells, though nothing in themselves, perform the important service of interrupting the solid and liquid phases, and so making the whole mixture softer and lighter. Ice cream without air cells is very difficult to scoop and bite into.

The relative proportions of the ingredients can of course vary a great deal, and together with the preparation technique, will determine the quality of the resulting ice cream. Federal standards call for a minimum of 10% milkfat and 20% total milk solids (fat plus protein, minerals, and lactose), with allowed maximums of 0.5% stabilizers (large carbohydrate molecules sometimes used to bind water and inhibit the formation of large ice crystals) and 0.2% emulsifiers (which improve the clumping of fat globules in the foam). In addition, there is a maximum of 100% *overrun* allowed: that is, the mix can be expanded up to double its original volume by incorporating air cells.

The two specifications most commonly varied are the contents of fat and air. Cheap ice cream has the minimum fat—10%—and the maximum overrun—100%—while expensive versions will have double the fat, half the air. Middle-range products split the difference. The more air ice cream contains, the fluffier and the warmer it seems (warmer because there is less ice and icy sugar solution to maintain the sensation of coldness), and the lighter it is. Federal standards specify that a gallon of ice cream must weigh at least

4.5 pounds (not including such optional ingredients as nuts and fruit). Some premium brands weigh close to double this minimum, and this is one reason for the huge disparity in price between these and the bargain brands. There may be as much actual substance in a pint of expensive ice cream as there is in a quart of the cheap. So hefting cartons is one crude way of estimating relative quality. Reducing the fat content can result in a coarser texture— there are fewer globules to separate the ice crystals, which can therefore grow larger, or to lubricate them on the tongue—and it will make the ice cream feel *colder* in the mouth than usual. This is one reason why ice milk and sherbet feel colder than ice cream (another is their higher sugar content, which causes them to freeze, and to melt, at a lower temperature).

Cheap ice creams also tend to use more artificial flavorings, stabilizers, and emulsifiers. Artificial flavorings are usually less subtle and complex than natural ones. Vanilla extract, for example, is imitated by vanillin, which is only one (albeit the most important) of its components. And emulsifiers and stabilizers, which are useful in masking the lower fat content, can impart a sticky, gummy quality and an odd aftertaste to ice cream. With a little concentration on temperature, texture, and the feel of an ice cream in the mouth, you can tell a great deal about what went into it.

Other frozen desserts generally have less fat, less air, and more sugar than regular ice cream. Soft ice cream is 3 to 6% fat, with an overrun of 30 to 50%; ice milk is about the same, but it is served colder (at about 10°F (−12°C), as against 20° to 22°F (−7° to −6°C) for soft ice cream). Sherbets contain small percentages of both milk fat and solids, while fruit ices generally have neither. Though ice milk, sherbet, and fruit ices all have the reputation of being low-calorie alternatives to ice cream, this is not necessarily the case. They make up for their lack of fat by including less air and as much as double the sugar (in the case of sherbets and ices, to balance the fruit's acidity); like ice cream they range around 200 calories per cup. Finally, there is a product as yet rare in this country, but less so in England —an imitation, nondairy ice cream made with vegetable oils, called "mellorine."

Making Ice Cream
There are three basic steps in making ice cream; preparing the mix, freezing it, and hardening it (letting it sit for a while after the freezing process). The first step is simply a matter of mixing the cream, sugar, flavorings, and other ingredients, if any, into a homogenous liquid. Heating helps to dissolve the sugar, and kills off bacteria that might otherwise survive in the frozen product. In the second stage, the liquid mix is brought from room temperature or above to a temperature well below the normal freezing point of water, where enough of the water solidifies to make the mix semisolid at least. The freezing process is an interesting one. The dissolved sugar, by getting in the way of water molecules that must bond together to form ice crystals, lowers the freezing point of the solution from 32° to about 27°F (0° to

−3°C). At this temperature the water molecules have slowed down enough that their mutual attraction becomes stronger than the disruptive influence of the sugar. As they crystallize, the water molecules are removed from the solution, which means that the *remaining* solution gets more concentrated with sugar: and so its freezing point is lowered even further. (The same concentration effect increases the boiling point in sugar solutions during candy making; see chapter 8.) It is clear from this trend that the liquid phase of ice cream will *never* freeze completely, though the lower the temperature gets, the less liquid remains. Ice cream is noticeably softer at 22°F (−6°C)—the typical temperature of soft ice cream—than at 0°F (−18°C) because half of the water is still liquid at the first temperature, but only 20% of it is liquid at the second. At 10°F (−12°C), the recommended serving temperature of regular ice cream, a half gallon will contain about a cup of liquid: a proportion that provides the right yielding, semisolid consistency.

Two aspects of the freezing process have a special influence on the texture of the final product. One is the number and size of the ice crystals produced. A few large crystals give a coarse, icy texture rather than a desirably creamy one. This can be avoided by cooling the ice cream mix rapidly with continuous movement. Rapid cooling ensures the simultaneous production of many "seed" crystals which, because they share the available water molecules among themselves, cannot grow as large as a smaller population could. And continuous stirring helps cool the mixture evenly, distribute the seed crystals, and prevent several crystals from growing into each other and forming a cluster that the tongue might notice (again, the same problems arise as sugar syrups cool into candies). The second influence on the final texture is the amount of air incorporated during freezing. Home ice-cream makers also produce the foam by stirring, and mostly once the mix has begun to thicken. Commercial producers can pump in compressed air at the end of the freezing stage. They also have an advantage in freezing rates: by using very low temperatures and small-diameter tubes that expose a large surface area to the coolant, they can freeze the mix in a matter of a few seconds or minutes. The home machine, by contrast, may take half an hour.

Hardening is the last stage in making ice cream. Once the mix has become good and thick, with a large number of small ice crystals and enough air, the ice cream is finished by freezing it more solidly. Because it is done without agitation, this stage is called "quiescent" freezing. All that happens is that more water is removed from the liquid phase and deposited onto ice crystals, leaving the various solid components less lubricated and so making the whole mixture firmer. Here again, the advanced technology of the commercial producers gives them some advantage. They squirt the semisolid mix into a cardboard package and then put it in hardening rooms, some of them kept as low as −50°F (−46°C), where the job is done in an hour or so. The rest of us must rely on brine or a freezer compartment that won't get below 0°F (−18°C), and so we must wait much longer for the proper consistency to develop as the ice cream approaches 10°F (−12°C).

Home Freezing

The odds in favor of producing a fine-textured ice cream at home are increased when a few helpful tricks are known. Preparations begin, of course, with the mix. It's not advisable to use straight heavy cream, even though it would give the richest result. Homemade ice cream tends to be on the dense side, and often can profit from more incorporated air. The mix will foam more readily the more protein it contains, and the more homogenized fat; and cream is not homogenized because large fat globules produce the stablest—not the lightest—whipped cream (since the ice-cream structure is frozen, stability is not a problem). Condensed milk or half-and-half will contribute some homogenized fat, and condensed or dry milk will increase the protein content of the mix without diluting it too much. One or more of these other milk products should be mixed with cream to obtain a softer, lighter solid. However, too much dry or condensed milk raises the lactose levels as well as the protein, and may lead to the formation of lactose crystals at low temperatures, giving the ice cream an unpleasantly gritty texture.

The job of freezing is accomplished with a rather vaguely defined mixture of ice and salt. Ice cream simply cannot be made without salt in the ice water. Salt does for the cooling bath exactly what the sugar does for the ice cream mix: it lowers the freezing point of the solution. Since the mix will not freeze above 27°F (−3°C), the surrounding bath must be colder than this, but pure ice water can never fall below the freezing point of water, or 32°F (0°C). If it did, it would be solid ice. Ice from the freezer is generally around 10°F (−12°C), or plenty cold enough to freeze the mix if it weren't for the combination of friction, warm mix, and warm air melting some of it into ice water with a much higher temperature. Since this is inevitable, salt is added to lower the freezing point of the water, and so the temperatures at which it can remain liquid (for the same reason, salt is used to turn icy roads into merely wet ones). The brine then absorbs heat from the mix, the colder ice melts and keeps the brine cold, and salt crystals dissolve to keep the brine from being diluted. Theoretically, one part of salt added to three of ice (by weight) can produce a liquid brine of −6°F (−21°C). In practice, our ice is not that cold to begin with, salt isn't *that* cheap, and we really don't want that cold a brine anyway, at least not during the first stage of freezing. A more usual ratio is one part salt to every eight of ice, and more often than not we just throw in a handful of salt for every few handfuls of ice. (Sherbets and fruit ices, because they contain more sugar and freeze at a lower temperature than ice cream, require a colder, saltier brine.)

Once the mix has been cooked up, it should be put in the refrigerator and allowed to cool down without much stirring. By bringing it down to 40°F (4°C) or so, we reduce the amount of cooling that has to be done in the brine, and by refraining from stirring, we avoid the possibility of churning some of the cream into butter. The mix is then put into the freezing container and surrounded by alternating layers of crushed ice and salt. The ice is crushed in order to multiply its surface area and so bring more cold matter into contact with the brine; this helps maintain lower temperatures. Coarse salt is preferred to table salt for several reasons. It is cheaper, less likely to fall through the ice and concentrate at the bottom as the process begins, and it dissolves more slowly, lowering the brine temperature more gradually. Too cold a brine will produce a coarse ice cream with too little incorporated air. On the other hand, a brine that gets cold only gradually and doesn't dip as far below 27°F (−3°C) will cool the whole mass of mix more evenly, so that many more crystals will form simultaneously, share the available water, and remain relatively small.

As it cools, the mix is kept in continuous motion to even out the temperature change and to incorporate air. It is usually recommended that about one third of the container be reserved for overrun. If the final ice cream reaches the brim, it will have increased its volume by half, or have an overrun of 50%. (Some of the expansion is due to the expansion of water as it becomes ice, but only a small amount.) If the ice cream is being made by hand, it should be cranked more vigorously as soon as the mix becomes noticeably viscous, since this is the point at which it begins to retain air well. Most of the earlier motion simply circulates the mix from cold exterior to warmer interior and back. Active churning as the liquid thickens also helps keep the size of the developing ice crystals down, and so maintains a smooth texture.

Once the ice cream is so stiff that motor or arm stalls, the home version of hardening begins. The mix container is sealed, and another few handfuls of salt are thrown into the brine to lower the temperature another few degrees. Because water is much denser than air, and so exchanges heat more rapidly, this technique is preferable to putting the ice cream into an ice box of similar temperature. In either case there is an advantage to dividing the ice cream among several small containers: the increased surface area will result in faster cooling. Whenever ice cream is transferred to other containers, be sure to prechill them to avoid surface melting and recrystallization.

Storing Ice Cream

Ideally, ice cream should be stored at fairly low temperatures, between −10° and 0°F (−23° and −18°C), to maintain its fine texture. It should also be well covered so that the fat doesn't pick up odors from the rest of the freezer compartment, and so that free moisture doesn't settle onto the surface and form large crystals. Some commercial cartons now include a single sheet of plastic wrapping that can be pressed directly onto the surface, however irregular it is; this is an excellent solution and one that you can easily imitate. The gradual coarsening in texture during storage is generally due to repeated partial thawings and freezings, whether from removing the ice cream from the freezer to serve it, or from fluctuations in the freezer temperature itself.

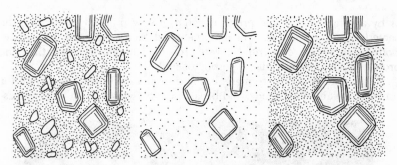

The growth of ice crystals during storage. Whenever ice cream is warmed slightly, the smallest crystals melt *(center)*. When the temperature drops again, the additional water is taken up by the surviving crystals, which get larger and larger *(right)*. A grainy texture results.

If the temperature rises even a little bit, some of the smaller ice crystals will dissolve, and the water so liberated is taken up by surviving crystals when the temperature declines again. The result: fewer, larger ice crystals. The lower the average storage temperature, the less noticeable the effects of this cycle. Partial thawing can also cause some crystallization of lactose and a consequent sandy texture. In this case, the higher the temperature, the less viscous the remaining water solution is, and so the easier it is for lactose molecules to move and find each other. Lactose crystals are distinguishable from ice crystals by their tendency to persist after the ice has melted, either on the tongue or in the dish.

FERMENTED DAIRY PRODUCTS

Curdling with Acid: Yogurt, Buttermilk, Sour Cream

Yogurt, buttermilk, and sour cream are, after cheese, the most common cultured dairy products in the West. They differ from cheeses in not being treated with rennet, the extract from calf's stomach. Instead, they are thickened, or curdled, solely by the action of acid-producing bacteria.

The first yogurt was probably eaten not long after the advent of dairying itself. If left alone, fresh milk quickly teems with lactic acid bacteria that sour it and eventually cause the casein micelles to aggregate. As long as fermentation does not go too far, the resulting thick texture and piquant taste is a refreshing change from plain milk. Similarly, the original buttermilk, the low-fat liquid left over from making butter, was contaminated by airborne bacteria during the churning, after which it thickened and soured slightly. Once these effects had been noticed and appreciated, they could have been controlled to some extent by using the same containers from batch to batch, or by adding some of the previous batch to new milk: in both cases, the bacterial culture would have been perpetuated. Sour cream and its relatives, clotted cream and *crème fraiche*, were probably late refinements of the same process applied to the high-fat portion of milk.

Today, these foods are made in more strictly controlled ways, both for convenience and consistency, and because the old ways do not work in pas-

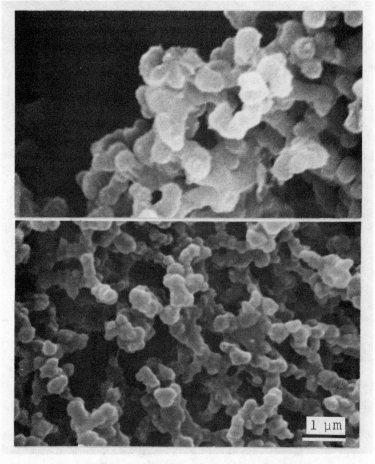

Top: yogurt made from unheated milk. Casein micelles form large, coarse clusters when bacteria produce acid. *Bottom:* yogurt made from milk preheated to 185°F (85°C). At this temperature, whey and casein proteins complex in such a way as to inhibit this clustering. The resulting yogurt has a finer, firmer texture. These and the following scanning electron micrographs of dairy products were kindly provided by Miloslav Kalab of the Food Research Institute, Agriculture Canada, in Ottawa. (From Kalab, *Scanning Electron Microsc.* 3 (1979): 261–72.)

teurized milk. As we shall see in more detail when we get to cheese, raw milk sours into wholesome foods because the acid-producing bacteria have certain advantages, including a head start, over less helpful microbes. Pasteurized milk, on the other hand, is nearly free of bacteria, and troublemakers have a better chance of prevailing. So where raw milk sours, pasteurized milk may spoil. In order to compensate for this shift in the ecological balance, we now add pure cultures of the desired bacteria to pasteurized milk.

Another innovation in the making of acid-coagulated milks is heating the milk to about 185°F (85°C), well above the usual pasteurizing temperature, holding it there for about half an hour, and only then cooling and inoculating it with bacteria. The fermentation is allowed to proceed until the desired acidity is reached, and then stopped simply by cooling the product down to refrigerator temperatures. The heat treatment does more than destroy the microflora. It denatures the whey proteins to some extent, unfolding the initially compact molecules into longer structures that increase the viscosity—thicken the texture—of the liquid. And in the case of yogurt, it results in a firmer gel that is less prone to separating into curds and whey. Apparently, when the lactoglobulin is denatured it reacts and forms a complex with one of the casein subunits, and this complex interferes with the aggregation of casein micelles. It doesn't *prevent* this aggregation, but it limits it to such an extent that the micelles seldom get close enough together to squeeze out much whey. The result is a finer matrix, one in which the liquid and solid phases are well integrated. The aggregation of casein micelles is caused, of course, by the bacterial secretion of lactic acid, which lowers the pH from about 6.5 to about 4.5, thereby causing the micelles to shed their mutually repellant negative charge. And with nothing to keep them apart, their chemical similarity naturally brings them together.

Yogurt

Yogurt can be made from whole or skim milk, and dried milk solids are sometimes added to provide more casein for the solid matrix. A mixed culture of *Lactobacillus bulgaricus* and *Streptococcus thermophilus* is added when the milk cools to 113°F (45°C), and the temperature is maintained at about 110°F (43°C) for four hours. If the bacterial culture is initially dry and therefore dormant, the incubation period may be several times as long. The two kinds of bacteria consume the lactose as an energy source, and excrete lactic acid as a waste product (the same waste product, by the way, that collects in our muscles during strenuous exercise and by interfering with further energy production may make them feel heavy and weak). The coccus (sphere-shaped bacterium) outgrows the bacillus (rod-shaped) until the accumulating acid slows it down, at which point the hardier bacillus takes over.

This ancient food has only recently caught on in the United States, where consumption has multiplied by a factor of 35 between 1955 and 1980.

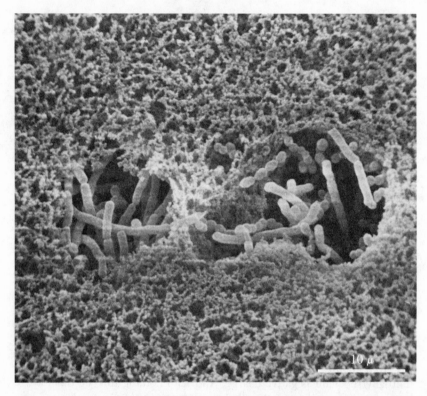

Yogurt and its bacteria. The bead-shaped denizens of this lair are lactic streptococci, and their more numerous rod-shaped companions the lactobacilli.

The first commercial American yogurt was produced by the Armenian Columbosians in 1931, and the Spanish Carasso family brought Danone, as it was then named (after Daniel, the founder's son) to New York in 1942. Dannon first put fruit in its yogurt in 1946 as a novelty and to balance the as yet unaccustomed sourness. The claim that yogurt is an especially healthful food dates from the turn of the century, and is based on the mistaken belief that the lactobacilli populate the intestine and, as they do in milk, suppress the growth of harmful microbes. Unfortunately, *Lactobacillus bulgaricus* cannot survive in humans (see chapter 11).

Buttermilk

Though some buttermilk is still drawn off butter today and is slightly soured during the churning, most is cultured from skim milk in order to obtain consistent results. It is incubated longer than yogurt, from 12 to 14 hours, because it is kept about 40°F cooler during the fermentation. Lacking actual contact with butter, it must derive a buttery flavor from the bacterium *Leuconostoc citrovorum*, which converts citric acid, a minor by-product of the lactose-fermenting bacterium, into diacetyl, a volatile molecule characteristic of butter flavor. The optimum temperature for the production of diacetyl is 68°F (20°C), and above this other, less appealing by-products accumulate. *Streptococcus lactis*, which grows well at 86°F (30°C), is used instead of the high-temperature bacteria found in yogurt. The culture temperature is kept at about 70°F (21°C) to develop the best flavor, and a rather slow fermentation is the result. Buttermilk is valued for the rich, tangy flavor it can impart, especially to baked goods.

Sour Cream and Cottage Cheese

Sour cream can be made either with or without the help of bacteria. Buttermilk starter can be cultured for 24 hours in cream, or vinegar can be added to the cream and the mixture allowed to stand until it has curdled.

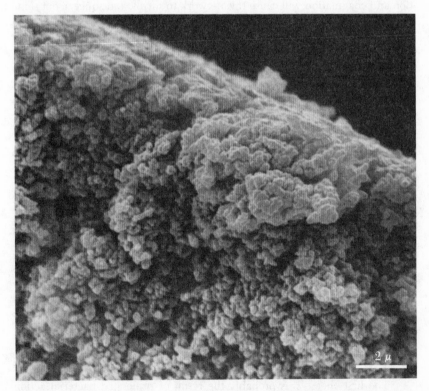

Close-up of a cottage cheese curd, showing the compressed casein network. (From Kalab, *Scanning Electron Microsc.* 3 (1979): 261–76.)

Cottage cheese is produced by two methods: one uses rennet, and the other lactic bacteria. In order to obtain the lumpy, dense curds, the skim milk is not subjected to the initial high-temperature treatment. Cooking and pressing after fermentation concentrate the curd and remove the whey, and washing removes much of the acid: hence the blandness of cottage cheese compared to other fermented products. It is called cottage cheese for its similarity to other unaged "farmhouse" cheeses.

Cooking Causes Curdling

Perhaps the most important thing to know about cultured milk products in the kitchen is that they are susceptible to curdling when made into sauces or added to other foods. Plain milk or cream is relatively stable when heated, but the extended heat treatment and abnormally high acidity have already caused partial aggregation of the casein in cultured products. All the water can still find a place in the extended network of casein micelles and whey proteins, but anything that the cook does to further the process of aggregation and coagulation will cause this network to shrink and squeeze out some of the liquid. You know that this has happened when you see distinct particles of protein—curds—floating in the separated liquid. Heat, salt, more acid, and even vigorous stirring all can cause curdling. The key to maintaining a smooth texture is gentleness. Heat gradually and moderately, and stir slowly.

CHEESE

Cheeses differ from the other cultured milk products principally in the extent to which curdling and fermentation are allowed to proceed. The coagulation of casein is furthered by the use of special enzymes, the whey is physically pressed out of the curd, and the bacteria are allowed to work long enough that fat and protein, as well as lactose, are broken down into simpler molecules. The compressed curds contain nearly all the casein and a good deal of the fat.

Cheese takes up about one tenth the volume of its original milk, and because it is drier and more acidic, it is much more resistant to spoilage. The usefulness of this transformation would quickly have become apparent to early agricultural societies, which could not otherwise have saved a surplus of milk for drier times. But there is really no good estimate of how old cheese making is. The earliest evidence known so far, a residue found in an Egyptian pot, dates from 2300 B.C. We do know that a wide variety of milks—from the familiar cow, sheep, and goat to the mare, water buffalo, and yak—have been made into cheese in many different parts of the world.

The first cheese was probably the result of prolonged bacterial action beyond the point at which a homogenous, yogurtlike texture is reached. It makes sense that rennet, an extract from the fourth, or true stomach of a milk-fed calf, would have been discovered as an especially efficient curdling agent after that organ had been used as a bag for carrying milk. When this discovery was made is not known, but it certainly preceded Roman times. It is clear from Columella's *Rei rusticae* ("On Rustic Matters," about A.D. 65) that a number of protein-precipitating substances were used in the ancient world. Milk, he said,

> should be curdled with rennet [*coagulum*] obtained from a lamb or kid, though it can also be done with the flower of the wild thistle or the seeds of the safflower, and equally well with the liquid which flows from a fig-tree if you make an incision in the bark while it is still green.

In fact, Columella's basic outline of the whole process is not much different from current practice. The curdling was done in fresh milk kept warm by the side of a fire. After coagulation was completed, the whey was pressed out, the curds sprinkled with salt—which both dehydrates them and protects them from spoilage—and the fresh cheese put in a shady place to harden. Salting and hardening were repeated once more, and the ripe cheese was then washed, dried, and packed for storage. According to Columella, it could then be shipped overseas without spoiling.

Early Appreciations of Cheese

It would appear that during the Dark Ages, the appreciation of fine cheese was preserved in the religious houses (as was true of so many other aspects of civilized life). About 50 years after Charlemagne's death in 814, an anonymous monk at the monastery of Saint Gall wrote a biography of him that includes this fascinating anecdote. Charlemagne was traveling, and found himself at a bishop's residence at dinnertime.

> Now on that day, being the sixth day of the week, he was not willing to eat the flesh of beast or bird. The bishop, being by reason of the nature of the place unable to procure fish immediately, ordered some excellent cheese, white with fat, to be placed before him. Charles . . . required nothing else, but taking up his knife and throwing away the mold, which seemed to him abominable, he ate the white of the cheese. Then the bishop, who was standing nearby like a servant, drew close and said "Why do you do that, lord Emperor? You are throwing away the best part." On the persuasion of the bishop, Charles . . . put a piece of the mold in his mouth, and slowly ate it and swallowed it like butter. Then, approving the bishop's advice, he said "Very true, my good host," and he added, "Be sure to send me every year two cartloads of such cheeses."

The word we have translated as "mold" is *aerugo* in the Latin—literally, "the rust of copper." Partisans of Brie cheese, which has an external coat of mold, have translated it as "skin," and they claim Charlemagne as one of Brie's illustrious admirers. But the adherents of Roquefort, a sheep's-milk cheese, have a much stronger case. It is veined with blue-green mold, which is much the same color as weathered, or "rusty" copper. It seems that Charlemagne was not alone in requiring instruction on the subject of cheese. A late medieval compendium (from about 1400) of maxims and recipes, known as *Le Ménagier de Paris*, includes this formula "To recognize good cheese":

> Not at all white like Helen,
> Nor weeping, like Magdalene.
> Not Argus, but completely blind,
> And heavy, like a buffalo.
> Let it rebel against the thumb,
> And have an old moth-eaten coat.
> Without eyes, without tears, not at all white,
> Moth-eaten, rebellious, of good weight.

According to the belated testimony of Thomas Fuller in his *History of the Worthies of England* (1662), the British Isles owe their knowledge of cheese making to the Roman occupation. By Fuller's time, Cheshire cheese was renowned. And by the time that Daniel Defoe made his *Tour Through England and Wales* early in the 18th century, both Cheddar and Stilton were well established. Cheddar, named for the Somerset town near which it was made, was remarkable for being a cooperative venture: the milks of all the cows in the area were mixed together to make the cheese. Said Defoe, "By this method the goodness of the cheese is preserved, and without all dispute, it is the best cheese that England affords, if not, that the whole world affords." A somewhat less admiring picture was drawn of Stilton cheese, named for a town in Huntingdonshire, "which is called our English Parmesan [actually, it is a cow's-milk version of Roquefort], and is brought to table with the mites, or maggots round it, so thick, that they bring a spoon with them for you to eat the mites with, as you do the cheese."

Predictably, a different view prevailed across the Channel. According to the French *Encyclopédie* (1751–76), Roquefort was the *"premier fromage de l'Europe,"* with Brie, Maroilles, and Gruyère not far behind. Regardless of particular rankings, the place of cheese in the two countries' diets was the same: for the poor, fresh or briefly aged types were staple food, sometimes called "white meat," while the rich enjoyed a variety of cheeses, mainly as a novelty. So Thomas Fuller wrote of Cheshire cheese that "Poor men do eat it for hunger, rich for digestion." And the *Encyclopédie* article on *"Fromage"* used the same idea to explain why it was that well-aged cheeses are served toward the end of the meal:

And so cheese that is almost putrefied, a condition in which it is sometimes eaten, should pass less as a food than as a seasoning, *irritamentum gulae* [a provoker of hunger or gluttony], which often stimulates to advantage the play of a stomach already filled with diverse foods, and which one can therefore eat with success at the end of the meal. It is principally this fact which is involved in the verse known to all the world: *Caseus ille bonus quem dat avara manus* [from the 11th-century School of Salerno: "That cheese is good which is served with a sparing hand"].

According to one historian, the Greeks too ate cheese at the end of the meal, but in order to renew their thirst for wine. In any case, even those who could survive without cheese have found it a necessity for aesthetic reasons. Early in the 19th century, the gastronome Brillat-Savarin wrote that "a dinner which ends without cheese is like a beautiful woman with only one eye."

Aversion to Cheese

If the characteristic aroma of cheese provokes hunger in some people, it provokes disgust in others. The 17th century saw the publication of at least two learned treatises (by Martin Schook, a Dutchman, and Thomas Sagittarius, a German), "*de aversatione casei*," or "on the aversion to cheese." And the author of "*Fromage*" in the *Encyclopédie* noted that "cheese is one of those foods for which certain people have a natural repugnance, of which the cause is difficult to determine." Today, with our knowledge of the role of bacteria and molds, the source of this repugnance is less obscure. The fermentation of milk, like that of grains or grapes, is essentially a process of limited, controlled spoilage. We allow certain microbes to degrade the original food, but not beyond the point of edibility. At the same time, the abundance of these microbes discourages the growth of other, potentially harmful ones. The period of fermentation for yogurt and fresh cheeses is so short that only the lactose is degraded, and then only to the relatively innocuous lactic acid. In cheese, on the other hand, fats and proteins are also broken down into sometimes highly odorous molecules. These same molecules are also produced during uncontrolled spoilage, as well as by microbial activity on moist, warm, sheltered areas of human skin: hence the oft-noted resemblance between the smells of cheese and feet. An aversion to such signs of decay has the obvious biological value of steering us away from spoiled food and possible food poisoning. Though the exact nature of this aversion is still a mystery (see the discussion of smell in chapter 12), it definitely exists, and it must be overcome in order to enjoy cheese. Once acquired, the taste for partial spoilage can become a passion that must express itself in paradoxes. The French call a particular grape fungus the *pourriture noble*, or "noble rot," for its influence on the character of certain wines, and the Surrealist poet

Leon-Paul Fargue is said to have honored Camembert cheese with the title
les pieds de Dieu—the feet of God.

Modern Developments

Until 1851, cheese was made the way it always had been, mostly: by
farmers with more cows than were needed to supply the demand for fresh
milk. A few years before 1851, the son of one Jesse Williams of upstate New
York married and took over another farm. He considered his father to be a
better cheese maker, and so he sent his milk to his father's farm for that
purpose. In 1851, Williams began to do the same for other farmers in the
region. And fifteen years later, there were 500 such specialized cheese and
butter factories in the state. "Associated dairying" had been born. At the turn
of the century pure microbial cultures became available, and soon after that
commercial preparations of rennet. Fungal enzymes (from species of *Mucor*
and *Endothia*) are now sometimes used in the curdling phases instead of
rennet. And since 1935, most cheese has been made from pasteurized milk,
a practical necessity, given the scale of factory operations. Today, most
cheese produced in the developed countries is factory-made. Even in France,
which established a *"Fromage appellation d'origine contrôlée"* in 1973 to
certify that traditional methods have been used in the areas specified, less
than 20% of the annual production qualifies. In the United States, the market
for process cheese, a mixture of aged and green cheeses blended with emulsifiers and pasteurized, is now larger than the "natural" cheese market. Even
"imitation pasteurized process cheese" has made an appearance: it replaces
milk fat with vegetable oils.

Making Cheese

The process of making cheese can be divided into three fundamental
steps. The first is the precipitation of casein into curds. Bacteria that produce
lactic acid are infused into the warm milk to obtain an adequate acidity for
the action of rennet, and to crowd out less desirable organisms. Then, rennet
is added, which causes the casein micelles to aggregate, trapping fat globules
and whey in the protein network. The second stage is the concentration of
the curds. Any free whey is drained off, and the curds are cut, pressed,
cooked, and salted to remove much of the rest. The final stage is the "ripening," or aging of the "green" curds. It transforms the initially bland and
either crumbly or rubbery curds into a smooth substance with a pronounced
and complex flavor. Ripening is mostly a matter of molecular breakdown
caused by the enzymes of microbes, both the original starter bacteria and
special ripening organisms. One way of classifying the thousands of cheeses
that are still made today is by the characteristic ripening organisms and their
location. Blue cheeses, for example, are ripened from within by veins of
mold, Brie and Camembert from without by surface molds, Cheddar and
Swiss from within by the evenly dispersed starter bacteria. Another useful

CHEESE MAKING

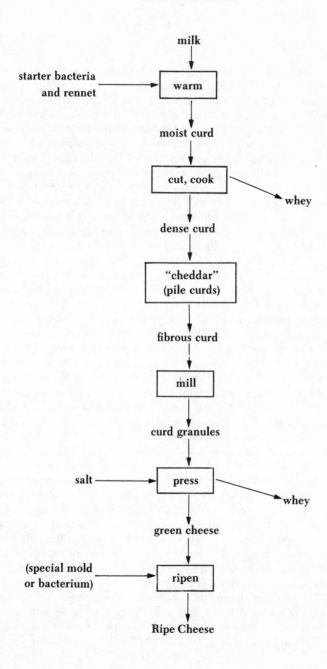

consideration is the water content, which is primarily a function of the methods used in concentrating the curds.

The Milk

Cheese making begins, of course, with milk, and this basic material itself can vary greatly in composition. Sometimes it is used straight, sometimes it is skimmed, sometimes it is enriched with cream; and each formulation will result in a different texture and somewhat different flavor (the less fat, the fewer fatty acids available, and the blander the aroma). The species of the source is also important. The milk of goats and sheep has a greater concentration of short-chain fatty acids in the fat than does cow's milk, and these highly odorous molecules give their cheeses—*fromages de chèvre* and Roquefort among them—their special pungent quality. Even particular breeds of cows eating particular feeds in particular climates will give their own particular milks. In his *Complete Encyclopedia of French Cheese,* Pierre Androuët claims that "sublime" cheeses are possible when the animals can feed on newly sprouting grass, meadows in full flower, or the second growth of autumn. Verifiable or not, it is a nice idea.

Today, most cheese is made from pasteurized milk, principally because of the scale of factory production. An individual farmer may be able to start curdling his hundred gallons only minutes after they leave the cows, but a factory uses thousands of gallons of varying quality from several different sources in a single batch, and the milk must be transported miles before the starter bacteria and rennet can be added. If even a small amount of the milk is contaminated somewhere along the line, undesirable bacteria may be able to compete with the starter bacteria, and the cheese may spoil. Soft cheeses are especially liable to such spoilage on account of their high moisture content and short ripening period, and they have been known to cause food poisoning. To minimize this danger, federal regulations provide that cheeses aged less than 60 days at temperatures lower than 35°F (2°C) must be made with pasteurized milk. Nearly all mass-produced cheese follows this rule, whether hard or soft, no matter how it is ripened, although "farmer's" or "farmhouse" cheese generally comes from raw milk. The only major exception to the rule is "Swiss"-style cheese; its characteristic holes fail to develop properly in curds whose casein has been preheated.

There is a general feeling that cheeses made from pasteurized milk are inferior to the others, that pasteurization, which kills most of the milk's bacteria, is a kind of technological interference with an otherwise natural process. What effect does it actually have on the quality of cheese? Apart from a slight loss of vitamins and the destruction of microbes, there are two notable changes. The action of the rennet is slightly retarded, apparently because the denatured lactoglobulin forms a complex with the casein subunit that rennet attacks, thereby protecting it from that attack. The result is a somewhat moister, weaker curd than is usually obtained from raw milk, one that must be ripened about twice as long before the full flavor and texture of the cheese develops. But because contemporary tastes run to milder flavors in any case, the ripening period is not significantly extended in practice. The second important effect of pasteurization is that it inactivates various enzymes in the milk, whether native or microbial. The loss of these enzymes does seem to make a qualitative difference in the final flavor of the cheese, since among them are molecules that break fats down into odorous fatty acids. While these are the same enzymes that quickly turn raw homogenized milk rancid, their slower, less drastic action in ripening cheese is desirable. To be sure, enzymes from the added bacteria do liberate fatty acids, but apparently not in the same proportions. And so the cheese made from raw milk may have a fullness and complexity of aroma that the other lacks. The more romantic idea that the native flora of milk is crucial to the flavor of cheese does not seem to be true.

While pasteurized milk has become fairly common in cheese making, homogenized milk has not, because that treatment confers few benefits. It does cause greater retention of both moisture and fat in the curd, and a finer, closer texture, but these qualities make for slow ripening. However, they are valued enough in unripened cheeses and some soft and blue cheeses that these are usually made with homogenized milk.

Starter and Rennet

The first step in cheese making proper is the addition of the starter, a pure culture of lactic *Streptococci* and *Lactobacilli*. These bacteria ferment lactose to lactic acid and reduce the milk's pH to the proper range for rennet to coagulate the casein. If the cows have recently been treated with antibiotics or have eaten treated feed, the bacterial growth can be inhibited enough to slow the subsequent curdling or even make it impossible. Ordinarily, the incubation time is about an hour at 85° to 105°F (29° to 41°C). These kinds of bacteria are natural inhabitants of milk once it has left the cow, and they are the reason raw milk turns sour; the fact that they outgrow the other microbes that fall into this nutritious liquid is what made primitive cheese making possible without simultaneous spoilage (pure cultures are used today to maintain greater control over this important stage). These bacteria outgrow the others for several reasons: lactose can be more rapidly metabolized than fat or protein; most bacteria don't thrive in acidic conditions; and many don't do well without plenty of oxygen (the small supply in milk is quickly exhausted). These particular fermenters of lactose are preferred not just because they thrive in milk, but also because they excrete very little besides lactic acid and carbon dioxide, and so leave behind a sharp but clear, clean taste. Microbes that use proteins and fats as fuel excrete highly odorous, sometimes noxious wastes. The usual starter culture works like a relay team during ripening. The *Streptococci* decline once the acidity gets to be too much for them, and the hardier *Lactobacilli* continue the work into regions of even lower pH.

Once the starter bacteria have churned out enough acid, the rennet is stirred into the vat. Rennet has been prepared for centuries by soaking the fourth, or true, stomach of a milk-fed calf in brine. Gastric extracts from other animals and even plant juices have also been used, as Columella records, but rennet has the unique property of causing coagulation without much actual digestion of the casein into smaller molecules. Plant proteases (protein-breaking enzymes) and the gastric secretions of *weaned* animals digest more than they coagulate, and this results in a soft, weak curd that may even disintegrate. The active substance in rennet is a single enzyme, *rennin*, which somehow disables the stabilizing subunit of casein, and so causes the normally separate micelles to clump together in the presence of dissolved calcium. Rennin is remarkably efficient. In pure form, one part will coagulate 5 million parts of milk. Commercial rennet, which is part brine and contains small amounts of other enzymes as well, works at a ratio of about 1 to 4500. During the drawn-out process of ripening, these trace enzymes probably act on proteins and fats and contribute to the flavor of the cheese.

Once the rennet or fungal enzyme has been added, the curd is allowed to form or "set" into a gel for an hour or two. The temperature used ranges from 70° to 95°F (21° to 35°C), and depends on the kind of cheese being made. Low temperatures result in soft, jellylike curds, high temperatures in rubbery curds, and medium temperatures in a firm coagulum: desirable materials for soft, hard, and semihard cheeses, respectively.

The milk gel is now treated in order to expel the whey that has been trapped in the casein network, and to develop the texture of the green cheese. The first step is to cut the gel into small chunks, thereby exposing the interior and allowing more liquid to drain out. Fine cutting releases more whey than coarse cutting, and so produces a drier, harder cheese. The second step is to cook the curds at 105° to 130°F (41° to 54°C) while agitating them. The heat, together with the continued action of rennet, causes the destabilized micelles to lose their separate structure and to fuse into tiny filaments of protein. The curd becomes denser and more compact, and still more whey is released. Again, soft-cheese curds are cooked at a lower temperature than hard-cheese curds, and some soft, crumbly cheeses like blue cheese and Roquefort are not cooked at all. Once the pieces of curd have become dense and somewhat rubbery, the stirring is stopped, and they are allowed to fall to the bottom of the vat. At this point, the whey is drained off. It used to be that this liquid, which contains proteins, salts, vitamins, and some fat, was saved and curdled with high heat and acid. Ricotta was originally a whey cheese, and the Norwegian Mysost still is. Once a necessary economy, true whey cheeses are now rare (ricotta is usually made with whole milk, and resembles cottage cheese). In the age of factory cheese making, whey has become a serious water pollutant; when released into rivers, it encourages the overgrowth of algae and so depletes the deeper water of oxygen. Whey is now frequently saved and added to other foods, including animal feed, and used to make growth media for antibiotic-producing microbes.

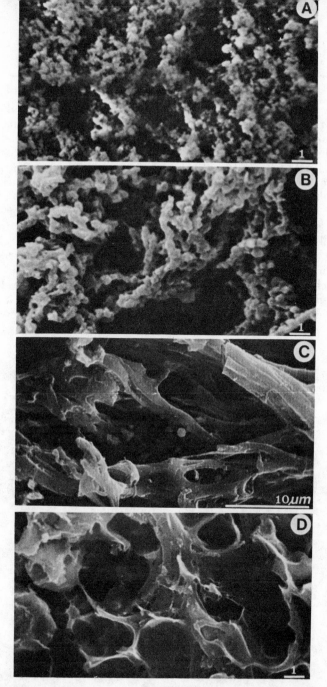

The development of Cheddar cheese seen through the electron microscope. *(A)* The casein micelles clot under the influence of acid and rennet. *(B)* The coagulated micelles form coarse chains during cooking. *(C)* Under the pressure of cheddaring, the chains fuse into smooth fibers. *(D)* The grain of the protein matrix is lost in the finished cheese because the curd is milled into small pieces after cheddaring. (From Kalab and Emmons, *Milchwiss.* 33 (1978): 670–73.)

As the small pieces of curd fall to the bottom of the vat, they matt together into a large, irregular mass. This mass is then usually cut into blocks which are piled on top of each other and allowed to sit for a period from a few minutes to two hours. Such "cheddaring" squeezes more whey out of the curd and forces the fine filaments of casein closer together. The physical pressure, the heat, and an ever-increasing acidity—the starter bacteria are still multiplying—change the texture of the curd from a rubbery elasticity to what is called the "chicken breast" condition. The casein filaments merge into larger fibers which, because of the vertical force exerted by the piled curd, all run horizontally. This "grain," similar to the grain of muscle fibers in meat, combined with the general consolidation of protein, produces a curd that can be pulled into tender strings. The curd has "mellowed." At this point, cheeses of the provolone and mozzarella types are kneaded and stretched in order to accentuate the fibrous structure and compact it further; they end up with a shiny surface and, like kneaded dough, they can be

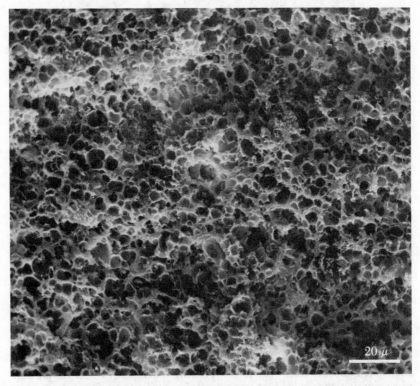

This view of mature Gouda shows the random orientation of the casein matrix typical of cheeses that are milled before final pressing into shape.

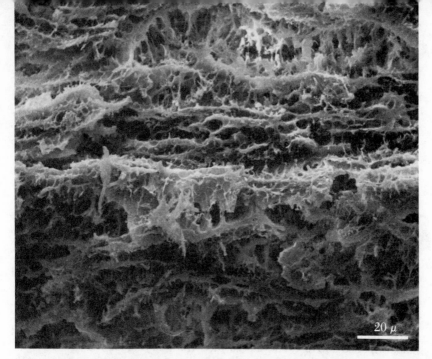

Mozzarella is one of the few cheeses that are not milled, and its protein network retains the grain of the "chicken-breast" stage. "String" cheeses can be peeled into long fibers for the same reason.

molded into various shapes. Some cheeses are not cheddared, and therefore do not develop as fine and smooth a texture as the others. Once cheddaring and mellowing are complete, the curd is milled into small granules in preparation for the final steps in cheese making. Kneaded curds are not milled, and retain the grain developed during cheddaring.

The next steps vary according to the kind of cheese being made. All cheeses except the softest—the cream and cottage kinds—are salted, either by direct addition to the milled curds, or by soaking the formed cheese in brine or rubbing its surface with granular salt. In the last two techniques, the salt slowly diffuses through the liquid phase of the cheese to its center. Beyond its contribution to taste, salt has two important effects on the cheese. It helps further dehydrate the curd, so that more whey can be removed by pressing the formed cheese; and it slows but does not altogether stop the activity of the starter bacteria, and thereby controls the rate at which ripening will occur. The growth of spoilage microbes is similarly suppressed. A few cheeses—the Greek feta for example—are salted heavily enough to bring all microbial activity to a halt, and so are called "pickled"; they clearly do not profit from a ripening period.

The method of incorporating salt varies from cheese to cheese. It is added directly to Cheddar curds, while Swiss types are brined, sometimes for as long as two weeks, and hard cheeses like Parmesan are rubbed. This last technique has the advantage of drying the surface into a very hard rind that protects the rest of the cheese from damage during handling, though there is also the disadvantage of making a certain fraction of the cheese inedible. The degree of salting must be carefully controlled: too little and the cheese will end up too moist, quick to ripen, and liable to spoil; too much and it will be dry and take forever to develop a ripe aroma.

Ripening

Either before or after salting, the milled curds are packed into molds and then pressed to remove the last free moisture. The curds of hard cheeses are pressed harder than those of soft cheeses. The resulting "green" cheese is then ready for ripening. This stage takes place in special storage rooms with carefully controlled temperature and humidity—basically, artificial caves. The temperature is kept fairly low, around 50°F (10°C), to ensure that microbial growth is slow and steady, and so avoid uneven maturing and the undesirable chemical by-products characteristic of rapid growth. The relative humidity must be kept high—80% for hard cheeses, 95% for soft and mold-ripened—to prevent the surface from drying out or the ripening organisms from failing. Cheeses that are ripened from within are frequently washed or brushed to remove undesirable microbes from their hospitable surfaces, unless they have been coated with wax, cloth, or some other protective layer. Most cheeses are turned over often to assure even ripening.

Ripening is a very involved process in which the microbes present in the green cheese slowly change its chemical composition and so its texture and flavor. As we have said, the change is overwhelmingly in the direction of dismantling complex organic molecules into simpler, smaller ones: lactose into lactic acid and carbon dioxide, fats into fatty acids, proteins into smaller chains of amino acids, individual amino acids, and even ammonia (NH_3). Each of these breakings-down contributes a multitude of flavor molecules to the final cheese. And the fibrous protein structure, already disrupted by milling, is smoothed out by the partial degeneration of casein. A waxy, malleable texture is the result in most cheeses. Because soft cheeses are moister and so more hospitable to bacteria and molds, they tend to ripen faster and undergo more extreme chemical changes than do hard cheeses, and are often held at lower temperatures to avoid overripening. For example, a mature Camembert, aged only one or two months, will have about half of its nitrogen-containing compounds, nearly all of which were originally in the casein matrix, converted into small, water-soluble compounds. And of this half, perhaps a quarter will be the simplest such molecule: ammonia, or NH_3. In fact, a distinctly noticeable, biting odor of ammonia in a soft cheese like Camembert or Brie is a sign of overripeness, while it is seldom even detectable in the aroma of hard cheeses that have been aged for a year or more. In very small amounts, ammonia contributes to the full aroma of cheese, but when it can be singled out it is reminiscent of floor cleaner, and may even seem to burn the nasal passages. A general definition of ripeness in cheese might be that point at which the chemical breakdown caused by microbial enzymes has developed a full, balanced range of volatile compounds, without any of these compounds overwhelming the others.

At the green stage, most cheeses are very much the same. It is the particular organism most responsible for ripening that gives a cheese its characteristic flavor. The original starter bacteria are generally also the ripening agents in hard and semihard cheeses like Parmesan, Cheddar, and Gouda. An additional strain of bacterium, *Propionibacter shermanii,* is included in the starter of Swiss-type cheeses; it lives on the lactic acid excreted by the other bacteria and gives off prodigious amounts of carbon dioxide gas, which collects in large pockets, or "eyes," as well as propionic acid, which contributes to the flavor. Roquefort, Stilton, and other "blue" cheeses owe their sharp aroma to veins of the bluish-green mold *Penicillium roqueforti,* whose growth is encouraged by running skewers into the inoculated green cheese and so releasing accumulated carbon dioxide, and admitting oxygen.

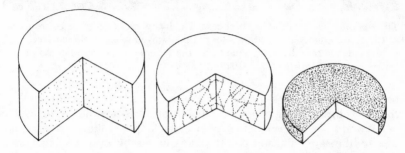

The three principal ways of ripening cheese: throughout, with the original starter bacteria *(left);* with discrete veins of mold *(center);* and from the surface, with molds or bacteria *(right).*

Camembert and Brie are ripened from the outside by a coat of white *Penicillium camemberti* mold, whose enzymes slowly penetrate the interior and transform it from a chalky, bland solid into a rich, creamy semiliquid. It is because these cheeses are ripened from the outside that they are formed into relatively thin wheels; any thicker, and the outer regions might liquefy while the center remained green. Brick and Limburger are examples of cheeses ripened by surface bacteria rather than molds.

Originally, the microbes characteristic of traditional cheeses infected the green cheese by happy accident—the Roquefort caves, for example, harbored one predominant mold, the Brie region another—and so particular cheeses could be made only in particular regions. In the late 19th century, a Danish brewery succeeded in isolating and growing pure cultures of individual yeast strains, and soon the same techniques were being applied to

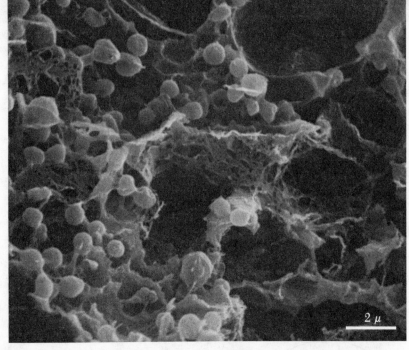

The ripening bacteria in Gouda cheese. The lacework at center is the remains of the membrane from a fat globule. Fat is removed from all these dairy products when samples are prepared for the scanning electron microscope. (From Kalab, *J. Dairy Sci.* 62 (1979): 1352–64.)

bacteria and molds. By the early 20th century, cultures of *Penicillium roqueforti* and other ripening organisms could be prepared and shipped anywhere in the world. Today, even in the original regions, pure cultures are generally used for the starter and added to the curds, or injected or sprayed into or on the green cheese, to assure consistent results. In only a few operations are the naturally occurring microflora allowed to do the work alone.

Penicillium sounds familiar, of course, because it is the genus of mold that gave the world its first antibiotic: penicillin. This fact gives us one clue about the ways in which the standard ripening organisms established their dominant position. Many ripened cheeses appear to inhibit the growth of other microbes, including potentially harmful members of the genera *Clostridium, Bacillus,* and *Staphylococcus.* This broad antibiotic activity not only serves the ripening bacteria and molds well, but may also help delay spoilage of the cheese. (In the 1950s, a group of polypeptides from *Streptococcus lactis,* collectively dubbed "nisin," were patented as antispoilage additives for cheese and bread.) It is unlikely that cheese has any distinctly medicinal effect, since the quantities of microbial products involved are very small. The antibiotics used in medicine are highly purified and concentrated. However, there is this intriguing bit of information: several different bacteria- and mold-ripened cheeses seem to slow the growth of tooth-decaying *Streptococci* when we eat them.

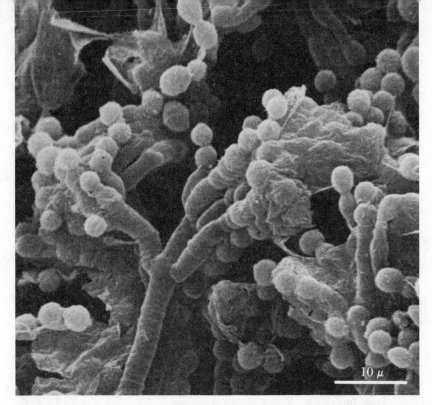

The luxuriant growth of *Penicillium* mold in blue cheese. Visible here are the fruiting structures that bear spherical spores at their tips.

Coating and Coloring

Once the cheese has ripened—or, in the case of starter-ripened types, once it has been molded and pressed—it is coated in order to prevent moisture loss, spoilage, and physical damage. Cloth, wax, fat, foil, and plastic are all used for this purpose. Some cheeses have also been artificially colored for centuries with dyes, which are added to the milk along with the starter bacteria or rennet. Carrot juice and marigold petals were among the first materials used, and annatto or achiote, the crushed seed of a tropical tree, has colored Leicester, Cheshire, and Cheddar cheeses for 200 years (the Spanish found the Mexicans reddening their chocolate beverage with it in the 16th century). Today, both bixin, the carotenoid pigment of annatto, and synthetic β-carotene, which also occurs naturally in plants, are the preferred dyes for cheese and other dairy products. Artificial coloring heightens the orange hue contributed by vitamin A from cows that have fed on green plants, and conceals its absence in the winter milk of hay-fed animals.

Pasteurized Process Cheese

Finally, a few words about two relatively recent contributions to the world's treasury of dairy products. Process cheese is essentially a material

CHARACTERISTICS OF SOME COMMON CHEESES

Type and Composition	Months Aged
Very Hard: 30% water, 26% fat, 36% protein	
Parmesan	8–36
Romano (sheep's milk)	8–24
Hard: 30–40% water, 30% fat, 25% protein	
Cheddar	6–24
Emmental, Gruyère (Swiss)	3–10
Provolone	3–6
Semihard (semisoft): 40–50% water, 30% fat, 20% protein	
Brick	3
Edam	3–12
Gouda	3–12
Munster	1–3
Port du Salut	1–2
Roquefort (sheep's milk)	2–5
Stilton	4–6
Soft: 50–75% water, 25% fat, 17% protein	
Bel Paese	1–2
Brie	1–3
Camembert	1–2

Unripened:				
Cottage and Ricotta (skim): 80% water, 4% fat, 13% protein				
Cream	50	35	8	0
Mozzarella (originally water buffalo's milk)	40	30	25	0
Neufchâtel	50	23	18	0
Pasteurized Process	40	28	25	0

Cheeses of the same name can vary widely in composition and in the way they are made. Typical ranges are given for ripening time and moisture content, and averages for fat and protein (the percentages will not add neatly up to 100).

that allows the immediate use of some green cheese, which therefore will not tie up space and money as it sits in a ripening room for several months. A blend of aged and green cheeses, (often Cheddar, Swiss, or brick) is ground, stabilized with emulsifiers, pasteurized, and packed while still hot. The pasteurization, of course, effectively halts aging, but because it does not kill all bacteria, salt and preservatives are added to improve the shelf life. Cheese spread is made in much the same way, with added moisture and a vegetable gum stabilizer giving the desired softness. Neither of these products has anything like the full flavor and fine texture of genuine aged cheese. In today's United States, however, more cheese is marketed as process cheese than as "natural."

Cooking with Cheese

The behavior of cheese when heated depends primarily on its proteins. Beyond a certain temperature, the casein will coagulate and separate from the fat and water into tough, stringy masses. Hard, well-ripened cheeses can tolerate higher temperatures than soft cheeses because more of their protein has been broken down into small, less easily coagulated fragments. The degraded protein of hard cheese is also more readily dispersed in a sauce. On the other hand, the more water and fat a cheese contains, the more easily it can be blended into a liquid that is composed mainly of these two materials. Once successfully made, a cheese sauce must be treated gently; if overheated, the casein will separate into stringy or grainy masses, and the rest of the sauce will become thin and soupy. In Swiss fondue, a sauce made for dipping and kept bubbling at the table in a chafing dish, alcohol in the white wine base lowers the boiling point enough that the protein should not curdle.

CHAPTER 2

Eggs

Eggs, their botanical counterparts the seeds, and milk are among the most nutritious foods on earth, and for much the same reason. Unlike meats or vegetables, they are all *designed* to be foods, to support chick embryo, seedling, and calf until these organisms are able to exploit other sources of nourishment. Eggs are one of the more versatile foods we have; they take well to a great variety of cooking techniques and combinations with other ingredients. Aside from their nutritiousness, characteristic flavor, and yolky richness, eggs are valued for two special qualities: the ability to bind other liquids into a moist, tender solid, and the ability to form a remarkably light, delicate foam. While the eggs of ducks, geese, and other birds are often used in some parts of the world, the chicken egg is by far the most common in our kitchens, and will be the exclusive object of our attention.

THE CHICKEN AND THE EGG

Over the centuries there have been several clever answers to the question, Which came first: the chicken or the egg? The Church Fathers sided with the chicken, pointing out that according to Genesis, God first created the creatures, not their reproductive units. The Victorian Samuel Butler awarded the egg priority in significance, if not in time, when he said that a chicken is just an egg's way of making another egg. This remark anticipates one modern interpretation of genetics, which sees organisms as the means by which particular sets of genes assure their survival, not the other way around. The commonsense answer, of course, is that neither preceded the other, that neither could exist without the other. About one point, however, there is no dispute: eggs existed long before chickens did.

As an entity, the egg is about as old as sexual reproduction in many-celled organisms, which evolved around a billion years ago. Generally defined, an egg is the female reproductive cell, together with enough nutrients to support at least part of the embryo's development once it has been fertilized by the male cell, and a membrane to protect it from changes in the external environment. The first, single-celled forms of life reproduced asexually, simply dividing themselves in two, and the first sexual organisms—probably a type of alga—would fuse and exchange genetic material directly into each other. Specialized egg and sperm cells became necessary when multicellular life evolved and this simple transfer was no longer possible. The first eggs were released, fertilized, and hatched in the oceans, and the protective membrane could be relatively simple because the egg's environment was the same mild salt solution as its parent's.

As animal life developed and diversified, it made important adaptations to new environments. The amphibians could move and breathe on land, but had to return to the water in order to reproduce; their eggs would dry up in the open air. Some time during the Carboniferous period, around 250 million years ago, the earliest fully land-dwelling animals, the reptiles, developed a self-contained egg with a tough, leathery skin that prevented fatal water loss. The eggs of birds, animals that arose some 100 million years later, are a refined version of this reproductive adaptation to life on land. (Mammals, including humans, make use of an alternative strategy: the embryo is retained inside the mother's body until its development is largely complete and it can breathe air on its own.)

Eggs, then, are millions of years older than birds. *Gallus domesticus*, the chicken more or less as we know it, is only a scant 4 or 5 thousand years old, a latecomer even among the domesticated animals (sheep and goats go back twice as far). Its background is, however, more exotic than most. The chicken's immediate ancestors were several types of jungle fowl native to Southeast Asia or India, where it was first bred. No one knows to what extent eggs, rather than bird meat, were a motive for domestication, but the choice was fortunate in one respect. Some birds will lay only a set number of eggs at a time, no matter what happens to them. Most birds, however, including the chicken, will lay until they *accumulate* a certain number; if the eggs are removed from the nest before that number is reached, they will continue to lay indefinitely. Over a lifetime, these "indeterminate" egg layers will produce more eggs than the "determinate" layers.

It appears that the chicken became a prized animal for largely nonculinary reasons as it slowly spread from East to West. The male of the species has long been noted for its aggressiveness; such words as "cocky," sexual slang, and even the feminine noun "coquette" embed this association in our everyday language. Cockfighting is one of the oldest sports known; it dates back at least to the 5th century B.C. in India, and was quickly adopted in

ROMAN EGG RECIPES

Pan-cheese: Take milk and the right pan, mix the milk with honey as for other milk dishes, add five eggs for a pint, three for a half-pint. Mix them in the milk until they make one body, strain into a dish from Cuma, and cook over a slow fire. When it is ready, sprinkle with pepper and serve.

Egg-cake made with milk: Take four eggs, a half-pint of milk, a cup of oil, and so mix them that they make one body. Throw a little oil into a thin pan, make it boil, and pour in your preparation. When it has cooked on one side, turn it onto a dish, moisten with honey, sprinkle with pepper, and serve.

—from Apicius (A.D. first or third century)

Persia, Greece, and Rome. The name of its arena, the cockpit, is still with us, and despite being outlawed in most states and many countries, the sport continues to thrive all over the world. For reasons that may or may not have had to do with its virility, the cock was considered by many cultures to be a good means of divination and a valuable sacrifice to the gods. According to Plato's *Phaedo,* Socrates's enigmatic last words as the hemlock numbed his limbs regarded a debt to the god of healing: he said to Crito, "We owe a cock to Asclepius." The less flamboyant hen and her eggs remained in the cultural background, but we know that eggs were commonly eaten in the West from Roman times on. Apicius gives recipes for *ova frixa, elixa, et hapala*—fried, boiled, and "soft" eggs—and early versions of an omelet and a custard. French and English recipe books from about 1400 include directions for making baked eggs, and their omelets and custards are quite close to the modern style.

Technology and the Egg

The chicken led a largely unnoticed career until the 18th century, when reports of elaborate Egyptian hatching ovens impelled the French naturalist René de Reaumur to write a treatise on hatching chicken eggs. The problem, of course, is heat, and Reaumur suggested various strategies: one might use areas near fireplaces or baking ovens, or even beds of fermenting dung (one of the oldest means; Aristotle and Pliny both reported that the Egyptians hatched eggs in dung). Then in the middle of the 19th century, a rapprochement between England and China brought a few specimens of a previously

LATE 14TH-CENTURY EGG RECIPES

Arboulastre: [First, prepare mixed herbs, including rue, tansy, mint, sage, marjoram, fennel, parsley, violet leaves, spinach, lettuce, clary, ginger.] Then have seven eggs well beaten together—yolks and whites—and mix with [the herbs]. Then divide in two, and make two *allumelles*, which are fried in the following manner. First, you heat your frying pan well with oil, butter, or whatever fat you like; and when it is well heated, especially toward the handle, mix and cast your eggs upon the pan, and turn frequently with a paddle over and under; then throw some good grated cheese on top. And know that it is done thus because if you mix the cheese with the eggs and herbs, when you fry the *allumelle*, the cheese which is underneath sticks to the pan. . . . And when your herbs are fried in the pan, shape your *arboulastre* into a square or round form, and eat it neither too hot nor too cold.

—from *Le Ménagier de Paris*

Erbolate [Herbed Eggs]: Take parsel, myntes, saverey, and sauge, tansey, vervayn, clarry, rewe, ditayn, fenel, southrenwode. Hewe hem and grinde hem smale. Medle [mix] hem up with ayren [eggs]. Do butter in a trape [dish] and do the fars [filling] thereto, and bake, and messe forth.

Crustade Lombarde: Take gode creme, and levys of Percely, and Eyroun, the yolkys and the whyte, and breke hem ther-to, and strayne thorwe a straynoure tyl it be so styf that it wol bere hymself. Than take fayre Marwe and Datys [marrow and dates] y-cutte in ij or iij and Prunes and putte the Datys an the Prunes and Marwe on a fayre Cofynne y-mad of fayre past and put the cofyn on the oven tyl it be a lytel hard. Thanne draw hem out of the oven. Take the lycour [liquid] and putte ther-on, and fylle it uyppe and caste Sugre y-now on, and if it be in lente, let the Eyroun and the Marwe out and thanne serve it forth.

(Say the strange-looking words from these early English recipes out loud, and they should be recognizable.)

unknown Chinese breed, the Cochin, to Europe and the United States. These showy, spectacular birds, so different from the run of the barnyard, touched off a chicken-breeding craze comparable to the Dutch tulip mania of the 17th century. During this "hen fever," as one observer of the American scene called it, poultry shows were very popular, hundreds of new breeds were developed, and the chicken took on new prominence among farm animals. Of course, appearance was not the only consideration in breeding, and by the end of the century the white Leghorn had emerged as the champion layer, and descendents of the Cornish the best meat bird. Soon a mere handful of varieties would account for the vast majority of chickens raised in the United States.

The 20th century has seen the general farm lose its poultry shed to the poultry farm or ranch, which has in turn been split up into separate hatcheries and meat and egg factories. Separate automated hatcheries freed the laying hens from the distraction of reproducing and so improved efficiency (newborn chicks are quickly sexed and the males simply discarded without having wasted any feed). The laying stock was carefully bred to increase output and eliminate such interference as broodiness, the hen's inclination to stop laying and sit stubbornly on its nest until its eggs hatch. Optimum feeds and laying conditions were developed. Closely confined hens don't waste energy on unproductive movement, and when it was found that short days cause hens to stop laying in winter, controlled artificial lighting became standard. Economies of scale dictated that production units be as large as possible, but large, confined flocks of genetically homogenous animals led to an increased vulnerability to disease, and vaccines and antibiotics became necessary. Today's laying hen is born in an incubator, eats a diet that originates largely in the laboratory, lives and lays on wire and under lights for about a year, until she lays less regularly, and produces between 250 and 290 eggs. As Page Smith and Charles Daniel put it, in their *Chicken Book*, the chicken is no longer "a lively creature but merely an element in an industrial process whose product [is] the egg."

The chicken has not been the only casualty of this development. As the egg business was starting up early in this century, poultry farming was considered the ideal occupation, the clear road to the American dream. The first technical innovations made it possible for small families to run poultry farms pretty much independently. Writing about her own experience in *The Egg and I* (1945), Betty MacDonald called the chicken business "the common man's Holy Grail." Thousands of people joined the quest, especially on the West Coast, where the climate reduced the cost of heating and cooling. But most of these new ventures quickly failed. Eggs were cheap, the new high-yield feeds expensive, the technology changing constantly. By the 1950s, only the larger, more advanced operations could afford to invest in air conditioning, conveyor belts, lighting systems, and so on. Before World War II, most laying flocks numbered 400 hens or less, while today, the majority are over 3000, and there are many farms with hundreds of thousands, and even a million hens.

The industrialization of the chicken has brought benefits, mainly economic, and these should not be underestimated. Both chicken and eggs are today the bargains among animal foods. The per capita consumption of chicken has nearly tripled in the United States since 1935, and the U.S. contributes some 70 billion of the estimated 390 billion eggs produced yearly in the world. For the city dweller, there has probably been a gain in egg quality, both in freshness and uniformity, over the days when small-farm hens ran free and laid in odd places. Prompt gathering, inspection, refrigeration, and transport have made it possible for good eggs to reach the urban consumer with less deterioration than was likely in the more relaxed, more humane past (an egg deteriorates more in one day at room temperature than it does in a week under constant refrigeration). The question is whether we can enjoy good cheap eggs without turning their creator into a biological machine. Short of the solution proposed by Page Smith and Charles Daniel—let each consumer raise his own chickens—it is difficult to see a positive answer.

THE COMPOSITION OF EGGS

How an Egg Is Made

The very commonness and familiarity of the egg lead us to overlook its remarkable progress from ovary to nest. All animals put a great deal of work into the business of reproduction, but the hen does more than most; her

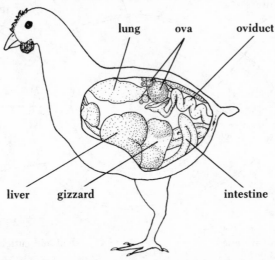

The hen's reproductive tract. The mature ovum, with its large supply of yolk, is overlaid with albumen proteins, membranes, and a shell during its day-long passage through the oviduct.

"reproductive effort," defined as the fraction of body weight deposited every day in the embryo and supporting tissues, is fully 100 times greater than a human's. The process begins with the several thousand immature egg cells each hen is born with in her single ovary. When the hen is of laying age, these ova begin to mature one at a time; if two mature simultaneously, a double-yolked egg will result. The most obvious element of maturation is the accumulation of yolk materials—mostly fats and related compounds, with some protein—which are synthesized in the hen's liver. This takes several weeks together, but is most rapid in the week or so before the ovum is released. If the hen feeds only once or twice a day, rather than continuously, her yolk will show a distinct layering of pigment. During this time the yolk comes to dwarf the egg cell itself, containing as it must the nutrients for 21 days of incubation in the nest.

The construction of the rest of the egg is initiated by ovulation. The egg cell, riding on the completed yolk, is released from the ovary and enters the funnel-shaped opening of the oviduct, a tube about 2 feet long. If the hen has mated in the last few weeks, sperm will be stored at the upper end of the oviduct, and will fertilize the egg cell. After about 15 minutes in this region, the egg cell and yolk move into a portion of the tube called the magnum, whose walls secrete the proteins of the albumen, or white of the egg

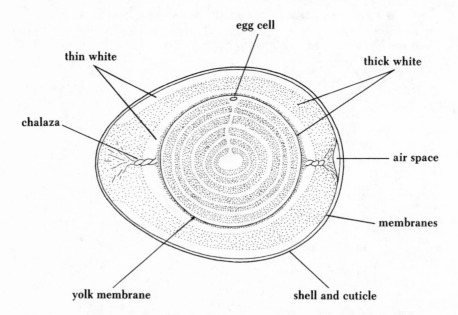

The structure of the egg. Layering of color in the yolk is caused by the hen's periodic ingestion of grain and its pigments. As the yolk grows, the egg cell moves from center to surface, where it can be fertilized easily.

("albumen" comes from the Latin *albus*, "white"). In the space of a few hours, the four layers of albumen, alternately thick and thin in consistency, are laid down over the yolk. The first thick layer is twisted by folds in the magnum wall to form the two *chalazae* (from the Greek for "hailstone" or "small lump"), or the cords which anchor the yolk to the shell and keep it centered. The next section of the oviduct encloses the yolk and albumen in two tough but very thin membranes, which are attached to each other everywhere except for one end, where the air pocket will later develop to supply the hatching chick with its first gulps of air. These membranes, which take about an hour to form, serve mainly as a barrier against bacteria.

The next 5 hours are spent in "plumping" the egg, or pumping water and salts from the oviduct walls through the membranes and into the albumen. The final stage, the making of the shell, occurs in the uterus, or shell gland and takes about 14 hours. The shell is about 4% protein and 95% calcium carbonate, materials which again are secreted by the surrounding tissue. Much of the calcium needed is removed from honeycomblike structures in the hen's own bones. The principal function of the shell, of course, is mechanical protection, though the embryo does derive some calcium from it. Since the enclosed embryo is alive and, although it cannot breathe, needs a constant supply of air, the shell is porous and allows oxygen to pass in, carbon dioxide to pass out. The shell color is a matter of genetics, and not of the hen's feed; Leghorns lay white eggs, Rhode Island Reds lay brown, robins blue, and so on. The last coating to surround the egg is a waxy cuticle, which slows the loss of water. The completed egg is expelled, about 25 hours after leaving the ovary, on the signal of prolactin, the same hormone that causes uterine contractions and milk secretion in mammals.

Aging and Egg Quality

The moment the egg leaves the hen, it begins to change in several ways, almost all of which have deleterious consequences for the cook. The principal change is one in pH; both the yolk and albumen solutions get more alkaline with time. This is because the egg "breathes," or expires carbon dioxide, which, dissolved in the internal liquids, is a weak acid. The yolk jumps from a slightly acidic pH of 6.0 to a more nearly neutral 6.6, while the albumen goes from a slightly alkaline 7.7 to 9.2 and sometimes above. (Incidentally, egg albumen and baking soda are the only alkaline ingredients to be found in the kitchen.) Because the albumen proteins repel each other more strongly at high pH and so tend not to form clusters that can deflect light, the white of an older, more alkaline egg is clearer than that of a fresh egg.

Most of the change in pH occurs in the first few days after the egg is laid, though it can be slowed down. It is commercial practice to coat the shell with mineral oil after the protective cuticle is lost during washing; the oil retards water loss and generally improves the storage life of the egg.

The change in pH would be of academic interest only were it not correlated with two other, more obvious changes. The proportion of thick white to thin, initially about 60% to 40%, decreases, and the yolk membrane weakens. This means that the cook must deal with an egg that is runnier in the pan and more likely to suffer a broken yolk. It is not known whether the thinning of the albumen is directly caused by the rise in pH. The yolk membrane is stretched as water slowly passes through it from albumen to yolk.

Because it is these physical characteristics that suffer most during storage, egg quality is defined in terms of albumen thickness and yolk roundness (a weak membrane allows the yolk to sag into a flattened disc). But these measurements can be made only by breaking the egg and making it unfit for sale, and so certain secondary indicators of quality are examined instead. Shortly after it is laid, the egg is *candled,* or placed in front of a light source that reveals its physical condition (electronic eyes are conducting more and more of these tests). Cracks in the shell and harmless but unattractive blood spots on the yolk and "meat spots" in the whites are quickly detected. Twirling the egg makes the yolk move, and gives some indication of its tendency to flatten and of the albumen's thickness.

Candling also reveals the size of the air cell, and the smaller this is, the better the egg. As the egg loses both carbon dioxide and moisture with time, its mass shrinks, and the empty space at the wide end of the shell gets larger and larger. In a fresh egg, this space is about ⅛ inch deep and the diameter of a dime. If it is substantially larger, then the wide end of an egg put in a bowl of water will rise well above the narrow end. This test is centuries old. Around 1750 Hannah Glasse gave two ways of determining the freshness of an egg, which would have been an important talent in households where it might be laid in an odd corner of the yard. One is to feel how warm it is—probably less than reliable—but the second indirectly measures the air cell: "[Another] way to know a good egg, is to put the egg into a pan of cold water; the fresher the egg the sooner it will fall to the bottom; if rotten, it will swim at the top."

Egg Grades

The grading of egg quality, which is not mandatory by law, is independent of the different sizes available. Generally, only the two top grades, AA and A, are seen in stores, with AA having a somewhat thicker white and stronger yolk. Because candling is not a foolproof process, up to 20% of a graded lot of eggs is allowed to be below grade. It is commonly thought that egg grades are an indication of freshness, but this is obviously not the case; eggs are graded shortly after being laid, and all eggs, AA or B, deteriorate with time. Higher grade eggs simply start out better than the others.

If you are going to use a particular carton of eggs fairly soon and will be boiling or scrambling them or making them into a custard, then the higher grade will not be worth the extra price. But if you go through eggs slowly, or like your poached and fried eggs neat and compact, or are planning to make a meringue, soufflé, or cake, then you might be better off with

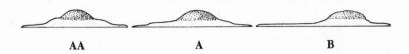

| AA | A | B |

Three different grades of eggs. The AA egg has a high proportion of thick white and a firmly contained yolk. The A egg has less thick albumen and a weaker yolk membrane. The B egg is likely to spread all over the pan, and its yolk is easily broken. Grading has nothing to do with egg freshness.

the premium grade. The lower grade's thinned egg proteins and a weakened yolk membrane, which can allow some fat to leak into the white, may lower the white's foaming power substantially. In any case, slow deterioration in quality goes on as long as eggs are stored, so prompt use is best. In the meantime, they should be kept refrigerated, and preferably in the carton or some other container rather than in an open rack. Otherwise the porous shell allows the egg to absorb strong odors and lose moisture and carbon dioxide more rapidly. Once broken open, eggs should be used promptly. Deprived of their highly effective barriers to bacterial infection, eggs become quite susceptible to spoilage.

Because the nutritional value of eggs varies only with their size, it is not an important factor in judging their quality. Larger eggs, of course, have more food value than small ones (jumbo, extra large, and large eggs weigh 30, 27, and 24 ounces per dozen respectively). A single large egg provides 6.5 grams of protein, or about 13% of the recommended daily intake for adults, as well as 80 calories, and good amounts of iron, phosphorus, thiamine, and vitamins A, D, E, and K. The disadvantage of eggs as a staple food is their high cholesterol content. Despite beliefs to the contrary, neither shell color nor the fertilization of the ovum has any nutritional significance. A fertilized egg contains only a microscopic embryo whose development has been arrested by refrigeration.

The Yolk and Albumen

The yolk and albumen are the parts of the egg used in cooking. They are both very complicated materials, and food science has a long way to go in explaining the intricacies of their composition and behavior. But we do know enough to make sense of traditional egg recipes and cooking techniques.

The yolk accounts for just over one third of the egg's weight after shelling. It is about 50% water, 34% lipids—fats and related substances—and 16% protein, with traces of glucose and minerals. Of the lipids, about two thirds are ordinary animal fats; one quarter, phosphorus and fatty acid complexes, including the emulsifier lecithin; and about one twentieth, cholesterol. The yolk contributes most of the cholesterol, about three quarters of the calories, and most of the vitamin A, thiamine, and iron of the whole egg. Its yellow coloration—*yolk* comes from the Old English for "yellow"—is caused mainly by pigments called xanthophylls. These are chemical relatives of carotene, a precursor of vitamin A, but are not themselves capable of being transformed into that vitamin. So you can't judge the nutritional value of a yolk by its color.

The disposition of materials in the yolk is not quite what might be expected. The continuous liquid phase is not fat, but a water solution of several proteins, called the livetins, which become the blood serum proteins in the chick. The livetins behave much as the albumen proteins do. Floating in the livetin solution is a variety of small bodies, the most conspicuous being the yolk granules and yolk spheres. The granules are intricate complexes of lipid-protein particles and threads about 1 micron, or 40 millionths of an inch in diameter. Fully three fourths of the yolk's protein and fat are contained in these tiny complexes. The spheres have much the same chemical composition, but are larger, on the order of 0.1 millimeter, or 0.004 inch, and contain several smaller bodies within them. The spheres are barely visible, if you look at a thin film of yolk on a white plate. Our everyday experience certainly gives us indirect evidence of the yolk's granular structure: a hard-boiled yolk comes out soft and crumbly, unlike the white's smooth, uninterrupted protein network.

The white accounts for nearly two thirds of the egg's shelled weight, and most of that is water. About 10% of the albumen is protein, with only traces of minerals, glucose, and lipids. The protein, as one would expect, is almost perfectly balanced in the amino acids necessary for animal life, and is, in fact, used as a standard for measuring the value of other protein sources, including meat (whole egg rates a 94, meat a 75). As we have seen, the albumen comes in two consistencies, thick and thin, with the yolk cords being a twisted portion of the thick. The difference is due to a particular protein component, ovomucin, which somehow organizes the otherwise weakly viscous solution of the other proteins into a laminated structure with more body.

There are six major albumen proteins, each with somewhat different properties, and each interacting with the others in ways that are largely unknown. Among the most populous are ovalbumin, conalbumin, and a small group known as the globulins, which together account for about 85% of the protein mass and behave in roughly the same way, as we shall see. Several of the minority population, however, are worth mentioning. Lysozyme, one of the globulins, is interesting for the fact that it is also present in human saliva; in egg or mouth, it has the useful property of weakening bacterial cell walls and so reducing the chances of infection. Avidin is an antagonist of the vitamin biotin, and makes the consumption of raw eggs in large quantities a bad idea (if it wasn't already). In the egg, this biotin-binding activity may have the purpose of discouraging bacterial growth. Biotin deficiency can cause disorders of the skin and digestion, but cooking eggs eliminates the problem because avidin is inactivated by heat. A third protein, ovomucoid, inhibits the digestive enzyme trypsin, and so will lower the nutritional value of all the egg proteins unless it too is denatured by cooking.

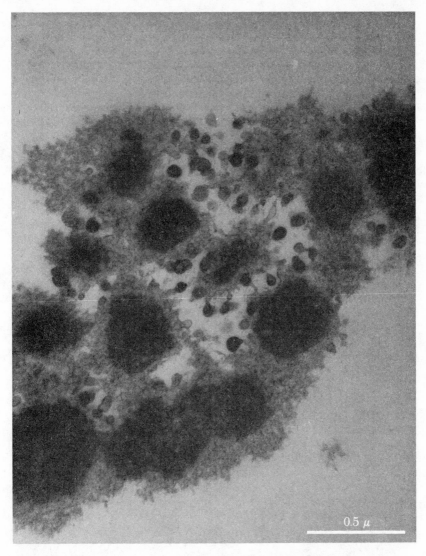

A single yolk granule teased apart and photographed through the electron microscope. It is a complex assemblage of various proteins, lipids, and lipid-protein combinations. (From Chang et al., *J. Food Sci.* 42 (1977): 1193–1200. Copyright © by Institute of Food Technologists; with permission of W. D. Powrie.)

EGG COOKERY

In addition to their intrinsic virtues, eggs have three remarkable talents that are exploited by the cook: they can thicken liquids into solids, introduce the light texture of a foam, and stabilize oil-and-water sauces with the emulsifiers in the yolk. The matter of emulsification is discussed in chapter 7, together with the (in this case) undesirable coagulation of egg proteins. Here, we will talk about useful coagulation and foaming.

The Chemistry of Coagulation

Protein Bonding

The behavior of eggs in the kitchen is mostly a matter of protein chemistry, and in particular the chemistry of coagulation. The albumen proteins are long chains of amino acids that are folded up onto themselves in a compact, roughly globular shape. Each protein molecule is held in this shape by various kinds of bonds between different parts of its chain. Different molecules are prevented from bonding to each other because in the native chemical environment of the albumen, each one accumulates a net negative charge: and particles with like charge repel one another. Both of these conditions—the bonding that shapes a molecule, and the mutual repulsion of different molecules—are very easily disturbed by changes in acidity, salt content, temperature, and even by bubbles of air. Such changes can cause the protein molecules to bond together into a solid mass: that is, to *coagulate*.

Take as an example the changes that occur during cooking. When eggs are heated, the increased energy of all the molecules in the albumen breaks some of the shaping bonds in the proteins, and the individual protein molecules begin to unfold. This unfolding exposes more of each molecule's length

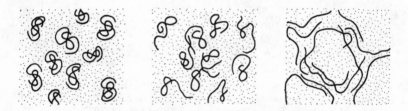

Egg proteins begin as folded chains of amino acids *(left)*. As they are heated, the increased motion of their component atoms breaks some bonds, and the chains begin to unfold *(center)*. The unfolded proteins then begin to bond to each other. If not carried too far, this results in a solid meshwork of long molecules that contains many small pockets of water *(right)*.

to the others, and so makes bonding between different molecules more likely. As the temperature of the egg rises, then, the proteins unfold, bump into parts of each other, and bond to each other. Eventually, the initially separate, globular molecules dispersed in water form a mass of partly extended, intricately interconnected proteins, with water molecules held in the interstices of the mass. The liquid has become a solid, and because the proteins have clustered together densely enough to deflect light rays, the clear matter has become opaque.

The particular qualities of this new solid, or gel—hard or soft, dry or moist, lumpy or jellylike—depend on the degree to which the proteins have bonded to each other. A mild interaction results in a relatively open structure that can retain a good deal of water, while more extensive bonding makes for a denser mass.

Overdoing It: Water Loss

The point of most egg cookery is to bond a liquid, whether the egg itself or a mixture of egg and other liquids, into a moist, delicate solid. This means avoiding such extensive coagulation that the proteins squeeze out the liquid they are supposed to retain. Either a rubbery texture or curdling—solid lumps floating in the wrung-out liquid—is the unhappy result of overcooking egg dishes. To understand this, we must descend once more to the molecular level. Water is held both within and around the coiled proteins by relatively weak hydrogen bonds, which are easily broken as the energy of both water and protein molecules increases. Only the stronger ionic and covalent bridges between proteins last for any amount of time as the solution is heated. Heat, then, automatically favors protein-protein bonds over protein-water bonds. And once two proteins join at a single point, their respective lengths are kept in close proximity, and further bonding is therefore more likely. The more the system is heated, then, the closer and more completely the proteins squeeze together, and the less room there is in which they can retain water. Dishes to which liquid has been added—scrambled eggs, for example—separate into two phases, the excess water and solid lumps of protein, while boiled or fried eggs lose water in the form of steam and get progressively more rubbery.

Most of the egg proteins coagulate readily when heated, but ovomucin, the protein that makes the thick white thick, is more resistant than the others. This is why the thin layer that runs away from the yolk as an egg is fried whitens and solidifies first. The yolk cords, or chalazae, are especially rich in ovomucin and may remain liquid long after the rest has set (one reason, perhaps, why many people tend to overcook scrambled eggs). The thin egg white begins to coagulate at about 145°F (63°C); at 150°F (66°C) neither layer of white will flow as a liquid, and the yolk proteins, the livetins, begin to set. At 160°F (71°C) the entire white has a firm but tender consistency. Continued heating makes the egg less tender.

Added Ingredients

As often as not, eggs are cooked along with other ingredients, whether sugar and milk in the case of custards, or salt and a little liquid—cream or sherry—in the case of scrambled eggs. Diluting eggs with other liquids raises the temperature at which coagulation begins, not for any subtle reason of chemistry, but simply because the protein molecules are isolated from each other by many more water molecules, and they must move that much more rapidly in order to link up at a noticeable rate. Sugar has the same effect, and for the same reason. One tablespoon of sugar surrounds each protein molecule in a large egg with a screen of several thousand sucrose molecules. Salt and acids, however, have the reverse effect and actually promote coagulation. Acid—cream of tartar, lemon juice, or the juice of any fruit or vegetable—lowers the pH of the egg, and thereby lowers the mutually repelling negative charges on individual protein molecules. The lower this charge, the smaller the repulsion, and the less force required to overcome it; the protein molecules can then bond together more easily. Salt, by dissociating into positive sodium and negative chloride ions, disturbs the electrical environment in its own way, again encouraging coagulation. So, diluting or sweetening eggs delays coagulation; salting or acidifying them accelerates it.

Cooked Eggs

Boiled Eggs

Boiling is often taken as a measure of minimal competence in egg cookery, and there are really only two things to keep track of with this method: the cooking time, which will determine the consistency of the white and yolk, and the water temperature, which actually should be short of the boiling point. Boiling water is turbulent enough to smash the eggs into each other and the pan walls, cracking the shells, allowing some albumen to leak out, quickly overcook, and release unpleasant amounts of hydrogen sulfide gas (of which more in a moment). Cooking time is a matter of taste and of experience, either yours or a cookbook author's. Soft-boiled eggs are cooked only long enough—typically 3 to 5 minutes—to get the white solid while keeping the yolk liquid. Hard-boiled eggs are cooked four or five times as long, to ensure that the yolk has fully solidified as well. Tender whites can be obtained by cooking eggs at no more than a simmer—about 185°F (85°C)—for 25 to 35 minutes. This relatively gentle heat keeps the whites from hardening excessively while the energy slowly makes its way to the yolk. The timing will vary not only with preferred texture, cooking technique, and the size of the egg, but also with altitude, because water boils at lower temperatures the higher you go. You can tell hard-boiled eggs from raw ones by the fact that boiled eggs are easily spun around on their sides and stopped. Because the liquid insides of a raw egg move somewhat independently, such an egg will spin only reluctantly, and may start moving again if it is stopped quickly and then released.

There are two peculiarities associated with hard-boiled eggs. One is the occasional difficulty encountered when peeling the egg. It turns out that peelability is affected by the pH of the egg white, and so by the egg's freshness. If the pH is below 8.9—in a fresh egg it is closer to 8.0—then the inner membrane tends to adhere to the albumen, whereas at the figure typical after three days of refrigeration, around 9.2, the problem no longer exists. Exactly what the chemistry involved is, no one knows, though some cooks claim that salt in the cooking water helps. Baking soda, which is alkaline, might seem an obvious aid, but for the fact that it can give a slightly soapy taste to foods if it isn't neutralized by acid. In any case, peeling trouble can always be balanced by the pleasant knowledge that the eggs must be very fresh. Carefully piercing the air cell with a pin may allow some of the cooking water to leak between the membranes and so ease their separation from the egg, but this also tends to make the egg crack. Modern science generally agrees with the empirically derived advice given by the 14th-century *Ménagier de Paris:* "Item, whether they be soft or hard, as soon as they are cooked put them in cold water: they will be easier to peel."

The other oddity about hard-boiled eggs is the occasional appearance of a greenish-gray discoloration on the surface of the yolk. The color is caused by a harmless compound of iron and sulfur called ferrous sulfide, which is formed only when the egg is heated, especially in the case of an extra-alkaline less-than-fresh egg. The yolk contains a good deal of iron, and the albumen a good deal of sulfur, primarily on side chains in the protein ovalbumin. When that protein is heated, some of its sulfur atoms are liberated and react with hydrogen ions in the albumen to form hydrogen sulfide (H_2S). In minute quantities this gas lends a characteristic and pleasant odor to cooked eggs and meat, but in larger quantities is the odor we associate with rotten eggs. As the gas forms, it diffuses in all directions, and some reaches the surface of the yolk, where it encounters iron and reacts to form the dark particles of ferrous sulfide (FeS). The way to minimize the discoloration is to minimize the amount of hydrogen sulfide that reaches the yolk. First, cook the eggs only as long as is necessary to set the yolk. Then plunge the cooked eggs immediately into cold water. This lowers the pressure of the gas in the outer regions of the white, since cool gases exert less pressure than hot, and cool protein loses less sulfur than hot. Consequently the hydrogen sulfide diffuses away from the yolk, toward the region of lower pressure at the surface. Finally, peel the eggs promptly: this also helps pull the gas away from the yolk.

Poached Eggs
A poached egg is made by carefully sliding the raw egg from a bowl into a pan of simmering water, where it slowly sets. The tricky thing about poached eggs is producing a nice, compact shape without overcooking the white. Fresh, Grade AA eggs are clearly the best candidates for this kind of

cooking, since they will have the largest proportion of thick white and will spread the least. Don't shell the eggs until just before they are to be cooked; if left to sit in the open air, their quality declines rapidly. Make sure the water is close to, but not at, the boiling point—turbulent water will tease the thin albumen away—so that the outer regions will coagulate quickly and restrain the rest. Then finish cooking with the water at a simmer to avoid overcooking the outside while the inside heats through. Adding some salt, lemon juice, or vinegar to the water will speed coagulation of the albumen surface and help preserve the egg's shape, but these ingredients also have strong flavors.

Fried Eggs

Like poaching, frying calls for fresh eggs and careful handling in order to prevent the white from spreading out to flapjack dimensions. Temperature control is more difficult in frying. Boiling water cannot exceed 212°F (100°C), and covers most of the albumen, heating it evenly from all sides, while the heat of a frying pan is more intense and is applied only to the bottom of the egg. The cook must steer a middle course between using too little heat, which allows the egg to spread before setting, and too much, which hardens the bottom while leaving the rest of the egg liquid (eggs done "sunny side up," or cooked only on one side, are especially tricky). According to food scientists, the ideal temperature range for frying eggs is 255° to 280°F (124° to 138°C). This is fine if you have a calibrated electric skillet; if you don't, this range can be approximated by heating the pan until butter sizzles without turning brown. Basting the top of the egg with butter from the pan, or putting the lid on and steaming the top with the egg's own moisture will help even out the heating of the sunny side, but at the cost of dimming the sun. These techniques cause the thin layer of white covering the yolk to coagulate.

Scrambled Eggs and Omelets

Scrambled eggs and omelets are both made by mixing the yolks and whites together—quite thoroughly in the case of omelets—and sometimes adding milk or water to produce a moister, somewhat softer mass; the coagulated proteins can hold somewhat more liquid than they do in the plain egg. The amount of liquid added is fairly critical. If so much is used that the holding capacity of the proteins is reached or exceeded, then even the slightest overheating will cause the liquid to be squeezed out and form a separate puddle. For scrambled eggs, the recommended amount is between 2 and 5 teaspoons per egg, while the omelet, which is cooked more rapidly, will only hold 2 or 3. Many cooks add no liquid at all to either dish, but take great care that the eggs do not dry out. Scrambled eggs are cooked slowly over low heat to avoid overcoagulation, and stirred and scraped occasionally to break the coagulating mass into large, soft lumps and to distribute the heat evenly. They should be removed from the stovetop while slightly underdone, since they retain heat and will continue to cook for some time on their own. Ordinary, or French omelets are cooked over high heat in a thin, intact sheet which is turned onto itself while the upper surface is still partly liquid. "Puffy" omelets are actually stovetop versions of the soufflé. Omelets have been known in England since about 1600, and in France centuries before that. The word has gone through various forms—*alemette, homelaicte, omelette* (the standard French)—and derives ultimately from the Latin *lamella*, "thin plate."

Custards

Custards are sweet, very moist, tender gels of egg protein. The quiche (the French version of the German *Kuchen,* "little cake") is simply an unsweetened custard. The basic custard contains an egg, 1 cup of milk, and 2 tablespoons of sugar, although more eggs can be used, and often extra yolks are added for greater richness. The ingredients are mixed together thoroughly, set in a slow oven, usually in a water bath, which moderates the heat (the walls of the custard pan can never exceed 212°F, or 100°C) and baked until set. A creamy rather than solid custard is made by stirring continuously during stovetop cooking to prevent the proteins from bonding into a fully solid mass. Though the milk contains some protein, it does not contribute much to the gel structure. More significant is its complement of salts, without which the diluted egg proteins are unable to gel: it is impossible to make a thick custard by replacing the milk with pure water. The most important effect of milk and sugar is to thin out the egg proteins. The milk alone increases the volume of liquid by nearly a factor of 4, and as we have seen, 1 tablespoon of sugar contains several thousand molecules for every molecule of egg protein. Because the egg proteins are so outnumbered, the coagulation temperature is elevated by about 50°. It is very easy to overcook them, and end up with grainy curds in a sweet soup.

Many cooks have known the temptation to raise the temperature after a custard has been nearly an hour in a low oven with no sign of setting. But there is good reason to resist. The lower the rate of heating, the greater the safety margin between setting and curdling. According to one study, a custard cooked by immersing the dish in boiling water thickened at 190°F (88°C) and curdled at 195°F (91°C). The same recipe cooked over water that is brought to the boil *after* the dish is in place thickened at 180°F (82°C) and curdled at 190°. The temperature gap between success and failure is doubled, and this gives the cook more leeway to get the custard out of the oven in time. There is another advantage to setting the proteins at the lower temperature—chemical reactions speed up as the temperature rises. The egg proteins are coagulating perhaps hundreds of times faster at 190° than they are at 180°: for a rough analogy, think of trying to stop on a dime driving a car going $\frac{1}{10}$ mile per hour, against driving at 60. So the slower the custard is cooked, the less likely the cook is to overshoot "done" and hit "finished."

Use a low oven temperature, never neglect to use the water bath, which slows the transmission of heat from oven to egg, remove the custard as soon as possible after setting, and bring its temperature down in a cold water bath (like scrambled eggs, a custard will continue to cook with the heat it retains). A skewer is the traditional tester; it will emerge from a set custard relatively dry. Once the proteins have bonded to each other, they are less likely to stick to anything else.

EGG FOAMS

The egg's talent is not exhausted by setting liquids into solids, as the light, delicate meringue, soufflé, and angel cake attest. Egg whites make excellent foams—their volume can be increased as much as 8 times by beating—and their secret is the remarkable teamwork of the albumen proteins. Yolks are not only poor foamers, but also ruin albumen foams.

Technically speaking, a foam is a dispersion of gas in a liquid; more loosely defined, it is a relatively stable mass of bubbles. All the foams we see in everyday life—sea spume, soap suds, meringue, or the head on beer—consist of tiny pockets of air (carbon dioxide in the case of beer) surrounded by a thin film of water with various substances dissolved in it. It is the dissolved substances that are the key to foams; plain water alone, or any pure liquid for that matter, will not foam, no matter how much it is agitated. This is because all liquids have a significant surface tension, a force that makes a liquid assume a compact shape. All the molecules of a liquid exert an attractive force on each other. Without such a force, they would fly apart and constitute a gas, not a liquid. Molecules at the surface of a liquid—at the interface, say, of water and air—are more attracted to each other than they are to the air, and so are pulled back into the liquid. This imbalance of forces minimizes the number of water molecules, and so the area of water, exposed to the air. The geometrical shape with the least surface area for a given mass of liquid is a sphere (hence the shape of dewdrops). Single large masses are more stable than small, numerous droplets (let two droplets touch each other, and they will coalesce). Because their liquid walls are only a few molecules thick, bubbles expose a huge surface area of liquid to the air, and they simply will not form unless the surface tension of the liquid is greatly reduced. This is the job of the dissolved or dispersed molecules in all froths: they interrupt the matrix of forces in the liquid enough to make bubbles that are at least momentarily stable. In the egg, these molecules are mostly proteins.

Stabilizing the Foam

The cook, of course, wants to make foams that last longer than suds, and the egg white is able to do this in two ways. Because the albumen is a thick, viscous solution, it drains more slowly out of bubble walls than does a thin liquid. All of the egg proteins, because they are large molecules that get in

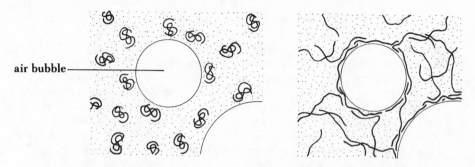

air bubble

The proteins in egg white produce a long-lived foam by unfolding at the interface between liquid and air and bonding to each other. A solid network of reinforcement is the result.

the way of each other and of the water that surrounds them, probably contribute to the noticeable viscosity of the white and yolk. The second way in which egg white stabilizes foams is by introducing a kind of reinforcement into the bubble walls, a culinary equivalent of quick-setting cement. Eggs contain several such materials, all proteins. Some are unfolded and coagulate to some extent simply by being forced into contact with air. This is apparently true of conalbumin, the globulins, and ovomucin. Here is what happens: as the liquid is beaten and air bubbles incorporated into it, the proteins in the bubble wall are subjected to an imbalance of forces. One part of a large protein molecule is attracted to the water molecules through hydrogen bonds, while another bordering on the pocket of air feels no such attraction. This imbalance causes the normally compact molecule to unfold, with its hydrogen-bonding side chains immersed in the water film, and its water-avoiding, or hydrophobic, side chains sticking up into the air pocket. Once unfolded, the protein molecules bond with each other and thus form a delicate but definitely reinforcing network, a solid lattice that holds the water in place while shielding it from the pockets of air.

Uncooked foams, those in chiffon pie and mousses for example, owe their lightness entirely to the egg proteins that can be "surface-denatured" readily. A different protein, however, is responsible for maintaining a foam structure when it is cooked, either alone as a meringue, or as the leavening of a soufflé or an angel cake. Because gases expand when heated, the air cells in the foam grow in the oven. They would eventually pop through the tenuous network of surface-denatured proteins, and the whole mass would collapse, if it weren't for ovalbumin, the protein that makes up over half of the albumen proteins. Ovalbumin doesn't unfold much when the egg is beaten, but it does coagulate readily when heated. It forms a solid network in the bubble walls that resists collapse even when the hot air escapes and, in the case of the naked meringue, when much of the water itself evaporates. Put simply, ovalbumin makes it possible for the cook to transform a liquid foam into a solid one.

The remarkable foaming power of egg white is the result of the combined activities of the various proteins. Ovomucin and the globulins increase the viscosity of the albumen, and so make possible a finely divided foam that is slow to drain. Together with conalbumin, they stabilize the foam at room temperature; ovalbumin preserves its structure in the oven. These roles have been confirmed by experimentally isolating the proteins and trying to make foams with them individually and in different combinations. A solution of pure ovomucin foams at room temperature, but coarsely; the addition of globulins reduces the size of the air cells and improves the volume. (Duck eggs are known to be an inadequate substitute for hen's eggs in making foams, and it turns out that they contain very little lysozyme, one of the globulins. Supplement them with lysozyme, and they do much better.) Successful ovomucin-globulin foams collapse in the oven, and while ovalbumin cures that problem, ovalbumin itself forms a foam at room temperature only after a very long period of beating, much longer than that required by ovomucin. The division of labor is not absolute. Some ovalbumin probably helps with the initial foaming, and the other proteins do eventually coagulate in oven heat. But these proteins are specialists to a large extent, and without any one of them, egg whites wouldn't be as versatile as they happily are.

Beating Is Tricky

This is not to say that meringues are easy to make. For one thing, it is quite possible to beat the albumen either too much or too little, with the common outcome of poor volume, premature collapse, and, in the case of overbeating, some curdling. When whipping begins, large pockets of air are trapped in the albumen, and at the interface between liquid and air the process of surface denaturation begins. For some time, however, the foam will be fairly coarse, and the extent of coagulation insufficient to support it. If the beating is stopped early, the foam will be coarser and lower in volume than it might be, and liquid that contains uncoagulated protein will slowly drain away, causing bubbles to break and the foam as a whole to collapse.

Overbeating, on the other hand, causes too extensive coagulation. Coagulation is the bonding of one protein molecule to another, and the more such bonds a particular molecule participates in, the fewer sites are left available to form hydrogen bonds with water molecules. At a certain point, the water-holding capacity of the protein will decline far enough that the foam will leak liquid—this time containing much less uncoagulated protein—clump up, and lose volume. The ideal beating period lies between these two extremes: enough to produce a protein film strong enough to hold the foam up, but not so much that the water is squeezed off. It turns out that this optimum stage comes *before* the maximum volume is reached, although there is no obvious way of recognizing either of these points. Most cookbooks say to continue beating until the foam is stiff enough to stand up in well defined peaks. It is probably a good idea to try to recognize this stage as soon as it has been reached and then stop well before the danger point.

Fats Interfere

No matter what the ultimate use of the foam, whether as a meringue or in a soufflé, whenever the greatest possible lightness or volume is desired, the whites are always beaten alone. Cookbooks insist that not a speck of yolk or any other fatty material be allowed in the bowl, and for good reason. A single drop of yolk in the white of an egg can reduce the foam's maximum volume by as much as two thirds. The effect of other fats and oils is less drastic, but great enough that bowls made out of plastic should be avoided. Plastics are polymers of hydrocarbon chains, which also form the bulk of all fat molecules, and so plastics tend to retain traces of fatty material on their surfaces.

Fats reduce the volume of albumen foams by getting in the way of coagulation. One of the ways in which protein molecules bond to each other is through the attraction between nonpolar hydrocarbon regions exposed by foaming. It appears that some of the yolk lipoproteins (combinations of protein and fatty molecules) bind to these areas, thereby preempting some protein-protein bonds and so weakening the coagulated lattice. Another possibility is that fatty materials collect at the water-air interface and interrupt it—just as a drop of oil will collapse soap bubbles—even before the coagulation begins. In any case, yolk and other fats are incompatible with the creation of a good volume of albumen foam. Once the foam is formed, however, it can be folded gently into the yolks or other fatty mixtures and do a fine job of leavening. It doesn't matter if the foam slowly degenerates in the mixture, because the important task—filling the mixture with more air bubbles than it could accommodate alone—has already been accomplished. This is exactly how soufflés and sponge cakes are made.

The Effects of Added Ingredients

Albumen foams are almost always made with other ingredients, and these generally have pronounced effects on the beating process and on the final product. Acid is often added in order to stabilize the foam, and sugar and salt to improve its flavor. Acid, usually in the form of cream of tartar (the solid salt of tartaric acid) has been found to have no effect on the volume of foam produced, but it does make the foam less prone to overcoagulation and the resultant lumpiness, drainage, and collapse. Only a very small amount of acid is necessary to make an obvious difference. Most recipes call for about $\frac{1}{16}$ teaspoon per white, which changes the pH of the albumen from about 9 to about 8 (which is still distinctly alkaline). This ten-fold increase in the number of hydrogen ions may lower the reactivity of some side groups on the protein molecules and so limit the extent to which they can bond to each other. (More acid can even limit coagulation during cooking: see the discussion of sauce béarnaise in chapter 7.)

Salt, however desirable its taste, increases the whipping time and decreases the foam's stability. Salt probably plays a role analogous to that of the undesirable yolk lipoproteins. The positive and negative ions compete for bonding sites on the unfolding ovomucin molecules, thereby reducing the number of protein-protein bonds and so weakening the latticework.

Sugar has a mixed influence on whipped albumen. On one hand, it delays foaming, and reduces its maximum volume and lightness. This happens for the same reason that sugar raises the temperature of egg coagulation in custards: the sugar molecules simply outnumber and get in the way of the proteins, and this slows the rate at which the latter bond to each other. This delay is a distinct disadvantage when the whites are whipped by hand— added sugar can double the beating period—but it may be an advantage when using an appliance, whose efficient motor brings with it the danger of overbeating in a matter of seconds. More clear-cut is sugar's enhancement of the foam's stability in the oven. Sugar helps prevent drainage and collapse by forming hydrogen bonds with water molecules and so delaying evaporation from bubble walls until after the proteins, especially ovalbumin, have had time to coagulate and reinforce the initial lattice. The meringue maker can avoid the disadvantages and exploit the advantages of added sugar by beating the whites without sugar, and then whipping some in just before baking.

Extra water is an ingredient seldom called for in recipes, but when added in small amounts, it will increase the volume and lightness of the foam, though there is a slightly greater likelihood that some liquid will drain off. The more liquid there is, of course, the larger the surface area that can be exposed to the air, and so the more bubbles that can be formed. At the same time, however, the proteins are diluted, and at a certain point they will be spread too thin to stabilize the foam. Albumen diluted by 40% or more of its volume in water cannot produce a stable foam.

Whipped Egg Dishes

Meringues

The meringue, which has been made since at least 1700, is the simplest of egg foam concoctions, in the sense that it contains only egg white, sugar, and traces of acid and salt. The proportion of sugar depends on the desired final texture. Typically, 2 tablespoons are added per white for a soft meringue, 4 for a hard one. The hard meringue is often used as the base for some kind of dessert sauce or fruit, and profits from the additional strength provided by the hardened sugar syrup. In any case, fine confectioner's sugar is preferred to ordinary granulated sugar because it dissolves much more rapidly. Undissolved granules give a meringue a gritty texture and can cause beads of syrup to form during baking.

The temperature of the meringue ingredients is not critical, as it is with whipped cream, but room temperature is preferred to the 40°F (4°C) typical of refrigerators because the surface tension of the albumen is lower and air is easier to beat in. The bowl should be scrupulously clean of any yolk or oil. The whites are put in the bowl and whipped until somewhat foamy, the salt and cream of tartar are beaten in, and whipping is continued. If a fast electrical beater is used, the sugar is slowly added after the foam first appears, while if the beating is done by hand, it is added near the end. As beating proceeds, air pockets are forced into the albumen; ovomucin, conalbumin, and globulins collect at the air-water interfaces, unfold, and coagulate, and the foam begins to stiffen. The initially clear liquid becomes opaque as it fills up with air cells and threads of coagulated protein, and gradually takes on a dull, dry surface and semisolid consistency. When the foam is stiff enough to stand on its own, even in relatively thin "peaks," beating is stopped. Overbeating can result in overcoagulation and irreversible clumping. If the foam gets lumpy, then there is nothing to do but start again with a new bowl of egg whites.

Meringues and Copper Bowls

For better than two centuries now, a quiet culinary tradition has specified the use of copper utensils in making meringues. One early trace of this tradition is a 1771 illustration in the French *Encyclopédie* that shows egg whites being beaten in a copper bowl. The practice must have been a matter of common professional knowledge, since it appears to go unmentioned in French cookbooks until Henri-Paul Pellaprat's *L'Art culinaire moderne* of 1936, which directs the cook to "Put the whites in the copper bowl reserved [*affecté*] for this use." Today, it tends to be the fancier books that call for copper; most recipes settle for the cheaper expedient of cream of tartar.

Only recently have people tried to explain why copper bowls should be used and what exactly they do for meringues. A side-by-side comparison

Detail of "*Pâtissier*," or "The Pastry Cook," from the *Encyclopédie*; the illustration was first published in 1771. The boy at right wields what the accompanying key calls "a copper bowl for beating egg whites and mixing them with the dough from which biscuits are made."

demonstrates that copper does indeed make a difference: it produces a creamier, yellowish foam that is harder to *over*beat—to turn into lumps and liquid—than is the snowy-white, drier foam made in ceramic or stainless steel bowls. Like cream of tartar, then, copper stabilizes the foam and gives the cook more leeway in the beating. On account of this similarity in effect, some cookbook writers have suggested that copper imparts an "acidity" to the whites, while others claim that the metal helps set up an electric field which somehow affects the proteins. Neither of these ideas sounds at all likely, and attempts to find a drop in albumen pH or a voltage between bowl and whisk have failed.

It has been reported that portions of several of the albumen proteins—ovomucin, the globulins (including lysozyme), and conalbumin—are retained in the bubble walls of beaten egg whites, and so are thought to be responsible for the stability of the uncooked foam. Ovomucin has generally been credited with being the principal stabilizer, and has gotten most of the attention from food scientists. But it happens that conalbumin has the very interesting property of being able to bind metal ions to itself. This protein was first described in 1900, and in the mid-1940s an investigation of its anti-bacterial properties disclosed the fact that conalbumin binds iron. This discovery led directly to a search for an iron-binding protein in human blood, and *transferrin*—so named for its important role in transporting iron throughout our body—was found in 1946. When it was determined that human transferrin, egg conalbumin, and milk lactalbumin are homologous proteins—for all practical purposes, the same molecule—the latter were renamed "ovotransferrin" and "lactotransferrin," though "conalbumin" is still commonly used for the egg protein. Over time, it was shown that when the pH of their environment is above 6, all the transferrins can bind a number of different di- and tri-valent metal ions, including iron, zinc, manganese—and copper.

For the meringue maker, the most interesting part of the transferrin story dates from the late 1950s. P. R. Azari, a graduate student in chemistry at the University of Nebraska, was trying to analyze the metal-binding mechanism of conalbumin. When the protein picks up metal ions—either one or two per molecule—the resulting complex is colored: salmon-pink in the case of iron, yellow in the case of copper. Azari's plan was to break up the molecules with enzymes and attempt to isolate the colored fragments for detailed analysis. But the procedure didn't work: even after long periods of incubation with the enzymes, little or no digestion occurred, and the protein-metal complexes remained intact. Azari soon discovered that the complex is much more stable, much more resistant to denaturation, than the protein is by itself. And this stability extends to surface denaturation, the unfolding of molecules in bubble walls. When conalbumin has bound metal ions to itself, it becomes much more difficult to unfold.

The relevance of this property to whipping egg whites is evident. Suppose that conalbumin can pick up enough copper ions from the bowl's surface to form a significant number of the more stable protein-metal complexes. If so, then some of the molecules involved in creating the raw foam will now be resistant to unfolding. This would make the mass of egg white harder to denature to the point of overcoagulation and curdling, and so more tolerant of overbeating. By stabilizing the conalbumin fraction, the copper bowl would stabilize the entire egg white foam.

To test this hypothesis, it was necessary to find a way of determining whether the protein-metal complexes are indeed formed when egg whites are beaten in a copper bowl. This turned out to be a relatively simple task. The complexes are colored, which means that they have a characteristic spectral "fingerprint": they absorb light in very particular regions of the electromagnetic spectrum. And when Stanford University biologists Winslow Briggs, Joseph Berry, and Sharon Long helped me run samples of copper-beaten egg whites through a spectrophotometer, we saw the fingerprint of the copper-conalbumin complex. Similarly, when we added iron salts to egg whites beaten in a ceramic bowl, the iron-conalbumin pattern showed up. These results were consistent with the theory that the copper bowl exerts its stabilizing effects on whipped egg whites by providing ions to the metal-binding protein conalbumin.

These findings suggested a further development in our culinary practice. If several different metals can complex with conalbumin, then why limit ourselves to copper, which is expensive and (in large quantities) can have toxic effects? However, formation of the iron-conalbumin complex in egg white does *not* result in a more stable foam, and zinc has only a small influence. Cooks have not been arbitrary in their choice of metal! Perhaps the iron complex is *too* stable, is not denatured by whipping and so cannot participate in the foam at all. Or perhaps copper does more than stabilize conalbumin. P. R. Azari suggests that it may react with sulfur-containing groups along other proteins and interfere with general coagulation in the bubble walls. Whatever the explanation, this striking culinary phenomenon may have interesting things to teach biochemists about how proteins react with each other during denaturation.

Cooking Meringues

Once the foam has formed, it is baked to coagulate the ovalbumin and other proteins, dry out the sugar syrup to the point where it is more solid than liquid, and so leave the structure strong enough to stand up for hours and even support a topping of some kind. Soft meringues are typically baked at 350°F (177°C) for 15 minutes. This crisps the surface but leaves the center moist and chewy. Hard meringues are baked at 225°F (107°C) for an hour or two, or are put into a hot oven with the heat turned off, and allowed to dry for several hours. They come out uniformly dry and crisp. Undercooking results in incomplete coagulation of the ovalbumin and may cause syrup leakage, beading, or collapse. Baking at too high a temperature, on the other hand, may coagulate the proteins so rapidly and fully that the water squeezed out may not be able to evaporate before beading up into tears of syrup on the surface. Weeping can also be a problem in humid weather, when the sugar in a well-dried meringue absorbs moisture from the air and dissolves into droplets of syrup, or makes the confection sticky and limp. In such conditions, the meringue should either be served very soon after baking or sealed in an airtight container.

Cakes

Angel cakes and sponge cakes both derive their light texture from foamed egg whites, but flour and, to a lesser extent, milk are responsible for their strength and stability. Sponge cake batter also includes egg yolks and sugar, and angel cake just sugar. The whites are beaten separately, and then are gently and quickly incorporated with a minimum of stirring into the rest of the mix. Foam and mix are put in the same bowl, and then "folded" into each other by repeatedly pushing a spatula down through the center, drawing it up the side of the bowl, and folding some of the lower layer over the top. This technique preserves as much of the foam as possible, where ordinary stirring would break many of the fragile air pockets. As the cake bakes, the air cells expand to raise the batter, and the flour starch gelatinizes, helping the heat-coagulable proteins reinforce the bubble walls.

Soufflés

Soufflés can be thought of as very light relatives of the foamed cakes. The name of the dish, which goes back at least to the 18th century, is the past participle of the French verb meaning "to blow," "to breathe," "to whisper"—meanings suggestive of the soufflé's notorious fragility. If the oven door is opened as the air cells are expanding and the soufflé rising, the air pressure and temperature can change enough to collapse the whole structure. And when the soufflé is removed from the oven, the trapped air and steam begin to cool and contract, inevitably bringing down the delicate web of proteins and starch.

For soufflés, the beaten egg whites are folded into a base more like a sauce than a batter; there is about half the amount of flour per egg in the soufflé as there is in a sponge cake. In fact, a white sauce—milk and flour—is frequently mixed with the yolks. Grated cheese or purées of meat, fish, or vegetables go into "savory" soufflés, and sugar, fruit, chocolate, or liqueurs into "sweet" soufflés. Soufflés must be cooked at a temperature high enough to set the proteins before the foam has reached its maximum expansion and begins to fall, but low enough to heat the interior without first burning the outside. A 400°F (205°C) oven will produce a creamy center when the surface is done, while 325°F (163°C) gives a more uniformly solid result.

Whipped Yolks

Egg yolks are also beaten in some culinary procedures, but because of their high fat content, and the fact that the yolk proteins are not easily surface denatured, they foam less effectively than the albumen. Zabaglione, a warm, richly frothy mixture of yolks, sugar, and Marsala wine, is the only well-known whipped yolk dish. More common is the beating of yolks and sugar over low heat before they are added to the base for sweet soufflés, foam cakes, and other desserts. This process is known as "ribboning," because the mixture gets viscous and trails off the spoon in a thick band. Ribboning dissolves the sugar, and, by incorporating some air into the mixture, disperses the yolk fats and emulsifiers evenly, which in turn helps disperse the fats in the dessert base.

Of course, eggs are used in many other dishes, often to contribute their yolky flavor or to help reinforce a foam. Various breads and cakes, and French and Italian styles of ice cream are examples. But these are minor roles. Egg yolks are, however, the crucial ingredient in the emulsified sauces, which include mayonnaise and béarnaise. This stellar role is reviewed in chapter 7.

CHAPTER 3

Meat

Meat has a special place in the diet of the Western world. For most of us, it is the main attraction of a meal. And this prominence has attracted the intense scrutiny of several interested groups. Nutritionists tell us that we eat more meat than is healthful; anthropologists debate whether a literal thirst for blood is an inescapable fact of our evolutionary history; scientists in the food industry try to find simple chemical combinations that will mimic meat's flavor and make inexpensive meat substitutes more palatable; cooks carry on a running argument about the proper method of roasting, a process capricious enough that Brillat-Savarin wrote, "We can learn to be cooks, but we must be born knowing how to roast." Some issues have been resolved more satisfactorily than others. Luckily, the behavior of meat in the kitchen is pretty well understood.

MEAT IN THE HUMAN DIET

By the word *meat* we mean the body tissues of animals that can be eaten as food, anything from frog legs to calf brains. However, we usually make a distinction between the "meat" animals—cattle, sheep, pigs—and poultry and fish, and between meats and *variety* meats, organs like the liver, kidneys, intestine, and so on. Meat proper is muscle tissue, whose function is to move some part of the animal. But the word has not always had this meaning, and its evolution suggests a gradual shift in the eating habits of the British people. In the *Oxford English Dictionary*'s first citation, from the year 900, *meat* meant solid food in general, in contrast to drink. A vestige of this sense survives today in the habit of referring to the meat of nuts. It was not until 1300 that *meat* was used to denote the flesh of animals, and not until even later that this definition displaced the earlier one, as animal flesh became preeminent in the English diet, in preference if not in absolute quantity. (The same transformation can be traced in the French word *viande*.) As one late sign of this preference, consider Charles Carter's *Compleat City and Country Cook*, published in London in 1732, which devotes some 50 pages to meat dishes, 25 to poultry, and 40 to fish, but only 25 to vegetables and a handful to breads and pastries.

For a while, it was thought that the privileged place of meat in our diet was simply a sign of our basic carnivorous nature: early man, after all, was a big game hunter. But our understanding of human evolution has become somewhat more refined than that. Research on the patterns of wear in fossilized teeth suggests that early protohumans ate far more fruit than anything else. It was not until *Homo erectus, Homo sapiens'* immediate predecessor, that we find even a fully omnivorous diet of fruits, nuts, shoots, and meat. How has such a huge shift in our eating habits come about?

In the first place, why do we like meat at all? Two very different answers are worth noting. According to the legend preserved in Genesis, meat eating is one consequence of the Fall. In the beginning, Adam and Eve are given "every herb bearing seed" and all fruits but one, for their meals; animals are not even mentioned as a source of food. After the Flood, however, God says to Noah, "Every moving thing that liveth shall be meat for you; even as the green herb have I given you all things." This expansion of our diet is not a reward, but rather is a recognition of the fact that "the imagination of man's heart is evil from his youth." As the French scholar and diplomat Jean Soler points out, Cain's murder of Abel appears to set a precedent for taking the lives of animals, and our eating habits continue to reflect this primal crime. So at the core of Western culture lies the notion that meat eating is a sign of human weakness and cruelty. This idea lives on today in various secular forms.

An alternative to this moral view is the nutritional theory offered by the school of "cultural materialists," notably the anthropologist Marvin Harris, who sees the practice of cannibalism—an extreme form of meat eating—as a consequence of the high quality of meat protein compared to plant protein. Because the biochemistry of most animals is pretty much the same as ours, their tissues supply us with the number and proportions of amino acids that we need, while the very different tissues of plants are not as helpful. In part for this reason, it is not a simple matter for vegetarians to maintain a well-balanced diet. Harris argues that cannibalism has been resorted to when the depletion of game or other sources of high-quality protein threatens the health of a group (though others say that the cultural system of which cannibalism is usually a part is too complex to be explained through biology alone). From this sort of perspective, meat eating seems to be a manifestation of the most basic biological needs of the human organism.

It certainly makes sense that a preference for the most concentrated, most complete source of protein available would have been advantageous for our ancestors. But this does not mean that the biblical allegory lacks its own kind of truth. Nourishment that entails the destruction of other creatures, though it is nature's way, has been morally troubling to many people.

The Domestication of Animals

Early man was a hunter-gatherer, dependent for the most part on what his immediate environment offered him for food. Hunts were used to supplement the less dangerous and complicated gathering activity, which could turn up insects and small animals as well as vegetable matter. Primitive hunting tools are known from about 15 million years ago, and big game hunting from about a million years ago. During these times, man was what is called an "opportunistic" meat eater: he ate meat when he could obtain it, and that was relatively seldom. The great turning point in human dietary history came during the Neolithic era, beginning about 10,000 years ago, when man domesticated a few plant and animal species and gave birth to agriculture. The knowledge accumulated during his millennia as a gatherer, and which made gathering more efficient—the growing seasons, the proper time to gather, the most prolific plants, the best growing conditions—was turned to the active control of crops. Strains of barley, wheat, and some legumes—peas, lentils, broad beans and chick peas—were now cultivated.

The domestication of animals came at about the same time, and probably out of necessity. Sheep and goats would have been attracted to the large, stable fields of grain, dogs and pigs to the garbage heaps of growing settlements; they had to be brought under some control or they would have interfered with food production. The ruminant animals—goats, sheep, and cattle—can digest cellulose, and so could feed on stalks without competing for man's valuable grain. In effect they could further increase the food yield of a given field. With the development of agriculture, then, humans were no longer limited to opportunistic meat eating, though grains and other plant foods would still make up the major part of their diet. The cultivation of plants and animals meant a significant narrowing in the range of foods man ate, and meat became one of the choices he concentrated on.

The History of Meat Consumption

Little is known about the human diet in ancient times. Homer's heroes eat nothing but roasted meat, and the Greek and Roman privileged classes are infamous for their indulgence in such unlikely morsels of flesh as the wombs of virgin sows. But the masses, whose dinners were not celebrated in literature, probably lived on gruels and breads made from grain. When we reach the beginning of the modern age, we are on sounder footing, and the researches of *Annales* historian Fernand Braudel and his colleagues seem fairly sure on several points. Human diet from about A.D. 1400 to 1800 was essentially vegetable, for economic reasons; at that time, agriculture fed from 10 to 20 times the population that could be supported by animals grazing on

THE DOMESTICATION OF ANIMALS

Animal	Estimated Date B.C.	Area of the World
sheep	9000	Middle East
dog	8400	Eurasia, North America
goat	7500	Middle East
pig	7000	Middle East
cattle	6500	Middle East
guinea pig	6000	South America
horse	3000	Russia
chicken	2000	India

the same acreage. The great exception to this general rule, which holds for Africa, Asia, and pre-Columbian America, is Europe, which had begun to eat a great deal of meat in the Middle Ages (the time when the word *meat* was evolving). The reason for this seems to have been Europe's relatively low population density, and the availability of vast areas for pasturage. From the 17th century on, however, a steadily increasing population cut into meat production, so that while the upper classes never went without, the lower frequently did.

In the 17th and 18th centuries, the majority of people on the European continent rarely took more than a quarter of their total calorie intake from meat. Often it was closer to 10%, with cereals in the form of bread and gruel making up the deficit. As meat became more of a luxury, the lower classes began to depend heavily on salted meats and on fish; salt cod from Newfoundland became an important commodity in the 16th century. This trend of increasing scarcity was reversed in the 19th century, when the development of artificial meadows, "scientific" stock raising, and imports from the Americas boosted the meat supply.

From the beginning, Americans have enjoyed a relative abundance of meat made possible by the extent and richness of the continent. In colonial days, before livestock became established here, settlers lived on game animals, and had so much meat that visitors from Europe were astonished and a little disgusted. In the 19th century, as the country became urbanized and more people lived away from the farm, salted meats became the rule. While the staple food in Europe was bread, in America it was salt pork. (The fact that salt pork was stored in barrels gave rise to such expressions as "scraping the bottom of the barrel" and "pork barrel politics.") Per capita meat consumption was huge. In the decade of 1830 to 1840, it averaged 178 pounds annually, a figure not matched again until the 1960s and 1970s.

In the 1870s a wider distribution of fresh meat, especially beef, was made possible by the cumulative impact of several advances, including the growth of the cattle industry in the West, the introduction of cattle cars on the railroads in 1867, and the development of the refrigerated railroad car by Gustavus Swift and Philip Armour. No longer did cattle drives of better than a thousand miles take their toll on the animals; no longer did meat have to be salted to make it from Chicago to New York. Upton Sinclair's account of a Chicago slaughterhouse in *The Jungle* (1906) helped to put a crimp in the meat industry's expansion, but it soon recovered. Since the early 1970s we have actually been eating less and less red meat, while our dependence on poultry has tripled in the last 30 years. In 1980, the per capita consumption of red meat in the United States was 150 pounds. Add to that 63 pounds per year of poultry and 14 of seafood, and the grand total is 227 pounds per year, or nearly ⅔ pound per day. With one fifteenth of the world's population, the United States eats one third of the world's meat.

This allocation of resources is possible only in wealthy societies like our own, because meat is a much less efficient source of food than plant protein. Our stock animals are fed grain to speed their growth and fatten them up, and it takes much less grain to feed a person than it does to feed a steer or chicken in order to feed a person. Even today, with advanced methods of production, it takes 2 pounds of grain to get 1 pound of chicken meat, and the ratios are 4 to 1 for pork, 8 to 1 for beef. It is only because we have a surplus of vegetable proteins that we can afford to depend on animals as a major source of food. It is easy under the circumstances to forget that during the span of human life on the planet, meat has been a luxury reserved for the few, and in many countries still is.

Research in nutrition and medicine over the past few decades suggests that we may be paying for our high consumption of meat and other animal products with our health. Americans eat about twice as much protein every day as they actually need, and most of the protein comes from red meat, eggs, and dairy products. The problem is not so much the excess of protein in the diet as the large amounts of saturated fats and cholesterol that accompany it. These substances have been implicated in the development of heart disease. With heart disease the leading cause of death in the U.S. today, Americans would be better off eating less and leaner meat than they do.

THE STRUCTURE AND QUALITIES OF MEAT

Research on muscle as a biological system is voluminous. Workers in the field come to it from very different angles: they may be interested in the basic phenomenon of cell movement, or athletic training programs or slaughterhouse technology or the treatment of muscular dystrophy. Even among food scientists there is a wide range of viewpoints, from the immediately practical to the visionary. A few years ago, one symposium member saw in current research the hope for muscular tissue culture: the growth of whole muscles from a single cell in the laboratory, without all the wasted resources that go to make up the rest of the animal! One thing the world can probably do without is test-tube tenderloin, but most cooks can do with an understanding of how meat is put together, what makes it tender or tough, red or brown or white, flavorful or bland.

Meat is composed of three basic materials: water, protein, and fat. On the average, lean muscle tissue is about 75% water, 18% protein, and 3% fat, but the cuts we buy in the store are usually packaged with the surrounding fatty tissues. The following table gives an idea of the relative compositions of common meats as they are prepared for the consumer, with dairy products included for comparison. The eventual texture and taste of the cooked meat will depend on the amount of fat and water in the tissue, and on the particular kinds and configurations of the proteins. Roughly, 50% of the protein is accounted for by the filaments that contract the muscle, 30% by the oxygen-storing pigments and various enzymes that serve the muscle fibers, and 20% by the connective tissues that hold the muscles together.

THE COMPOSITION OF MEATS

	% water	% protein	% fat
beef	60	18	22
pork	42	12	45
lamb	56	16	28
turkey	58	20	20
chicken	65	30	5
fish	70	20	10
milk	87	4	5
hard cheese	37	25	31
soft cheese	70	15	7

Tissue Types

Muscle Fibers

Muscle tissue is made up of long, thin cells, or muscle fibers, bound together by thin sheets of connective tissue. The individual fibers can be as long as the whole muscle, and some exceed a foot in length. The bundles of fibers are organized in groups to form an individual muscle, which is contained and anchored to bone by other connective tissues. The longitudinal structure of the muscle bundles accounts for what we call the "grain" of the meat: it is much easier to cut or chew in the direction of the fibers than across them. For this reason, we carve less tender meats *across* the grain, so that we can then chew *with* the grain.

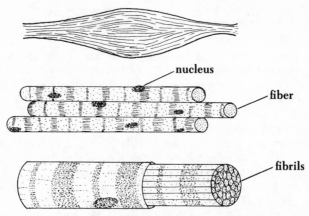

nucleus

fiber

fibrils

The structure of muscle tissue. Each muscle *(top)* is composed of many individual cells, or fibers. A single muscle cell contains many nuclei. The fibers are in turn filled with many fibrils *(bottom)*, or complexes of actin and myosin, the proteins of motion. Muscles grow by increasing the number of fibrils in each fiber, and not by generating whole new cells.

The job of any muscle is to perform work by contracting itself when it receives the appropriate signal from the nervous system. This movement is made possible by the two major proteins found in muscle, myosin and actin. These proteins take the form of long, chainlike molecules that lie parallel to each other. When an electrical impulse is received from the associated nerve and brings into play various ions and regulatory molecules, the actin and myosin filaments slide past each other, and then by means of *cross-bridging*, or forming mutual bonds, lock into place. The change in relative position

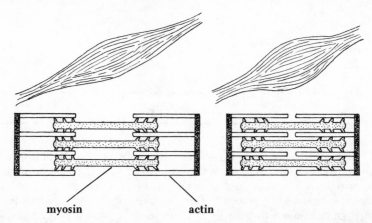

myosin   actin

A schematic view of actin and myosin in action. When a muscle contracts, the filaments of actin and myosin slide past each other and decrease the overall length of the complex. The contraction is sustained by cross-bonds between the two proteins which prevent them from sliding back to their original position.

shortens the muscle cell as a whole, and the formation of cross bridges maintains the tension by holding the filaments in place. The actin and myosin filaments now form one complex molecule, actomyosin. Each muscle cell will have a cross section of from 500 to 2000 actomyosin complexes, or fibrils.

Actin and myosin are the most versatile molecules of motion. These same proteins propel amoebas, and keep the contents of all plant and animal cells in constant circulation. Animals have mastered movement by concentrating, organizing, and harnessing these fundamental, even primitive structures.

Connective Tissue

Connective tissue is the physical harness of the muscles. The significance of the connective tissue lies not in the cells themselves, but in the substances they secrete. What we call connective tissue is for the most part not living tissue, but protein structures surrounding the live cells that made them. It is found between individual muscle fibers, between fiber bundles, between whole muscles, and between muscles and other kinds of tissue; it also makes up the blood vessels that supply the muscle with oxygen and nutrients and carry away waste products. The diaphanous membranes found between different sections of chuck steak or between the muscle and skin of a chicken

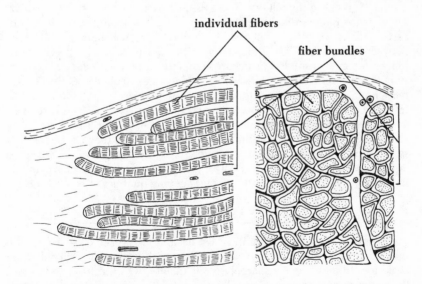

Muscle fibers are grouped in bundles and held in place by connective tissue. The particular makeup of connective tissue has an important influence on the texture of a given meat.

breast, the tendons attaching the muscles to the chicken leg bone, and the gristle that surrounds veins and arteries in lower quality cuts are all different kinds of connective tissue.

Connective tissue is made up of three basic proteins: collagen, elastin, and reticulin. Elastin is the yellowish main component of blood vessel walls and ligaments. As its name implies, it is elastic. Reticulin is fibrous and concentrated in the spaces between muscle cells. Elastin is especially tough, and neither it nor reticulin is weakened by heat. But collagen is, and this makes it a very important factor in the texture of cooked meat. Collagen is the major structural component of the simplest of many-celled animals, the sponges, and accounts for some 30% of the protein in the human body. It is found in the skin and tendons as well as in between muscle cells and muscles, and it is a large part of the matrix in young bones that is later filled with hard minerals. The name comes from the Greek for "glue producing," referring to the fact that when it is heated in water, insoluble collagen is transformed into a *gelatin,* a soluble, gummy solution that can be used for glue (hence the consignment of skin-and-bone horses to "the glue factory") as well as a thickener for soups and desserts (culinary gelatin is made from pig skin). This transformation can help make some tough meats tender, as we shall see.

Fat

Fat tissue surrounds muscle tissue and is incorporated into it. In the first instance it is called deposit or adipose tissue, and in the latter the term "marbling" is used to describe the pattern of white splotches in the red matrix of muscle. Even when it is not readily visible, there will be some fat tissue between muscle bundles, and fat molecules are present in the cell membranes and cytoplasm. Fat tissues are composed of fat cells, which are mostly storage tanks for fat molecules held in a network of connective tissue; they help insulate the animal and are a source of reserve energy. Unsaturated fats tend to be soft at room temperature because their structure is less regular, and so less orderly than that of saturated fats (see chapter 13). Pork, lamb, and poultry contain fewer saturated fats than beef, and they are discernibly softer, especially just out of the refrigerator. Only about 20% of fish fat is saturated, which is why it is normally liquid. As we shall see, fats have important influences on the texture, taste, and keeping qualities of cooked meat.

The Texture of Meat

Cooking can make tough meat tender and vice versa, of course, so the initial texture of meat is no guarantee of how it will come out. Still, what the cook does to meat depends largely on its quality, which can be understood by way of its composition. Generally, this rule holds true: meat will be tough when the muscle fibers are coarse rather than fine, when there is a lot of connective tissue, and when the meat is lean. We'll take each of these points in turn.

Muscle fibers are small in diameter when the animal is young and its muscles little used. As it grows and exercises, the muscles enlarge—not by increasing the number of fibers, but by increasing the number of actomyosin filaments within the individual fibers. The number of muscle cells stays the same, but they get thicker. And the more filaments there are to cut through, the tougher the meat. Stockmen have known this basic fact for a long time, and therefore try to minimize their animals' activity. It's unlikely that Longhorns were at all tender after a four-month cattle drive just to get to market. Within a single animal, the tenderest meat will come from those muscles that are least used. The tenderloin, for example, is a muscle that runs along the back of the animal and gets little action, while the shoulder is used continually in walking and standing, and is relatively tough. Turkey legs are a lot tougher than turkey breasts, for the same reasons.

So the lesson here is that young animals and little-used muscles make for the tenderest meat? Well, not quite. Unfortunately, there are counter-vailing trends in the other two categories. The younger the animal, the less fat there is likely to be in the meat, especially in the form of marbling. And it turns out that fat contributes to the tenderness of meat by acting as a "shortening" agent, much as it does in pastry (see chapter 6). When it is melted during cooking, fat penetrates the tissue and helps separate fiber from fiber, lubricating the tissue and so making it easier to cut across or crush. Without much fat, otherwise tender meat becomes dry and resistant. Deposit fat is much less successful in penetrating the tissue than marbled fat, hence the value placed on good marbling in steaks. A second unfortunate correlation is that the younger the animal and the more finely fibered the tissue, the more connective tissue is found in the meat. Veal, for example, contains about twice as much collagen as year-old calf meat. But the subject of collagen is a tricky one, because that protein can be denatured and dissolved into gelatin, and the collagen of young animals is much more easily denatured than older connective tissue. Again, more of this when we get to the subject of cooking.

In addition to fiber size and fat and collagen content, texture will also be affected by the muscle's shape, size, function, and enzyme and acid contents. Speaking very generally, however, younger and little-used muscles are usually the most tender.

Why Fish Is Delicate

The typical flaky texture of fish is clearly related to that animal's unique anatomy. While the muscles of mammals and birds are composed of very long fibers arranged in longitudinal bundles, fish muscle consists of segments ("myotomes") of rather short fibers which are separated by large sheets ("myocommata") of very thin connective tissue. There is very little connective tissue in fish—about 3% of its weight, as opposed to 15% in land animals—and what there is is very fragile and easily converted to gelatin. The combination of sparse, weak connective tissue and short muscle bundles

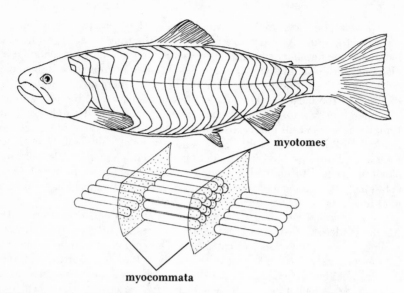

myotomes

myocommata

Unlike the muscles of mammals or birds, fish muscle is arranged in layers of short fibers—the myotomes—which are separated by very thin sheets of delicate connective tissue—the myocommata.

results in the great tenderness of fish, and its troublesome tendency to fall apart altogether during cooking.

Muscle Types, Color, and Flavor

Have you ever wondered why chickens and turkeys have white and dark meat? Why veal is pale, beef heartily red, fish snowy? The explanation is fascinating, and tells us something about human muscle and activity as well. It turns out that the pink or red coloration of meat is not due primarily to blood and its oxygen-carrying hemoglobin—something like half of the animal's blood is removed after slaughter—but to oxygen-*storing* myoglobin, which is located in the muscle cells proper and retains the oxygen brought by the blood until the cells need it. Tissues with different colors, then, contain different concentrations of myoglobin. The muscles that require a lot of oxygen have a greater storage capacity than those that need little, and are consequently darker red. To some extent, oxygen use can be related to the general level of activity: muscles that are exercised frequently and strenuously need more oxygen. Chickens and turkeys do a lot of standing around, but little if any flying, so their breast muscle is white, their legs dark. Game birds, on the other hand, spend more time on the wing, and their breast meat may be as dark as their drumsticks.

This explanation is fine as far as it goes. But why is it that many fish, which are pretty constantly on the move, have pristinely white flesh? To explain this, we must turn our attention back to the individual muscle cells.

Fast and Slow Fibers

Red and white muscles differ in more fundamental ways than color. In the last few years, it has been determined that one of the basic muscle proteins, myosin, exists in several different forms. A single molecule of myosin consists of two very long intertwined chains of amino acids attached to four shorter chains. Both the long and the short chains in white muscle myosin differ from those in red muscle myosin. Different types of myosin have been found side by side in the very same muscle cell, and both red and white cell types are found in most muscles. Individual cells and whole muscles are seldom if ever all red or all white. The important factor in determining muscle type is the relative populations of myosins and cell types.

For the sake of this discussion, however, we will stick to the extreme cases of red and white muscle. The two myosin types and the corresponding cells seem to specialize in their work. What we have been calling red cells are more frequently called "slow-twitch fibers," and white cells, "fast-twitch fibers." Slow fibers are designed to meet the need for slow, continuous activity, and fast fibers for intermittent bursts of work with long rest periods in between. These two functions involve different energy supplies. Slow cells burn fats, and contain many of the cell bodies called mitochondria, which specialize in using fats as fuel. This system has an absolute need for oxygen,

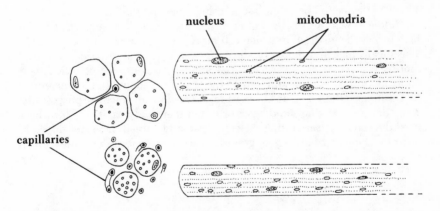

nucleus   mitochondria

capillaries

Fast muscle cells *(above)* are larger than slow cells *(below).* Because they are designed to burn carbohydrates rather than fats, and can do so temporarily in the absence of oxygen, fast cells contain less myoglobin (hence their light color), are less well supplied with blood, and contain fewer fat-burning mitochondria than slow cells.

and the myoglobin-rich slow cells are well equipped to meet this need. Fast cells, on the other hand, have few mitochondria and instead burn glycogen, a carbohydrate, right in the cell fluids. This process can take place without oxygen, if need be, although a waste product, lactic acid, will accumulate and limit the cell's endurance: hence the fast cell's fitness to work only in short bursts.

It is now thought that the first generation of embryonic muscle fibers consists entirely of fast cells, some of which are later converted into slow cells by various influences. The kinds of nerve endings and signals associated with the tissue, chemical stimuli from nearby tissues, and patterns of muscle use (exercise) all affect the nature of the muscle. For example, the whale, which undergoes prolonged oxygen deprivation during its dives and which moves almost constantly, has muscle that is nearly black, so heavily concentrated is its oxygen-storing myoglobin in preponderantly slow fibers. Frogs and rabbits, which make quick, sporadic movements and use very few of their muscles continuously, have very pale muscle consisting mainly of fast fibers. Humans are somewhere in between, although there is a great deal of variation. On average, our skeletal muscles are half fast, half slow fibers. But long-distance runners may approach 80% slow cells, and sprinters, 75% fast cells. Athletic training, which comes relatively late in the development of the body, doesn't seem to affect the ratios of fast to slow cell types, but does improve the efficiency of one or the other depending on the program. In the case of runners, inherited anatomy may very well be destiny.

Why Fish Is White
The distinction between fast and slow fibers provides a neat solution to the riddle of why fish flesh is often brilliantly white. The important factor here is that the medium in which fish move, water, is much denser than air. This means that in order to produce a given increment of speed, fish need more sheer power than do terrestrial animals, because water resists their effort more than air would. On the other hand, the density of water means that fish can attain a neutral buoyancy—can for all practical purposes be weightless—by storing some lipids or gas in their body cavities. They don't need the elaborate supporting skeletons that land animals have developed in order to counteract the force of gravity. As a consequence fish devote a much

Cod have the habit of resting on the ocean floor. Their body *(left)* contains less dark muscle than that of the mackerel *(right)*, which is a more active hunter.

greater proportion of their body weight, from 40% to 60%, to muscle tissue than do other vertebrates. And for the same reasons, their muscle has tended, over time, to become much more clearly differentiated into fast and slow types than is true of terrestrial animals. In order to take quick evasive or offensive action, fish need to develop high power very quickly, while slow cruising in a practically weightless environment takes very little effort. The great preponderance of fish muscle—usually from 75% to 90%—is therefore almost pure fast tissue, used only for the occasional bursts of high speed that punctuate a life of slow-speed cruising. Slow tissue is generally limited to the areas just under the skin or near the fins that are used at any speed. (The pink tone of salmon and trout flesh is caused by astaxanthin, a carotenoid pigment derived from insects and crustaceans that the fish feed on.) The majority of fish muscle, then, is an emergency power pack, while the minority cell population is worked much harder, or at least more often. This arrangement is possible only because fish need not worry about supporting their weight; a land animal could never survive carrying around a huge mass of muscle that was used only in extreme situations. Fish muscle, then, is so abundant and so white as a result of mechanical adaptations to the marine environment. The same is true of crab, lobster, and shrimp.

Activity and Flavor

Generally speaking, well-exercised, tough meat is more flavorful than tenderer, less exercised meat. Think of the difference between veal and beef; between tenderloin and flank steak. The stronger flavor apparently has something to do with the cell contents peculiar to the slow tissue, and particularly with the higher concentration of fat. Though this phenomenon is not very well understood, it is well-known: 150 years ago, Brillat-Savarin made fun of "those gastronomers who pretend to have discovered the special flavor of the leg upon which a sleeping pheasant rests his weight." Today in Brittany it is customary to ask for a chicken "qui a bien couru," one that has had the run of the farm, because poultry raised on wire is thought to be insipid. However it was not always thus: one food historian has recorded the preference of 17th-century Frenchmen for the chickens of Mans "raised in coops or baskets suspended from the wall, immobilized in the dark." Such chickens had a more delicate flavor than common, run-of-the-farm birds. Less than overpowering flavors have the advantage of being more versatile: the French praised veal as the chameleon of their cuisine.

SLAUGHTER, AGING, AND STORAGE

Thus far we have mainly considered live muscle, but live muscle must be converted into cuts of meat fit for the individual cook, and the way this is done will affect the quality of the meat. Some attention must be given, then, to the practices of slaughtering and aging.

Slaughtering

By a fortunate coincidence, the most humane methods of slaughter are those that result in the highest quality of meat. It has been recognized for centuries that any kind of stress on the animal immediately before slaughter—whether fasting, duress in transport, or fear—has an adverse effect on the final product. This is due to elementary body chemistry. When muscles are active, they consume their own energy stores, mostly the carbohydrate glycogen. The waste product of this process is lactic acid, which can be carried away by the blood or further oxidized while an animal is still alive. For a time after an animal is killed, the muscles continue to work at maintaining body temperature, and since blood is no longer flowing, lactic acid accumulates. If, however, an animal has been under stress prior to being slaughtered, its resulting muscular tension will have depleted the glycogen supply. The muscles will therefore accumulate less lactic acid after death; and the meat will have a lower acid content than it would if its glycogen supplies had been normal. Such meat is called "dark-cutting," a condition first described in the 18th century. Although it has the same nutritional value as normal meat, dark-cutting meat is unattractive, gummy in texture, and tends to spoil faster, since acid conditions inhibit the growth of many bacteria and molds. It pays, then, to treat animals well before the slaughter, and the meat industry has learned this lesson thoroughly. In November, 1979, the *New York Times* reported that a Finnish slaughterhouse had evicted a group of young musicians from a nearby building because their practice sessions were resulting in dark-cutting meat.

Commercial slaughter is carried out as untraumatically as possible. The animal is stunned, usually with a blow or electrical discharge to the head, and then is hung up and bled from one of the major blood vessels. A good deal of research has gone into stunning practices. It has been found, for example, that a square wave, rather than the house current sine wave, and a high rather than a low frequency, result in less muscle tension and internal bleeding. The animals are bled because blood is an excellent medium for the growth of microorganisms, and removing it decreases the risk of spoilage. About half of the blood in a given animal is removed by bleeding; the rest is retained mostly in blood-rich tissues like the heart and lungs.

The slaughter of the animal has important effects on the chemistry and even the structure of its muscle tissue. Lactic acid accumulates, lowering the pH of the tissue and in turn promoting the unfolding, or denaturation, of some proteins in the fibers (collagen and elastin, however, are unaffected). The denaturation of the protein molecules frees the water that was previously bound up in their interstices (see page 596); this causes fluid loss, or "weep," from the carcass, and dries out the meat to some extent. The actin and myosin filaments bind together to form actomyosin, the complex characteristic of muscle contraction, and so produce the condition of *rigor mortis*,

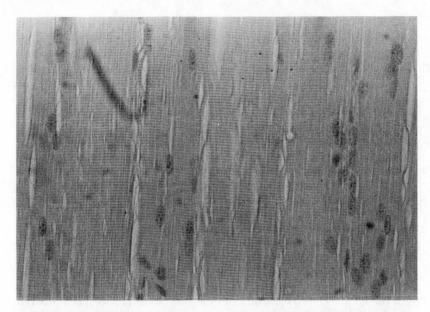

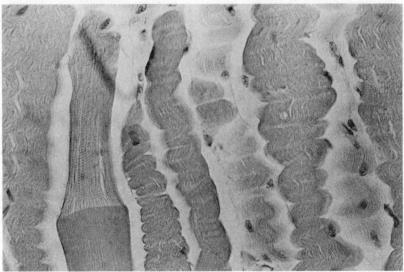

Microscopic view of rabbit muscle 30 minutes *(top)* and 6 hours *(bottom)* after slaughter. As the chemical makeup of the cells changes, the fibrils contract and the muscle clenches in rigor mortis. Rigor eventually passes (see photograph on page 110). (From P. C. Paul and H. H. Palmer, *Food Theory and Applications.* Copyright © 1972 John Wiley and Sons, Inc. Reprinted by permission.)

when the muscles clench and cannot be relaxed. If the muscle is in a contracted position before rigor sets in, the actin-myosin bonding will be all the stronger, and the meat will be tougher. This is one reason that carcasses are hung up in such a way that all their muscles are extended. As the chemical conditions in the cell continue to change and the actomyosin filaments are affected, the muscle eventually relaxes again to some extent. Rigor passes in about a day in beef, and after 6 hours or so in pork and chicken.

A last effect of slaughter has to do with color. When it is doing its job of storing oxygen, the principal pigment of muscle, myoglobin, is bright red. Once the animal is slaughtered, it is deprived of oxygen and the pigment reverts to its unoxygenated form, which is purplish (just as hemoglobin is blue in our veins and red in our arteries). When meat is cut and a surface exposed to the air, myoglobin at the surface quickly picks up airborne oxygen, and reverts to the pink-red color we associate with good fresh meat. Butchers call this color "bloom," and to preserve it they will usually wrap meat loosely in paper or package it in oxygen-permeable film. The brownish-gray color characteristic of less-than-fresh meat results from the oxidation of myoglobin to the metmyoglobin form—not the simple addition of oxygen, but the removal of an electron from the iron atom. This chemical reaction is enhanced by bacterial activity, extreme temperatures (including cooking), and low oxygen and high salt concentrations. Whether the conditions are especially favorable or not, the conversion of myoglobin to metmyoglobin proceeds continuously and cannot be reversed by the cook. The older a particular piece of meat, the grayer it will be.

Aging

Like cheese and wine, meat benefits from a certain period of "aging," or slow chemical change, before it is consumed. Its flavor improves and it gets more tender. Exactly what happens during aging is not known, but the general impression is that the muscle's own enzymes are the principal agents. As lactic acid accumulates in the tissue after slaughter, it begins to break down the walls of lysosomes, the cell bodies that store protein-attacking enzymes. As a result, these enzymes, whose normal function is to digest proteins in a controlled way for use by the cells, are liberated and attack the cell proteins indiscriminately. Flavor changes are probably the result of the degradation of proteins into individual amino acids, which generally have a strong flavor. It is not clear whether these same enzymes also tenderize the meat by breaking up the actin-myosin complex; the evidence is contradictory. Whatever the cause—enzymes, continuing mechanical tension, an unfavorable balance of acids and minerals—the muscle filaments do undergo some disintegration, and the tissue becomes softer. Connective tissues, however, are not much affected although some collagen is degraded. Stored at 34° to 38°F (1° to 3° C), a temperature range that limits the growth of microbes, beef profits from 10 days to 3 weeks of aging, and lamb from a

Top: the heme group, a carbon-ring structure at the center of both hemoglobin and myoglobin that holds oxygen for use by the body's cells. The protein portion of these molecules, the globin, is a long, folded chain. *Bottom:* three different states of the heme group in uncooked meat. In the absence of oxygen, myoglobin is purple *(left)*. Myoglobin that has picked up a molecule of oxygen gas is red *(center)*. When little oxygen is available, the iron atom in the heme group is readily oxidized—relieved of another electron—and the resulting metmyoglobin is brownish *(right)*. Metmyoglobin is characteristically found in meat that is several days old.

week. In the 19th century, beef joints would be held until the outside was literally rotten; the French called this *mortification*. Today, most beef is cut up to separate the loin and ribs for brief aging in the cooler, while the less choice chuck and round are shipped to market immediately. With cooler space at a premium and storage tying up assets, extended aging is not common. What aging does occur is usually a matter of refrigerated transit from packer to market.

Fat Is a Limiting Factor

Fats as well as muscle are affected by the sudden stoppage of blood circulation. Since the blood no longer supplies either white cells to control infection or antioxidizing agents, the fat tissue becomes susceptible to infection and rancidity. Oxidation is the replacement of an oxygen ion for a hydrogen ion in the three carbon-hydrogen chains of the fat molecule. This change destabilizes the molecule, makes it possible for other odd chemical fragments to find a place along the chain, and so gives rise to unpleasant flavors. Unsaturated fats, on account of their geometry, are particularly susceptible to these reactions (see chapter 13). Fat oxidation is one of the limiting factors in the aging process and in storage generally. Meats with high proportions of unsaturated fats—fish, poultry, pork, lamb, and veal—cannot be kept as long as beef.

Storage

Meat, like most foodstuffs, can be rendered inedible by several different conditions, including the action of light, oxygen, enzymes in the meat itself, and microorganisms. The first three problems are, for the most part, negligible today: we keep meat in a dark, cool place, well wrapped, and the enzymes do little damage below the normal body temperature. (The major exception to this rule is fish, which, being cold-blooded, have enzymes designed to be most effective at the average temperature of their home waters—sometimes as low as 40°F (4°C). So eat fresh fish fast, before it eats itself.) Today, our primary battle is with bacteria and molds.

Microbial Contamination

The muscles of healthy livestock are usually free of bacteria, though there are diseases that can infect animals and thereby humans. Salmonella is the most prevalent of these, and worm infections, of which the best known is trichinosis, are also a problem. Thorough cooking is the best safeguard against these diseases. Meat spoilage, however, is caused by bacteria and molds introduced during processing. The animal's hide is a prime source of contamination. The skin prevents bacteria from penetrating into live muscle, but this means that it harbors great numbers of microbes that can be transferred to the muscle when the carcass is cut up. The meats that we eat skin and all—poultry, for the most part—are especially prone to spoilage because, despite postslaughter washings, many bacteria remain. For example, a typical piece of pork in the supermarket will have a few hundred bacteria per square centimeter, while a piece of chicken may have 10,000 in the same area. This simple statistic is the reason for cleaning hands, knives, and cutting boards thoroughly with hot, soapy water after handling chicken and before handling other foods, and for cooking poultry well, especially its surfaces. Such precautions have not always been observed, and one shudders at the hygienic consequences of this late 14th-century recipe for skinned, roasted peacock: "when he is rosted take hym of, and let hym coole awhile, and take and sowe hym in his skyn, and gilde his combe, and so serve hym forthe."

Bacteria and molds break down the surface layer of meat by digesting the proteins and carbohydrates into a liquefied film, producing hydrogen, carbon dioxide, and ammonia gases in the process. And they convert residual glycogen into acids that further speed the denaturation of the proteins. Bacteria can also discolor meat by oxidizing myoglobin into the brownish metmyoglobin, or even converting it into yellow and green "bile" pigments. Finally, bacteria produce putrid odors by converting proteins and amino acids into highly noxious substances, including mercaptans, a group that includes one of the active ingredients of skunk spray, and hydrogen sulfide gas, the characteristic constituent of rotten egg smell. Spoiled meat is generally more offensive than other rotten foods precisely because it contains the protein necessary to generate these compounds.

Most of our storage techniques are designed to counteract microbial growth in some way. Bacteria and molds generally thrive in moderate temperatures, high humidity, and low salt concentrations, and there are accordingly several different antimicrobial strategies: refrigeration, dehydration, and salt curing, as well as such drastic treatments as antibiotics and radiation.

Cold Storage

Refrigeration is the method of meat preservation most familiar to us. It probably began simply by storing foods in dark, underground chambers that were fortified when possible with stream and lake ice. It has been known since the 16th century that salts lower the freezing point of water, and the 18th century managed to reach −27°F (−33°C) by exploiting this fact. Large-scale freezing in the meat industry arrived around 1880, and in 1923 Clarence Birdseye founded the first company devoted to quick-freezing, which minimizes damage to foods. The logic behind refrigeration is obvious: both bacteria and enzymes in the meat itself work less actively at low temperatures. Even so, spoilage continues at a slow rate. Take that piece of chicken with 10,000 bacteria per square centimeter; even at 40°F (4°C), an average refrigerator temperature, the surface will become slimy, indicating a 10,000-fold population increase, in only 6 days. And fish enzymes and microbes may actually prefer 50°F to 75° (10°C to 24°), since they are accustomed to the cold.

Freezing, of course, greatly extends the storage life of meat by bringing biological processes dependent on water to a halt, though chemical reactions like oxidation continue slowly. It is usually recommended that home freezers operate at 0°F (−18°C), though most are closer to 10° or 15°F (−12° or −9°C). There have been some remarkable claims made for naturally frozen flesh: Russians say they have found edible mammoth meat 20,000 years old deep in the ice of Siberia, and Yukon-frozen horse-bone marrow 50,000 years old was reportedly served at an exclusive New York dinner party.

Freezing is an extreme treatment that inevitably damages the tissue. It results in *drip*, the loss of fluid rich in valuable proteins, vitamins, and salts that causes a viscous pool of liquid to form when the meat thaws. What happens is this: as the temperature in the tissue reaches the freezing point, ice crystals begin to form between the muscle cells, where the concentration of salts is lowest. The crystals intrude on the soft cell membranes and may puncture them, thereby exposing the muscle proteins to the external fluids. As the growing ice crystals take up more water from the tissue, the concentration of salts and enzymes within cells increases. This denatures membrane and other cell proteins, and liberates yet more water to be taken up by the ice. As a result, the water-holding capacity of the tissue is lowered, and fluid is released from the damaged cells when the temperature is raised again. The significance of the fluid loss, of course, is that the cooked meat will come out drier, and so tougher, than it should. For some reason, fish muscle is especially liable to come out dry and stringy after freezing. Drip can be minimized by increasing the *rate* of freezing: the faster that water freezes, the smaller the crystals that are formed, and so the less damage done to cell membranes. The colder your freezer, the faster it will freeze food.

Another side effect of freezing is *freezer burn*, that familiar discoloration of the meat surface after it has been in the freezer for a while. This is caused by the sublimation of ice crystals from the surface of the meat—the equivalent of evaporation at below-freezing temperatures—which in effect dries out the tissue (see page 636). The concentration of water in the meat is much higher than in the surrounding air of the freezer, and so the surface will tend to dry out if given the chance. You are then left with a patch of freeze-dried meat on the surface of otherwise normally moist tissue, and both its taste and texture are adversely affected. Avoid freezer burn by wrapping meat as tightly as possible with water-impermeable film.

A final problem with freezing is that it promotes the oxidation of fats, especially the unsaturated fats in chicken and pork. For the same reason that freezing denatures proteins, it also encourages the alteration of fat molecules, and this results in rancid flavors. These too can be minimized by quick freezing and careful wrapping, but they cannot be eliminated altogether. Beef, whose fat is largely saturated and relatively stable, can be kept frozen for years, but pork, poultry, and fish are limited by fat oxidation to freezer lives of a few months.

Heat Treatment
In 1810 the Frenchman Nicholas Appert discovered that if he sealed food in a container while heating it, the food would be preserved. This was the beginning of canning, a form of partial cooking which, when done properly, is quite effective: canned meat 114 years old has been eaten without

distaste, if not exactly with pleasure. In the technique called pasteurization, food is heated enough to stop the growth of microorganisms while causing the least possible damage to the food itself. In the more severe treatment called sterilization, all microorganisms are killed regardless of the often drastic effect on the food. Neither process finds much use in the average kitchen today. Cooking itself, of course, kills most bacteria and can be used in conjunction with refrigeration or freezing to keep food edible.

Dehydration

The technique of controlling moisture to preserve meat has been known at least since ancient Egypt. The eastern American Indians made a highly nutritious and portable food called *pemmican* out of dried meat, fat, and berries. Large-scale dehydrated meat production was developed during World War II for logistical purposes, and more recently, freeze-drying has become an important commercial process. Like canned meats, though, dehydrated meats are rarely used outside of emergencies or special situations like campfire cooking.

Salt Curing

Salt has been used to inhibit microbial growth for thousands of years, and was especially important in the days before refrigeration. It works by creating such a concentration of dissolved ions outside of the bacteria and mold cells that water inside the relatively dilute cells is drawn out across their membranes. The microbes dry up and either die or slow down drastically.

Originally, meat was soaked in a strong brine solution or covered with whole grains of salt, which were known in England as "corn"—hence the term "corned beef." Today bacon is presliced and passed through the curing solution, and is ready in less than a day. Salt tends to dry out the meat as well as the microbes by drawing fluids from the tissues, and freeing bound water by denaturing the proteins. Cured meat therefore contains less water than uncured meat and proportionally more fat.

Salting today is done mainly for aesthetic purposes. We have come to like the taste of ham and bacon, and we salt pork for that reason, not to improve its storage life. As a result, meats are now treated with much milder cures than are necessary to prevent microbial growth, and so they are more susceptible to spoilage than they used to be (and are still sometimes thought to be).

Curing also has an effect on color. We all know from experience with hams, sandwich meats, and so on that cured meats retain a bright pink-red color even after cooking. This is due to the presence of a chemical group called *nitrite* (NO_2^-), which reacts with the purplish myoglobin pigment to form pink nitrosomyoglobin. This compound is quite stable during changes in temperature, but is sensitive to oxygen and light. For this reason cured meats are often vacuum-packed and exposed as little as possible to light (or to the consumer's eye). In addition to bringing about the color change, nitrite is a very effective antibacterial agent, it retards fat oxidation, and it contributes flavor. Originally a trace mineral in the curing salt, nitrite was intentionally added to meat in the form of a nitrate salt, saltpeter, beginning in the 16th or 17th century. Today, pure nitrite is put into the curing brine. In recent years, nitrite has come under suspicion as a carcinogen (see chapter 10 for details).

Smoke Curing

Smoking is another venerable preservation technique which is actually a kind of slow, low-temperature cooking. But it is also a chemical treatment. Smoke is a very complex material, with upward of 200 components that include alcohols, acids, phenolic compounds, and various toxic, sometimes carcinogenic substances. The toxic substances inhibit the growth of microbes, the phenolics retard fat oxidation, and the whole complex imparts the characteristic flavor of burning wood to the meat. Salt curing and smoking are often combined to minimize the fat oxidation that salt encourages. A recipe for ham that has come down to us in the Latin cookbook of Apicius uses this double treatment: it directs the cook to salt the meat for 17 days, dry it for 2 in the open air, oil it and smoke for 2 days, and then store it in a mixture of oil and vinegar.

Chemicals and Radiation

Two distinctly modern contributions to meat preservation are still being evaluated: chemical preservatives, including antibiotics, and radiation. Chemicals have been used often and heavily in the twentieth century when good hygienic practices were difficult; that is, during the World Wars. Today, relatively few chemical additives are permitted in fresh meat. Antibiotics are, however, used a great deal in this country, a practice that is coming under increasing criticism. Nearly all poultry, about 85% of the hogs, and 60% of the cattle raised in the U.S. are fed antibiotics, usually tetracycline or penicillin. Close to half of all the antibiotics manufactured here go into animal feeds. This practice reduces the incidence of disease in the animals, and makes it possible to raise large numbers in very small areas. Traces left in the tissues after slaughter may also slightly delay spoilage. The problem is that this widespread deployment of drugs may be resulting in the rapid evolution of drug-resistant bacteria that will be more difficult to control in both animals and humans. Partly on these grounds, several European countries ban the indiscriminate use of antibiotics in animal feed, and there is a movement in the United States toward the same position.

Radiation—the sources used are gamma rays, soft X rays, and cathode rays—kills microorganisms by ionizing chemicals in the tissue, and these ions in turn destroy proteins, both bacterial and meat. This is the limiting factor in the process: not enough meat protein is damaged to hurt the nutritional value of the food, but radiation does have adverse effects on *flavor*. It can cause the formation of hydrogen sulfide gas and mercaptans, some of the most odiferous and odious chemicals known to man, and ironically also the products of bacterial action. The ions that result from radiation also hasten the oxidation and thereby the rancidity of fats. Tests have shown that low doses of radiation can extend the shelf life of beef to three weeks, although this also involved carefully controlled temperatures and elaborate wrapping.

Although little used at the moment, there is reason to believe that irradiation may become more widespread in the near future. Both government and industry showed interest in food irradiation in the early forties, but industrial participation in research tapered off to nearly nothing when it became clear that there were major problems to be overcome. The Food and Drug Administration made it even less attractive when it officially classified irradiation as an additive rather than a process, and so subjected the technique to more stringent guidelines than would otherwise have been the case. In recent years, however, the World Health Organization and the International Atomic Energy Agency have declared food irradiation to be a process, after research showed that under proper conditions no residue was left in the tissue by the treatment. It is thought that the F.D.A. may follow suit.

MEAT COOKERY

We cook meat for four basic reasons: to make it safe to eat, easier to chew and to digest (denatured proteins are more vulnerable to our enzymes), and to make it more flavorful. Fortunately, the issue of safety is a minor one today. We all know to cook pork to at least a medium doneness (an internal temperature of 150°F (66°C) gives a 13° safety margin) to kill any trichinae present. Aside from this, all that is called for is sensible hygienic care in handling the food. In the rest of this chapter we will concentrate on the physical and chemical transformations of meat during cooking, and their effects on our impressions of texture and taste. A few words, though, about meat cookery and nutrition. One of the inevitable consequences of cooking meat is the loss of some juice, whether to the roasting pan or the stew liquid. Generally, the longer or hotter meat is cooked, the more juice will be squeezed out of it. These fluids are made up primarily of water, fat, and soluble vitamins and minerals. The B vitamins, one of meat's nutritional specialties, are carried off in the watery juices. The loss of fats results in a reduction in the meat's calorie content and removes some vitamin A, which is fat-soluble and responsible for the yellowish color of fat in older meats. High temperatures will destroy some thiamine, but cooking doesn't really hurt the meat's nutritional value unless it is grossly overdone. If you recoup the fluid losses by making gravy from the drippings or a sauce from the braising liquid, the nutritional decline will be slight.

Cooking and Flavor

The substances responsible for the flavor of meat have been the object of long and intensive research. The French gastronome Brillat-Savarin accepted the claim of his early 19th-century contemporaries that they had found it to be a water-soluble chemical, which they dubbed "*osmazome*," or sauce-scent. He called this discovery "the greatest service rendered by chemistry to alimentary science," and the term was still current when Fannie Farmer wrote the first version of her cookbook in 1896. Today we know that the problem is much more complicated; there are probably hundreds of constituents which account for the particular flavor of any food. Heat creates the characteristic flavor of cooked meat in two basic ways. First it damages the cell membranes and mixes the cell contents. Fats come in contact with water-soluble compounds, and reactions between free amino acids, sugars, minerals, fats and related substances, and enzymes can all contribute to flavor. Second, intense heat favors "browning reactions," exceedingly complex chemical changes that involve mostly proteins and carbohydrates (see chapter 14). Browning reactions are a boon to the food industry, which is always looking for short cuts to palatable products. Several patents have been awarded for meat "essences" produced by the hit-or-miss method of throwing a few likely compounds into a test tube and heating them up to browning-reaction temperatures.

It seems that Brillat-Savarin and his confreres were not entirely wrong: there is a water-soluble fraction of meat responsible for a basic "meaty" flavor, one that does not vary much from one kind of animal to another. Some research suggests that the flavor peculiar to a given animal species may derive from water-soluble compounds contained in the fat tissue that are transformed into fat-soluble compounds by cooking. An odd experiment has been performed that demonstrates how influential and variable animal fats can be in determining meat flavor. Several fats were heated in a vacuum, to make certain typical cooking reactions impossible by limiting the supply of oxygen, and then they were sampled by the investigators. It was found that lamb fat still smelled like lamb, but beef fat smelled like apples, and pork fat like cheese.

The flavors we have been talking about are generally diffused throughout the meat. But we are all familiar with the fact that roasted, broiled, or fried meats develop a crust that is much more intensely flavored than the rest of the meat. This is because browning reactions are greatly accelerated at high temperatures. The interior of the meat can never reach temperatures higher than the boiling point of water, or 212°F (100°C), until all the water is cooked out of it, at which point it would resemble shoe leather. The outside surface, however, is quickly dried out, and so can reach the temperature of the surrounding cooking medium—perhaps 300° or 400°F (149° or 205°C)—at which both flavor and color quickly develop. In addition, the dry surface attracts moisture from the center of the meat, thereby concentrating more browning-reaction participants in the crust. Brillat-Savarin, among others, called this action of the frying pan the "surprise," perhaps for its rapidity or its sudden effect on the palate.

Leftovers

While cooking is necessary to develop the characteristic flavors of meat, at the same time it promotes certain chemical changes that lead to so-called "warmed-over flavors" and limit the storage life of meat and even its appeal on being rewarmed (of course complex, strongly-flavored dishes may actually improve with time and reheating; here, we are talking about the meat component alone). The flavor of cooked beef, for example, has been found to change noticeably after only a few hours in the refrigerator. Fats are the principal source of off-flavors in cooked meats. The heating process releases reactive compounds from the muscle cells that encourage oxidation of fatty substances—especially phospholipids and unsaturated fats—located within the muscle tissue itself. One major culprit is ferrous (iron) ions from the denatured pigments, myoglobin and hemoglobin. Meats with a greater proportion of unsaturated fats—poultry and pork—are more susceptible to warmed-over flavor than beef and lamb. If you know in advance that you will be storing some or all of a meat dish, there are several ways to minimize fat oxidation and the development of warmed-over flavor. Avoid using iron or aluminum utensils, and put off salting until the food reaches the table; metal ions and salts accelerate oxidation. Pepper seeds and skin, onions, and potato peelings all seem to inhibit it, if you can find a way of working them into the recipe. Finally, use an oxygen-impermeable film to cover the meat (polyethylene is gas permeable), and try to eliminate any air pockets in the package. Fried chicken is a particular problem because the coating retains a thin air layer over the whole surface. Finally, try to eat the leftovers as soon as possible.

Cooking and Texture

Our understanding of the effects of cooking on meat texture is less primitive than our understanding of flavor production. Before we can discuss how cooking techniques influence texture, however, we must define the term a bit more fully. The texture of a food is a function of its physical structure: the way it feels to the touch, and the ease or difficulty with which it is broken down, by knife or tooth, into manageable pieces. Generally, we like meat to be tender and juicy rather than tough and dry. This means that in cooking we want to minimize fluid loss and toughening of the meat fibers themselves, while maximizing the conversion of tough collagen in the connective tissue to water-soluble gelatin. There are, however, ways in which we can tenderize meat *before* cooking as well.

Tenderizing

The most straightforward way of tenderizing meat is to damage it physically: cutting, pounding, grinding are all ways of breaking down the structure of the muscle bundles. Grinding, of course, is the most extreme method, since it breaks down the tissue into tiny shreds. Pounding and cutting are generally useful only for fairly thin cuts, and grinding creates an entirely different sort of texture: ground sirloin is no substitute for a tender steak. Another often-used technique is a chemical one: the application of certain plant enzymes or acid marinades. The enzymes have been used for hundreds of years. In pre-Columbian Mexico, meat was wrapped in papaya leaves before cooking, and today a derivative of those leaves, the enzyme papain, is available commercially, diluted in salt and sugar. The transition from leaf to canister came in 1941 with the discovery that several plants, including the fig, the pineapple, and some fungi, produce protein-digesting enzymes that can break down muscle and connective proteins in meat. The problem with tenderizers, however, is that only the surface of the meat is fully exposed to the enzyme, so that it can get mushy on the outside while the interior remains unaffected. The distribution can be improved somewhat by poking holes in the meat with a fork, but this is as much a mechanical as a chemical technique, and will cause greater fluid losses during cooking.

The enzyme itself, because it digests protein, will lower the water-holding capacity of the meat, and so increases fluid loss. If you do use tenderizers, you should know that they do not accomplish much until they reach a temperature of 140° to 175°F (60° to 79°C), so there is no point in letting them sit on the meat at room temperature. The enzyme is inactivated when the boiling point is reached. Marinades containing wine or vinegar can also tenderize the surface of meat—their acid denatures the surface proteins—but again the result will be drier meat. In general, there are at present no really satisfactory ways of tenderizing meat chemically.

The Importance of Fat

The fat content of meat has two different effects on texture. First, it lubricates meat fibers, slipping among the cells and making them easier to pull apart. And together with the flavor compounds they carry, fats play an important role in stimulating the flow of saliva. Food scientists who have studied the subjective sensation of juiciness find that it consists of two phases: the initial impression of juiciness as one bites into the meat, and the continued presence of moisture as one chews. The first phase is determined by the actual moisture content of the meat, while the second is the result of salival stimulation. This connection helps to explain why it is that meat from young animals—veal, for example—is often tender at first, but seems to turn dry as it is chewed: it has very little fat and is less flavorful than mature meat. Cooks have compensated for this deficiency by serving such meats with piquant sauces or stuffing them with cheese or some other, richer meat.

What Heat Does to Muscle

Cooking has a drastic effect on the muscle fibers. They begin to shrink in width as soon as cooking begins, and start to shorten when the tissue reaches 130°F (54°C). By about 170°F (77°C), the cells have shrunk as much as they can, and are beginning to develop cracks and breaks. These physical changes are manifestations of chemical changes in the actin and myosin proteins, and of weakening in the cells' membranes that allows some cell contents to escape. As the filament proteins heat up and move around more and more energetically, the forces holding them in their native configuration are overcome, and at about 100°F (38°C) the molecules start to uncoil. Once they begin uncoiling, they expose more of themselves to each other and bond together, and so the filaments coagulate, or form solid masses where before they were held in a delicate association.

There are two clear signs that coagulation is occurring. One is that the meat begins to look more and more opaque, whether it is white or red: the threads of coagulated filaments and other proteins block light rays more thoroughly than the initially dispersed arrangement (this is also why egg white is clear when raw, but opaque when cooked). The other sign is that the meat begins to exude juice. Although it is 75% water, raw meat leaks very little when it is cut, but rare meat is quite juicy. The reason for this sudden appearance of fluid is that the coagulating proteins are squeezing out the water that used to separate them from each other and that used to be trapped in their coiled structure. It's the same thing that happens when you twist the strands of a wet rope more tightly together, or squeeze a wet towel. In any case, moisture leaves the tissue as the proteins coagulate, and by the time the meat reaches 170°F (77°C) most of the liquid that can be freed by this process has been released. Well-done meat, because it has already given up its juices, does not exude them or seem as moist when chewed as rare meat does. The combination of fiber contraction and fluid loss explains why meat shrinks during cooking, and becomes denser and less jellylike.

Judging Doneness by Color

We usually define the doneness of red meats not by their temperature but by their inner color. We can do so because the pigment myoglobin is also a protein, and its changes parallel those of the fiber proteins. Up to about 140°F (60°C) myoglobin remains unaffected and its color stays red, but temperatures from 140° to 160°F (71°C) disturb its structure: it loses the ability to bind oxygen, and the iron atom at its center gives up an electron, thereby forming a new, tan-colored compound called a hemichrome. In this temperature range, the color of the meat will run from deep red to light shades of pink. By 175°F (79°C), enough hemichrome has accumulated to produce a light brown-gray shade. As the fiber proteins get more solid, dense, and dry, the pigment gets more and more drab. This is why rare meat is pink and juicy, well-done meat brown-gray and dry.

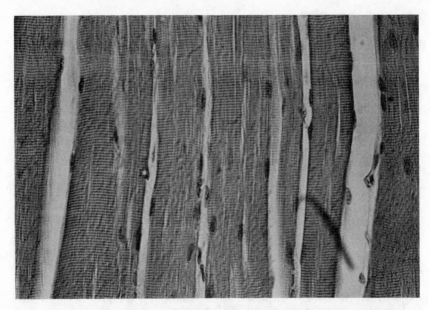

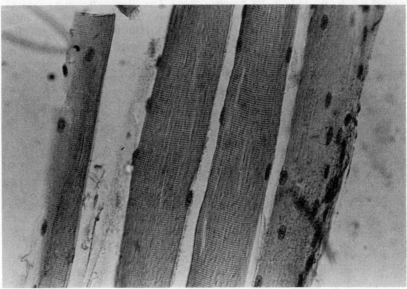

Top: rabbit muscle 24 hours after slaughter. The fibers have relaxed, and they and the fibrils now stand out more distinctly from each other. The horizontal bands indicate separate contractile units in the fibrils. *Bottom:* cooked rabbit muscle. As the proteins have coagulated, the cells have lost liquid and shrunk in both diameter and length (notice the crowding of horizontal bands). The fiber at left has lost its pliability and has broken into two pieces. (From P. C. Paul and H. H. Palmer, *Food Theory and Applications.* Copyright © 1972 John Wiley and Sons, Inc. Reprinted by permission.)

$$O_2 \qquad\qquad O_2 \qquad\qquad NO \qquad\qquad CO$$

N⟍ ⋮ ⟋N   N⟍ ⋮ ⟍N   N⟍ ⋮ ⟋N   N⟍ ⋮ ⟍N
Fe^{+2} Fe^{+3} Fe^{+2} Fe^{+2}
N⟋ ⋮ ⟍N   N⟋ ⋮ ⟍N   N⟋ ⋮ ⟍N   N⟋ ⋮ ⟍N

globin   denatured globin   denatured globin   denatured globin

(red)   (tan)   (pink)   (pink)

The effects of cooking on meat pigment. *From left to right:* red, oxygen-carrying myoglobin; the oxidized, denatured form characteristic of well-done meat; the pink form that results from heating nitrite-cured meat (NO is nitrous oxide, a product of nitrite); and the pink form that sometimes develops in uncured meats in an oven where carbon monoxide (CO) accumulates.

There are a couple of common exceptions to this sequence, however. The red-pink cooked color of cured meats is caused by a stable but denatured form of the nitrite-myoglobin complex. And the pink juices that sometimes run out of a well-cooked chicken are a complex of myoglobin and the carbon monoxide or nitrous oxide gases that are produced by the gas flame or electrical element.

The Necessity of Compromise

Given what happens to the muscle fibers in meat as they heat up, one lesson seems obvious: the cook should stop cooking before all the juice is lost and the fibers dry out and toughen. The problem is that the collagen in connective tissue requires more extensive cooking in order to be converted into soft gelatin. This process is in some ways the reverse of protein coagulation in the muscle fibers: tightly wound fibers of pure collagen come apart into more vulnerable single molecules as they heat up. The "shrink temperature," the temperature at which the ordered structure of this protein collapses, is about 140°F (60°C) in meat (a mere 105°F (41°C) in fish), and it is only at temperatures higher than this that the individual molecules separate from one another into the loose association we call gelatin. The conversion of collagen to gelatin is most rapid at temperatures close to the boiling point of water, or 212°F (100°C). So if you cook meat rare, just to the point where the fiber proteins are coagulating, exuding juice, and are still tender, the tough connective tissue will be mostly unaffected. But if you cook the meat well enough to gelatinize the collagen, then the fibers will be dry and dense. The only way to get tender fibers and connective tissue at the same time is to cook the meat so long that the fibers begin to disintegrate altogether. But meat treated in this way is crumbly and mealy rather than firm and juicy. Clearly the art of cooking meat lies in compromise.

The cook has several very different techniques with which to fashion this compromise. Meat cookery includes all the basic cooking methods:

THE EFFECTS OF HEAT ON MEAT

Temperature				Protein-bound	
(°F)	(°C)	Pigment	Fiber Proteins	Water	Collagen
100	38	red	uncoil, begin to coagulate	begins to flow	
120	49				
140	60				begins to dissolve
		pink			
160	71		mostly coagulated		
		gray-brown		flow ceasing	
180	82		forming denser association		
200	93				dissolving rapidly
220	104				

among the dry methods, roasting (today, really baking), broiling, and frying; among the moist methods, boiling, steaming, braising, and stewing. The important difference between these two groups is not moisture per se, but their characteristic temperatures and rates of heating (see chapter 14). Dry methods generally use temperatures well above the boiling point of water, a common example being the 350°F (177°C) oven. Moist methods are necessarily limited to a maximum temperature of 212°F (100°C); you can't get water hotter than that without a pressure cooker. On the other hand, dry methods, which transfer heat across a layer of air (except in the case of frying), are much less efficient than moist methods, because water has a much higher heat capacity than air. So it will actually take about the same time to boil a large piece of meat as it will to roast it, even if the oven temperature is double the boiling temperature. And each different combination of cooking temperature and time has its own balance of tenderizing and toughening influences.

The Searing Question

There is one misconception about meat cookery that still enjoys great popularity, even though it has long since been discredited. Does the gist of this description of cooking sound familiar?

Thus as the exterior pores contract, the moisture contained in the object cannot escape any more, but is imprisoned there when the pores close.

This quotation comes not from a blurb for convection ovens, but from Aristotle's treatise on meteorology (Book 4). The theory has changed little except for the terminology—today we would say that the food's juices are "sealed in" by high temperatures, keeping it moist and tender. To see how this idea came into prominence and lives on is to see how recipes can be perpetuated more by theory than by result.

Not much is said in the oldest surviving cookbooks about the precise methods for boiling or roasting foods, perhaps because there were no precise methods, or because early manuscripts were meant for other professional cooks who would already have known the basics. One of the first "modern" expressions of the idea of trapping juices came from Sir Kenelm Digby, a 17th century naval commander, scientist, philosopher, and writer, who left a treatise on cooking. He directed that the capon or chicken be set on a spit, heated through, basted with butter, and sprinkled with flour: "This by continuing turning before the fire will make a thin crust, which will keep in all the juyce of the meat." In time, retaining the meat's juice would become the overriding goal of many cooks, at least in England and America. France remained impervious to the controversy until late in the 19th century. The great chefs before that time generally put larded paper over the meat to prevent burning during the roasting, and then removed the paper and browned the meat at the very end to give the roast *"une belle couleur."* For them, juice retention was not a concern.

In the 18th and early 19th centuries, most cookbooks in English ignored the problem of fluid loss, and recommended beginning the roast well away from the fire and moving it closer to brown it at the end. The turning point in England and America comes in the middle of the 19th century, and a single man, the eminent German chemist Justus von Liebig, was responsible. His book *Researches on the Chemistry of Food* was translated into English in 1847, and an American edition appeared one year later. In it, Liebig summarized recent research in muscle chemistry and then spelled out its practical implications. Chemists working all over Europe had demonstrated that any sort of muscle tissue, whether fish, flesh, or fowl, has several basic constituents. Since these substances are also a part of human muscle, Liebig reasoned, then they had to be elements of human nutrition necessary to replace worn-out materials. But some of them are water soluble and so easily lost from the meat during cooking. It would be best to minimize this loss in order to maintain the nutritional value of the meat.

This could be done, Liebig said, by heating the meat quickly enough that the juices are immediately sealed inside. He explained what happens when a piece of meat is plunged into boiling water, and then the temperature reduced to a simmer:

> When it is introduced into the boiling water, the albumen immediately coagulates from the surface inwards, and in this state forms a crust or shell, which no longer permits the external water

to penetrate into the interior of the mass of flesh. But the temperature is gradually transmitted to the interior, and there effects the conversion of the raw flesh into the state of boiled or roasted meat. The flesh retains its juiciness, and is quite as agreeable to the taste as it can be made by roasting; for the chief part of the sapid [flavorful] constituents of the mass is retained, under these circumstances, in the flesh.

And if the crust can keep water out during boiling, it can keep the juices in during roasting, so it is best to sear the roast immediately, and then continue at a lower temperature to finish the insides. Conversely, if one wants to make a nutritious soup—in the last century you could buy a concoction called Liebig's Extract, made under his supervision—the meat is put into cold water and heated slowly, thereby drawing all the nutritious substances into the liquid. After this treatment, according to Liebig, the cooked solid tissue itself has lost much of its nutritive value.

We know today that most of this is simply not true. The water-soluble components are minor products of muscle metabolism and nutritionally negligible. Any crust that forms around the surface of the meat is not waterproof. And the solid hunk of long-boiled meat still contains most of the protein. But in its day, Liebig's account answered the unspoken need for some rational, systematic approach to cookery. Its influence was swift and, in English-speaking countries, nearly universal. If you didn't follow Liebig's advice, you at least had to argue openly with him. The revolution is most obvious in Eliza Acton's *Modern Cookery for Private Families*, which, in its second edition of 1845, gives traditional directions for boiling (begin with cold water) and roasting meat. In the next edition, published post-Liebig in 1855, the subtitle has become "In Which the Principles of Baron Liebig and Other Eminent Writers Have Been as Much as Possible Applied and Explained," and the directions for roasting and boiling are now pure Liebig. Another good barometer of the change is Mrs. Sarah J. Hale's 1857 *New Cook Book*, whose recipes are obviously pre-Liebig, but whose introduction quotes Liebig in footnotes and contradicts the recipes.
        There were those who resisted the new wave, saying that the techniques of the "modern professors" resulted in unpleasant flavors and tough meat. But such voices were a distinct minority, and the Liebig method lasted for better than fifty years. One of its last major proponents was Fannie Merritt Farmer in her original *Boston Cooking School Cook Book* (1896). It fell, finally, with the accumulation of more knowledge about the composition of animal tissue, and with more careful analysis of Liebig's own claims. Late in Liebig's life, his collaborator on the Beef Extract withdrew from the project, announcing that its nutritional value had been greatly exaggerated. But even after Liebig's rationale for the early-searing method had been disproven, the method itself lived on under various guises, often rather eccentric. A prime example is Auguste Escoffier's authoritative *Guide Culinaire* of 1902. Escof-

fier applied a baroque imagination to the theory of albumen coagulation and came up with this remarkable narrative of what happens to meat during braising. First the meat is browned in a frying pan.

> The point of browning is to drive back into the interior the juices that strive to escape, and to produce around the meat a sort of armor, which thickens from the surface to the center as the cooking progresses.
>
> Under the influence of the heat of the liquid that bathes the meat, the fibers contract and force the juices to move toward the center. Soon, the heat arriving at the center forces the compressed juices to decompose and to release the excess of water which they contain, and which does not delay in vaporising, distending and dissociating the fibers directly submitted to its action.

And so on. The wonderful *panache* is enough to convince you that Escoffier has been inside a pot roast to witness what he calls the "process well-known in physics under the name of Capillarity."

The brief decline in the Liebig technique's popular standing began early in this century, when the invention of the meat thermometer made it possible to measure the temperature within cooking meat, and when the federal government made funds available for research in home economics. In 1930, a study done at the University of Missouri found that meat roasted at a constant temperature actually lost *less* fluid than initially seared samples. This information was picked up and disseminated by the meat industry, and so we read in the first (1936) edition of *The Joy of Cooking* that the "old-fashioned" roast takes an initial searing, while the "modern" method sponsored by "our national packers" calls for the constant and relatively low temperature of 325°F (163°C). By the third edition (1953), the *Joy* gives the modern method top billing, and so implies that it is the preferred choice. A new orthodoxy was taking hold. But not very strongly, and not for long. In the 1964 edition, the first after Mrs. Rombauer's death, the searing method makes a comeback, and in the 1975 edition Marion Rombauer Becker wrote that "we have found the flavor markedly superior when the meat is sealed by high temperature at the outset—and the difference in shrinkage is negligible."

Notice how the grounds of the argument have shifted since Liebig's time. The issue is no longer nutritional value or juiciness, but taste. And here we are on firmer ground. We *do* know for a fact that whether done early or late, searing does not seal, but it does brown: it won't prevent flavor from escaping, but it *creates* flavor via the complex browning reactions. And because it has become a matter of taste, today's experts have come to different conclusions on the matter. Some recommend searing all but the toughest meats, others are devotees of the constant method. So there is a good reason to sear meat, but it has nothing to do with nutrition or juiciness. The many recipes and ads that perpetuate Liebig's theory probably do so because the image it evokes is vivid and appealing.

The Cook's Choices

The choice of a particular method of cooking should depend on two basic considerations: the kind of meat involved, and your own personal tastes. Obviously it makes a difference if you are cooking a roast rather than a steak, pork rather than fish, ribs rather than round. And if you can decide whether you prefer rare or well-done meat, chewy or crumbly texture, mild or savory exterior, then your choice of method becomes even clearer. The important thing is to be able to match your material and your predilections with the treatment most likely to do justice to both.

Kinds of Meat

Three qualities of meat should be kept in mind when judging the best way to treat a particular cut. One is the intrinsic texture of the fibers, which we have seen to be determined by the location of the muscle or muscles in the body, and by the animal's age and activity. Adelle Davis suggested a way of understanding why certain areas in the animal are likely to be tender or tough, at least for quadrupeds. Get down on all fours and walk and "graze"

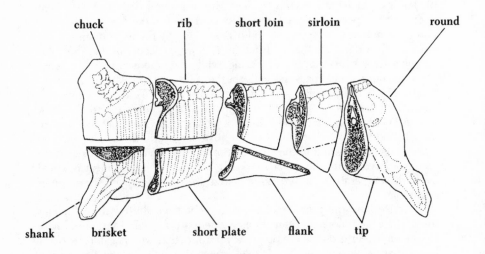

The cuts of beef. The rib, short loin, and sirloin are generally the tenderest cuts and do well with dry heat. With some exceptions, the other cuts are best cooked with moist heat to soften the connective tissue, and are frequently ground into hamburger.

AGES OF MEAT ANIMALS AT SLAUGHTER

Beef:
Veal	Less than 3 months
Calf	3 to 24 months
Steers (unsexed males), some Heifers	Less than 42 months

Lamb: — Less than 14 months

Pork: — 6 to 7 months

Poultry:
Game hens	5 to 7 weeks
Broiler, fryer (both sexes)	9 to 12 weeks
Roaster (both sexes)	3 to 5 months
Capon (unsexed male)	Less than 8 months
Hen, stewing chicken, fowl	More than 10 months
Roasting ducks (both sexes)	Less than 4 months
Young turkey	Less than 15 months
Mature turkey	More than 15 months

for a while; you will notice that the neck, shoulders, chest, and "legs" take a beating, while the back is more relaxed. As a general rule, the further away the cut is from hoof or horn, the more tender the meat. And remember that generally, younger animals—and veal, lamb, and pork all come from younger animals than beef does—will have tenderer because finer, less-used muscle fibers.

The second quality is fat content. Young animals, though tender, usually have less fat, and so will tend to seem drier than more mature cuts. This is especially true of veal compared to beef. In the federal grading system that was introduced in 1927, color, firmness, and marbling are the three factors that determine the quality of a particular carcass. Beef must have a certain degree of marbling in order to qualify as Prime. Today, Prime beef is getting rarer, partly because consumers are buying leaner meat, and partly because the cost of an extra month's grain feeding is becoming prohibitive. It turns out that lamb and pork are well-larded even when young, and are not as much of a problem as veal. Still, the general rule holds: a readily visible pattern of fat in the meat will mean a moister texture, although also higher calorie and cholesterol counts and less protein.

Finally, there is the connective tissue. Again, look at the meat: if you see a lot of membranous tissue, there will be a lot of collagen. Parts of the body where several muscles are bound together and anchored to a bone will be relatively tough. Among commonly used cuts of beef, the rump contains the most collagen and elastin. Young animals tend to have a higher proportion of collagen in their muscles than mature animals, but it is more easily converted to gelatin than older collagen, which develops more extensive cross-linking among its molecules.

Fish and Variety Meats

As a consequence of their very different muscle organization, fish and shellfish constitute a special category of meat. Because their connective tissue is fragile, the muscle fibers short, and their fat content relatively low, they generally should be cooked as little as possible, only to the point that the muscle proteins coagulate. Beyond this point, the tissue tends to dry out and either toughen or disintegrate. Those organs of land animals that are not muscles and contain little connective tissue—liver is probably the most commonly used—should also be treated gently; long cooking will dry them out and toughen them. Hearts, gizzards and other muscular variety meats should be treated like the lean, well-exercised tissues they are, and should be cooked slowly and long.

Techniques for Cooking Meats

Broiling

Broiling subjects the meat to the heat source, whether it be gas flame, glowing charcoal, or electrical element, with the mediation of only a few inches of air. It is a very hot process—gas burns at 3000°F (1649°C), the electrical element glows at about 2000°F (1093°C)—and this limits its usefulness; the whole piece of meat must be cooked through before the outside surface is burned. For this reason broiling is traditionally limited to relatively thin and tender cuts like chops, steaks, small poultry parts, and fish. Meat with a lot of connective tissue is less suited to broiling because collagen does not have the time or reach the temperature necessary to be converted to liquid gelatin. The one great advantage of broiling is that the very high temperatures at the meat surface are ideal for browning reactions; thus broiled meats have a characteristically intense flavor. A minor drawback is that these same temperatures are high enough to burn fat, and small conflagrations can result when the melted drippings spit toward the heat source. Avoid this by using a proper broiling pan, which allows the melted fat to drain under a metal shield.

Frying

Pan frying, or sautéing, depends largely on the conductive transfer of heat, the direct movement of energy from the pan to the meat. Enough fat or oil is added to the pan to prevent the meat from sticking. Frying is a fairly hot, rapid cooking method, and so is limited to the same thin, tender materials best suited for broiling. Frying temperatures are significantly lower than broiling temperatures, however, primarily because the smoke point of the fat must not be exceeded if burnt flavors are to be avoided. Different fats have different smoke points, and the choice of fat determines the maximum cooking temperature. Unclarified butter cannot be heated above 250°F (121°C) without burning, while animal lard can reach 400°F (205°C) and vegetable oils close to 450°F (232°C). Vegetable shortening, because it contains added emulsifiers, is limited to about 370°F (188°C).

Because heat is conducted very efficiently from pan to meat, the meat surface tends to be browned very quickly—in a matter of a minute or two—and so, to prevent the surface from being toughened while the inside cooks, the heat is usually reduced after the initial browning. If the pan is covered, water vapor is trapped and a process more like braising results (see below).

Roasting

With roasting, or more accurately oven baking, we return to the searing controversy traced above. Roasting depends on infrared radiation from the oven walls and on convection, or the transfer of heat via currents of air, which has a much lower heat capacity than either water or oil or metal. It is a relatively slow method, well-suited to large pieces of meat that take time to cook through to the center. You will want to sear the meat, either in an initially superhot oven or in a large frying pan, if you prize the brown crust and its intense flavor, and don't mind if the meat is somewhat drier as a result. But if juiciness is your desideratum, then skip the initial browning. The choice of temperature is more complicated, and depends on the quality of the meat.

Recall that connective tissue requires long cooking at high temperatures to be converted into gelatin, while muscle proteins, once they have coagulated at about 160°F (71°C), will only get drier and tougher with time, until the fibers themselves begin to disintegrate. If you roast at a relatively high temperature, say from 350°F (177°C) up, then for a given doneness your meat will have been cooked for a relatively short time. As a result, the muscle proteins will not have been overcooked, but the collagen will have gelatinized less as well. This is an appropriate strategy, then, for cuts with little connective tissue, such as rib roasts. With low temperatures, say from 250° to 300°F (121° to 149°C), the meat must remain in the oven much longer, on the order of an hour per pound rather than 20 or 30 minutes. The collagen will gelatinize fairly completely, but the muscle tissue itself will become mealy rather than firmly tender. Because the heating process is more gradual, there is less extreme coagulation in the tissues, and less fluid will have been squeezed out in the process, although the longer time spent in a dry oven may cause increased evaporation of moisture. The low-temperature method is appropriate for less tender cuts, such as the chuck and rump. Lamb and pork, because they come from relatively young animals and have high fat contents, are much less affected by changes in cooking temperature than is beef. Because bone conducts heat more readily than moist tissue, bony roasts will cook faster than boneless ones.

There is some disagreement about the value of basting, which does interrupt and slow down the cooking process. Except for very lean cuts and areas like the chicken breast, which are easily dried out, the fat content of the tissue can probably keep the meat moist by itself. Stuffed poultry should be cooked at about 325°F (163°C), because lower temperatures allow bacteria in the stuffing to flourish for too long, and higher temperatures shorten the cooking time enough that the stuffing may never cook through.

Braising

A word about roasts that are not roasts. In days past, as the first *Joy of Cooking* attests, there have been roast recipes that call for adding stock to the pan, for covering the roast with foil, or both. It was thought that added moisture, or a moisture trap, would help preserve the juiciness of the meat. Because they are covered containers, today's low-temperature slow-cooking crock pots have the same effect. In fact, such techniques, because they *speed* the cooking process—steam is a more efficient heating medium than air— actually increase the fluid loss, and tend to produce drier roasts. Moreover, because they depend on moisture, these methods are not properly called roasts, but braises.

Braise is a term borrowed from France, where it originally referred to a closed pot sitting on top of, and covered by, charcoal; today, one can braise in the oven or on the stovetop. Braising is cooking with a small amount of water in a closed container; it differs from stewing in using less water and larger cuts. The pot roast, made from the chuck or rump of beef, is perhaps the most common braised meat. As a technique, the braise is characterized by a relatively low temperature, the boiling point of water, and yet a more efficient transfer of energy than is possible in dry heating. In practice, this means that a large piece of meat will reach temperatures near this maximum relatively quickly, and can be kept there for prolonged periods without any danger of the outside burning. Such treatment is exactly what is called for in order to convert collagen to gelatin. Food scientists have compared collagen conversion in a rump roast when it is roasted well done, which would take an hour or two, and when it is braised for 30 minutes and for 90 minutes. The roasted sample had 14% of its collagen gelatinized, the 30-minute braise 11%, and the 90-minute braise 52%. The obvious conclusion is that a long slow braise is the best way to tenderize meat that is tough on account of its connective tissue content. For higher grade meats with little connective tissue, however, long braising makes the meat much tougher than it would be if it were cooked quickly to a lower final temperature. A short braise— just enough to reach the desired doneness—is appropriate for tender cuts like chops. Most recipes call for browning the meat before braising; this is advisable not to seal in juices, but to provide flavor by triggering the browning reactions that require high, dry heat. If you brown the meat in a different pan, be sure to use some of the braising liquid to deglaze it by dissolving the flavorful browning reaction products left behind.

Stewing

The stew involves more liquid, smaller pieces of meat, and generally longer cooking times than the braise. In a stew, we want the flavor of the meat to permeate the large amount of liquid, so we tend to let the meat be cooked well beyond the point at which it is done. In both cases, though, the end point in cooking is rather fuzzily defined as "fork tender," and so timings are not terribly precise.

Microwave Cooking

Microwave cooking is neither a dry nor a moist technique, but electromagnetic. Very high frequency radio waves cause electrically asymmetrical molecules, especially water, to vibrate, and these molecules in turn knock into other molecules—proteins, fats, and so on—and impart their energy to them. The water in the tissue is heated up, and in turn heats up the tissue itself. Because radio waves penetrate organic matter, the meat is not cooked from the surface inward, but simultaneously to a depth of a couple of inches. The water molecules are excited faster than they can transfer their energy to the muscle proteins, so that the meat may "cook" for a relatively short time, and then sit for a while with the oven off as its temperature rises to the desired doneness. In practice, though, the radiation may penetrate unevenly, and a large piece of meat may have to be shifted around to compensate for this. Since the air in the oven is not heated, microwave ovens do not brown the outside surface, and this important disadvantage has led some manufacturers to add a conventional electrical browning element to the oven. Microwave cooking is very fast—this, of course, is its great selling point—but for that reason it tends to result in greater fluid loss than conventional means. On the other hand, it has the compensatory virtue of being very effective in solubilizing collagen—why, we don't know. Microwave roasting would seem to be especially handy for tougher cuts of meat, while conventional methods are more appropriate for higher-quality cuts like leg of lamb and rib roasts.

A Note on Carving

Whether it be a leg of lamb or a turkey, the main attraction of your meal may be done a disservice when you carve it, and not just visually. Let the roast sit after removing it from the heat. It will continue to cook for a while because the outside layers are still warmer than the inside and they go on transferring heat to the center. For this reason, you should take the roast out when it is 5 degrees or so short of the desired end temperature. Then there is the old bugaboo of fluid loss. A roast that is cut when fresh from the oven will leak much more juice than it would if it were first allowed to sit for 15 minutes. Apparently some protein coagulation is partly reversible, and the water-holding capacity of the tissue increases as the temperature evens out and begins to drop. Finally, use a good sharp knife. Sawing away has the effect of compressing the tissue and, like a squeezed sponge, the roast will ooze much of its delicious liquid away.

CHAPTER 4

Fruits and Vegetables, Herbs and Spices

PLANTS AS FOOD

From his earliest days as a species, man's primary source of food has been plants. Indeed, they are the ultimate nourishment for all animals, including strict carnivores, because they are at the beginning of the food chain. Unlike animals, plants can synthesize organic materials from the minerals, air, and sunlight, and so they are the true origin of the proteins, carbohydrates, and other complex molecules necessary to animal life.

The feeling that plants are our original and proper food is a deep-rooted one. In the Golden Age described by Greek and Roman mythology, the earth gave of itself freely, without cultivation, and man ate only nuts and fruit. And in Genesis, Adam and Eve spend their brief innocence as gardeners:

> And the Lord God planted a garden eastward in Eden; and there
> he put the man whom he had formed. And out of the ground
> made the Lord God to grow every tree that is pleasant to the
> sight, and good for food. . . . And the Lord God took the man,
> and put him into the Garden of Eden to dress it and to keep it.

Only after Cain kills Abel does the human race eat meat. Many individuals and groups, from Pythagoras to the present, have chosen to eat only plant foods, often for reasons of principle. Even today, for the great majority of the earth's population, which lives in poverty and has no choice, plants constitute the primary, sometimes the sole source of nourishment.

The first plants to be domesticated, about 10,000 years ago, were the grains and legumes, the richest sources of protein and carbohydrates in the plant world; and so the most important nutritionally. The plants that give us what we call vegetables and fruits generally came under cultivation much later. They contribute mainly vitamins, minerals, and some carbohydrates to the diet, and have always been more of a supplement than a staple food. To this day fruits and vegetables are usually served as side dishes or desserts or as accessory ingredients in a main course.

Herbs and spices are used primarily to imbue other foods with their flavor. They have little nutritional contribution to make because their strong flavors limit our consumption of them. Yet spices have played a remarkable social and economic role in history. For millennia they were valued as a retardant and disguise of spoilage, as medicines, and as a symbol of wealth. Attempts to break monopolies on the spice trade were the immediate cause of the first voyages to the New World.

To some people, herbs are common garden products and spices exotic, often tropical plants. To others, herbs are the green leaves and stems, and spices the roots, seeds, and nuts of any flavoring plants. The first definition has the disadvantage of being geographically relative, so we will adopt the second.

Fruit or Vegetable?

The definitions of fruit and vegetable are more complicated, and reveal some of our buried attitudes toward plant foods. Both words derive from the Latin. *Fruit* originally meant any plant used as food, and gradually came to mean the edible layer that surrounds seeds. In the 18th century, as the systematic science of botany arose, the word was given a technical sense which is still in force: a fruit is the organ derived from the ovary and surrounding the seeds. Plant anatomy replaced edibility as the criterion. Also in the 18th century, and perhaps as a result of the narrowed range allowed to *fruit*, *vegetable* makes its first appearance as a word meaning plant foods eaten along with meat or other parts of the meal; originally it meant simply a plant, as opposed to an animal or inanimate object.

Of course, the technical definition is not commonly in force today. Green beans, eggplants, cucumbers, and even corn kernels are all fruits anatomically, but we call them vegetables. Common usage is based not on plant anatomy, but culinary custom: we employ apples and oranges in one way, corn and eggplants in another. No less an authority than the United States Supreme Court has based its august opinion on this consideration. Late in the last century, a New York food importer had claimed duty-free status for a shipment of tomatoes from the West Indies. He argued that tomatoes were fruits, and so under the regulations of the time, not subject to import fees. The customs agent disagreed, and imposed a 10% duty on the shipment he defined as vegetables. Constitution and statute offering no guidance on this question, the Court decided on the grounds of linguistic custom. Tomatoes, held the majority, are "usually served at dinner in, with, or after the soup,

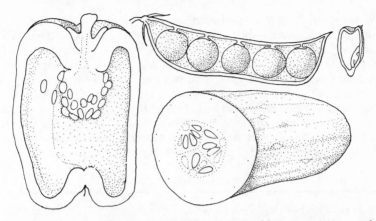

Despite the fact that we call and treat them as vegetables, peppers, pea pods, cucumbers, and even corn kernels are anatomical fruits. Each derives from the flower's ovary and contains one or more seeds.

fish, or meat, which constitute the principal part of the repast, and not, like fruits, generally as dessert." Ergo the tomato is a vegetable, and the importer had to pay the duty.

Fruit is distinguished from vegetables in another important way, one that is also reflected in our language. Consider the following common expressions. "That job is a real plum." "That's a peach of a dress." "Life is just a bowl of cherries." "She's the apple of her father's eye." Now another group of sayings: "That show was pure corn." "Your offer isn't worth a hill of beans." "He's nuts!"

By and large, we refer to fruits when we want to convey praise, and to other plant products in order to disparage. Of course, there are exceptions: figs, raspberries, and the word *fruit* itself can be used to express derision, while *carrot* can mean a reward or inducement (at the cost of turning its object into a quadruped, however). The lemon is perhaps the most obvious exception, but in fact it proves the rule, as we shall see. Even the etymology of the terms reflects the special status of fruit. *Vegetable* comes from the Latin verb *vegere*, meaning to animate or enliven. *Fruit*, on the other hand comes from *frui*, meaning to enjoy, to delight in, to have the use of something. This root is largely evaluative and imputes desirability to its object. So both etymology and the figures of speech that parallel it implicate fruit with pleasure, and vindicate the judgment of most children and probably not a few adults: vegetables may be good for us, but it is fruit that *tastes* good.

Culinary fruits are distinguished from other foods by a single trait: sweetness, or a favorable balance between sweet and sour. Studies of human newborns and other mammals indicate that, of the four basic taste sensations, only sweetness is innately preferred. And the plant products that we call fruits and treat as such—they have been eaten at the end of the meal at least since the Greeks—generally have a much higher sugar content, and often more acid, than those we call and treat as vegetables. Most temperate-zone fruits are from 10 to 15% sugar, and tropical fruits like bananas, figs, and dates from 20 to 60%. And now we see why *lemon* means "dud": its 1% sugar content fails miserably to match its acidity.

We will generally abide by the everyday definition of fruit in this chapter. In discussions of plant structure and function, however, the word will refer to the ovary-derived organ, regardless of its flavor.

Historical Background

How long has the Western world been eating the plant foods we eat today, and in the way that we eat them? Briefly, only a very few common vegetables—broccoli, cauliflower, brussels sprouts, celery—have not been eaten since before history, although it was only in the 16th century that the *variety* of foods we now know became available to any single culture. Recognizable salads go back to the Middle Ages, and boiled vegetables in delicate sauces to 17th-century France.

Spices Changed the World

Beyond being the primary food source for all animals on the earth, plants—and especially the spice plants—have played a decisive role in the shaping of history in the last five centuries. The great European hunger for the exotic spices was largely responsible for the development of Italy, Portugal, Spain, Holland, and England into major sea powers during the Renaissance, for the discovery of the New World, and for advances in astronomy, timekeeping, and the science of magnetism, all important to accurate navigation. Vasco da Gama, Columbus, John Cabot, Magellan, and many other less successful explorers were looking for a new route to the cinnamon-, clove-, nutmeg-, and pepper-rich Indies, in order to break the expensive monopoly of Venice and of the southern Arabians, who had been geographic and economic middlemen in the spice trade for millennia. They failed in that quest, but succeeded in opening the "West Indies" to European exploitation.

The New World was rather disappointing in its yield of spice: it held none that the Europeans were looking for, and only three of its novelties would catch on—allspice, red pepper, and vanilla. But its wealth of new vegetables, including the common bean, corn, squashes, tomatoes, potatoes, and sweet peppers, eventually transformed the cuisine of the Old World. Also transformed was the worldwide distribution of flora. European colonists introduced their native plants to the Americas, planted East Indian spices in the Caribbean, brought American novelties back to Europe. The result: an unprecedented geographical dispersal of economically important plants. As unnatural as this may seem, we shall see that this trend is quite consistent with the nature of plants.

Prehistory and Early Civilizations

What of eating habits before this revolution? Our knowledge of customs before the Middle Ages is limited by incomplete or ambiguous evidence, reliance on authorities who wrote long after the fact, and uncertainties of translation, but a few ideas seem commonly accepted. Many plants had been under human cultivation long before the rise of any civilization, by the unsophisticated but slowly effective means of gathering useful plants and spilling

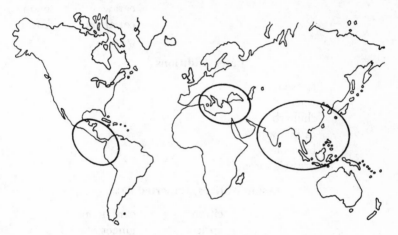

The three principal regions in which food plants were domesticated

their seed near fertile refuse heaps. Along with the protein-rich cereals and legumes, root and gourd vegetables, which were protected by the soil or a tough outer skin, were probably the most important to prehistoric man. Primitive Europeans seem to have relied on wheat, beans and peas, turnips, onions, radishes, and cabbage. In Central American villages, winter squash, tomatoes, avocados, beans, and corn were staples around 3500 B.C., while Peruvian settlements relied heavily on the potato. Millet, wheat, rice, coconuts, and bananas were most important in Asia. Mustard seed flavored foods in Europe and in Asia, where ginger may also have been used. Red pepper was probably the chief spice in the Americas.

By the time of the earliest Mediterranean civilizations in Sumer and Egypt (about 3000 B.C.), most of the plants native to that area that we eat

VEGETABLES, FRUITS, AND SPICES USED IN WESTERN EUROPE

Mediterranean Area Natives, Used B.C.

mushroom	onion	apple	basil	dill
beet	cabbage	pear	marjoram	parsley
radish	lettuce	cherry	oregano	fennel
turnip	artichoke	grape	mint	bay
carrot	cucumber	fig	rosemary	caper
parsnip	broad bean	date	sage	fenugreek
asparagus	pea	strawberry	savory	garlic
leek	olive		thyme	mustard
			anise	poppy
			caraway	sesame
			coriander	saffron
			cumin	

Later Additions

spinach
celery
rhubarb
cauliflower
broccoli
brussels sprouts

Asian Natives, Imported B.C.

citron	cardamom
apricot	ginger
peach	cinnamon
	turmeric
	black pepper

Imported Later

yam	lemon	clove
water chestnut	lime	mace, nutmeg
bamboo	orange	tarragon
eggplant	melon	

New World Natives, Imported 15th–16th Centuries

potato	kidney bean	pineapple	allspice
sweet potato	lima bean		red pepper
pumpkin	sweet pepper		vanilla
squashes	avocado		
tomato			

today were already in use (see chart). And by 2500 B.C., spice trade with the East had already been established. There are records of a barge expedition from Egypt to the lower Red Sea from about this time, and it was probably not the first. Around 1200 B.C., Ramses III proudly recorded huge offerings of cinnamon, a product of Ceylon.

Greece and Rome

With the Greeks and Romans we begin to see the outlines of our own cuisine. The Greeks were fond of lettuce, and habitually ate fruit at the end of meals. Pepper from the Far East was in use around 500 B.C., and quickly became the most popular spice of the ancient world. It flavored the cereal gruel that was the staple of the masses, it hid the flavor of spoilage, and to some extent, slowed spoilage by inhibiting microbial growth.

In Rome's heyday, vegetables were regularly served as hors d'oeuvres and fruit as dessert, and lettuce was served at both the beginning and end of meals. The art of grafting, which had also been practiced by the Greeks, was advanced enough that the natural historian Pliny could say at the beginning of the Christian era, "This part of life has long since reached its summit; everything has been tried" (Book 17). Records show that the Romans had discovered or developed 25 varieties of apple and 38 of pears. Fruits were preserved whole by immersing them, stems and all, in honey, and the pioneer gastronome Apicius (he lived in the first century, but the earliest manuscript dates from around A.D. 300) gives a recipe for pickled peaches. Among vegetables, the carrot was disdained and the cabbage very popular, and there was a long-standing dispute about the virtues of garlic. Finally, the Romans may be considered partly responsible for the spice competition of 1500 years later, for in their cuisine, spices became an essential ingredient for the first time. From the recipes that survive, it would seem that no food was served without liberal application of several strong flavors. Apicius's cumin sauce, to be eaten with shellfish, is typical.

ROMAN HERBS AND SPICES

Cumin sauce, for shellfish: pepper, lovage, parsley, mint, aromatic leaf [e.g., bay], malabathrum [a Middle-Eastern leaf], plenty of cumin, honey, vinegar, liquamen [a fermented fish paste, probably similar to our anchovy paste].

—from Apicius (A.D. first or third century)

The Piquant Middle Ages

As the Romans conquered Europe, they brought tree fruits, the vine, and cultivated cabbage with them, as well as their heavy spice habit, which may have been exacerbated by the temptations of conspicuous consumption. One historian has it that a renewed interest and indulgence in spices followed from the visit of Haroun al Raschid, the caliph of Islam immortalized in *The Thousand and One Nights*, to the Emperor Charlemagne around 800. In any case, sauces of the time match or even outdo Apicius in piquancy. The recipes we give come from two nearly contemporaneous (late 14th-century) sources, one an anonymous compilation from the court of Richard II, and the other the work of Guillaume Tirel, known as Taillevent, cook to Charles V. A comparison of the two recipes for green sauce would suggest that the English had more native herbs to choose from, or at least appreciated them more than the French.

MEDIEVAL HERBS AND SPICES

Sauce Cameline, for meats: France: ginger, mace, cinnamon, cloves, grain of paradise [similar to cardamom], pepper, vinegar, bread [to thicken].

England: Ginger, cloves, cinnamon, currants, nuts, vinegar, breadcrusts.

Verde Sauce: France: parsley, ginger, vinegar, bread.

England: parsley, ginger, vinegar, bread, mint, garlic, thyme, sage, cinnamon, pepper, saffron, salt, wine.

—from Taillevent (France, ca. 1375),
and the court of Richard II (England, 1390)

The English also seem to have been more fond of garden vegetables at this time, or to have had a better supply, although for most of the common people, vegetables were a rare luxury. Taillevent includes only one vegetable dish in his group of recipes for Lent, a puree of white beets and cress which is fried and boiled in almond milk. The English manuscript, on the other hand, has recipes for beans and spinach (boiled and then fried), one for herb fritters dipped in honey, and a salad and compote, for which we print the directions. Salads in late medieval England are recognizable, if a bit heavy

VEGETABLES AT THE COURT OF RICHARD II

Salat: Take parsel, sawge, garlec, chibollas [scallions], oynons, leek, borage, myntes, porrectes [young leeks], fenel, and ton tressis [cress], new rosemarye, purslarye [purslane]; lave, and waishe hem clene; pike hem, pluck hem small with thyn honde, and myng hem wel with rawe oile. Lay on vynegar and salt, and serve it forth.

Compost: Take rote of parsel, pasternak of rasens, scrape hem, and waishe hem clene. Take rapes [turnips] and caboches [cabbages] ypared and icorne [cut]. Take an earthen pane with clene water, and set it on the fire. Cast all thise therinne. Whan they buth are boiled, cast thereto peeres and parboile hem wele. Take thise thynges up, and lat it kele on a fair cloth. Do thereto salt, whan it is colde, in a vessel. Take vynegar, and powdor, and safron, and do thereto. And lat alle thise thynges lye thereinne al nygt other al day. Take wyne greke and honey clarified togider, lumbarde mustard, and raisons, corance al houl [whole currants]; and grynde powdor of canel [cinnamon], powdor douce, and aneys hole, and fenel seed. Take alle thise thynges, and cast togyder in a pot of erthe, and take thereof whan thou wilt, and serve it forth.

on the onion family, with fresh herbs taking the place of lettuce. The compote employs heat, alcohol, sugar, salt, and spices to create an inhospitable environment for microbes, and so to give the dish a relatively long pot life. Fruit was seldom eaten raw, in part because the influential Arab physician Galen thought the practice unhealthy.

New World, New Foods

The Renaissance brought competition to find free access to the East, the discovery of the Americas, and the infusion of new plant foods into Europe. Corn, the common bean, squashes, the tomato, hot and sweet peppers, the potato, peanut, avocado, and pineapple were all brought back and met with various receptions. The "French" bean was an immediate success, while outside of Italy the tomato was grown only as an ornamental, and corn and potatoes became staple foods for livestock and the poor. The 16th century saw the successful cultivation of sugar cane and other East Indian spices in the West Indies, and the subsequent rise of fruit preserves and jams (raw fruit was still not much trusted).

The 17th and 18th centuries were a time of assimilating the new foods and making advances in the culinary arts. Cultivation and breeding received new attention; Louis XIV's orchards and plantings at Versailles were legendary. A few new plants made their entrance: watermelon, okra, and black-eyed peas were carried by the slave trade from Africa to the colonies. Vegetables, once the produce of gardens, were becoming items of commerce; Adam Smith wrote in his *Wealth of Nations* (1776) that turnips, carrots, cabbages, and such "things which were formerly never raised but by the spade ... are now commonly raised by the plough." And for some reason, salads were suddenly the object of great admiration. French hygienists of the period recommended that fruit be eaten at the beginning of a meal, and salad at the end, because the salad, one wrote, "moistens, refreshes, frees the stomach, encourages sleep, enlarges the appetite, tempers the ardors of Venus, and appeases the thirst." And John Evelyn, one of the charter members of England's Royal Society and an enthusiastic gardener, wrote a book-length disquisition on the subject, called *Acetaria: A Discourse of Sallets* (1699). In it we learn that lettuce is now the primary ingredient: "by reason of its soporifous quality, lettuce ever was, and still continues the principal foundation of the universal tribe of Sallets, which is to cool and refresh, besides its other properties" (which included beneficial influences on "morals, temperance, and chastity"). Evelyn goes on to give an analysis of the proper composition of a salad, and we quote it for its admirably strict standard.

> We have said how necessary it is that in the composure of a sallet, every plant should come in to bear its part, without being over-power'd by some herb of a stronger taste, so as to endanger the native sapor and virtue of the rest; but fall into their places, like the notes in music, in which there should be nothing harsh or grating: And though admitting some discords (to distinguish and illustrate the rest) striking in all the more sprightly, and sometimes gentler notes, reconcile all dissonancies, and melt them into an agreeable composition.

Vegetable Cuisine, Lent, and the French
Perhaps the most significant development of the 17th century was the elevation of vegetables into dishes that were important in their own right, rather than being mere garnishes. Much later, the English cookbook author Ann Bowman suggested that because Catholic France was forced to rely heavily on vegetables for fast days and during the 40-day abstinence of Lent, "it is, therefore, from French cooks we have derived our happiest directions for cooking vegetables." England, on the other hand, after a period of strict

observance during the Middle Ages, had become more and more relaxed about dietary prohibitions after the split with Rome in the 16th century, and with the Revolution of 1648–49, the Lenten laws quickly became obsolete.

France's first great culinary writer, Pierre François de La Varenne, chef to Henri IV, recorded the first elaborate recipes in which vegetables were the centerpiece, and indirectly suggested that religious practice was at least partly responsible for their development. In the section devoted to general cooking of his book *Le Cuisinier françois* (1651, new edition 1712), he gives recipes only for artichokes, asparagus, mushrooms, and cauliflower. But in the section on meatless cuisine there are instructions for cooking peas, turnips, lettuce, spinach, cucumbers, cabbage (five ways), chicory, celery, carrots, cardoons, and beets, as well as the other four. And the recipes are recognizably modern. For example, compare the procedure for "Asparagus in a Fragrant Sauce" to our hollandaise.

> Choose the largest asparagus, scrape them at the bottom, and wash. Cook them in some water, salt them well, and do not let them overcook. When done, let them drain, and make a sauce with some good fresh butter, a little vinegar, salt, and nutmeg, and an egg yolk to bind the sauce; take care that it doesn't curdle. Serve the asparagus well garnished with whatever you like.

(Eggs and dairy products had been allowed in meatless cooking for about 100 years.)

For a long time, English cooks didn't catch on to this delicate touch with flavorings. In his 1726 book, *England's Newest Way in All Sorts of Cookery*, Henry Howard gives a recipe for a fricassee of mushrooms that calls for cream, white wine, gravy, pepper, mace, nutmeg, anchovies, thyme, and shallots; the sauce is thickened with egg yolk, butter, and flour, and the dish garnished with scalded spinach. Gradually, the Anglo-American tradition moved away from long procedures and heavy flavoring to the simple boil-and-butter method. This shift reflects the strongly domestic base of our cuisine, in which quickness and ease are among the highest considerations.

Meanwhile, in France, La Varenne's lead is followed. Antonin Carême, the most influential chef and culinary writer of the nineteenth century, claimed in his *Art of French Cooking in the 19th Century* (1835) that "it is in the confection of the Lenten cuisine that the chef's science must shine with new luster." Carême prided himself on his large repertoire of vegetable entremets, which included all the materials of La Varenne and added broccoli, truffles, eggplant, sweet potatoes, and potatoes, these last fixed *à l'anglaise, dites, Mache-Potetesse* ("in the English style, that is, mashed"). Of course, such mastery tends to undermine the whole point of Lent. In his *366 Menus* (1872), Baron Brisse wrote, "No need for a sad glance toward Lent. A peasant, after having been a domestic at his lord's, said: 'I can't eat meatless—it's too expensive. One must have game fish, Geneva trout, Rhine carp, first fruits and vegetables of the season, Champagne wine, and truffles.' Our man was right: are the meatless meals of our Lenten enthusiasts really meals of abstinence?"

The Influence of Technology

The 19th and 20th centuries have been remarkable mainly for great shifts in the eating habits of urban populations. The industrialization of Europe led to an influx of people into the cities, which had limited market facilities and were isolated from much of the farmland. As a result, the working class ate very little fresh produce, apart from the easily stored root and tuber vegetables that had been so important in prehistory. This situation was alleviated to some extent by the development of rail transportation, then canning around 1850, and refrigeration a few decades later. The use of refrigeration on freighters brought dependable shipments of bananas and coconuts to temperate climates for the first time, as well as increased supplies of less exotic produce. Around the turn of the century, vitamins—their concentration in plant products and their nutritional significance—were discovered, and fruits and vegetables were canonized as one of the four food groups that should be eaten at every meal.

Today, breeding is big business, although it is now more often directed at improving yield, uniformity, and durability in transit than at making improvements in flavor. And small home gardens are enjoying a revival, partly in response to the decline in quality of mass-produced, far-shipped supermarket produce. Nevertheless, per capita consumption of fruits and vegetables is substantially lower compared to 40 years ago, when we in the United States ate 154 pounds of fruit and 265 pounds of vegetables. In 1979, each of us on average ate 83 pounds of fresh fruit, and 56 processed; of vegetables, 104 pounds fresh, 55 canned, and 11 frozen.

THE NATURE OF PLANTS

The evolutionists continue to argue about the number of "kingdoms" in the huge diversity of living things on the planet, but plants and animals are assuredly two of them, and the two most advanced results of several billion years' evolution. They are radically different, and have diverged over one fundamental issue: food. Plants are generally "autotrophic," or "self-nourished"; given a supply of water, minerals, oxygen, carbon dioxide, and sunlight for energy, they can survive and even flourish, entirely independent of other organisms. Animals, on the other hand, cannot synthesize from such primitive materials the complex proteins, carbohydrates, and other molecules necessary to life. Animals are "heterotrophic," or "other-nourished," and depend on the ingestion of other organisms, plant or animal, to meet their needs for energy and tissue-building supplies.

How did these two strategies arise? We will never know for sure, but a generally accepted model is the following. The planets formed about 4.7 billion years ago, and some hundreds of millions of years later, the earth had developed an atmosphere and oceans from gases and vapors trapped in its interior. The action of ultraviolet radiation, electrical discharges in the atmosphere, radioactivity, and heat on the materials of the earth's surface led to the formation of complex molecules. And somehow out of this primeval soup there developed self-enclosed and reproducing cells. Initially, all of them probably survived by breaking down the complex molecules around them, thereby releasing the energy stored in the chemical bonds. In the process, though, the supply of these molecules would be depleted. In the resulting conditions of scarcity, cells that could reverse this strategy and build up complex molecules from simpler ones would be at a competitive advantage. And the development of these autotrophic cells would in turn ensure the continued existence of heterotrophs, which could then live off of the dissolved contents of dead cells ("saprophytes": modern-day examples are the mushrooms), or the materials stored in other living cells (parasites).

Several different types of autotrophs probably developed, including bacteria that could use sulfur, nitrogen, and iron compounds to produce energy by chemical manipulation. But perhaps the most important development in the history of life was a bacterium capable of capturing the energy in sunlight and storing it in sugar molecules. Chlorophyll, the green pigment we see all around us, is the compound that makes this conversion possible, according to the equation

$$6CO_2 + 6H_2O + light \rightarrow C_6H_{12}O_6 + 6O_2.$$

The bacteria that came eventually to rely on chlorophyll gave rise to the green plant—and, indirectly, to the animals as well. Before photosynthesis, the earth's atmosphere contained little oxygen, and the sun's ultraviolet rays penetrated all the way to the earth's surface, with the result that living organisms were restricted to habitats several feet below the ocean surface, where the high energy radiation could not reach. The blue-green algae liberated vast quantities of oxygen via photosynthesis, and in the upper reaches of the atmosphere the gas was converted to ozone, which blocked ultraviolet light. Land life was now possible. Our own existence, then, both as heterotrophs and as oxygen-breathing dwellers on the earth's surface, is largely predicated on the greenery we walk through and cultivate and consume every day of our lives. Sublime thoughts for the salad course!

Why Plants Don't Have Muscles

This divergence in feeding strategies helps make sense of many other obvious differences between plant and animal. Land-dwelling autotrophic organisms need access to the soil for minerals and trapped water, to the atmosphere for gases, and to the sun for energy—sources which, once located, are rather reliable. Plants have developed a straightforward, economical structure that depends on this reliability: roots penetrate the soil to reach stable supplies of water and minerals; leaves maximize their surface area to capture sunlight and exchange gases with the air; and stalks support leaves and connect them with roots. Unlike the animals, which depend for their survival on finding other organisms, plants did not need to develop the power of locomotion.

Defense Mechanisms

While the plant may have no need for eye, muscle, or brain, it has been forced to develop special strategies for which the mobile, sensitive animal has little need. One such area is defense. The animal can detect possible danger and react in two basic ways: fight or flight. But stationary plants? They compensate for their limited "behavior" and physical deterrents (thorns and bark, for example) with a remarkable diversity in biochemical makeup. It has been estimated that fully 80% of all known naturally occurring compounds are to be found in the plant kingdom. At last count, this is some 12,000 molecules, and the job of cataloguing is nowhere near done. Most of these chemicals have no apparent role in the plant's metabolism, the primary, day-to-day processes of energy production and growth. These "secondary" compounds very often have a defensive role to play. They are distasteful, poisonous warning signals that discourage heterotrophs of all sorts—bacteria, fungi, animals—from consuming them.

Plants engage in a wide range of passive chemical warfare. A partial list would include irritating compounds like poison ivy oil and capsaicin, the "hot" in hot pepper; toxic and bitter alkaloids like strychnine, caffeine, and nicotine; the cyanide compounds found in lima beans and citrus and apple seeds; substances that interfere with the digestive process, including tannins and inhibitors of digestive enzymes; and goiter-inducing chemicals found in the cabbage family. Perhaps the most ingenious weapons in the arsenal are the juvabiones and ecdysones. These are molecules closely related to insect growth hormones, which control the larva's maturation into adulthood. When ingested by the insect, these plant products either delay or speed up the insect's maturation, with the result that its reproductive stage is either never reached or bypassed in favor of continued physical enlargement. Finally, there are also defenses against crowding by other plants. These are usually growth inhibitors, which may be washed by rain from the leaves to the ground (as in the walnut and eucalyptus) or secreted directly from the roots (as in apple trees and wheat).

Why, with all these nasty herbivoricides in plants, is the world not littered with the corpses of their victims? Because animals have learned to recognize potentially harmful plants and to avoid them. The senses of smell and taste are chemical senses that can detect general classes of compounds in very small concentrations, and animals have developed innate responses to the bitter taste typical of alkaloids and cyanide, to the sweet taste of nutritionally important sugars. Sweetness is pleasing, bitterness (except for acquired tastes like quinine water and coffee) is not. One scientist has even conjectured that the reptiles' relative insensitivity to bitterness may have contributed to the decline of the dinosaurs: the rise of the alkaloid-containing Angiosperm (flowering plant) coincides with their sudden and mysterious disappearance.

Beyond this aversion to the taste of poisons, some insects and other animals have developed specific detoxifying mechanisms that permit them to have an otherwise inedible plant all to themselves. Gray squirrels can eat the deadly Amanita mushroom, the koala bear can eat eucalyptus leaves, and monarch butterfly caterpillars store up the poison in milkweed, which in turn acts as a deterrent to insect-eating birds. Humans have learned to cope in different ways. Many toxins can be destroyed by heat or leached out of the plant by boiling water, and this was surely one factor in the development of cooking in prehistoric times. It is also well established that cultivated varieties of wild food plants—cabbage, lima beans, potatoes, and lettuce are all examples—are less toxic to humans and domestic animals than the wild types.

Secondary compounds are not universally harmful. In fact, hundreds have turned out to be among the most useful chemicals known. Plants were our only source of medicinal compounds for thousands of years, and the synthetic products we have today are often based on natural models. Willow bark, used by American and European natives to relieve pain, gave us the precursor to aspirin. Digitalis, quinine, and morphine are all plant products. The sex hormones called estrogens were first manufactured from a related compound in yams. The Chinese have developed a male contraceptive from gossypol, a constituent of cotton. Rubber, resins, scents and pigments, spice flavors, and natural pesticides like pyrethrum and rotenone are all secondary compounds. The irritating oils of mustard, ginger, and black and red pepper were among the first chemical preservatives; today these same spices are powdered together and sold as a footwarming concoction for skiers and winter backpackers. And tannins from leaves, bark, and galls (the protuberances caused by an infestation of wasps) have been used for thousands of years to tan animal hides into leather (synthetic tanning chemicals are now the rule).

Plant Reproduction: Flowers and Fruit

A second area of special need for plants, after defense, is reproduction. The higher plants and animals reproduce sexually, that is, by fusing genetic material from male and female organs, usually from different individuals. The mobile members of the world do it in basically the same way: male and female seek each other out. Stationary plants, however, must depend on middlemen to join male and female. The result, among land plants, is pollen, borne either by the wind or by animals. To encourage animals to help out, advanced plants have developed the flower, a structure made of modified leaves and designed to attract, by color, shape, or scent, a particular aide who, while collecting nutritious nectar or pollen for food, will also spread the pollen from one plant to another.

Once pollen and ovule have been brought together and the seeds have formed, the next generation must be seen off to a good start. The female animal simply searches out an advantageous location and deposits her young there. But, again, plants need help. If all seeds fell straight from the plant to the ground, they would have to compete with each other and with their overshadowing progenitor for the limited resources of a small area. Most seedlings would die, and the population would grow very slowly, if at all. So successful plant species have tended to develop mechanisms for dispersing their seeds over as wide an area as possible. These mechanisms include seed appendages that catch the wind, containers that pop open and spray their contents in all directions, structures that catch on passing fur, and structures that can hitch a ride *inside* passers-by by tempting them to eat.

Fruit is in essence a device of seed dispersal, the result of a long coevolution between plant and animal. One needs food, the other a transportation service, and fruit is the compromise, the medium of exchange. Different animals call for different inducements, and while we are a long way from being able to say exactly how particular fruit characteristics have evolved, some generalizations are possible. Reptiles are not normally climbers, and fruits that appeal to reptiles are usually borne near the ground or dropped at maturity. Birds are sensitive to color contrasts and can easily reach heights, and typical bird fruits are accordingly brightly colored and remain attached to branches. Fruit bats are color-blind, attracted to their own odor, and have trouble negotiating leaves; fruits designed for them are drab, smell musty, and hang exposed below the foliage. Mammals are generally color-blind but have a good sense of smell, and possess teeth; their fruits develop odors and can have tough skins. Primates, relative latecomers to the mammal family, can climb and see colors and have invaded the birds' territory.

Of course, the fruit will not have served its purpose if the seeds are consumed along with it. They escape this fate in several ways. Seeds and their coating may be too large and hard to be eaten with the fleshy covering (peach, avocado), or small and numerous enough that some will be spilled during feeding (tomato, watermelon). They may be distasteful, even poisonous, like the cyanide-bearing apple and citrus seeds, or they may be constructed so that they pass through an animal's digestive tract uninjured, and finish up in a freshly manured location. The wild tomato of the Galapagos Islands has developed a seed coat so tough that it requires the good offices of the tortoise to digest it away and make germination possible.

Unlike the rest of the plant, fruit is *meant* to be eaten. This is why the complex of taste, odor, and texture is so well matched to our predilections as mammals. And so it is that, in everyday language, fruits tend to stand for or suggest desirability. But fruit is meant to be eaten only when the seeds—its whole reason for being—are mature and viable. This is why ripening occurs. Leaves, roots, stalks can be eaten at any time, and the earlier the tenderer, but we must wait for fruit to indicate to eye and palate that it is ready to engage our services.

Geographical Range

Flowers and fruits, like humans, are relatively recent innovations on our planet. Life arose about 4 billion years ago, but flowering plants have been around for only about 200 million, and been dominant for about 50 million, or about 1% of the time life has existed. An even more recent development is the "herbaceous" or annual habit. Most of our food plants are not long-lived trees, but small, fast-growing plants that produce their seed and die in one growing season. This short life cycle makes it possible for plants to invade and establish themselves rapidly in new areas—the plants we call "weeds"

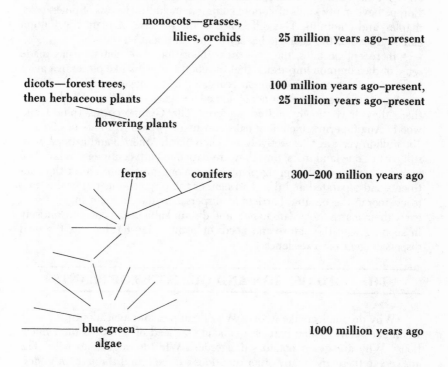

monocots—grasses, lilies, orchids   25 million years ago–present

dicots—forest trees, then herbaceous plants   100 million years ago–present, 25 million years ago–present

flowering plants

ferns   conifers   300–200 million years ago

blue-green algae   1000 million years ago

A rough sketch of plant evolution, showing the periods during which the more familiar forms have flourished. The conifers—pines, firs, and their cone-bearing relatives—shared the earth with dinosaurs, early mammals, and the long-established insects. The first manlike apes arose 25 million years ago, and the human race itself is perhaps 2 million years old. Like us, the herbaceous dicots and monocots on which we rely for food are newcomers to the planet.

are especially masterful at this—and so give them greater flexibility in increasing their range and surviving inclement seasons. The herbaceous habit has worked to our advantage as well: crops grow to maturity in a few months, plantings can be changed from year to year, and many of our vegetables come from parts of these plants that, were they toughened to survive winters, would be inedible. Herbaceous plants became widespread only in the last few million years, just as the human race was emerging. In a real sense, many of our food plants are our evolutionary partners, although with agriculture we have taken their evolution into our own hands.

The relative ages of plants help us to understand an important but often forgotten fact: most of our food plants were originally restricted to isolated parts of the world. We have seen how the geographical isolation of spice plants changed world history, and how the discovery of the New World changed the culinary life of Europe. On the other hand, North America and Europe have many kinds of flora in common, including forests of pines, oaks, maples, and chestnuts. This difference in the patterns of plant distribution can be traced to the geologic history of the continents.

In recent decades, the geological theory of plate tectonics has made sense of the common impression that the continents look like pieces in a giant jigsaw puzzle. Around 200 million years ago, it is believed, they were joined together into one supercontinent, named Pangaea, or Whole-Earth. Since then, they have slowly shifted position. The Gymnosperms (pines) and woody Angiosperms (including oaks and maples) date from about 300 and 150 million years ago respectively, or when North America and Europe were still part of one land mass; thus they are native to both continents today. The younger herbaceous flowering plants came along much later, after the continents had separated and they had neither the physical means nor the time to overcome the oceanic barriers to dispersal. It took human ingenuity to carry them across the oceans to new and distant habitats: an outcome entirely in keeping with the nature and needs of plants. Man has become the seed dispersal agent par excellence.

THE COMPOSITION AND QUALITIES OF PLANTS

Why do onions make us cry? Why do apples and avocados turn brown when cut open? Which part of a cabbage or head of lettuce is most nutritious? Why are green potatoes dangerous? Why do vegetables wilt? The answers to these and many other questions can be found through an understanding of the structural and chemical makeup of plant tissues.

Structure

The Cell

The overall sensory qualities of plant foods are determined by the contents of individual cells. Like animal cells, plant cells consist of a nucleus

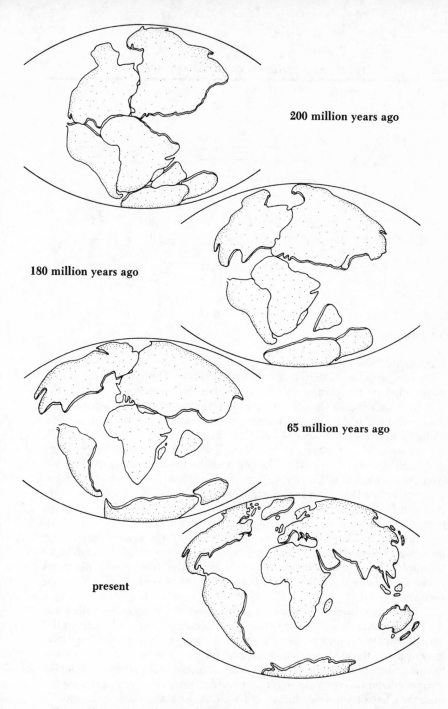

200 million years ago

180 million years ago

65 million years ago

present

Movement of the continents in the last 200 million years. Pines and many other familiar forest trees evolved when North America and Europe were in contact, and so are native to both continents. Most annual food plants evolved after they had split apart, and so had to be introduced to one continent or the other by humans.

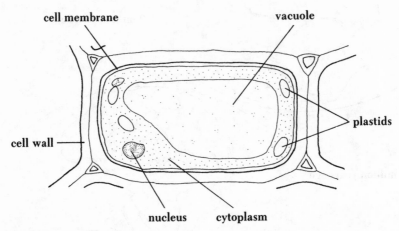

A typical plant cell

surrounded by cytoplasm—a solution of water, lipids, proteins, and many other compounds necessary to growth and function. The cytoplasm is in turn surrounded by the plasma membrane, which separates the contents of the cell from other cells. Smaller membrane-contained areas called vacuoles store enzymes, starch, sugar, proteins, certain pigments, and waste or secondary compounds in a water-based fluid. Plant cells differ from animal cells in two ways. They possess a rigid *cell wall*, which surrounds the membrane and lends structural support to the cell and the tissue of which it is a part. And they possess *plastids*, discrete structures within the cell which perform a variety of functions: they may contain chlorophyll and do the work of photosynthesis (chloroplasts), contain yellow or orange fat-soluble pigments (chromoplasts), or store starch, protein, or fat (leucoplasts).

Broadly speaking, the texture of plant foods is largely dependent on the thickness of the cell walls. Color is determined by the relative numbers of chloroplasts and chromoplasts, and sometimes by the water-soluble pigments in vacuoles. Taste and nutritive value are fixed by the contents of the storage vacuoles.

There are three basic kinds of cells in plants. *Parenchyma* cells form the fundamental living tissue of fruit, leaf, stem, and root, and perform the various duties of photosynthesis, nutrient storage, and originating new growth along the stem or root. Animals have no close equivalent to parenchyma cells. *Collenchyma* cells, which are analogous to cartilage in animals, are specialized parenchyma cells that provide support in young, growing tissues; they have somewhat thicker cell walls and usually occur in strands or cylinders. *Sclerenchyma* cells, comparable to bone cells in animals, are tissue-strengtheners found in mature areas, where they occur in masses. They have very thick cell walls and often contain *lignin*, the characteristic component of wood. Nutshells, the stone that surrounds a peach seed, the seed coats of beans and peas, and such economically important fibers as hemp,

Three types of plant cells. Parenchyma cells *(left)* do general duty.
Collenchyma *(center)*, with their thickened cell walls, give structural support
to young tissue. Sclerenchyma *(right)* are little more than cellulose and lignin
and strengthen older parts of the plant. The more supportive or protective
tissue there is in a fruit or vegetable, the tougher it will be.

flax, linen, and sisal are all mainly sclerenchyma cells. The lignified fibers
that make old vegetables stringy, and those gritty, grainy bodies that give
pears their characteristic texture, are sclerenchyma growths.

Tissues

Tissues are groups of cells organized to perform a common function.
The "ground" or fundamental tissue is the primary mass of cells. Surrounding and infiltrating the ground tissue respectively are the dermal and vascular systems, which correspond to our skin and circulatory system. The cells
that form the outer surface of the plant, the boundary between organism
and environment, can be either *epidermis* or *periderm*. The epidermis is
usually a single layer of cells that secretes waxes, a fatty material called cutin,
and other substances, in order to form a protective water-impermeable coating. Periderm is a corky material that replaces the epidermis when root and
stem are enlarged by secondary growth. Our culinary experience of periderm is usually limited to the skins of potatoes, beets, and so on; periderm
on a stem vegetable is a sure sign of toughness.

The vascular tissue is the system of cells that transports nutrients from
root to stem to leaf and back again. The work is divided between two sub-
systems: *xylem*, which conducts water and minerals from the roots to the rest
of the plant, and *phloem*, which conducts sugars down from the leaves. Vascular tissue also takes on support duties, and will contain sclerenchyma fibers
for that purpose.

The last basic type of plant tissue is the secretory system, usually isolated
cells that occur both on the surface and within the plant. They correspond
to our oil and sweat glands. Secretory cells produce various secondary compounds, often either to attract or repel animals—flower nectar and poison
ivy oil are two examples. The large mint family, which includes other com-

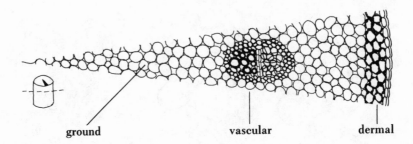

ground   vascular   dermal

The three kinds of plant tissue in a stem. Fibrous vascular tissue and thick dermal layers are common causes of toughness in vegetables.

mon herbs like thyme and basil, is characterized by glandular hairs on stem and leaf that contain the aromatic oils.

Plant Organs

Although it is often impossible to tell exactly where one stops and another begins, we recognize six distinct plant organs: the root, the stem, the leaf, the flower, the fruit, and the seed. We will take a closer look at seeds in chapter 5.

The *root* anchors the plant in the ground, absorbs and conducts moisture and nutrients to the rest of the plant, and stores food supplies. The fleshy part of the carrot, radish, beet, and yam is a specialized storage root whose main component is starch. The typical root has a simple structure: a central cylinder of vascular tissue surrounded by a storage area made up of parenchyma cells with large vacuoles. Our root vegetables develop this storage area in different ways, and so produce different patterns. In the carrot, the central core pushes outward, maintaining the two-sectioned pattern. The

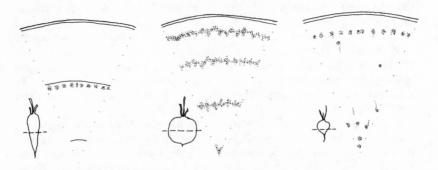

The patterns of vascular tissue in carrots, beets, and radishes

beet produces concentric layers of storage and vascular tissue, and its cross section looks much like that of a tree trunk. In the yam, radish, and turnip, parenchyma cells develop into both new vascular and storage tissues, with the result that the two are intermingled and no clear pattern emerges.

The *stem*'s principal function is to conduct nutrients between the root and the rest of the plant, but it also provides support for the food-synthesizing organs, the leaves, and it stores nutrients as well. Stems, like roots, can develop into specialized storage organs; some common examples are the potato, sweet potato, and water chestnut. The proof that, contrary to common sense, the potato is not a root, is the fact that it gives rise to stem buds and sets of leaves from the eyes, which are arranged in a tight spiral around the long axis. Only stems can do this.

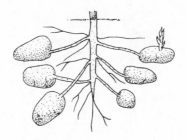

The potato is the swollen tip of an underground stem, and gives rise to new stems.

Leaves specialize in the production of high-energy food molecules— sugars—via photosynthesis, a process that requires exposure to sunlight and good supplies of carbon dioxide and water. The leaf accordingly differs in structure from the other organs. It contains very little storage tissue, and no secondary protective growth that would interfere with its access to light or air. As a result, leaves are the most fragile and short-lived parts of the plant, and are continuously being shed and re-formed. To maximize light capture, the leaf is flattened out into a thin sheet with a large surface area, and the "palisade" parenchyma cells are heavily populated with chloroplasts. To promote gas exchange, the leaf interior is filled with thousands of tiny air pockets, which increase the number of cells exposed to the air. This structure accounts for the fact that leafy vegetables shrink much more when cooked than roots and stems do. The spongy interior is collapsed by the heat, and the entire structure loses volume. Some leaves are as much as 70% air by volume.

An exception to the rule against storage tissue in leaves is the onion family (tulips and other bulb ornamentals are exceptions as well). The many layers of the onion (and the single layer of a garlic clove) are leaf bases that surround very small stem shoots. They store water and starch during the

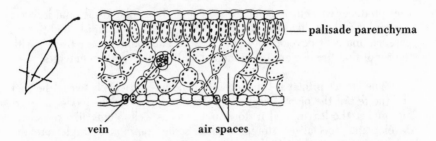

palisade parenchyma

vein   air spaces

Cross section of a leaf. The dark bodies crowded inside cells are chloroplasts. Because photosynthesis requires a continuous supply of carbon dioxide, many cells within the leaf are directly exposed to the air. Leaf tissue is accordingly light and spongy. (Exceptions to this rule are plants that have adapted to very harsh climates.)

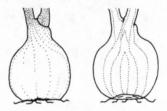

The bulb of the onion consists of leaf bases swollen with stored water and carbohydrates for the next year's growth.

plant's first year of growth for use during the second, when it will flower and produce seed.

The *flower* is the plant's reproductive organ. Here the male pollen and female ovules are formed; here too they unite (usually pollen from one flower joins the ovule in another) and develop into seed embryos. The flower is generally quite small compared to the rest of the plant and is rarely eaten by itself. Broccoli, for example, is mostly stem tissue with a few very small flower buds attached. And cauliflower is not a proper flower, but one gone haywire. All that white "curd," as it is called, is a degeneration of the stem tip into a sterile mass of undeveloped buds. The only true flower regularly consumed for itself is the artichoke, a kind of thistle. Even in this case we eat only the base, or heart, which is an outgrowth of the stem, and the bases of the surrounding protective leaves, the bracts. The part we call the "choke" and throw away is the mass of immature flowerets, which turn a beautiful violet-blue when allowed to develop.

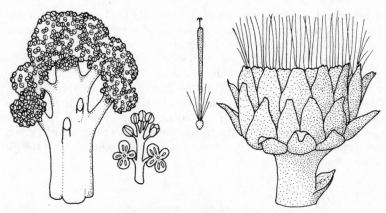

Broccoli and the artichoke are masses of flowers. The actual flowers of the artichoke are the tough, stringy innards we throw away as the "choke."

The *fruit* is the organ derived from the flower's ovary or adjacent tissues, and it contains the seeds. Its function, as we have seen, is to extend the seeds' geographical range by taking advantage of the eating habits and mobility of animals. It is the only part of the plant that is meant to be eaten and that has no support, nutritional, or transport responsibilities to the other organs. Consequently, the fruit consists almost entirely of storage parenchyma tissue filled with carbohydrates useful to animals. The changes in fruit qualities that signal the seeds' maturity—the phenomenon of ripeness—will be discussed below (page 161).

Texture

The texture of fruits and vegetables depends on two factors: the nature of the cell walls and the abundance of water in the tissues.

Plant cell walls, unlike animal cell membranes, are rigid and not very flexible. Their major component, and in fact the most abundant plant product on earth, is cellulose. Like starch, cellulose consists of a chain of sugar molecules, but a slight difference in bonding makes cellulose invulnerable to human digestive enzymes and very resistant to any chemical degradation. Even ruminant animals and termites depend on digestive tract bacteria to make cellulose available to their systems. This remarkable stability makes cellulose valuable to such trees as the sequoia and bristlecone pine, some specimens of which are several thousand years old. It is quite valuable to the human race as well. About 10 billion tons of cellulose are produced by plants every year, and we use it, either barely purified or heavily processed, for a vast range of materials, including cloth, paper, cellophane, celluloid, rayon, acetate fibers, and such plastic products as Ping-Pong balls, steering wheels, and eyeglass frames. Cotton is 98% cellulose, and lumber 30 or 40%. Cellulose accounts for about a quarter of our livestock feed. It becomes most visible to us in the winter as hay, a stubble field, or the delicate skeletons of weeds. It would be hard to imagine life without it.

When a plant cell divides, microscopic fibrils of cellulose are formed and laid down in the plane of the new wall. They provide the framework for other substances that then fill in the interstices. If you think of the cellulose fibers as the metal reinforcing bars in a building, then these other substances would be the concrete. They too are long polymers of various sugar molecules. One component is the hemicelluloses, a vaguely defined group of compounds that are broken apart more easily than cellulose; they are partly dissolved during cooking.

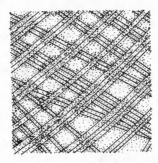

Plant cell walls are made up of a framework of cellulose fibers embedded in a mass of amorphous materials, the hemicelluloses and pectic substances.

The other important component is the pectic substances, which play a prominent role in fruit softening. As ripening progresses, insoluble protopectin is converted by enzymes into soluble pectin, and the cell walls are weakened. Like the hemicelluloses, pectin is lost from fruit and vegetable cell walls during cooking.

Thus crisp, even hard plant material can be softened to some extent by leaching out parts of the cell walls. Ripening does this naturally, cooking artificially.

One last cell wall component must be mentioned even though it is not often encountered in food, and for good reason. Lignin, which like cellulose is resistant to degradation, is produced during the plant's secondary growth (growth not at the shoot tip, but in older tissues) and functions solely as a strengthening agent. It is the major constituent of wood, and is frequently found in the walls of sclerenchyma cells. Most vegetables are harvested well in advance of appreciable lignin formation, but occasionally we must deal with woody root and stem vegetables. For this kind of toughness there is little remedy, except to peel away the lignified areas.

The second important factor in plant texture is the inner water pressure, or *turgor*, of the individual cells. Water is the principal constituent of living tissue, and is found in the cell vacuoles, the cytoplasm, and even combined with the sugar-based molecules of the cell wall. Though these three areas are separated by the cell and vacuolar membranes, the membranes are water-permeable. Water molecules can diffuse across them to maintain equilibrium in the relative amounts of water and dissolved molecules and ions. A drop in water content outside the cell will draw water from the cell, while an increase will result in a flow of water into the cell.

When a cell reaches its limit in water content, the vacuole swells and presses the cytoplasm against the cell membrane, which in turn presses against the cell wall. The wall is somewhat flexible, and bulges a little to accommodate the swollen cell. The mutual pressure exerted by many cells against each other results in an erect, rigid tissue, or what we would judge a crisp texture. As we bite into a turgid vegetable, we experience an initial resistance and then a burst of juice as the walls rupture. But if the tissue has lost water, the vacuoles have shrunk, the cell membranes have drawn away from the walls, and the vegetable has become limp and flaccid.

This system of water relations is one reason for the design of modern refrigerators. The crisper is a confined compartment that generally maintains a higher humidity than the rest of the refrigerator does. Vegetables left on open shelves will lose water to the relatively dry air and wilt, while in the crisper they remain turgid longer.

The relative importance of cell wall structure and turgor in determining texture differs from food to food. Leaf vegetables suffer more noticeably from water loss than do other vegetables, and watermelon more than apples, whose cell walls are stronger.

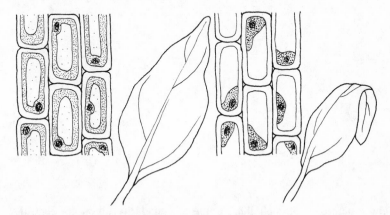

Plant tissue that is well supplied with water is filled with fluids and mechanically rigid *(left)*. Loss of water causes cell vacuoles to shrink, the cells become partly empty, and the tissue weakens *(right)*.

Color

There are three families of plant pigments, each with different functions in the plant's life, and each with different behaviors in the kitchen. The effects of cooking on color will be discussed later on.

Chlorophyll

The green chlorophylls are the most common and familiar family of pigments. Chlorophyll a is bright blue-green, and is about twice as common as the olive-colored b form. Chlorophyll is the molecule that traps solar energy and makes it available to form sugar out of water and carbon dioxide,

The structure of chlorophyll a. For the sake of clarity, all but the central atoms have been omitted; there is a carbon atom at all other junctions of lines. Chlorophyll b differs only in the makeup of the upper right-hand corner of the molecule. Chlorophyll is remarkably similar in structure to the heme group in meat pigments (compare the drawing on page 99).

in the process known as photosynthesis. It is concentrated in discrete cell bodies called chloroplasts, and is never found in the cytoplasm.

The chlorophyll molecule has a ring structure similar to the heme molecule in animal blood. Instead of iron, it has a magnesium atom at the center, and it trails a long hydrocarbon tail, called the "phytyl" group, at one end. The phytyl group gives the molecule as a whole the property of being soluble only in fats, and not in water. But both the central atom and the tail group are rather easily removed from the molecule during cooking, and chlorophyll is therefore susceptible to both color change and leakage into the surrounding water.

Carotenoids

Carotenoids are so named because the first member of this family to be chemically isolated came from carrots. These are the pigments responsible for the yellow and orange colors in fruits, vegetables, flowers, and autumn leaves, as well as the red color of tomatoes, watermelon, pink grapefruit, apricots, and bell peppers (most red colors in plants are caused by anthocyanins: see below). There are many different carotenoids, and they are always found in mixtures, never singly. Chloroplasts always contain carotenoids, in the ratio of about one to every three or four chlorophyll molecules. Carotenoids play an indirect role in photosynthesis by trapping certain wavelengths of sunlight and funneling the energy to the chlorophyll system. In addition, they absorb excess solar energy and help prevent the chlorophyll system from being overloaded and destroyed. Like chlorophyll, the carotenoids are fat soluble, but they are often found in special pigment bodies, or chromoplasts, as well as in photosynthetic chloroplasts. And they are relatively unaffected by cooking.

Some carotenoids—in particular, two kinds of carotene—have a nutritional as well as aesthetic significance; they are converted to vitamin A in the human intestinal wall. Strictly speaking, only animals and animal-derived foods contain vitamin A; fruits and vegetables contain only its precursors. But without these pigment precursors there would be no vitamin A in animals either.

Anthocyanins

This third major class of plant pigments is a subgroup of the phenolic compounds, which include tannins. Their only known function is to give color to flowers and fruits (the word comes from the Greek for "blue flower"), and they are responsible for most of the red, purple, and blue in plants, including many berries, red grapes, apples, red cabbage, radishes, and eggplant. A related group, the anthoxanthins ("yellow flower") are pale yellow compounds found in potatoes, onions, and cauliflower. Unlike the chlorophylls and carotenoids, the anthocyanins are water soluble and located in cell vacuoles. Like chlorophyll, though, they are easily and radically changed during cooking: try the experiment of sprinkling a little baking soda on some crushed rose petals or a spoonful of (cheap) wine. We'll go into this trait in more detail below.

Browning

A different kind of phenolic compound is responsible for the discoloration of fruits and vegetables when the tissue is cut or bruised. We are all familiar with the way apples, bananas, pears, eggplants, avocados, and raw potatoes turn brown when they are sliced or bitten into. This discoloration is the work of an enzyme known as polyphenoloxidase, which oxidizes phenolic compounds in the tissue and causes them to condense into brown or gray polymers. A similar kind of enzyme acting on a similar kind of compound is responsible for the "browning" of humans in the sun. Polyphenoloxidase is also present in mushrooms, apricots, pears, cherries, peaches, and dates, but it is missing from citrus fruits, melons, and tomatoes. These fruits will eventually turn brown simply by the action of oxygen in the air, but the chemical reactions are much slower without polyphenoloxidase, and it takes days for their color to change. In intact tissue, the enzyme and the phenolic compounds are segregated, but when cells are disrupted, they come into contact with each other and react to form the dark pigments.

Enzymatic browning can be discouraged by several means. Chilling the food below about 40°F (4°C) will slow the enzyme down. Even better, boiling temperatures will destroy it, but this means cooking and so altering the flavor and texture of the fruit or vegetable. Chloride ions inhibit the enzyme, so that salt solutions will retard discoloration, but again at the cost of flavor. Immersing the cut pieces in cold water limits the enzyme's access to oxygen and slows browning somewhat. Various sulfur compounds combine with the phenolic substances and block their reaction with the enzyme; they are often applied commercially to dried fruits. The single handiest method for the cook to use is the old standby, lemon juice. The enzyme works only very slowly in highly acidic conditions. Apparently malic acid, which is found in apples and grapes, is even more effective than the citric acid of lemons, but those fruits are not as concentrated a source of acid as lemons are, and their own distinctive flavors would interfere more with the food to be protected.

Another acid that inhibits enzymatic browning is ascorbic acid, or vitamin C. In fact, vitamin C was first isolated precisely because it has this property. Around 1925 the young Hungarian biochemist Albert Szent-Györgyi became interested in plant chemistry when he saw a similarity between the darkening color of damaged fruit and that of patients suffering from adrenal gland disorders. When he began to investigate why certain kinds of plants do *not* turn brown, he found that the juice of nonbrowning plants could delay the discoloration of browning ones. He isolated the substance that delayed browning and analyzed it.

It was an acid and it seemed to be related to an unknown sugar which I called "Ignose," the substance itself being called "Ignosic acid." But the editor of the journal to whom I sent my paper did not like jokes and rejected the name. "Godnose" being no more successful, we agreed that the child's name should be "hexuronic acid." Later, with advancing knowledge of its structure it had to be rebaptized in haste and it is now called ascorbic acid (sometimes, cevitamic acid) because it is identical with Vitamin C and prevents scurvy. In this way I became a father without wishing it, the father of a vitamin.

Szent-Györgyi thought he was on to something when he found great quantities of ascorbic acid in the adrenal gland, but *that* lead did not pan out.

Flavor

The overall flavor of plant foods is a composite of several different factors: the astringent effect of tannins, the balance between sweet sugars and sour acids, and the odor of many different aromatic compounds, including alcohols, esters, ketones, and volatile acids. When it comes to judging quality, the acid-sugar balance is very important for fruits, while for vegetables, which contain relatively little acid or sugar, the mixture of aromatic components is most distinctive.

Astringency

Astringency is neither an aroma or a taste, but a description of how the mouth feels when a particular kind of chemical comes in contact with it. It is that dry, puckery, constricting sensation that follows on a sip of strong tea or an assertive red wine, or a bite into less than ripe fruit. It is caused by the class of phenolic compounds called tannins, so named because they have been used since prehistory to tan animal hides into tough leather by complexing with proteins on the skin surface. In the plant, tannins are defensive compounds that counteract bacteria and fungi by interfering with *their* surface proteins. They also deter herbivores by virtue of their astringent effect on the mouth and their interference with digestion. The sensation of astringency is caused by the "tanning" of proteins in the saliva and mucous membranes of the mouth; lubrication is reduced and the surface tissues actually contract. Tannins are most often found in woody tissue and immature fruit, and only rarely in the herbaceous parts we eat as vegetables.

Sweet and Sour

Sugar is the primary material stored in fruit, which uses it for energy and for the synthesis of other organic materials. The average sugar content of ripe fruit is 10 to 15% by weight; at the extremes, the lime has less than 1%, the date more than 60%. In most fruit, sugars supplied by photosynthesizing leaves are stored as starch, which is then converted back into sugar as ripening commences.

A development parallel to the conversion of starch into sugar is the decline in acid content during ripening. Taken together, these two trends explain why unripe fruit is generally sour and ripe fruit sweet. There is a handful of different organic acids—citric, malic (characteristic of apples), tartaric, oxalic—that account for the acidity of most fruits and vegetables. All plant tissues are acidic to some degree; none have a pH above the neutral 7.

Odor

The subject of odor is less than clear-cut. The so-called essential, or volatile, or aromatic oil of a plant is defined as the set of all the compounds that can be distilled from the plant and that contribute to its characteristic aroma. These compounds have more to do with our impression of a food than its literal taste, since taste is limited to four possibilities—sweet, sour, salty, bitter. A given plant part has upward of 100 aromatic components, and most plant parts have most of these in common. The important factor is the relative concentrations, the set of predominant substances, rather than the whole list.

The function of these aromatic oils is probably defensive; most are present in minute quantities, have no primary role in the plant's growth, and are generally toxic to bacteria, insects, and animals to some degree (hence the early use of some spices as preservatives). And yet, despite their origin as chemical warning signals, animals have developed positive responses to some of them, and find them attractive. Our spice rack is basically a collection of our favorite essential oils. For an interesting theory about why spices are such an important and characteristic element of regional cuisines, see chapter 12.

Only a few families of plants have strongly scented oils. The two most important are the carrot family, or Umbelliferae (the Latin refers to the characteristic parasol-like mass of flowers), and the mint family, or Labiatae (referring to lip-like lobes in the flower). The carrot family includes 3000 species, and contributes anise, caraway, coriander, cumin, dill, fennel, and parsley to the kitchen. The 3200 species of the mint family give us basil, mint, marjoram, oregano, rosemary, sage, and thyme. The generosity of these two groups becomes clear when we learn that the 19,000 species of the daisy family or Compositae (yes—composite flowers), the largest family of plants outside of the orchids, are represented in the spice rack only by tarragon and, for herb-tea drinkers, camomile.

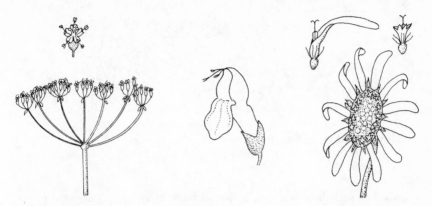

Left: the umbrellalike flower mass of the carrot family. *Center:* the prominent lip on a flower of the mint family. *Right:* many individual flowers are massed together in the "flower" of the daisy family.

The essential oils of the carrot family are concentrated both in the seeds and in cavities between cells in the leaf and stem; those of the mint family in specialized hairlike oil glands that cover both leaf and stem. Seeds, of course, are already relatively dry when they are harvested, but even herbs are usually dried before use, for a couple of reasons. Dry organic material is less likely to spoil, and drying, by breaking down the cell structure, makes it possible to extract more oil. Dry herbs are therefore often a more potent source of flavoring than fresh ones. But because the oils are volatile, they slowly evaporate, and in time even dried herbs and spices lose their flavor. Storing them whole rather than ground, and in air- and light-tight containers will minimize evaporation and oxidation and thereby extend their shelf life.

The volatile oils in fruits are usually concentrated in the skin, again often in specialized oil glands. The citrus fruits are a good example: we are all familiar with the sudden fragrant haze that appears when an orange skin is torn. The globular oil glands are especially obvious in grapefruit. These oils are often flammable, as squeezing an orange peel toward a candle flame will confirm.

Unlike herbs, vegetables have relatively mild odors when raw, and tend to develop stronger ones when heated or cut up. The effect of heat is due in part to the fact that volatile substances become more volatile as the temperature is raised. But there are two other important factors. One is the action of previously sequestered enzymes on disrupted tissue, which can result in the formation of new volatile compounds. The other is the chemical reactions that heating can bring about. In both cases aroma precursors, themselves odorless, are transformed into aromatic substances. The classic examples of these two possibilities are the onion and cabbage families, both blessed with similar sulfur-based compounds.

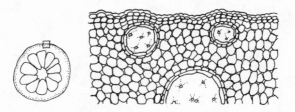

Oil glands in the skin of a citrus fruit

Onions and their relatives, as we all know from experience, are mild-mannered until we cut into them, at which point they immediately become quite odiferous and may even make our eyes water. The defensive advantage such a property has for the plant is obvious. Garlic was the first of these vegetables to be investigated thoroughly. Its tissue contains an odorless precursor, derived from the common sulfur-containing amino acid cysteine, which is stable in normal conditions. When we disrupt the tissue's cells, this compound is brought into contact with an enzyme that converts it into molecules of ammonia, pyruvic acid, and a mildly garlicky but unstable compound. This in turn breaks down into diallyl disulfide, the major and powerful constituent of garlic odor.

Chives, leeks, and onions all follow this general pattern, but end up producing different disulfide compounds, and so have distinct odors. The volatile substance in onions that makes the eyes water—the *lachrymator*—and the one that burns the tongue both seem to arise from another cysteine derivative that is rearranged by enzymes when the cell contents are mixed. The lachrymator apparently irritates the eyes by dissolving in their fluids and forming sulfuric acid. It is a relatively unstable compound and quickly decomposes in the onion tissue to less obnoxious forms. Unfortunately for the cook, the onion's production of the lachrymator naturally explodes during the most intimate phase of their relationship: chopping and dicing.

Despite this impressive chemical arsenal, the onion family is quickly subdued by cooking. Not only are its odor compounds driven off by high temperatures, but some of them appear to be converted into another complex molecule that is 50 to 70 times sweeter than a molecule of table sugar.

The members of the cabbage group—mustard, brussels sprouts, cauliflower, broccoli, turnips—are also characterized by sulfur-containing compounds, some similar to the odor precursors in onions, but the enzyme that would convert them into aromatic compounds is inactive at the pH of living tissue, and so they do not contribute to flavor. Instead, the important volatile components in this group are the so-called mustard oils, or isothiocyanates, of which more than thirty are known. In the live plant, they are bound to sugar molecules and are inactive, but when the tissue is broken, enzymes remove the sugar and liberate oils with a sharp, burning flavor and often eye-watering effects. These oils can range from the relatively mild, characteristic cauliflower or broccoli variety, to the powerful type found in mustard seeds. The disabling, even lethal chemical weapon called mustard gas is a synthetic preparation of isothiocyanate derivatives.

Members of the cabbage family are cooked more often than not, and the rather persistent odor that results is both familiar and annoying. It was probably for this reason that cabbage and cauliflower were the first vegetables to have their volatile output analyzed as they cooked, back in 1928. The mustard oils and cysteine derivatives break down to form various odiferous compounds, including hydrogen sulfide (typical of rotten eggs), ammonia, mercaptans, and methyl sulfide; eventually these may react with each other to form especially powerful trisulfides. The longer the vegetable is cooked, the more of these molecules are produced, and the flavor of the cooked vegetable gets stronger, not weaker, as time passes. For example, the amount of hydrogen sulfide produced in boiled cabbage doubles in the fifth through the seventh minute of cooking. So as the pungent bite of the raw vegetable disappears, it is replaced by an ever stronger odor. This fact, together with the general rule that fewer nutrients are lost in shorter cooking periods, is good reason not to overcook this particular family.

Nutrition and Other Benefits

On the average, fruits and vegetables contribute 1% of our daily intake of fat; 7% protein; 10% calories; 20% niacin, thiamine, and iron; 25% magnesium; 35% vitamin B_6; 50% vitamin A; and 90% of our vitamin C. In other words, we depend on them primarily for vitamins; grains, legumes, dairy products, and meat are all more important sources of proteins and calories. There are, of course, some significant exceptions to this generalization. Potatoes yield more energy per crop acre than cereals, and contain about 2% by weight of biologically useful protein. The banana, with its high sugar content, is also a useful energy source, and the avocado is anomalously well-stocked with oils, which it produces instead of sugar.

Vitamins and Color

There is a useful guideline for estimating the relative nutritiousness of vegetables. Plants manufacture vitamin C from sugars, which are supplied by the leaves as the product of photosynthesis. So the more light a plant gets, the more photosynthetic activity there is, the more sugars are produced, and the more plentiful is the vitamin C. In addition, the more light a plant gets, the more pigments—both chlorophylls and carotenoids—are needed to handle the energy input, and so the darker the coloration of the leaves and stem surfaces. Carotene, we recall, is the precursor to vitamin A. So the darker the color of a leafy vegetable, the more vitamin C and vitamin A precursor it is likely to contain. The light-colored inner leaves of lettuce and cabbage contain a fraction—perhaps one thirtieth—of the carotene found in the darker outer leaves. Unfortunately, the outermost leaves are often weather-beaten and insect-ridden, tough and scarred to the point of inedibility.

As for fruits and fruit-vegetables, their skins often contain higher concentrations of vitamins and minerals than the pulp. Citrus peels—again, the least attractive part of the fruit—have vitamin C concentrations 5 to 7 times higher than the juice.

Fiber: A Mixed Blessing

Plant foods contribute most of our fiber, or undigested bulk material, in the form of cellulose and other cell wall materials. A great deal has been said in recent years about the value of fiber in preventing intestinal diseases, basically by stimulating the intestinal wall and speeding the passage of food through the system. There is probably some truth in this—actual evidence is scanty—but there is also evidence that fiber can reduce the nutritional value of foods eaten with it by shielding them from the digestive enzymes (see chapter 11 for more details).

While we're on the subject of plant cell wall components and the intestine, a word about pectin, the substance which, because it is more soluble than its precursors, is partly responsible for the softening of ripe fruit. Some writers justify the old saw "An apple a day keeps the doctor away" by pointing out that apples, along with citrus fruits, are a rich source of pectin. And pectin, it turns out, is a general intestinal regulator that is quite effective in counteracting diarrhea. The trade name Kaopectate is based on the names of its two main ingredients, kaolin (clay) and pectin. Pectin is used in many medicinal preparations, and is sometimes added to baby food. It is also the thickener of choice for jellies and jams, as we shall see (page 171).

The Danger in Some Common Fruits and Vegetables

In a way, this section is unnecessary. Millennia of plant selection and breeding have given us staple foods that are quite safe to eat; a poisonous staple food would be a contradiction in terms. But this does not mean that all vegetables, fruits, and spices are good for you prepared any old way, or in any amount. As we have seen, almost all—if not absolutely all—plants contain secondary compounds meant to discourage animals from eating them. And even though many plants, including cabbage, lima beans, potatoes, lettuce, and squashes, have been altered by human manipulation to decrease their side effects (and in the process have been rendered more vulnerable than their wild ancestors to disease and insect pests), they are still their parents' offspring. While this should not be cause for any anxiety, there is a place for informed prudence, especially around a few particular foods.

Green Potatoes: Alkaloids

Alkaloids are alkaline-behaving, nitrogen-containing complexes that appeared in evolutionary history about the same time as the mammals, and which seem to be especially effective against them. Almost all known alkaloids are poisonous at high doses, and most disrupt animal metabolism at lower doses; this is the attraction of caffeine and nicotine for the sleepy and jumpy.

Among commonly eaten plant foods, only the potato seriously threatens us with alkaloid poisoning. The production of alkaloids in the potato tuber—especially in small or immature ones—is stimulated by exposure to light and to either very cold or fairly warm storage temperatures. Small amounts of the alkaloids solanine and chaconine are normally present in the potato and contribute to its characteristic flavor. But the higher levels that result from mishandling can be toxic, and there are several recorded instances of serious group poisonings caused by bad potatoes.

Fortunately, there are warning signals. Potatoes also produce chlorophyll when exposed to light, so green tubers are automatically suspect. And a burning, pepperlike sensation on the tongue also indicates high levels of alkaloids. These substances are not destroyed by heating, and so must be physically removed from the vegetable. Most of the alkaloids are concentrated within $\frac{1}{16}$ inch of the surface, and peeling a slightly green potato deeply will make it safe to eat. Potato sprouts are rich in these alkaloids and should be thoroughly excised before the potato is cooked.

Fruit Seeds and Lima Beans: Cyanogens

Cyanogens are inactive, sugar-cyanide complexes that give rise to hydrogen cyanide, a very potent inhibitor of the respiratory system, when they are reached by an enzyme that liberates the cyanide. This occurs when cyanogen-containing plant tissue is damaged. Among common foods, cyanogens are found primarily in the seeds of the rose family—apple, pear, peach, apricot, plum, almond, and citrus fruits—as well as in lima and kidney beans, the yam, sweet potato, and bamboo shoots. (Oddly, lettuce, celery, and mushrooms contain the appropriate enzyme but not the cyanogen.) The seeds do not leak cyanogens to the surrounding fruit, and in fact the pulp may somehow break them down; it has been noticed that peach seeds from stones that had split open contained no cyanogens.

Kirsch, a liqueur distilled from crushed cherries, contains a very small amount of cyanide, and in the past, preserves and ratafias were flavored by steeping peach pits in the liquid. It is very seldom that deliberate consumption of fruit pits is a problem, although there have been instances of people who have roasted and eaten a few dozen apple seeds as a novelty and died of cyanide poisoning, and children will sometimes eat them absentmindedly. The occasional swallowed seed is no danger, but you should make a general policy of avoiding them.

Lima beans, on the other hand, *are* deliberately consumed, and for this reason many countries, including the United States, restrict commercially grown varieties to those with the lowest cyanogen contents. Lima beans used in Java and Burma can have 20 to 30 times the concentration allowed in Western countries. Though not entirely free of cyanogens, our lima beans are no danger to us. And residual toxin can be driven off by the simple expedient of boiling water. Hydrogen cyanide is a gas, and escapes the cooking liquid as soon as it is formed—as long as the pot is not covered. This is also the way in which manioc and bamboo shoots are made safe for consumption in parts of the world where improved varieties are not available.

Nutmeg: Toxic Oils

Plant oils, which we have seen to be responsible for the flavor of many foods, especially the spices, also contain toxic compounds, some of which are also the flavor compounds. Today spices are valued only for their flavor, but in earlier times they enjoyed high repute as stimulants and cure-alls, and with some reason. In the past, when spices were a symbol of wealth and a disguise for spoiled food, they were used in much larger quantities: doses at which the toxic oils would have had noticeable effects.

Among the more prominent toxic oils are thujone and myristicin. The first is a major component of the spice wormwood, and it is the reason that wormwood is little used today. As an ingredient in the liqueur absinthe, it was implicated in the appearance of brain cortex lesions among heavy absinthe drinkers, and absinthe was banned in several countries around 1915 (Pernod, anisette, pastis, and ouzo are today's safer approximations). Myristicin, however, is still with us. It is found primarily in nutmeg and mace, although it has also been detected in small quantities in black pepper and carrot, parsley, and celery seeds. Until the nineteenth century, nutmeg and mace were thought to be effective against nearly every ailment known. Even now nutmeg has the unjustified reputation of bringing on menstruation and spontaneous abortion. Myristicin is a strong hallucinogen, with such side effects as severe headaches, cramps, and nausea. Nevertheless it has been a popular drug among prison inmates and the poor because it is cheap and easily obtainable. In his autobiography, Malcom X tells of using it in a Boston jail, and the great jazz saxophonist Charlie Parker is reputed to have eaten nutmeg with milk or cola. But there is no need to worry about the nutmeg you might scrape onto your eggnog or pudding. The average dose required to produce hallucinations is on the order of two whole nutmegs.

Many other toxic oils are scattered through the plant world. Sassafras contains an oil similar to thujone, and as a result sassafras extract is no longer allowed in root beer. Even the ubiquitous menthol, derived from the peppermint plant, caused heart irregularities in a woman addicted to peppermint candy. But there is little cause for alarm. Normal, everyday eating habits simply don't include significant doses of toxic plant oils.

The Cabbage Family: Goitrins

The goitrins are thiocyanate compounds peculiar to the cabbage family that can indirectly cause health problems by interfering with the body's uptake of iodine, an element crucial to the operation of the thyroid gland, which controls our metabolism. Enlargement of the thyroid, a condition known as goiter, can result. The goitrins are a problem, however, only for those people on a marginal diet to begin with, and who have a persistent iodine deficiency. Eating a lot of cabbage or cauliflower can aggravate a preexisting thyroid problem, but generally can't initiate one.

Beans: Protease Inhibitors and Lectins

The classes of molecules called protease inhibitors and lectins are found primarily in seeds and are troublesome by virtue of their interference with digestion. Both are complex protein structures that are rendered harmless by boiling temperatures. The protease inhibitors, as their name implies, interfere with the work of protein-cleaving enzymes, by complexing either with the enzyme or with the food protein itself. They have the additional effect of overstimulating the pancreas, by forcing it to produce more and more enzymes. The lectins bind to receptor cells in the intestine and prevent the absorption of nutrients. Beans are the most common food group to contain protease inhibitors and lectins, although potatoes and bananas harbor small amounts of the latter. Laboratory animals on diets of raw bean meal actually lose weight because these compounds *increase* the animals' needs for various nutrients. Fortunately, few of us live on raw beans.

Antivitamins

Antivitamins are compounds that destroy, complex with, or mimic the structure of vitamins, in each case making them unavailable to the body. Soy beans contain a carotene enzyme; antithiamine substances are known in several berries, beets, red cabbage, and brussels sprouts; and vitamin A seems to interfere with vitamin K! Little is really known about the properties or significance of these substances, but one would guess that their effects, in a normal diet, are negligible.

Again, it should be emphasized that the point of this section is not to provoke concern. It is to give you some idea of the great complexity of plant composition and human nutrition, of the many potential problems presented by plant defenses, and of how well the millennia of genetic selection and our culinary tradition serve us in making plants safe to eat.

Fruit Ripening: A Special Case

Up to this point, we have been talking about the composition and qualities of plant foods in general. But culinary fruits differ significantly from vegetables because they undergo a marked *change* in composition and therefore in texture, flavor, and other qualities. Vegetables too change gradually; leaves decay, stems and roots become lignified and tough. But fruits progress from inedibility to deliciousness. How this happens, and why, is now our subject.

Technically, the fruit is a distinct organ that develops from the flower's female tissue, the ovary, and encloses the maturing seeds. Most fruits are simply the thickened ovary wall, or else they incorporate nearby tissues as well. The apple, pear, fig, and strawberry, for example, are all composed principally of the *receptacle*, or stem tip in which the flower parts are embedded, and the pineapple is a conglomerate of many individual flower-like units surrounding a central stalk. The fruit usually develops into three distinct layers: a protective skin, a seed coat, and an intermediate layer which is thin and dry in the nuts and grains, but thick and succulent in the fleshy

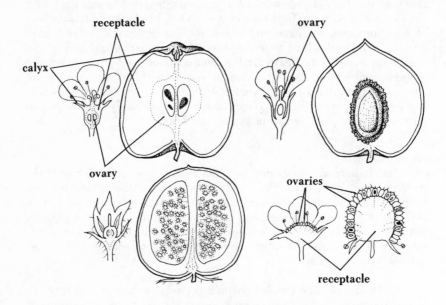

Four different fruits and the flower tissues from which they develop. The edible portion of apples *(upper left)* and pears derives from the flower base, or receptacle. Because the ovary is set below most of the flower parts, their remnants scar the top of the fruit. Stone fruits (peach, **upper right**; plum, apricot) derive from an ovary that sits above the base of the flower parts. The same is true of the fruits called berries, here represented by the tomato *(lower left)*, which have more than one seed. Raspberries *(lower right)* are not true berries, but rather multiple fruits that develop from many ovaries set in the same receptacle. Each little segment is a complete stone fruit.

fruits. Since the fruit does not support the plant structurally, like a root or stem, or nutritionally, like a root or leaf, it is composed primarily of storage cells, with a minimal complement of photosynthetic and strengthening tissue.

Fruit goes through four distinct stages of development. The first is fertilization of the female ovule by male pollen, an event that initiates the production of growth-promoting hormones and so leads to the expansion of the ovary wall. Some plants, called "parthenocarpic," can develop fruit without fertilization. Bananas, seedless grapes, and navel oranges are the most common carriers of this genetic trait, but many others can be tricked into setting fruit without seed by abnormal temperatures during flowering, or by such chemical manipulations as the direct application of growth hormones.

The second stage of fruit development is the multiplication of cells in the ovary wall. This phase is surprisingly short. In apples, it is finished a month after flowering, though it takes four months for the apple to mature. And in the tomato, cell division is virtually complete at the moment of fertilization (you can see the fully formed fruit at the base of the flower as soon as it opens). The only common exception to this rule is the avocado (a vegetable to us, but, with sugar added, a fruit in Brazil), whose cells continue to divide until it ripens.

Most of the noticeable growth during fruit maturation is a manifestation of the third stage, the expansion of an already fixed number of storage cells; and this growth can be remarkable. Melons at their most active put on better than 5 cubic inches a day, and pumpkins an average of 12 ounces, with daily gains of 20 ounces not unkown. Most of this expansion is due to the accumulation of water-based sap in the cell vacuoles. Mature fruit storage cells are among the largest in the plant kingdom, in watermelon reaching up to 350,000 times their original size, or about a millimeter in diameter.

As the cells enlarge, water and minerals from the roots and sugars newly synthesized in the leaves are transported to the fruit, where they are put to various uses. As each cell expands, its surface area increases, and new cell wall material, which takes the form of cellulose fibers embedded in a cementlike layer of pectic substances, must be laid down. Sugar, which is the cell's ultimate source of energy, is stored in the cell vacuole as sugar or in the form of organic acids, or is converted to more compact, less reactive starch granules (the avocado forms oil rather than starch). So-called secondary compounds, among them poisonous alkaloids and astringent tannins, are synthesized to deter infection or predation. And the many enzymes, hormones, and other intermediate molecules necessary to the life of the tissue are continually replenished.

The gradual work of cell enlargement goes on for weeks, even months, and, compared to what happens in the rest of the plant, it is not very remarkable. But the fourth stage of fruit development, ripening, is unique. Ripening is a sudden, rapid, and drastic change in the life of the fruit, which merges into its death. While cell expansion will stop if the fruit is cut off from the plant, detached mature fruit can go ahead and ripen as if by remote control, at times even faster than fruit left on the plant. (The anomalous avocado will not ripen *until* it has been picked—the tree appears to supply some inhibiting substance—with the result that "storing" avocadoes means leaving them on the tree.)

Ripening consists of several nearly simultaneous events. Skin color changes, usually from green to something else. Starch and acid contents decrease, and sugar content increases. The texture softens; secondary compounds disappear. A characteristically fruity odor develops. Translated to the consumer's perspective: fruit becomes sweeter, softer, and tastier, and it visually advertises these improvements.

The Action of Enzymes

The single largest cause of all these changes is the action of many different enzymes which break down complex molecules into simpler ones. Color shift, for example, is due primarily to the destruction of chlorophyll, together with some synthesis of other pigments. The membrane surrounding the chloroplast weakens and permits enzymes to reach and destroy the green pigment, which is normally so intense that it masks the presence of the other pigments invariably accompanying it.

Enzymes also convert starch back into sugar and so make the fruit more palatable. An extreme example of this is the banana, which goes from a ratio (by weight) of 25% starch, 1% sugar to 20% sugar, 1% starch after ripening (the missing percentage is used to provide energy for the ripening process). Several common fruits, however, do *not* store starch, and this fact has important practical consequences. Their final sugar content will depend entirely on the time they spend attached to the sugar-supplying leaves. Melons, citrus fruits, and pineapples will not get any sweeter after they are picked; in these cases, we are left at the judgment and mercy of the growers.

Even these fruits will, however, improve in texture after picking. Fruit softens because pectic enzymes convert the cell wall "cement" into more soluble forms, which then dissolve away, leaving the entire network weakened. For reasons that are not entirely understood, this change is so extensive in tree-ripened pears that they become unpleasantly mealy. Pears turn out best if picked (but not eaten) prematurely.

Not all secondary compounds follow this trend of being broken down by enzymes during ripening. Alkaloids do, but tannins bind to each other to form unreactive polymers. And fruity odors are a complex combination of the many volatile compounds that result from the general breakdown of tissue.

It was long thought that ripening was nothing more than the progressive breakdown of cellular structure, which resulted in the mixing of reactive compounds normally segregated from one another, and so in eventual chemical chaos. Today, however, it is thought that ripening is the final, carefully programmed stage of fruit development, a directed process that involves the synthesis of new materials, not just mere destruction. The key piece of evidence for this view is the role of ethylene in ripening.

Ethylene and Ripening

In the Caribbean islands, around 1910, it was reported that bananas stored near some oranges had ripened earlier than the other bunches. In 1912 California citrus growers noticed that green fruit kept near a kerosene stove changed color faster than the rest. What secret ripening agent did stove and fruit have in common? The answer came two decades later: ethylene, a simple hydrocarbon gas

$$\begin{array}{ccc} H & & H \\ \diagdown & & \diagup \\ & C = C & \\ \diagup & & \diagdown \\ H & & H \end{array}$$

which, when applied to mature fruit, triggers ripening. More recently, it has been determined that the fruit itself produces ethylene—a process called "autostimulation"—well in advance of ripening. It is not merely a by-product of cell disorganization, then, but a specific hormone that initiates this process in an organized way.

This knowledge of ethylene's effects has now found widespread commercial application. Bananas, tomatoes, and other fruit are shipped hard and unripe to reduce the damage caused by picking, packing, and transport and then are gassed with ethylene on arrival to ripen them for the counter. Although the flavor and texture of oranges do not change after picking, they are also treated to improve their color. But the control of ripening is hardly an invention of the twentieth century. In the 4th century B.C., Theophrastus, a student of Aristotle's and author of the first known botanical treatise, wrote of the Egyptian fig, "it will not mature fully unless it is cut and anointed with oil. . . . For the excess juice is drawn off by the wound, and the oil, like the sun, warms the open fruit and accelerates its maturation" (*On the Causes of Plants*, Book 5). This practice of "oleification" for the purpose of ripening figs about a week early continued well into this century in the Mediterranean region. Research at Yale has shown that long-chain fatty acids in the oil, which the investigators call "oleanimins," stimulate the fruit to produce the ripening hormone ethylene. Today, most fig growers say that the gain of a few days is not worth the trouble, and that the treated fruit is not as tasty. Consumers make the same remark about supermarket tomatoes. Ethylene per se is not responsible for their anemic flavor; removing the fruit prematurely from its source of nourishment is.

Exactly how ethylene works is not known. It probably increases the permeability of cell membranes, and triggers the synthesis of enzymes that are immediately responsible for the series of changes that constitute ripening. It is known that once ethylene reaches a certain concentration in the fruit, the cells suddenly begin to respire—to use up oxygen and produce carbon dioxide—from 2 to 5 times more rapidly than before. This respiration rate is a sign of furious biochemical activity. The cells are not simply dying away and disintegrating, but are living a last, intense phase of life. As it ripens, the fruit actively prepares for its end, organizing itself into a feast for our eye and palate.

SPOILAGE AND STORAGE

There are two primary causes of spoilage—those changes in taste, texture, and safety that make food inedible—in fruits and vegetables: infection by microbes, and action by the plant's own enzymes. Because plant surfaces are fairly dry and their cell contents acidic, molds rather than bacteria are the predominant problem; among the very acidic fruits, green colonies of the genus *Penicillium* are the most common. Mold spores are everpresent in the air, and their growth on fruits and vegetables will occur only when large numbers of spores are around—for example, in the vicinity of other moldy food—or when the plant is especially vulnerable to infection after suffering physical damage. Simple mechanical injury to the tissue that breaks through the skin and crushes a patch of cells will expose nutritious fluids to the contaminated air, and will mix enzymes with their target molecules. Uncontrolled enzymes destroy vitamins, develop off-flavors, soften cell walls, and produce the characteristic brown, bruised color. Even when plant tissue merely wilts, cell contents are concentrated by water loss to the point that they eat away at cell membranes; once again damaged tissue is the result. And, in the case of fruits, ripening itself is the beginning of inevitable decay.

The aim in storing fruits and vegetables, then, is to avoid or at least postpone the disorganization of plant tissue. Ideally, of course, we would consume our food fresh from the farm or garden. It has been shown that the composition of vegetables changes radically in only a few hours after harvest, simply because the cells continue to do their business despite the fact that they have been cut off from their food and water supply. Corn and peas, for example, lose up to 40% of their sugar in 6 hours at room temperature, either by converting it to starch or using it for energy to stay alive. Bean pods and such stem vegetables as asparagus and broccoli begin to use their sugar to make tough, indigestible, lignified fibers. The sweetness, flavor, and even texture of many vegetables deteriorates quite rapidly, then, and it's no sentimental illusion that corn picked one minute and dropped into boiling water the next just can't be matched.

Unfortunately, the ideal is impractical; there is almost always a lapse of time between harvesting food and using it. The inevitable deterioration can be minimized, however, by following a few basic guidelines. "One rotten apple spoils the barrel"—moldy fruit or vegetables should be discarded and the container, whether a refrigerator pan or fruit bowl, should be cleaned occasionally and thoroughly to reduce the mold population. Don't subject plant tissue to physical stress, whether it be dropping apples on the floor or packing tomatoes tightly into a confined space. Even rinsing in water is enough to make many berries more susceptible to infection by abrading their thin, protective epidermal layer with clinging dirt particles. On the other hand, soil harbors large numbers of microbes, and should be washed from the surfaces of less delicate fruits and vegetables. And stem and leaf vegetables should be kept moist enough to prevent wilting.

Beyond being saved from external stress, plant tissue must be saved from itself: as it ages, it inevitably declines in quality. The aim here is to slow or stop the metabolic activity of the cells. In the short term, this means refrigeration. But if the food is to be kept longer than a few days, more drastic measures must be taken; these can include freezing, canning, drying, making preserves, and pickling.

Refrigeration

Plant cells continue to live at lowered temperatures, but their biochemical activity is slowed. The same goes for spoilage microbes, which proliferate less rapidly. Most fruits and vegetables take well to this treatment; apples, for example, can be kept up to 6 months if the atmosphere is also controlled. But there are important exceptions. Not surprisingly, fruits that are native to tropical or subtropical climates do not do well in the cold. Bananas suffer cell damage and the release of browning and other enzymes, which is why their skin turns black in the refrigerator. Avocadoes will darken and fail to soften further if stored below 45°F (7°C). Citrus fruits develop spotted skins. All these fruits, as well as pineapples, melons, eggplants, squash, tomatoes, cucumbers, green peppers, and beans, keep best at the relatively high temperature of 50°F (10°C). Ordinary potatoes turn sweet at temperatures below 40°F (4°C) because, although their respiratory processes slow down, the conversion of starch to sugar does not, and excess sugar supplies build up (when they return to room temperature cold-sweetened potatoes consume the excess sugar). Most other vegetables, including carrots, cabbage, and lettuce, keep best right at 32°. Because the cell sap contains dissolved sugars and salts, it will not actually freeze until the temperature drops several degrees below the freezing point of water. Finally, small fruits keep well only for a day or two because the low temperature results in condensation on their delicate skin, which in turn favors the growth of mold.

Because few of us can provide ideal temperatures for the various foods we keep in our refrigerators, the best we can do is compromise.

A couple of other factors influence the success of refrigeration. One is humidity. The water content of the air in the refrigerator should be high enough to prevent wilting. Practically, this means keeping plant foods in restricted spaces—"crispers"—to slow down moisture loss to the compartment as a whole and to the outside. At the same time, the actual condensation of water on food surfaces is undesirable. A second necessity is air. Unlike most other foods, fruits and vegetables are still alive and respiring, and so consume oxygen. If oxygen is denied, then the plant cells switch to anaerobic respiration, which results in the accumulation of alcohol in the tissue. This in turn causes local cell damage and the brown cores and dry brown spots below the skin that we often see in apples and pears that have been in storage. Closed plastic bags should not be used to store fruits and vegetables for any extended time, even though they do prevent water loss. Paper and cellophane, which are moisture- and gas-permeable, are preferable.

Freezing

The more drastic process of freezing stops the respiratory processes cold, and thereby kills the tissue. Most, but not all, of the water in the cells crystallizes, and this stops most chemical activity. One side effect of freezing is the damage done to cell structure by the water crystals, which can puncture cell membranes and walls. The result is tissue that is less able to hold water and so is less crisp than the fresh original. But freezer damage is less extensive in plant tissue than in meat. The intercellular spaces, which are often large and which contain some water but very little else, freeze first, and then draw water out of the cells by osmosis. This in turn increases the concentration of dissolved substances within the cell, which further lowers the freezing point. Parts of the cell interior do not freeze at all on account of this continuous change in concentration. Still, concentrated enzymes and salts cause the cell membranes to deteriorate, and crystals can form within the cell and damage the tissue physically. The size of the crystals, and so the amount of damage done, can be minimized by freezing the material as quickly as possible to as low a temperature as possible. Under these conditions, many seed crystals form and draw water out of solution; at higher temperatures only a few seed crystals form and, to incorporate the same amount of water, they grow much larger and do more damage. Vegetables tend to be more acceptable after freezing than fruits because we usually soften them even further by cooking, while we prefer fruits uncooked and firm.

Although freezing temperatures greatly reduce enzymic and other chemical activity, they do not eliminate them, and the possibility of spoilage remains. (In fact, nonenzymic reactions may even be enhanced by the high concentration of dissolved material inside the cell.) As a result, vitamin content and color can suffer. The solution to this problem is *blanching*. In this process the food is immersed in rapidly boiling water for a minute or two, just enough time to inactivate the enzymes, and then just as rapidly immersed in cold water to stop further cooking and weakening of the cell walls. If vegetables are to be frozen for more than a few days, they should be blanched first. Fruit is rarely blanched because the resulting changes in flavor and texture are less acceptable in a food that is normally eaten raw. Enzymic browning in frozen fruit can be prevented by packing it in a solution of sugar and/or ascorbic acid. Sugar will also improve the texture of frozen fruit by being incorporated into the cell wall material, which becomes stiffer.

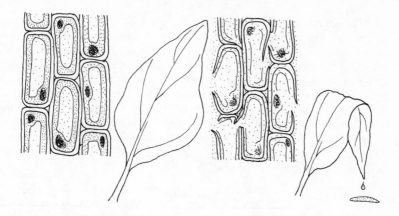

Freezing damages plant tissue when ice crystals and concentrated enzymes and salts damage cell membranes. When the fruit or vegetable is thawed, its cell fluids leak out, and with them go some crispness of texture.

Since frozen tissue no longer respires, its wrapping should be as air- and water-tight as possible. Surfaces left exposed to the relatively dry atmosphere of the freezer may develop freezer burn, the slow, patchy drying out caused by the transformation—sublimation—of frozen water molecules directly into vapor (see page 636). Burned patches have a rough texture and develop off-flavors as the local concentration of active molecules rises. It is generally agreed that frozen vegetables should be cooked immediately rather than thawed first. Thawing gives microbes and any remaining enzymes time to do some harm.

Canning

An alternative to freezing—at the opposite end of the thermal scale— is canning, which involves the rapid heating of hermetically sealed tin or glass containers. The heat destroys harmful organisms and deactivates enzymes, and the tight seal isolates the food completely from external contamination. Since the canned food is to be stored at room temperature after processing, all harmful organisms must be destroyed. Otherwise, once the temperature drops, they will resume their growth and spoil the food. Highly acidic foods—tomatoes and most common fruits—need the least severe treatment, usually about 30 minutes in a bath of boiling water, because their low pH inhibits the growth of most microbes. Most vegetables, however, are just on the acidic side of neutral with a pH of 5 or 6, and are much more hospitable to bacteria and molds; they must be more severely treated. Typically, vegetables are cooked in a pressure cooker at 240°F (116°C) for 30 to 90 minutes.

The archvillain of the canning process is the bacterium *Clostridium botulinum,* which, being anaerobic, thrives in the airless cans and produces its deadly nerve toxin, which causes botulism. The toxin itself is easily destroyed by mere boiling, but the bacterial spores are very hardy. Some have survived boiling for 5 hours. Unless they are killed by the extreme condition of higher-than-boiling temperatures, the spores will proliferate into active bacteria once the can is cooled down, and the toxin will accumulate. One precautionary measure is to boil canned produce after opening; this will destroy any toxin that may be there. But any cans that show evidence of bacterial activity—endplates bulging from the pressure of gases produced by bacterial metabolism—should be discarded.

Drying

Freezing and canning both slow or stop microbial growth by subjecting the plant tissue temporarily to extreme conditions, but culinary tradition has also developed methods of storage in which the food itself is made inhospitable to bacteria and molds. One such method is drying, the reduction of the tissue's water content to the point that nothing, or very little, can grow in it. This is probably one of the most venerable preservative techniques; the sun, fire, and mounds of hot sand have been used to dry food since prehistory. Today, home gardeners easily dry herbs, which require little attention and in fact should be protected from direct sunlight, since it tends to drive off flavorful volatile compounds. Fruits and vegetables, however, benefit from treatment to inactivate the enzymes that cause vitamin and color damage. Vegetables are usually blanched, and light-colored fruits are dipped or sprayed with a number of sulfur compounds that prevent enzymic browning. While sun drying used to be the most common treatment for prunes, raisins, apricots, and so on, forced hot-air drying is now widely used because its effects are more predictable.

Freeze-drying is actually a controlled version of freezer burn: moisture is not lost by evaporation, but by sublimation, the transformation of ice directly into water vapor. Although we think of freeze-drying as a very recent innovation, it actually goes back at least to the pre-Columbian Andes, the native ground of the potato. The Peruvian Indians would stomp on potatoes during the day to squeeze water out, and leave them out to freeze at night. Today, vegetables are more often freeze-dried than dried with heat, although their use in this form is pretty much limited to instant soup mixes, emergency rations, and camping foods.

Preserves

Another ancient technique for making plant foods, especially fruits, resistant to spoilage is to boost their sugar content to the point where microorganisms will be dehydrated by the osmotic pressure across their membranes. If the concentration of dissolved material is higher outside the microbe than it is inside, then water will be drawn across the cell membrane, and the microbe will be incapacitated or killed. Such products are commonly divided into three groups: preserves, which are whole fruits in a sugar solution that ranges in consistency from a thick syrup to jelly; marmalades and jellies, which are small pieces of fruit, or the fruit's juice, incorporated into a sweet gel; and jams, which are crushed or chopped fruit blended with a sweet gel into a fairly homogenous mixture.

Evaporation and condensation of water at the surface of preserves can result in patches hospitable to molds. It is to avoid surface mold that paraffin is often used to seal home preserves, and for the same reason such products are best kept refrigerated after opening.

Variations on this technique are many and difficult to date. Perhaps the first discovery was that when boiled down, and its own natural sugars thereby concentrated, fruit kept much longer. According to Apicius, many foods, including meat, were preserved in honey in classical times. The big boost for preserves came in the 16th century, when the Spanish began growing sugar cane in the West Indies. By 1800, sugar was cheap and plentiful enough for the middle classes to use it in preserves, and cookbooks of the period regularly include instructions for putting up fruit with pounds and pounds of sugar.

Sugar affects not only potential contaminants, but also the fruit being preserved, and especially its texture. If simply boiled, fruit quickly softens into mush. But cooked in a sugar syrup it tends to remain relatively firm and to maintain its shape. It will shrink as water is drawn out of the cells, but sugar molecules interact with the cell wall hemicelluloses and pectins, and become partly incorporated into the structure, making it firmer. Whole fruit preserves are best handled in this way, although hard fruit may need an initial boiling so as not to end up *too* stiff. Of course, added sugar affects the taste of the product, and some compromise is necessary between maintaining the natural taste and achieving the desired consistency.

Pectin Gels

Preserves get their smooth, semisolid consistency not from sugar, but from pectin, which can be extracted from the cell wall cement in plants by boiling. The long, stringlike pectin molecules bind a liquid into a solid by bonding to each other and forming a meshwork that traps the liquid in its interstices. Some fruits, including grapes and most berries, are rich enough in pectin to produce excellent gels on their own, while others, including apricots, peaches, and strawberries, need a supplemental source. Preserve makers can buy powdered pectin that has been extracted from apple cores and parings or from the intermediate white layer, or *albedo,* of citrus fruits. Or they can make their own extract by simmering lemon slices that have been lightly pared.

It turns out that a pectin mixture will gel only under certain conditions. The long pectin molecules ordinarily have a negative electrical charge in water, and so repel each other. Acid conditions reduce the electrical charge and so remove this barrier to bonding. Even in acid, however, pectin does not readily gel because its molecules are more likely to encounter and bond with water molecules than with each other. But if sugar is added, its highly water-attracting molecules take enough water out of circulation to allow the pectin molecules to reach each other. To thicken a preserve with pectin, then, we need acid and sugar: the two substances that fruits specialize in storing. Sounds like a cinch.

But as anyone who has tried knows, it's anything but a cinch. Making preserves is a tricky business because the necessary balance between pectin, acid, and sugar is a very delicate one. Food scientists have found that a pH between 2.8 and 3.4, a pectin concentration of 0.5 to 1.0%, and a sugar concentration of 60 to 65% are generally optimal, but you would have to be cooking in a well-equipped laboratory to measure the first two conditions (sugar content is easily measured by boiling point). And given the fact that the composition and quality of pectin vary with the fruit, and are difficult to determine and control even in commercial extracts, it seems much less surprising that some jams and jellies never gel at all, while other seize up into something resembling a gum eraser. Experience or a reliable guide remain indispensable in this branch of cooking.

The basic procedure for making preserves is as follows. The fruit, whole or cut up, is simmered for 10 to 20 minutes in a small amount of water to extract pectin. If the fruit is deficient in either pectin or acid (apples, pears, peaches), then lemon slices (minus the flavorful yellow portion of the rind) may be included in this initial step. Longer cooking is avoided to prevent extensive breakdown of the pectin molecules. An appropriate amount of sugar is then added, and the mixture kept at a rolling boil (sugar protects the pectin to some degree from damage) until the desired concentration is reached. If the mixture fails to gel when a sample is cooled in a spoon, more lemon juice or pectin extract can be added to adjust the balance of ingredients.

There is a special kind of pectin that will gel with very little sugar and less than the usual acidity. Calcium ions, which lack two electrons, will readily form bridges between negatively charged points on different molecules of this pectin, and thereby build up the desired meshwork. This kind of pectin is frequently used in low-calorie jellies and jams.

Pickles

Pickling is the preservation of foods by impregnating them with acid, which discourages the growth of most microbes. There are two basic techniques: simply adding acid, usually in the form of vinegar, and brining, also known as fermenting, in which acid-producing bacteria are encouraged to grow.

The first method is quite straightforward. The vegetable is cooked to a soft consistency or put in brine for a short time to draw out moisture that would dilute the vinegar. Then it is immersed in the vinegar, often with several spices. Sweet and sour pickles, fruit pickles, onions, cauliflower, peppers, and tomatoes are all commonly treated in this way, as are the mixtures called piccalilli, chutney, and ketchup (from a Chinese word for "pickled fish sauce"; in 17th-century England it meant a sauce fermented from mushrooms or walnuts, but now it is most often applied to pickled tomato pulp). Bacteria in these products are almost entirely inactive, but molds and yeasts can grow at low pH and contaminate the surface if it is not covered.

The process of making fermented pickles is more complicated. The vegetable is put in a brine solution strong enough to prevent the growth of undesirable bacteria, but weak enough to allow the growth and predominance of several species that produce lactic acid. Sauerkraut and dill pickles are the most common fermented products. In its first incarnation as a staple for the workers on the Great Wall of China, sauerkraut was simply cabbage covered with wine. Now it is made from fresh cabbage that is shredded and salted. The liquid drawn out of the tissue by osmosis is then supplemented with water to cover the vegetable completely. A salinity of about 2.25% is the goal. At the right temperatures, between 65° and 70°F (18° and 21°C), a bacterium called *Leuconostoc mesenteroides* grows and produces lactic acid and other minor compounds that contribute to sauerkraut's flavor. When the acid level reaches about 1%, this bacterium declines, and is replaced as the major population by *Lactobacillus plantarum*, which jacks up the acid content to as much as 2%, although a final level of 1.7%, reached after two or three weeks of fermentation, is considered ideal. Notice that no bacteria need to be inoculated into the initial brine; they are ready and waiting in the atmosphere around us.

Young cucumber pickles are made somewhat differently. An 8% brine flavored with dill and other spices is used to bring the pickles up to a final acidity of 1.0 to 1.5%. Vinegar is usually added at the beginning of the process to help prevent the growth of unwanted bacteria. Another method, employing a final brine of about 16%, does without the vinegar, but the pickles turn out so salty that they must be soaked in water before being eaten. Fermented pickles are trickier than those simply put up in vinegar; the wrong temperature or salt concentration can result in the wrong bacterial populations, soft or hollow pickles, or off-flavors.

COOKING FRUITS AND VEGETABLES

Cooking affects all the qualities of plant foods. Safety and edibility are generally improved: harmful compounds are either leached from the tissue (alkaloids, cyanogens), or inactivated (protease inhibitors, lectins). And in most cases, flavor is intensified because high temperatures make the aromatic molecules more volatile and so more easily detected. Overcooking, however, will result in the eventual *loss* of flavor as the volatile compounds are driven off or altered by the prolonged heat. Herbs and spices must likewise be cooked long enough to extract the flavorful oils from their tissue and diffuse them through the food, but not so long that they are lost. Long-simmered dishes like soups and stews may need some reinforcement late in the day. Among special cases, recall that cooking moderates the harsh flavor of onions but strengthens the odor of cabbage and related vegetables. Three other characteristics—color, texture, and nutritional value—are subject to some deterioration during cooking, and we will now see how these changes can be minimized.

How Cooking Affects Texture

We've seen that plant texture is determined by both the cell wall structure and the inner water pressure, or turgor, of the tissue. The application of heat, whether by boiling, baking, or stir-frying, tenderizes the food by weakening the cell walls and extracting water. First, heat denatures the proteins that make up the cell membranes, which thereby lose the selective permeability that regulates the cells' water content. Water leaks from the cells, the tissue loses turgor, and the plant becomes wilted and flabby. Even boiled vegetables, surrounded by water as they are, lose water during cooking, as weighings before and after will demonstrate.

Then there are the changes in the cell walls. While cellulose is not affected by heat, the hemicelluloses and pectins are. Some hemicelluloses dissolve. And the distribution of the pectic substances is altered: the amount of soluble pectin increases at the expense of the insoluble protopectins, and the walls lose still more of their "cement." The result: substantially weakened cell walls, and tenderer tissue.

The problem in cooking vegetables, of course, is how to make the tissue tender without making it too soft. Usually we take the commonsense approach of sampling the food during cooking and stopping when it is tender but still firm. In some cases color can also be used as an index of doneness; green vegetables are considered overdone if their color turns decidedly dull. One possible generalization is that leaf vegetables, with their relatively thin, exposed, and delicate layer of tissue, need only a minute or two of heating, while stem and root vegetables may require many times that amount. Experience and personal taste remain the best guides.

The Effects of Alkalinity

One interesting sidelight to consider is the pH of the water used for boiling. The loss of hemicelluloses and pectin is very much dependent on pH: hemicelluloses are more soluble in alkaline water, while pectin dissolves most rapidly at a pH below 4 or above 4.5. Cooking vegetables in distinctly alkaline water quickly makes them mushy, while neutral or acidic water will maintain the tissue's firmness for a longer period (the water vapor produced in steaming has a distinctly acidic pH of about 6). The pH of tap water varies from place to place and can be on the acidic or basic side, though boiling tends to make it slightly more alkaline by driving off various dissolved gases and concentrating remaining salts. All plant tissues are acidic, however, and when boiled they release acid to the surrounding water. Added acid in the form of vinegar or lemon juice will affect both the flavor and, if the vegetable is green, the color of the food. Of course, these considerations do not arise in the cases of stir-frying or baking, in which no water is used and the tissue is at the mercy of its native pH.

Starchy Vegetables

Potatoes, winter squashes, and other starchy vegetables constitute a special case in the matter of cooking and texture. For them, alteration of cell membranes and walls is less important than alteration of the starch granules. We examine starch granules in detail in the section on bread making in chapter 6. Here, it is enough to know that in raw tissue they are hard, closely packed agglomerations of starch molecules, and do not soften appreciably during cooking until the temperature reaches the "gelatinization range," which in the potato is from 137° to 150°F (58° to 66°C). At this point, the granules begin to take up water, which disrupts their compact structure, and they swell to many times their original size. Once swollen, they are tender and very easily damaged by physical pressure. In starchy vegetables, then, the primary task in cooking is to reach the gelatinization range in all parts of the tissue.

Several authoritative cookbooks concur in the view that in order to obtain the best French-fried potatoes, one should subject the potato strips to two cooking periods, first at a low temperature, around 325°F, and then at a higher one, perhaps 370°F (163° and 188°C). This odd-sounding technique does make sense. In frying foods, we aim both to cook them through and to brown the surface sufficiently to produce the characteristic fried flavor. But too high a frying temperature will brown the surface before the interior is cooked, while too low a temperature will take a long time and result in the absorption of more oil and perhaps in an overcooked interior. In the two-stage method, the first serves to cook the potato through without

 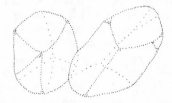

When compact starch granules reach their gelatinization temperature in water, they suddenly take up large quantities of water and swell into delicate, tender masses.

browning the surface: the typical direction is to fry the potatoes until limp, which is to say until the starch granules have gelatinized. We then let the strips cool to room temperature. By the second treatment, the potato strips have been precooked and covered with a film of gelatinized starch, which slows any further oil absorption. The second frying can then be done at a high temperature and stopped as soon as the outside is browned.

Cooking Fruit

Finally, a word about the texture of cooked fruit. Mushiness is one of its least appealing characteristics, and this can be remedied to some extent by loading the cooking syrup with sugar, which helps strengthen the disintegrating cell walls. In effect, an artificial sugar "wall" replaces the original cell wall. At the same time, the presence of sugar in the cells draws some water back into the tissue, to maintain an equilibrium in sugar concentration, and this restores some turgor to it.

What Happens to Color

One change in the color of leaf and stem vegetables as they are cooked has nothing to do with the pigment involved. That wonderfully intense, bright green that develops within a few seconds of throwing the vegetables into boiling water is a result of the sudden expansion and escape of gases trapped in the spaces between cells. Ordinarily, these air pockets dim and refract the color of the chloroplasts; when they are collapsed by the heat, we can see the pigments much more directly. But it usually takes more than a few seconds to cook vegetables, and some chemical alteration of pigments does result. The only major exception to this rule is the carotenoid group, which is chemically very stable. Carrots and tomatoes, for example, lose little if any color when boiled. Pressure-cooker temperatures, however, will cause the carotenoid pigments to change molecular form and so alter their color from reddish-orange to yellow-orange. But compared to the green chlorophylls and multihued anthocyanins, the carotenoids are the model of steadfastness.

Chlorophyll, the most common pigment, is susceptible to two basic changes in composition, which in turn affect its color. Chlorophyll's central magnesium atom can be nudged out rather easily by heat, and is readily replaced either by hydrogen or other metal ions. In acidic cooking water, the plentiful hydrogen ions replace the magnesium center, and chlorophyll a turns grayish-green, chlorophyll b yellowish. In acidic water that also contains zinc or copper ions, these metals compete successfully with the hydrogen for a place in the pigment, and the result is a bright green. The replacement of magnesium by hydrogen is by far the most common cause of color change in cooked vegetables, because heat—even from steam or the hot oil used in frying—damages cell structure and brings the plant's own

N   N   N   N   N   N

   Mg^{+2} H   Cu^{+2}

N   N   N   N   N   N

    H

     H

bright green   gray-green   bright green

**Replacement of the central magnesium atom in the chlorophyll complex
changes the color of the pigment. In acid conditions, hydrogen ions dull the
color *(center)*, while free metal ions like copper restore its brightness *(right)*.**

acids into contact with chlorophyll. This is why dull, olive-green vegetables
are so depressingly familiar.

Keeping the Green Bright

From this summary of pigment chemistry, two solutions to the problem
of color deterioration seem to leap out: cooking in alkaline water, to limit the
availability of hydrogen ions, or deliberately introducing metal ions into the
cooking water to restore bright green color in spite of the pH. It turns out
that these very remedies have been part of Western culinary tradition for
thousands of years. In the Roman Apicius, we read *"omne holus smarag-
dinum fit, si cum nitro coquatur."* "All green vegetables will be made emer-
ald colored, if they are cooked with *nitrum.*" Like our baking soda, the
natural soda that the Romans called nitrum would make the cooking water
alkaline. And in her English cookbook of 1751, Hannah Glasse gives this
advice: "Boil all your Greens in a Copper Sauce-pan by themselves, with a
great Quantity of Water. Use no iron pans, etc., for they are not proper; but
let them be Copper, Brass, or Silver." And English cookbooks of the early
19th century advise cooking vegetables and making cucumber pickles with
a ha'penny—a copper coin—thrown in to improve the color. All of these
practices survived in some form until the beginning of the 20th century.

Why is it that we don't avail ourselves of these tried and true touches?
Because each has unfortunate side effects. Long ago it was recognized that,
in large amounts, copper is a poison; we now know that it can affect both
liver and brain function. In 1753 Sweden outlawed the use of copper cooking
pots in its armed services, and in the early 19th century several pamphlets
appeared in England warning of the health hazards posed by pickles, beers,
bakery products, and candies that had been prepared in copper vessels. The
pleasure of eating bright green vegetables does not outweigh the risk to
health.

The use of nitrum, or baking soda, stopped about the same time as
copper's despite the earlier animadversions of such worthies as Tabitha

Tickletooth, who wrote in *The Dinner Question* (1860): "Never, under any circumstances, unless you wish entirely to destroy all flavor, and reduce your peas to pulp, boil them with soda. This favorite atrocity of the English kitchen cannot be too strongly condemned." Even Mrs. Tickletooth, however, was retrograde enough to allow a pinch when cooking cabbage and spinach, although she observed that soda makes the flower buds fall off of broccoli. In 1883, Mrs. Lincoln of the Boston Cooking School knew that the use of soda is unwise because soft water "extracts the soluble parts" of organic material, and by 1931 Mrs. Rombauer's *Joy of Cooking* noted that vitamins are destroyed by alkali.

Vegetables cooked in alkaline water turn mushy because the cell wall hemicelluloses and pectin are more soluble, quickly lost, and cause the rapid and extreme softening of the tissue. In addition, the phytyl group on the chlorophyll molecule is more easily removed, which means that more pigment leaks into the cooking water. The effects on vitamin retention are less than clear-cut, but they are certainly not positive. For several reasons, then, but primarily for one of aesthetics, soda is also a bad idea for bringing back the green.

Mrs. Lincoln knew, though 50 years later Mrs. Rombauer did not, that there is a simple, practical, and safe way to minimize chlorophyll damage. Put the vegetables into large volumes of rapidly boiling water, leave the pan uncovered for the first few minutes, and don't cook longer than 5 to 7 minutes (if necessary cut the vegetable into small, quickly cooked pieces). Lots of water will immediately dilute the plant's own acids, thereby minimizing their effects, and those acids that are volatile escape during the first few minutes of boiling; if the pan were covered, they would condense on the lid and fall back into the water. An abundance of water is dictated by yet another consideration. Plants contain enzymes that can destroy vitamins (see below) and alter pigments. Like the plant acids, the enzymes are mixed indiscriminately with other cell contents when the cell membranes are destroyed by heat. Chlorophyllase, which removes the phytyl group from chlorophyll, making it water soluble and more prone to alteration, is most active between 150° and 170°F (66° and 77°C) and is only destroyed when boiled. Large amounts of boiling water will suffer a relatively small temperature drop when the vegetables are added, and so the enzyme will have less time to work, less pigment will be lost to the cooking water, than if only a little water were used, or if the water started out cold and had to be heated through the 150°–170° range to reach a boil. This is especially important for spinach, which contains quite a bit of the enzyme, and less so for beans and peas. As we shall see, though, large volumes of water tend to increase vitamin losses.

Among other common techniques, steaming and stir-frying have the disadvantage of retaining the vegetable's acids rather than diluting them, while stir-frying has the advantage of very rapid cooking, thanks to the higher-than-boiling temperature of the pan surface and the traditional use of bite-sized pieces.

Red Wine into Blue

So much for the green. The usually reddish anthocyanins and their nearly colorless cousins, the anthoxanthins, are already water soluble, so there is no point in worrying about losing pigments to the cooking water; it's inevitable. Their most important characteristic from the cook's point of view is their sensitivity to pH—so sensitive that they have been proposed as pH indicators, like litmus paper—and to the presence of metal ions. Where chlorophyll gets duller or brighter according to these conditions, the anthocyanins change color completely. In acid solutions, they tend toward the red; around neutral pH, they are colorless or light violet; and in alkaline conditions a blue form predominates. You can easily demonstrate this startling change for yourself by grinding up some red cabbage, blackberries, concord grapes, even red autumn leaves or flower petals, and adding small amounts of baking soda or lemon juice. You might even try the recipe that comes down to us from Apicius. "To make white wine out of red wine. Put bean-meal or three egg whites into the flask and stir for a very long time. The next day the wine will be white. The ashes of white grape vines have the same effect." Both vine ashes (from whatever color grape) and egg white are alkaline substances. The color does change, although blue-gray seems more accurate than white. None of these substances, however, does anything for the wine's flavor.

This sensitivity to pH is seldom manifested in everyday cooking, where slightly acid conditions usually prevail. One exception is blueberry pancakes or muffins leavened with baking soda, which may turn the berry juices greenish (a combination of blue anthocyanins and yellow anthoxanthins). But some of these pigments do react with metal ions from our utensils. Some anthocyanins form grayish, green, and blue complexes with iron, aluminum, and tin respectively. The normally pale anthoxanthins found in potatoes and onions can form red, blue, green, or brown complexes with iron and aluminum, depending on other conditions; this is why carbon steel knives may discolor these vegetables. Beet pigments, the so-called betacyanins and betaxanthins, are somewhat different compounds and less susceptible to color change. Irreversible destruction of the anthocyanins can be caused by prolonged storage at high temperatures and the presence of oxygen, sugar, and ascorbic acid (vitamin C). A familiar example of this is the occasional dull brownish color of strawberry preserves. Acidic conditions and low temperatures favor the preservation of anthocyanins, and a little lemon juice inhibits the formation of pigment-metal complexes.

Nutritional Value

The most important contributions fruits and vegetables make to our diet are vitamins, especially A and C, and some minerals. Vitamins A precursors, the carotene pigments, are not water soluble, while C and the B vitamins and minerals are, and will leak into the cooking liquid. And carotenes are chemically stable, while thiamine is sensitive to heat and vitamin C to oxygen. Accordingly, loss of B and C vitamins increases as the cooking time and the volume of cooking water are increased. In addition, any treatment that exposes more vegetable tissue to the water—peeling or cutting into pieces— will result in higher losses. Up to a point, cutting into pieces can offset higher losses through the greater surface area by decreasing the cooking time; the smaller masses heat through faster. But chopping or dicing increases the surface area so much that the shorter cooking time cannot compensate for the increased rate of vitamin loss.

Substantial losses of vitamin C can be due not to leakage, but to enzyme action. In chopped or shredded tissue, a previously segregated enzyme now has access to the vitamin and destroys it. And when a vegetable is put into boiling water, up to 20% of the vitamin can be digested in one minute if it takes that long for the water to resume boiling. Like chlorophyllase, the vitamin C enzyme is most active at elevated temperatures, and is inactivated only by boiling. Such vegetables as potatoes, which are baked, and so warm up very slowly, lose much more vitamin C than they do when boiled, because the enzyme has more time to work. This is an excellent reason for the cook to put vegetables only into already boiling water.

Whereas large volumes of water are best for preserving the bright green color of vegetables, less water means more vitamin C. Studies have shown that more than twice as much vitamin C is retained in vegetables cooked in a small amount of water or steamed, compared to vegetables boiled in enough water to cover them completely. So to some extent the cook must choose between appearance and nutrition. It is possible, of course, to salvage the nutrients that leak out by making the cooking liquid into a soup. Minerals and B vitamins do survive, but vitamin C is easily oxidized and will not last long.

NOTES ON INDIVIDUAL FRUITS, VEGETABLES, HERBS AND SPICES

The remainder of this chapter is a compilation of facts about those plant-derived foods that are commonly found in our kitchens. Each entry is meant to place these foods historically, botanically, and, when this is of interest, biochemically. Nutritional content is mentioned only when a food is remarkable in some way. If many entries sound tentative, it is because there remains a great deal of uncertainty about the origins and precise classification of even the most familiar food plants.

FRUITS

Major Families

The Citrus Family: Lemon, Orange, Grapefruit, Lime

With the exception of the grapefruit and other recent hybrids, the members of the citrus family are native to Southeast Asia, and were first cultivated in India (our word *orange* comes from the Hindi), China, and Japan. The citron, now known only in candied form, was brought back to Europe by Alexander the Great in the 3rd century B.C., but it wasn't until the Middle Ages that the lemon, and in the 15th century the orange, made it to the West, where they were initially treated as ornamentals and spice plants. The seedless navel orange was known along the Mediterranean by the 17th century, and the loose-skinned mandarin or tangerine species had been long cultivated in China and Japan before Europeans found them in the 19th century. The grapefruit was born in the West Indies in the 18th century as a cross between the orange and the *pummelo*, a large citrus fruit that had been brought to the New World a few decades earlier. The Ruby variety with its bright red flesh was discovered as a chance "bud sport," or mutation, on a McAllen, Texas, farm in 1929. The original lime, variously called "West Indian," "Mexican," and "Key," was displaced in this country by the "Persian" or "Tahiti" lime, apparently a hybrid of the lime and the citron, around 1920. The Persian lime bears larger fruit and is more resistant to cold and pests, although it is said that the small, round Key lime—the hybrid is lemon-shaped—is more flavorful. The Key lime still predominates in the rest of the world. Other notable citrus hybrids are the Temple orange, a cross between mandarin and orange, and the tangelo, mandarin crossed with grapefruit.

albedo   carpel   juice vesicles

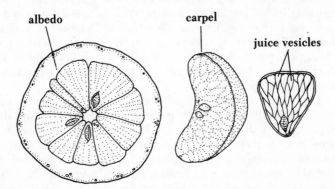

The anatomy of a citrus fruit

Each segment of a citrus fruit is a *carpel,* or a compartment of the ovary. The juice-containing vesicles grow from hairlike tubes on the segment membranes, which fill with water and dissolved substances as the fruit develops. Within each vesicle are many juice cells, their vacuoles full of sap. In the navel orange, the open end is the beginning of a second cluster of carpels, miniature segments nestled in the shallow cavity or navel of the main fruit.

Citrus fruits contain little starch and therefore do not sweeten much after picking. For this reason, a crop must reach the desired sugar-acid balance before it can be harvested. Color change is triggered by the arrival of cold temperatures while the fruit is still on the tree; in the tropics, oranges are always green. Fruits of commerce are usually treated with ethylene to improve their color, and coated with an edible wax to slow moisture loss. California oranges can be distinguished from Florida's by their relatively thick skins, which protect them from the West's dry climate. Citrus fruits are of course valued for their vitamin C content. As it happens, this vitamin is much more concentrated in the peel and albedo—the white layer just under the peel—than it is in the flesh. The juice of an orange contains only a quarter of the vitamin C in the whole fruit, and grapefruit juice an even smaller proportion. The color of orange juice is primarily due to carotenoid pigments with no vitamin A activity.

The Rose Family: Apple, Pear

These are the only two general names for the *pome* fruits (fruits with a characteristic compartmented core; the word comes from the Latin for "apple"), but today there are upward of 7000 varieties of apples and several thousand varieties of pears abroad in the world, most of them developed in the 18th and 19th centuries. Only about a hundred varieties of each fruit are grown commercially. Both apples and pears are natives of Europe and West Asia and were introduced to North America in the 17th century. There is evidence that the apple was already under cultivation in Neolithic times. The apple has long had a prominent place in Western culture as a symbol of temptation and instigation: the unnamed fruit in Genesis has traditionally been an apple; Paris gave Aphrodite a golden apple and thereby caused the Trojan War; William Tell shot an apple off his son's head; Snow White's stepmother used an apple to poison her, and Isaac Newton's legendary inspiration was a falling apple.

The pear has never been as popular as the apple, probably because it is not as easily stored. Pears are much more prone to tree and storm damage and have a relatively short shelf life. And they get mealy if allowed to ripen on the tree; to be at their best they must be picked before the onset of ripening. In contrast, dessert (sweet) apples are best when tree-ripened. Apples can be stored for many months under the right conditions, and this is perhaps one reason why they are not one of the fruits we tend to preserve. It has been generally observed that smaller apples have smaller cells and will keep better than larger ones. The only vitamin appreciably present in pome fruits is vitamin C, and this is located primarily in the peel.

The Rose Family: Apricot, Cherry, Peach, Plum

These *stone* fruits, or *drupes*, all come from a single genus in the rose family, *Prunus*, and contain a large seed surrounded by a hard coat composed mostly of lignin. The apricot and peach are natives of China, the cherry of Europe and West Asia, and the plum of the Caucasus. The apricot was cultivated as far back as 2000 B.C., and arrived in the Mediterranean area by way of Persia and Armenia in the first century A.D. About 90% of world production is now American, and 90% of that output is from California. The cherry was known to the ancient Greeks, and was brought to the New World by the earliest immigrants, but Europe continues to account for the bulk of world production. The peach was brought to Europe in Roman times, and is now grown primarily in the United States, South Africa, and Australia. There are two main varieties of peach, freestone and clingstone; their names describe how easily the fruit is separated from the stone. Pectic substances (part of the cell wall cement) become much more soluble in the freestone during ripening than they do in the cling. Freestones have almost disappeared from the market today because cling peaches are firmer, easier to transport, and more deeply colored (the orange pigment in apricots and peaches is a precursor of vitamin A). The dry, mealy, "woolly" texture of some supermarket peaches is caused by picking the fruit early and putting it into cold storage for two weeks or more; this treatment reduces the level of enzymes that convert insoluble protopectin into soluble pectin.

The nectarine is a fuzzless variety of peach, and this is the major difference between the two, although nectarines also tend to be smaller. The fuzzy trait is genetically dominant. Nectarines and peaches can develop from each other spontaneously as mutations from seed or as bud sports.

The Rose Family: Blackberry, Raspberry, Strawberry

There are several species of both the blackberry and raspberry, and all are members of the same genus, *Rubus*. At least two of the European raspberries have been cultivated since the 17th century, and today both these and American natives are grown commercially. These berries are actually aggregates of several small fruits: each tiny segment is a complete drupe, a fruit with a stony layer surrounding the seed. The strawberry, on the other hand, is a "false fruit," derived from the base of the flower rather than the ovary. The true fruits of the strawberry plant are what look to be the seeds on the surface of the berry. Each tiny *achene*, as they are called, is an individual fruit containing a single seed. The strawberry is naturally found in both Eurasia and the Americas, although two American species produce the largest fruits. It was planted in Roman gardens, and later cultivated by the

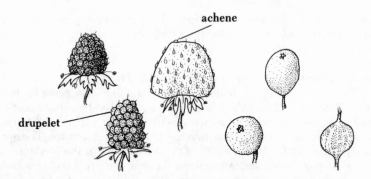

Left to right: blackberries and raspberries are multiple fruits, and the strawberry a "false" fruit: the "seeds," known technically as achenes, are the true fruits that correspond to the fleshy drupelets of the multiple fruits. Blueberries, cranberries, and currants are all true berries, or single fruits derived from the plant's ovaries.

French in the 14th century, as well as by the Indians of North and South America. Early in the 18th century, a French military engineer coincidentally named Frézier (the French for strawberry is *fraise*) was sent to report on Spanish colonies in Chile, and brought the large-berried native species back to Europe, where the breeding that has given us our modern varieties began.

The Squash Family: Melons

Of the melons commonly eaten in the United States, the watermelon stands apart from the others. This native of southern Africa was introduced to India in prehistoric times, was eaten in Egypt around 4000 B.C., came to the Americas with the slave trade, and is now cultivated all over the world. The other melons—musk, cantaloupe, honeydew, crenshaw, casaba—are more closely related members of the single genus *Cucumis*, the same one that includes cucumbers. The origin of these melons is obscure, and placed tentatively in either Africa or India. Their introduction to Europe seems to have been by way of cultivation by the Moors in Spain. The name *cantaloupe* comes from a former papal garden near Rome, Cantalupo, where the variety was developed; what Americans call cantaloupe is usually a musk melon, which is smaller and has a netted rind rather than a deeply ridged one (its orange color is contributed by a vitamin A precursor). Some varieties of melon grown in Asia are used as vegetables.

Anatomically, the flesh of the watermelon, in which the seeds are evenly dispersed, is placental tissue, while in the other melons it is derived from the ovary wall. Melons have essentially no starch reserves before ripening so they cannot get sweeter after picking (although their texture softens as the pectic substances in the cell walls get more soluble). For this reason, they should be left on the vine as long as possible. A good sign of ripeness is a clean break between melon and stem, rather than a cut in the stem itself.

Miscellaneous Fruits

Banana

The banana, native to India and Malaya, is a distant relative of ginger, turmeric, and cardamom, and has been eaten by man for several millennia. Its domestication, and the development of its seedless trait (the plant is propagated by way of shoots from an underground stem, or *corm*) remain obscure. Virtually all we know is that the banana arrived in Africa around A.D. 500, in Polynesia around 1000, and in the Canary Islands and the Americas in the 15th and 16th centuries when Europeans knew it as the "Indian fig" (botanically the fruit is a berry). It became a significant export item only late in the 19th century with the development of refrigerated shipping. Today, the banana remains an important staple food, with only about 20% of the world crop exported from the primary producing areas of Africa, Asia, and Central and South America, and the majority of that coming only from the Americas. Its importance stems from its unusually high carbohydrate content, which before ripening is almost entirely starch, and afterward, in the dessert varieties, almost entirely sugar—up to 20% by weight. About half of the world crop is eaten as we eat bananas, sweet and raw, while the other half is cooked as a starchy vegetable (both immature dessert varieties and other, nonsweetening varieties can be used in this way as "plantains"). The banana is notable for its fairly high potassium content. One biochemical oddity of the banana is that it synthesizes and stores in the peel the neurologically active compounds serotonin, dopamine, and norepinephrine. This fact encouraged some people in the 1960s to smoke dried banana peel in hopes of getting high, an idea that sounded like a joke on the face of it; it never seemed to catch on.

Blueberry, Cranberry, Currant

Blueberries and cranberries, two species of the genus *Vaccinium*, are found in the wild in northern Europe and North America, as are currants. The cultivated blueberry, a native of the American east, north, and northwest, has been purposely bred only since about 1910, when a U.S. Department of Agriculture scientist and a New Jersey cranberry grower developed 15 improved varieties. The commercial cranberry too is an American native, and has been introduced to Scandinavia and Great Britain. Half the annual cranberry crop in the U.S. comes from Massachusetts, where cultivation began in 1840. The species of currant grown commercially comes from northern Europe. It was cultivated before 1600 in the Low Countries, and brought to North America in colonial times. Black currants, which the French make into the wonderful liqueur crème de cassis, cannot legally be grown in many states because the plant is an alternative host of a serious tree disease, the blister rust of the white pine. Currants are notably rich in vitamin C.

Date

The date, the fruit of a kind of palm tree, is native to India. It was cultivated very early in the Middle East, and is now grown wherever desertlike conditions—high temperatures and low humidity—prevail. In many regions it is an important source of carbohydrates; up to 70% of its dry weight may be sugar. Although dates have large pits, they are not stone fruits: the bulk of the pit is not lignin, but hemicelluloses that act as a food reserve for the embryo. Dates are eaten both fresh and hard-dried in producing areas; most of us see soft, partly dried dates.

Fig

The fig is a native of Asia Minor that was imported into the Mediterranean area and used by the Egyptians 6000 years ago. In Greece and Rome it was an important part of the common man's diet, and in some places still is; like the date, it is valued for its sugar content. The Greek fig trade has given us our word *sycophant,* which literally means "one who shows the fig." It is surmised that the original sycophant was someone who tried to ingratiate himself with the authorities by informing on fig smugglers. In Mediterranean countries, giving or showing the fig is an obscene gesture involving the thumb. Figs were introduced to North America around 1600, and were first planted in California by the Franciscans around 1770, but commercial cultivation was not begun in this country until 1900.

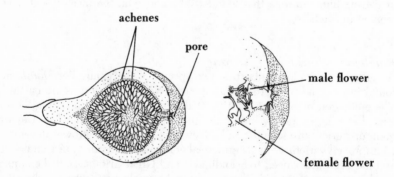

The wild fig, which contains both male and female flowers within the "fruit," actually the swollen flower base. Anatomically, the fig is an inverted version of the strawberry: it surrounds rather than supports the true fruits, or achenes. Fig wasps enter and exit via the small pore at the end.

The fig, like the strawberry, is a peculiar fruit, with its fleshy flower receptacle in this case surrounding, rather than supporting, the seedlike achenes that are the actual fruits. The common fig develops parthenocarpically, that is, without fertilization of the flowers, which develop *inside* the receptacle, and the achenes contain no embryos. The Smyrna (or Kalimyrna) fig must be pollinated in order to develop fruit, but this particular variety puts out only female flowers, and depends on the wild caprifig tree to pollinate it. Figs are not wind-pollinated, or even bee-pollinated—the anatomy of the receptacle discourages both—but *wasp*-pollinated, and only a certain kind of wasp, about ⅛ inch long, can do the trick. The grower of Smyrna figs must tie caprifigs containing wasp eggs to the branches of his Smyrna trees. The eggs hatch, the wasps develop, they exit the caprifig and pick up some pollen, and then visit Smyrna figs, pollinating them in the process. Luckily, they do not lay their eggs in Smyrna figs, which can be distinguished from common figs by their hard "seeds." This sophisticated technique, an artificial yet natural means of insemination, has been known since the Greece of Plato and Aristotle. Artistotle's student Theophrastus wrote in his *Inquiry Into Plants* (Book 2) that "caprification" was a remedy for trees that shed their fruit prematurely, though his understanding of the process was faulty: "Gall insects come out of the wild figs which are hanging there, eat the tops of the cultivated figs, and so make them swell."

Grape

Botanically, the grape is a berry, but common usage and history have set it apart from the other, relatively minor "berries" described above. A native of Asia Minor, the grape has been cultivated for 6000 years, perhaps first by the Egyptians. The Greeks and Romans developed new varieties, and the Romans introduced the vine to colder regions of Europe, including France. Today, there are about 60 species grown on 25 million acres worldwide, making the grape one of the most heavily cultivated fruit crops. Of course, only a small fraction of the annual production is used for table grapes. The great attraction of the grape is that its juice spontaneously ferments to a smooth liquor that keeps very well because of its high alcohol content and low pH (about 3). (Winemaking is discussed in Chapter 9). Nearly all the wine made in the world comes from varieties of a single European species, *Vitis vinifera,* as do such table grapes as Tokay, Ribier, Emperor, muscat, and Thompson seedless. American grape jelly, juice, and most northeastern wines are made with the distinctively flavored Concord grape, a variety of the American species V. *labrusca.* Sun-dried grapes, or raisins, have been prepared and eaten for many thousands of years. California is now the world's largest producer of raisins, which are made mostly from the Thompson seedless variety.

Kiwi

The kiwi fruit is a newcomer to American tables and to the world in general. It is the berry of a woody vine, *Actinidia chinensis*, native to the Yangtze valley of China. It seems never to have been cultivated until about 1900, when a British botanist collected a few plants and brought them to the West. It was first grown in New Zealand in 1906, and soon after became a commercial crop. The French nouvelle cuisine adopted the kiwi fruit as much for its striking appearance when sliced—a ring of small black seeds set in lime-green flesh—as for its delicate flavor, and today plantings are being established in California to meet the growing interest. Outwardly homely on account of its hairy, dull brown coat, the kiwi fruit is rich in vitamin C and has the virtue of keeping well for several months in cold storage.

Pineapple

A native of South America, the pineapple was introduced into Europe before 1535, where for a while it was a great sensation. Its form appeared in architectural ornament and coats of arms, and it was a popular fruit to grow under glass. It was brought to Hawaii in the 18th century and eventually became the state's major fruit crop; today the world's leading producers of pineapple are Hawaii, China, and Southeast Asia.

Botanically, the pineapple is a compound or collective fruit consisting of many berrylike structures fused together and surrounding a stem. Like melons, the pineapple contains no starch reserves and so will not sweeten further once it is picked; consequently the fruit must be harvested after the onset of ripening. Ripening is marked by the loss of some chlorophyll in the shell, but the growers say that the best sign of a ripe fruit is that it makes a solid sound when thumped, and has a small, compact leaf crown relative to the size of the fruit itself. Like figs, pineapples contain a protease, an enzyme that breaks down proteins, and this must be destroyed by boiling if the fruit is to be used in a gelatin preparation. Otherwise, the protease will simply digest the gelatin away, leaving a soupy, unappealing dish.

VEGETABLES

Fungi

Mushrooms, Morels, Truffles

Mushrooms, morels, and truffles are among the most primitive of our foodstuffs. They are related to the molds and yeasts, and are *saprophytic:* that is, they are unable to photosynthesize sugars, and must live on the decaying remains of other organisms. Many edible fungi are found only in symbiosis with the roots of trees; the fungus extracts sugars from the roots, and, in exchange, gives them soil minerals, especially phosphorus, which it can gather more efficiently than the tree can alone. Such fungi cannot simply be planted like other vegetables. The part we eat is only one stage in the organism's life, the fruiting body put forth by the fine, cottony underground mycelium for the purpose of making and disseminating spores. Fungi also differ from the higher plants in the chemical composition of their cell walls, which are made not of cellulose but of *chitin*, the carbohydrate-amine complex that makes up the outer skeleton of insects.

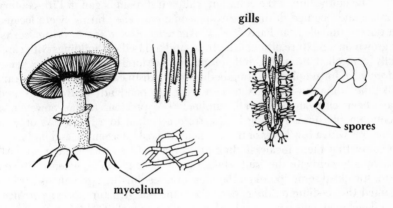

The anatomy of a mushroom

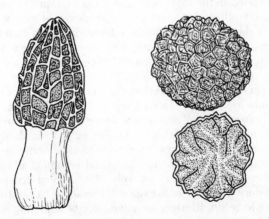

The morel, shown about one-third actual size, and the truffle in ordinary and cross-sectional views, about three-fourths actual size

Many different species of fungi have been and are eaten all over the world. Most of them are gathered from the wild, and only a few have been successfully cultivated. Traces of puffballs have been found in Stone Age settlements, and the Greeks and Romans considered mushrooms to be delicacies, although some of the earliest written references are concerned with poisonings rather than pleasure. A few common species produce protective toxins, and wild mushrooms should not be eaten until they have been positively identified as edible. In the Middle Ages, one species was commonly ground up with water and sugar in order to attract and kill flies.

The cultivation of the common white mushroom began in 17th-century France and boomed during the Napoleonic era. The "farms" were located in quarry tunnels near Paris. Today, *Agaricus bispora* (or *A. brunnescens*) is grown on a mixture of manure, straw, and soil in dark buildings with carefully controlled humidity and temperature. Morels, species of the genus *Morchella*, form hollow fruiting bodies with a distinctive honeycombed cap. Like the white mushroom, they can grow independently of tree roots, and have been cultivated in small numbers. The precious truffle, however, a dense, knobby mass ranging in size from a walnut to a fist, grows only in symbiosis with a few kinds of trees—including oaks, hazelnuts, and lindens—so cultivating them means finding an available forest, or planting one. All attempts to cultivate the truffle failed until the mid-1970s, when a French team sprinkled some spores on the roots of a seedling in a greenhouse, transplanted the seedling outdoors once the symbiotic structure, or *mycorrhiza*, had developed, and harvested truffles in the spring of 1978.

The several varieties of truffle (species of *Tuber*) never break above the ground, and must be smelled out by pigs, trained dogs, or goats. German scientists have recently discovered that truffles produce a musky chemical that is also secreted in the male pig's saliva and prompts mating behavior in the sow. The investigators suggest that "the biological role of this boar sex pheromone might explain the efficient interest of pigs in search of this delicacy." Human interest in truffles may also owe something to this hormone: men secrete it in their underarm sweat. Périgord in France is known for black truffles, northern Italy for white. These once common fungi are now so expensive (French production is less than a tenth of what it was a century ago, due mainly to the loss of forest lands and overharvesting) that they are used primarily in very small amounts to flavor sauces and pâtés.

The rich, almost meaty flavor of the fungi and their ability to intensify the flavor of vegetable dishes are largely due to an abnormally high content of glutamic acid, which makes them a natural version of monosodium glutamate. Mushrooms respire very actively after harvest compared to most produce, and during four days' storage will lose about half of their sugar and starch reserves to chitin. Refrigeration and airtight wrapping will slow them down, but these conditions also cause moisture to condense on their surface and so may speed spoilage. They should be used as quickly as possible after purchase.

Root Vegetables

Beets

Beets have been eaten by man since prehistory, and are native to a wide swath of Eurasia from Britain to India. In the 18th century, a white variety of beet began to be cultivated for sugar production. Up to 8% of its weight is sugar, an exceptional figure for a vegetable. The beet consists mostly of a swollen hypocotyl, or lower stem, although it is partly a root. The ability to metabolize the bright red pigment, betacyanin, is controlled by a single genetic locus; those people who have inherited two recessive genes pass the pigment in their urine.

The Cabbage Family: Radish, Turnip

The radish is a native of the eastern Mediterranean, and was cultivated by the ancients. Like the beet, it is a swollen hypocotyl, and only partly a root. Most of the enzyme responsible for its hot taste—it reacts with another substance to form a mustard oil—is located in the skin, and peeling will therefore moderate its effects. The turnip has been under cultivation for about 4000 years in Eurasia as a staple, starchy food. It consists of both hypocotyl and taproot, and is now used mainly by the poor and for animal fodder.

The Carrot Family: Carrot, Parsnip

The carrot is native to Afghanistan, and was known to the Greeks and Romans, although it was not widely used in Europe until the Middle Ages. Early varieties were red, purple, or black with anthocyanin pigments. A pale, yellow anthocyanin-less strain arose in the 16th century and became very popular, perhaps because it would not color sauces and soups. It was in 17th-century Holland that the familiar orange type, rich in carotene (the precursor of vitamin A), was developed. The carrot was brought by the colonists to the New World, where it escaped from cultivation to become the wildflower Queen Anne's lace. It has always been less popular here than in Europe, and has been commonly used in this country only since World War I. Anatomically, the carrot is the swollen base of the taproot.

The parsnip is native to Eurasia and was known to the Greeks and Romans; the variety known to us today was developed in the Middle Ages. It too is a taproot, and like the turnip was a very important staple food before the introduction of the potato.

Tuber Vegetables

Potato

The potato, a relative of tobacco and the tomato, is indigenous to Central and South America, from the southern United States to the tip of Chile.

It was cultivated more than 4000 years ago in mountainous areas, up to 15,000 feet, where corn cannot grow, and was a staple food of the Incas. The name comes from a Caribbean Indian word for the *sweet* potato, *batata*. Spanish explorers brought the plant to Europe around 1570, and England and Ireland had it by about 1610 and immediately accepted it as an important food crop. Because it was hardy and easy to grow, the potato was inexpensive and the poor were its principal consumers. Until about 1780, it was rigorously excluded from prudent French tables, where it was thought to cause leprosy, but a pharmacist named A. A. Parmentier, who had eaten it while a Prussian prisoner during the Seven Years' War, campaigned hard for it, and achieved the stunning coup of having Louis XVI wear a potato flower and serve the tuber at court. The courtiers immediately made it the new fashion, and the provinces soon imitated them.

The potato came to the United States as a food crop indirectly, via Ireland, in 1719; the first large area of cultivation was near Londonderry, New Hampshire. It quickly became established on all continents, in temperate, subtropical, and tropical climates, and is now the most important vegetable in the world, excluding only the tropical lowlands, where manioc, another large tuber (tapioca is made from its starch) is more easily grown. The hazards of dependence on the potato were made clear in the Irish potato famine of 1845–49, when a devastating blight struck, leaving one million dead and leading to the emigration of another million and a quarter. The spectre of this disaster was strangely invoked in 1939, when Germany accused Britain of airlifting Colorado potato beetles onto her potato fields in an attempt to subvert the Reich. In 1950 the U.S.S.R. claimed that the U.S. was doing the same thing in East Germany.

The potato is the swollen tip of an underground stem, and stores energy in the form of starch to support new stems that arise from the "eyes." The high starch content and not insignificant protein—upward of 3%—together with enough vitamin C to prevent scurvy if eaten in large quantities, make it an ideal food crop. The potato yields more energy per acre than do cereals, and it is a much more convenient crop to grow and harvest. Greened potatoes and sprouts harbor toxic alkaloids and should be deeply peeled or discarded (see page 159). Although the peel contains a disproportionate amount of nutrients considering its share of the weight, the rest of the tuber still accounts for most of the vitamins.

There are two characteristics of the potato that are familiar to nearly every cook and that food scientists have investigated with varying success. One is the rough division of potatoes into dry, "mealy" types that, because their cells tend to separate from each other when cooked, are best for baking and mashing, and moist, "waxy" types that, because their tissue is more cohesive, are best for cutting into scallops and chunks for potato salad. Exactly what accounts for this difference in behavior is not known, though mealy potatoes generally contain more starch. If you are unsure which kind you have and would like to know before cooking, you can tell by putting one in a brine made of one part salt to 11 parts water. Waxy potatoes will float, while mealy potatoes are denser and will sink.

The other characteristic is the tendency of some potatoes to develop a large, dark-colored region during cooking. "Stem-end blackening," as it is called (it occurs at the end that had been connected to the plant) is caused by the reaction of ferric (Fe^{+3}) iron ions (formed from ferrous (Fe^{+2}) ions during cooking) with phenolic substances in the tuber. Exactly why some potatoes are more susceptible than others is again not known; it is thought that soil and climate may be important factors. The cook can minimize stem-end blackening in boiled potatoes by making the pH of the water distinctly acidic. If you are having this problem, add ½ teaspoon cream of tartar for every pint of cooking water after the potatoes are half done.

Sweet Potato

The sweet potato is the true root of a member of the morning glory family and despite a resemblance to the yam is entirely unrelated to it. It is native to Central America, but may have spread to Polynesia before the arrival of the white man. Columbus brought the sweet potato to Europe on his first voyage to the New World, and by the end of the 15th century it was established in China and the Philippines. It is now cultivated in most subtropical areas of the world. It is remarkable for its 3 to 6% sugar content, which is increased by storage at warm temperatures and during the early stages of the cooking process (enzymes break down starch into its glucose units). The sweet potato provides more calories, minerals, and vitamin A (but less protein) than does the white potato.

Water Chestnuts

The water chestnut is a corm (a swollen underwater stem tip) of a plant in the sedge family, which is related to the grasses. It is a native of the Far

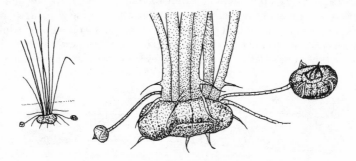

The water chestnut is a corm, an organ that stores food and gives rise to new stems.

East, and is cultivated primarily in China and Japan. The corm differs from the tuber in bearing scalelike leaves. There is also an unrelated plant whose seeds are known as water chestnuts in Central Europe and Asia.

Yam

The yam is the tuber of a plant that is related to the grasses and lilies. Among the earliest of the flowering plants, it had spread across the continents before they separated, and afterward evolved separately in tropical Asia, Africa, and America. It may have been cultivated as early as 8000 B.C. in Asia. The yam is valued for its high starch content, but some Old World varieties do contain a poisonous alkaloid, and must be peeled and then boiled or roasted to be safe. What are called yams in the United States are actually a variety of sweet potato.

Leaf and Stem Vegetables

The Lily Family: Asparagus, Leek, Onion

Plants in the lily family are relatives of the grasses, and have structures more like them than like other vegetable plants. Asparagus is native to temperate Eurasia and is unique in having no leaves, properly speaking, but rather *phylloclades*, or delicate photosynthetic branches. It was known as a delicacy in Greek and Roman times, and later, high demand for it led to the use of many substitutes, including young blackberry shoots and leeks, which were called "poor man's asparagus." It remains expensive today because the shoots grow at different rates, and must be harvested by hand. An odd side effect of asparagus has been known for centuries. As the learned Frenchman Dr. Louis Lemery put it in his *Treatise of All Sorts of Foods* (1702, in a contemporary English translation), "*Sparagrass* eaten to Excess sharpen the Humours and heat a little; and therefore Persons of a bilious Constitution ought to use them moderately: They cause a filthy and disagreeable Smell in the Urine, as every Body knows." From 1956 until 1980, it was thought that the excretion of odorous methyl mercaptan after eating asparagus was a dominant genetic trait; if you had the particular gene, you were a "stinker." But a recent study found that *all* asparagus eaters excrete methyl mercaptan; it is the ability to *detect* its odor that varies from person to person. The sul-

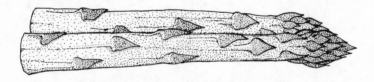

Asparagus and its peculiar branches, the phylloclades, which are clustered near the tip of the immature stem

fur-containing amino acid methionine is suspected as the precursor in asparagus.

Leeks were cultivated in prehistory in their native Mediterranean region, and were distributed across Europe by the Romans. The main body, like that of the scallion, consists of fleshy leaf bases. The leek has long enjoyed a lively extraculinary status. For better than 1000 years it has been the popular national plant of the Welsh. The legend goes that during a victory by the last Briton King, Cadwallader, over the Saxons in 640, the Welsh wore leeks in their hats to distinguish themselves from the enemy. Chaucer's Reeve refers memorably to the leek in describing those who are old in years and young in desire:

For in our wyl ther stiketh evere a nayl,	(will)
To have an hoor heed and a grene tayl,	(hoar head)
As hath a leek; for thogh oure myght be goon	(gone)
Our wyl desireth folie evere in oon.	(on and on)

Is the onion stem or leaf? Actually it is both. The concentric shells of onion tissue are the swollen bases of the previous year's leaves, and contain food reserves for the following year's growth, when the plant will flower. Ibsen's Peer Gynt and many others since have used the onion as an emblem of superficiality and emptiness; peel away the layers, they say, and you are left with nothing. A nice conceit, but not quite accurate: peel the leaf bases away and you are left with two stem buds, from which the second year's growth arises. At the center of the onion, then, is the beginning of a new life.

The onion is native to a broad region stretching from Israel to India, and has been cultivated since at least 3000 B.C. It and its flavorful relatives were especially missed by the children of Israel after they left Egypt, according to Numbers 11: "We remember the fish, which we did eat in Egypt freely, the cucumbers, and the melons, and the leeks, and the onions, and the garlic." The genesis of the onion's odor is discussed on page 156.

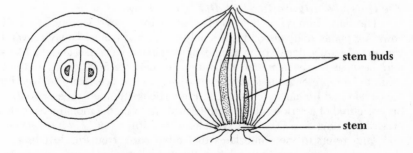

The onion in cross- and long-section, showing the stem buds which the surrounding leaf bases provide with food.

The Cabbage Family: Cabbage, Brussels Sprouts

Head cabbage, brussels sprouts, cauliflower, kohlrabi, kale, and broccoli are all varieties of one remarkable plant species, *Brassica oleracea*. The original wild cabbage is native to the Mediterranean seaboard, and this salty, sunny habitat accounts for the thick succulent leaves and stalks that make these plants so hardy. In such conditions it is very difficult to retain water, and the cabbage developed the waxy cuticle and thick, water-storing leaves characteristic of desert plants. The cabbage has been cultivated for about 2500 years, and was introduced to Britain in its domesticated form by the Romans, who believed it to be a prophylactic against the discomforts of high living. Said Cato,

> The cabbage surpasses all other vegetables. If, at a banquet, you wish to dine a lot and enjoy your dinner, then eat as much cabbage as you wish, seasoned with vinegar, before dinner, and likewise after dinner eat some half-dozen leaves. It will make you feel as if you had not eaten, and you can drink as much as you like.

The cabbage's ability to thrive in cold climates made it very popular in Eastern Europe. Sauerkraut, however, or the principle of pickled cabbage, was brought to Europe by the Tartars from China.

Anatomically, the cabbage is a large, main terminal bud. Brussels sprouts are small, numerous lateral buds along the main stem of the plant. This variant was apparently developed in northern Europe in about the 5th century, although the first clear record of its existence is from 1587. Brussels sprouts became a popular vegetable in Europe only after World War I. In recent years, Oriental species of the cabbage family, including the oblong-headed Chinese cabbage and nonheading leaves and stalks of bok choy, have become increasingly available in American markets.

The Daisy Family: Lettuce, Endive, Dandelion

The daisy family, or the Compositae, is the second largest family of flowering plants, and yet contributes only a few food plants. Today's several varieties of lettuce derive from an ancestor strain native to Asia and the Mediterranean; it has been under cultivation for close to 5000 years. Lettuce seems to be represented in some Egyptian art, and was certainly enjoyed by the Greeks and Romans. According to Pliny (Book 19), the Emperor Augustus was cured of a serious illness with lettuce, and raised a monument to the prescribing physician. The first syllable of its Latin name, *Lactuca*, means milk, and refers to the white latex that often oozes from the leaf base, a characteristic it shares with its relative, the dandelion. One variety of lettuce, *Lactuca virosa*, is grown for "lettuce opium": the latex contains triterpenoid alcohols that act as soporifics. Many herbalists claimed that even ordinary lettuce calmed the nerves and induced sleep.

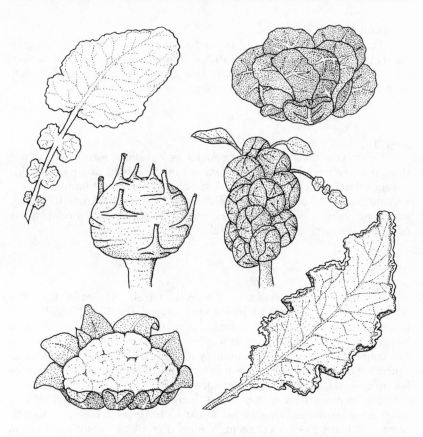

Some of the members of the remarkably various cabbage family. *Clockwise from upper left:* leaf of the wild cabbage, the terminal bud of head cabbage, the lateral buds of brussels sprouts, kale, the inflorescence of cauliflower, and the swollen stem of kohlrabi.

Endive and its relative chicory are native to India but were known to the Egyptians and the Greeks and Romans. The broad-leafed variety of endive is known as escarole. These plants are more bitter than most kinds of lettuce. The dandelion seems to be found naturally on all continents, although most varieties are native to Eurasia. It has occasionally been cultivated on a small scale, and has been used as a novelty and in emergencies as a green vegetable, probably since prehistory. The leaves can be eaten raw in salads or cooked, like spinach; the inflorescence is used to color and flavor dandelion wine.

Bamboo Shoots

These are the stem sprouts of bamboo, various members of the grass family that are native to tropical East Asia. Most bamboo shoots contain toxic concentrations of cyanogens, which cause cyanide poisoning, and so they must be cooked before canning or eating.

Celery

Celery consists of a bunch of petioles, or leaf stalks, rather than a main stem. It is first recorded as a food plant in France in 1623, and was probably developed either there or in Italy. A member of the carrot family, it has a very distinctive flavor and is especially valued for its crisp texture. It has little nutritional value. *Celeriac*, the starch-storing lower stem of a special variety of celery, is served as a boiled vegetable.

Rhubarb

The name rhubarb comes to us from the Greeks, who called this plant the vegetable of "barbarians [foreigners] beyond the Rha [Volga]." It is native to the southeast part of Russia, and is known to have been cultivated in Italy since the 17th century. It was introduced to America just after the Revolutionary War. Though technically it consists, like celery, of leaf stalks, rhubarb is quite acidic for a vegetable and is usually employed as a substitute for fruit in early spring pies and preserves.

The rhubarb has had a somewhat shady reputation since World War I, when Americans were encouraged to eat its leaves as a vegetable supplement, and many cases of poisoning resulted. For a long time it was thought that oxalic acid was the culprit, until it was realized that the stalks, which are safe, also contain significant amounts of this acid; so does spinach. The leaf toxin has not yet been identified.

Spinach

Spinach is indigenous to southwestern Asia, and is now cultivated all over the world. It was a relative latecomer to the European scene, not arriving until the late Middle Ages, but quickly became so popular that beet and turnip greens were sometimes used as substitutes for it. In the classic cuisine of France, spinach was likened to *cire-vierge*, or virgin beeswax, capable of receiving any impression or effect, while most other vegetables imposed their taste upon the dish. It was said to take a great chef to do justice to such a delicate vegetable. Despite Popeye, spinach is not remarkably rich in iron, though it is a good source of vitamin A.

Flower Vegetables

The Cabbage Family: Broccoli, Cauliflower
The cauliflower was known in Europe in the 16th century and introduced to England early in the 17th, probably from Cyprus. The solid head is a degenerate, sterile flowering structure. Growers obtain the uniformly white color by carefully covering the inflorescence with the outer leaves, blocking sunlight and preventing the formation of chlorophyll. Broccoli probably developed in Italy and followed cauliflower into culinary usage about a century later. Again, the edible part is the entire flowering structure, including the stalks and the many separate flower buds. It is unclear which of these prolifically floral versions of the cabbage gave rise to the other. Consumption of broccoli in the United States has been significant only since World War II. Broccoli leaves contain much more vitamin A than the buds and are worth eating.

The Daisy Family: Artichoke
The artichoke, a kind of thistle native to the Mediterranean region, was probably known to the Greeks, and was a delicacy in Rome, a fact of which Pliny professed to be ashamed: "thus we turn into a corrupt feast the earth's monstrosities, those which even the animals instinctively avoid" (Book 19). The name is a corruption, via Italian, of the Arabic *al'qarshuf;* for some reason the Latin word *cynara* did not survive, even in the Romance languages. The edible parts of the artichoke are the fleshy bases of the bracts, or protective leaves, and the heart, which is actually the base of the inflo-

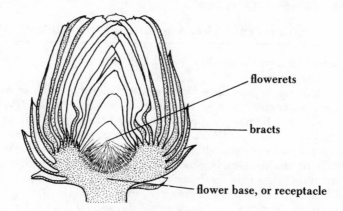

flowerets

bracts

flower base, or receptacle

The artichoke. Anatomically, the heart corresponds to the false fruit of strawberry and fig.

rescence. The "choke" is made up of the actual flowerets, which if allowed to bloom, turn a deep violet-blue. The still relatively minor consumption of artichokes in America began early in this century, when a group of Italian immigrants settled around Half Moon Bay on the California coast and planted a few hundred acres. In 1906 they made the first shipment to the East Coast.

A subject of recent controversy among food writers is the possible effect artichokes have on the taste of other foods, particularly wine. James Beard wrote in *Gourmet Magazine* (1971) that "there is a great feeling among serious wine drinkers that artichokes spoil the flavor of fine wines and therefore should be forbidden at great dinners." The first scientific report of this phenomenon dates from 1934, when 250 people at a dinner were given several taste tests, among them one involving artichokes with mayonnaise. After eating the artichoke course, 40% of the company said that water tasted normal, but 60% thought it seemed sweet. In 1954, an organic acid unique to artichokes was isolated and dubbed *cynarin*, and in 1972 another taste test was performed using cynarin and one other artichoke constituent. This time, only 6 out of 40 did not report a change in the taste of water. It may be that sensitivity to cynarin is genetically determined. The substance seems to stimulate the sweetness receptors in the taste buds of susceptible people, making everything taste sweet for a short period of time. When cynarin's effect was confirmed, there was talk about revolutionizing the prepared food industry by replacing sugar with cynarin. Magazine ads for Cynar, an Italian artichoke aperitif, have noted: "Yale University research has now confirmed Italian studies that artichokes make food that follows taste sweeter. This is why Cynar, made from artichokes, makes food tastier." Fortunately, there is more to tastiness than sweetness.

Botanical Fruits That Are Culinary Vegetables

The Squash Family: Cucumber, Pumpkin, Winter and Summer Squash

The cucumber (genus *Cucumis*) is a native of either India or Africa, and has been under cultivation for 4000 years. It was known to the Egyptians and the Greeks and Romans, and was brought to France in the 9th century and to England in the 14th. From the Middle Ages to the 18th century it was considered to be an unhealthful vegetable. Dr. Lemery remarks that they are "hard of Digestion, because they continue long in the Stomach," and so are appropriate only for "young Persons of an hot and bilious Constitution." Today, cucumbers are grown all over the world. True gherkin pickles are made from the fruit of a separate species.

The pumpkin and squashes (genus *Cucurbita*) are all of American origin, and many thrive in the arid climates of the southwestern United States and northern Mexico. Some of the 25 species have been cultivated for 9000 years, in particular the pumpkin and winter squashes, which may have been valued at first for their protein- and oil-rich seeds. Winter varieties—examples are acorn, butternut, and Hubbard squashes—are allowed to mature into hard, starchy fruits that keep well for months, while summer varieties— zucchini, crookneck—are eaten while soft and immature.

The Bean Family: Bean, Pea

There are primarily three species of beans eaten fresh either in their pods or out: the broad or faba bean; the French, kidney or common green bean; and the lima bean. The broad bean is indigenous to southwestern Asia and the Mediterranean, and was cultivated by the Egyptians, Greeks, and Romans. It was brought to the New World by explorers and is now popular in Mexico and Brazil as well as in Europe, the Middle East, and India. The common green bean originated in Central America, and was cultivated at least 7000 years ago in Mexico. The Spanish and Portuguese brought it to Europe in the 16th century and the Europeans, in turn, brought it to North America. Today it is consumed primarily in Europe and the Americas. The lima bean is native to tropical America and has been cultivated in Peru since 5000 or 6000 B.C. Its common name comes from Peru's capital city. Spanish explorers brought it home and then to the Philippines, Asia, Brazil, and Africa. It is now the main legume of Africa. Lima beans, especially those that are colored when dry, contain significant concentrations of cyanogens and are safest when boiled. More about dried beans and bean sprouts in the next chapter.

The pea is a native of western Asia. It was cultivated by the Greeks and Romans, and was an important dried food in the Middle Ages. Fresh peas were not eaten in Europe until the 16th century, when they became a great delicacy and luxury; both the wife of Louis XIV and the mistress of Louis XV recorded their infatuation with this innovation. Sugar peas and snow peas are varieties eaten very underripe for their pods.

The Nightshade Family: Eggplant, Pepper, Tomato

This remarkable family of plants also includes the potato, tobacco, the hallucinogenic jimson weed, and deadly nightshade. The eggplant (the name was first applied to a white-skinned variety) is a native of India, and was little known in the ancient Mediterranean. Arab traders introduced it into Spain and northern Africa in the Middle Ages, and its use was established in Italy in the 15th century, in France by the 18th. Today it is cultivated in almost all warm regions of the world, although India and the Far East remain the largest consumers. Any one who has fried eggplant knows that it seems to drink up several times its weight in oil. This is due primarily to the very spongy texture of its tissue: a high proportion of its volume consists of intercellular air pockets. The point is reached, however, when the heat of the pan and oil begins to collapse the structure, and then, like a squeezed sponge, it gives up much of the oil.

The sweet or bell pepper is a mild variety of the mostly pungent species of *Capsicum*. It is native to tropical America, and was domesticated long before the arrival of European explorers. It has been cultivated in Europe from the 16th century, and has now spread all over the world. All peppers will turn color—to shades of yellow or brown, purple or red—when ripe. Green peppers are simply picked before they are ripe, and are generally more common and cheaper in the market because they are more easily transported and keep better than ripe ones.

The tomato, which is indigenous to the Andes, was domesticated in Mexico (the name derives from the Nahuatl language) and brought to Europe in 1523 after the Conquest. By the end of the 16th century it was established in England, where the fruits were known as "love apples," but only as an ornamental plant. For better than two centuries it was a common foodstuff only in Italy. Perhaps on account of its known poisonous relatives, the idea persisted that the fruit of the tomato plant was poisonous (the foliage *is*), and the Latin binomial assigned it in the 18th century retains some of this doubt: officially, it is *Lycopersicon esculentum*, "edible wolf's peach." Large-scale consumption in Europe came only in the middle of the 19th century, and in America there seem to have been several waves of opinion about the tomato's edibility right up to the turn of the century. Today, we eat them in quantities second only to the potato. An odd anatomical fact: the highest concentration of vitamin C in the fruit is found in the jellylike material surrounding the seeds, which is made up of storage parenchyma cells from the placenta.

Avocado

The avocado has been cultivated in Central America for perhaps 7000 years. It was introduced to Jamaica in 1650, and to the Old World tropics and Florida and California in the 19th century. There are three general races of avocado, with hybrids among them. The two principal California-grown varieties are a Guatemalan-Mexican hybrid that stays green when ripe and a Guatemalan type that turns black at maturity.

The avocado is an anomaly among fruits in several ways. Cell division continues through the life of the fruit. Its fat content, at 20%, is about 20 times the average for other fruits: in the tropics it is known as "poor man's butter." Avocados are poor in vitamins A and C but rich in the B vitamins and minerals. Their sugar content *decreases* rapidly during ripening, and in Brazil, where the avocado is treated as a fruit, sugar is mashed into it for dessert. And the fruit will not ripen on the tree as long as the skin is unbroken; it must be cut from the tree for ripening to begin. Thus the best way to store avocados is to leave them on the tree, and this is commonly done for up to seven months. It seems that the leaves supply a hormone to the fruit that prevents ripening; harvesting the fruit cuts off the supply of this inhibitory substance and initiates the production of ethylene. The avocado is also unique among fruits in its inability to metabolize anaerobically. If deprived of oxygen (in a tightly twisted plastic bag for example), the ripening process is halted, and when oxygen is restored, the fruit will spoil. The avocado is highly susceptible to chilling injury, and if stored at refrigerator temperatures for very long it will become discolored and develop off-flavors. Avocado flesh turns brown very quickly from enzymic action. This process can be slowed down, although not stopped, by the addition of acids such as lemon juice or vinegar.

Corn

The corn we eat as a vegetable is a sweet variety of the grain that is picked "green," or before the kernels begin to dry out. The history and composition of this important grain are discussed in the following chapter.

Okra

Okra is native to tropical Africa or Asia and was cultivated by the Egyptians in the 12th century A.D., but there is no mention of it in ancient Western sources. It came to the southern United States with the slave trade in the 18th century and has been cultivated there ever since. A relative of the hibiscus, okra bears capsules that are used unripe as a vegetable. It is also valued for a mucilaginous substance, mostly long carbohydrate molecules, that helps thicken soups and sauces.

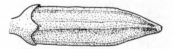

The okra capsule

Olive

The olive is native to the eastern Mediterranean region, and has been cultivated since at least 3000 B.C. for cooking, lamp, and cosmetic oils—it is about 18% oil by weight—and as a food. It was brought to the New World in the 15th century by the Spanish. Today, 90% of the world's olive production is used to make oil, and some 98% of the acreage is in the Mediterranean region. Olive oils are graded informally according to the method of extraction and the content of free oleic acid, which indicates the extent to which fat molecules have been degraded into their fatty acid components (free fatty acids lower the smoke point of oils). "Virgin" oil comes from the first pressing of the fruit, and is unrefined. Subsequent pressings extract substances that impart a harsh flavor to the oil, and refining is usually necessary to remove them. Olive oil simply labelled "pure" is probably a mixture of first and second pressings. Virgin oil is allowed up to 4% free oleic acid;

"fine," "superfine," and "extra virgin" are allowed 3, 1.5, and 1% respectively.

Anyone who has bitten into a raw olive knows that olives must somehow be processed before they are edible. Olives are usually pickled, and they contain a bitter glucoside called oleuropein (from the olive's botanical name, *Olea europea*) which is usually removed first. This has been done since Roman times by soaking the fruit in a lye solution and then washing it thoroughly. The watery, oleuropein-rich residue left after raw olives are pressed for oil—what the Romans called *amurca*—was used, so Cato and Varro tell us, as a weed killer, insecticide, and a lubricant for leather and axles. Today's Greek olives are as strong-tasting as they are because they have not been treated with lye to remove the oleuropein. They are either simply cured by packing in dry salt, or are pickled in brine, where they undergo a lactic fermentation. Green Spanish olives are picked before they are ripe, treated with lye, and then brined. California ripe olives are first dipped in a ferrous gluconate solution to fix the pigment, then treated with lye, and immediately packed in brine. Because they are not allowed to ferment for a few weeks, these olives have neither the pickled flavor nor the resistance to spoilage of the other kinds, and so must be sterilized in the can. The cooking makes some contribution to their characteristically mild flavor.

HERBS AND SPICES

Major Families

The Mint Family

With the single exception of sweet basil, which is indigenous to tropical Africa and Asia, all the herb plants in the mint family are natives of the Mediterranean region. The Labiatae are characterized by square stems—an everyday ornamental example is the coleus—and by hairlike oil glands on the surfaces of leaves and stems, which are the parts used as flavorings. The glands harbor the essential oils that make this family so valuable in the kitchen.

Basil was known and used by the Greeks and Romans, who had odd ideas about its properties. The Greek word *basilikon* means "royal," perhaps indicating that the herb was reserved for the king's use, but the Latin *basiliscus* refers to the basilisk, a fire-breathing dragon, perhaps in the belief that the plant was a charm against the beast. It is used to flavor the liqueur Chartreuse.

Marjoram is found in two forms: sweet marjoram, the more common and more delicate in flavor, and wild marjoram, which is usually sold as oregano. Both have been enjoyed since classical times.

Oregano is a word that comes from the Greek and means "mountain brightness," or "mountain joy." It was discovered in the fifties that this herb, now so commonly used in pizzas and other tomato-based foods, and yet little

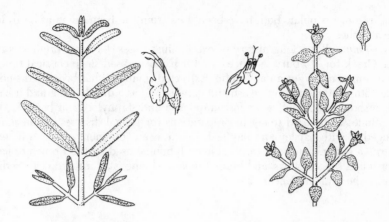

Rosemary and thyme, plants in the mint family. Family characteristics include square stems covered with hairlike oil glands, and flowers with a prominent lower lip.

known in the United States until the GIs returned from Italy, is actually two distinct plants from entirely different families. One is wild marjoram, and comes from several European countries; the other is a plant with similar but stronger aroma native to Mexico, from the Verbena family.

Peppermint is a more pungent offspring of *spearmint* that has been in use since biblical times. The main component of the mint's peculiarly refreshing active oil, menthol, is now one of the most widely used plant products: cigarettes, candy, liqueurs, toothpaste, mouthwash, cough drops, rubbing creams, and room deodorizers are only a few of its many applications today. Menthol is a remarkable substance. At low concentrations, it raises the threshold temperature at which the cold receptors in our skin begin to discharge: that is, it makes our warm mouth feel cool, and a cool drink feel cold. In larger doses it can be used as an anesthetic and even, like mustard, as an irritant.

Rosemary is an evergreen shrub that can grow up to 15 feet tall. The underside of its needlelike leaves is covered by a growth of oil hairs that make it look white. Its name comes from the Latin for "sea dew," and refers to the fact that it is often found growing along the sea coast. Its use in Rome was primarily medicinal; today it accompanies meats.

Sage was highly esteemed as a medicinal herb in Greece and Rome, and was used in England before the arrival of tea, to make infusions. Since the 19th century, it has been thought to be especially appropriate in pork dishes.

Summer and winter savory are two closely related species; the first is an annual, the second a woody perennial. Summer savory has a milder flavor and is more popular. Both have been used from the Romans on as herbs in meat dishes.

Thyme has a long history of extraculinary uses. Its name originates in the Greek for "to burn sacrifice," and it has been used since classical times as a fumigant and as an antiseptic. Extracted thyme oil has been used since the 16th century in mouthwash and gargles, and as a disinfectant, and it has been shown to be active against salmonella and staphylococcus bacteria. It continues to be used today in preparations for fungal diseases of the skin. Together with parsley and bay leaf, thyme is a component of the *bouquet garni*, a general purpose, removable herb bundle used by the French to flavor soups, sauces, stews, and braised meats. Thyme is an important ingredient in the liqueur Benedictine.

The Carrot Family

With some exceptions, this Mediterranean family, known botanically as the Umbelliferae, contributes its flavors not in its green growth but in its essential-oil-rich "seeds," which are not really seeds at all. What we call the seeds are actually complete fruits, even if they are dry.

Anise was used as a condiment by the Greeks, Romans, and Hebrews, and today is most often found in desserts and in liqueurs, including anisette, ouzo, and Pernod. Mexico is the largest producer. The botanically unrelated "star-anise" is the fruit of a tree in the Magnolia family; it contains a remarkably similar if harsher essential oil. Anethole, the major component of both oils, is also shared by anise's cousin fennel and the completely unrelated licorice.

Caraway seeds have been found in Swiss Lake dwellings of 8000 years ago. Today, it is one of the most popular spices in central Europe, used to flavor breads, cheeses, and liqueurs, as well as meat and vegetable dishes.

Chervil was used as an herb by the Romans; today, it is far more popular in France than in the United States. It is similar to parsley, though more

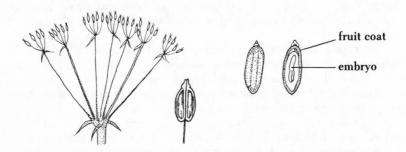

Generalized drawing of inflorescence in the carrot family. Anatomically, the "seeds" of dill, cumin, caraway, and so on are complete fruits.

delicate. Together with fresh parsley, tarragon, and chives, chervil is one of the *fines herbes*, a classic combination with which the French flavor soups and sauces.

Coriander has one of the longest recorded histories of the spices. There is evidence that it has been used since 5000 B.C., and it is prominently mentioned in Sanskrit records and Egyptian papyri. It became popular in 17th-century Paris, where it was the principal ingredient in Eau de Carnes, a concoction that could be used either as a liqueur or as a cologne. It has become a favorite spice in South America, and is called for, both as seed and as greens, in many Mexican dishes.

Cumin is mentioned several times in the Old and New Testaments, and references to it are found very early in English records. Today, it is used in Europe primarily to flavor breads and cakes, in Central and South America for sauces, and in many Indian dishes. Recently, the world's largest producer has been Iran.

Dill has been a constant of the culinary life of the Mediterranean since the Greeks, and it was also cultivated early on in India and Asia. Both the feathery leaves and the fruits of this plant are used. It is especially popular in India and Scandinavia.

Fennel was cultivated by the Egyptians; the name refers to a kind of fragrant hay. It is commonly used in sweets, liqueurs, and breads. The leaf stalks of one variety are eaten as a vegetable.

Parsley was used by the Greeks for medicinal purposes, and was made into a crown for the winners of the Isthmian and Nemean games. Homer mentions it as part of Circe's pleasant lawn in *The Odyssey*. Our word for it derives from the Greek *petroselinum*, "rock-celery," which apparently referred to its natural habitat. The Romans used it as an herb and introduced it into Britain. Today it is often used merely as a visual garnish.

The Ginger Family

The *Zingiberaceae* include three popular spices, all native to southern Asia.

Cardamom seeds were an important element of Greek trade with the East as early as the 4th century B.C. and may have been the most popular spice in Rome. They are most often met with today in cuisines from tropical countries, some Scandinavian baked goods, and in the cardamom-flavored coffee that is very popular in the Arab countries. Cardamom seeds are the third most expensive spice, after saffron and vanilla, and are produced primarily in India and Sri Lanka.

Ginger derives its name from a Sanskrit word that seems to mean "horn-shaped," perhaps referring to the shape of the whole spice. It is technically a *rhizome*, an underground stem that looks like a thickened root but in fact can give rise to both new roots and shoots. The Chinese were using ginger in the 6th century B.C., and it was brought to the Mediterranean by Arab traders sometime before the first century A.D. Ginger seems to have

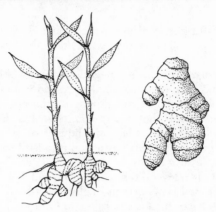

The underground stem that we call ginger root. Like the potato, a swollen
stem tip, this fleshy organ gives rise to new stems and roots.

been one of the more abundant spices in the cuisine of the Middle Ages;
practically every sauce recipe includes it. It was introduced to the West
Indies by Spain in the 16th century. Today we find it in two different areas:
Asian cuisines, and such sweet Western preparations as candies, desserts, and
drinks.

Turmeric, whose name is of obscure origins, is another underground
stem distinguished by its bright yellow color. In the Middle Ages it was called
"Indian Saffron" and was used as a substitute for that rarer spice, and as a
dye. It still colors Asian fabrics, and in the West is used as a dye in some
margarines and dairy products. Unlike ginger, which is often available fresh,
turmeric is usually sold in dried and powdered form. It is a major ingredient
of curry powders. These and plain turmeric should be stored in the dark
because turmeric is very sensitive to light.

The Onion Family: Chives, Garlic, Shallots

These "allies" of the onion, as they are often punningly called—the
onion is *Allium cepa*—have added a piquant touch to foods for thousands of
years, beginning in the Mediterranean region.

The *chive* is a mild, miniature version of the onion that forms compound bulbs, each surrounding a bud. Because the bulbs amount to so little,
only the green shoots are eaten.

Garlic carries a more potent flavor (its chemistry is discussed on page
156). The English word *garlic* comes from the Anglo-Saxon for "spear
plant," a good coinage for any of the onion family, which send up erect,
branchless shoots. Garlic bulbs are composed of individual bulblets, or *cloves,*
each with two leaves—a thin, dry coat and a thick, fleshy body—enclosing
the bud.

The *shallot,* unlike chives and garlic, which are species related to the
onion, is one variety of the onion species itself. It forms clusters of small

bulbs, and has a milder, subtler flavor than the other varieties. The name comes from the Greek *askalon*, which was the name of a trading town in South Palestine particularly associated with this variety.

The Cabbage Family: Mustard, Horseradish

Mustard is the seed of two members of the cabbage family native to Eurasia. It is used whole, ground, or mixed with various other ingredients into a paste (turmeric is a common addition in the United States). More mustard is consumed every year in this country than any other spice except pepper. The pungency of the seed, which develops only when it is broken and wetted, is caused by the reaction of enzymes with isothiocyanate compounds to form the so-called mustard oils. There is some evidence that mustard seed was chewed with meat in prehistory, perhaps to hide the flavor of spoilage. Both the Greeks and Romans used it as condiment and as medicine. According to Pliny (Book 20), mustard was put in smelling salts, chewed for toothache, and applied externally—like our mustard plaster—for various aches and pains (by irritating the skin, mustard oils and other "counterirritants" draw large amounts of blood to surface capillaries, and can thereby actually relieve inflammation in deeper tissue). Its name comes from the Latin *mustum*, or grape juice, with which the ground seeds were moistened into a paste.

Horseradish comes from the root of a plant in the cabbage family native to southeastern Europe, and is used mainly as a condiment because it too forms mustard oils when its tissue is broken.

Miscellaneous Herbs and Spices

Allspice, so named because it seems to combine the flavors of cinnamon, nutmeg, and cloves, is the dried, unripe berry of a tropical evergreen tree native to the West Indies and Central America. It is grown exclusively in the Western Hemisphere, and Jamaica is the largest producer. Spanish explorers discovered it in Mexico, and it was in use in London by 1601.

Bay leaves are taken from the bay laurel tree, an evergreen native of Asia Minor that grows to 50 and 60 feet. In classical times, laurel wreaths were bestowed as symbols of honor; hence our term *laureate*. According to Ovid, the laurel is thus distinguished because Daphne was metamorphosed into this tree rather than surrender herself to Phoebus, whom Cupid had wounded. Bay leaf is a common ingredient in Mediterranean cooking.

Capers are the pickled floral buds of a shrub commonly found in the Mediterranean region. Italy, Spain, and Algeria are the major producers. Capers are used to add a salty-sour note to sauces for fish and boiled meats.

Cinnamon is the dried inner bark of two related evergreen trees in the laurel family. One is true cinnamon, and the second, which is more strongly scented, is properly called *cassia*. The trees are native to Asia, and cinnamon was among the first commodities regularly traded from the East to the Mediterranean, at first along arduous land routes, and then by barge. There is evidence that cinnamon was used in Egypt around 3000 B.C., and it is commonly mentioned in the Old Testament. In Exodus, for example, God has Moses consecrate the temple with a mixture of myrrh, cinnamon, cassia, and olive oil. During the Middle Ages, it was second in popularity only to black pepper. Today, most of the "cinnamon" sold in the United States is actually cassia. The two are most easily distinguished by color: true cinnamon is tan, while cassia is a darker reddish brown.

Cloves are the dried floral buds of a tropical evergreen tree native to the Molucca Islands in the East Indies, so famous for their spices that they were known as the Spice Islands. The name comes from the Latin for "nail" and refers to the shape of the bud. Chinese courtiers used the clove to sweeten their breath in the presence of the Emperor around 300 B.C. and eugenol, the main constituent of clove oil, is still found in some brands of mouthwash. The spice was used by the Egyptians in the 2nd century A.D. and was known throughout Europe by the 8th century. The monopoly on the clove trade held by the Portuguese, who discovered the Moluccas while searching for cloves, and then by the Dutch, was broken by the French in 1770, when a diplomat managed to smuggle some seedlings out and into French colonies. Today, half of the world output (Tanzania is the largest producer) is consumed in Indonesia, where tobacco is smoked with cloves. It has been known for centuries that eugenol has local anesthetic properties, and it is still used to alleviate toothaches. The oil is also used in perfumes and bath salts.

Fenugreek, or "Greek hay," is the seed of a leguminous plant native to southern Europe and Asia; it was used by the Egyptians. The largest producers today are India, Egypt, Lebanon, and Argentina. Because it is a seed and a legume, fenugreek is rich in protein and in some countries is an important food, not just a spice. It is the principal flavoring in imitation maple syrup.

Licorice is a word that comes from the Greek *glykyrrhiza*, meaning "sweet root," and refers both to a plant in the legume family and to its extract. This native of the Middle East was used in Egypt 4000 years ago for medicinal purposes; today, it is consumed mostly in the form of candy and tobacco, which it sweetens and moistens. Licorice extract is made by boiling the yellow roots of the plant in water, and then evaporating the excess liquid. The blackish mass left behind has two important components: the essential oil anethole, which licorice shares with anise and fennel and which contributes the flavor characteristic of these three spices, and glycerrhetic acid, which is sweet. Not as sweet, however, as its precursor in the raw root, glycerrhizin, which is 50 times sweeter than table sugar. Before the advent of cheap sugar, licorice root was often cut into strips and chewed by itself.

Mace is the aril, or covering layer, of the nutmeg, which is the seed of an evergreen tree native to the Moluccas, the central Spice Islands. Mace is separated from the nutmeg by hand and dried. Nutmeg and mace were brought to the Mediterranean by Arab traders in the 12th century, and later

Mace is the thin membrane that separates the shell of nutmeg from the surrounding fruit.

became lucrative monopolies, along with cloves, of the Portuguese and then the Dutch. The Dutch were particularly ruthless in their efforts to eliminate competition; they massacred the inhabitants of one island and destroyed three quarters of the clove and nutmeg trees in order to limit production. In a brief period when they controlled the Moluccas, the British introduced the nutmeg tree to Singapore and the West Indies.

Nutmeg is the shelled seed of an evergreen tree native to the Moluccas; see the entry for *mace*. Like mace, it is primarily used in sweet beverages and desserts, preeminently eggnog and doughnuts. A component of its oil, myristicin, is a hallucinogen, and nutmeg has been used despite very unpleasant side effects as a cheap high. Mace and the seeds of the carrot family also contain myristicin, but in much smaller quantities.

Black Pepper, like white pepper, is the dried fruit of a tropical vine native to India. Black pepper found its way to the Mediterranean early in history, and was used by the Greeks and Romans; today, the largest producers are still India and Indonesia. It continues to account for fully one quarter of the world commerce in spices. Black pepper is made by picking the fruits green, letting them "ferment" in the sun for a while in order to develop a more pungent flavor, and then drying them. The blackened skin appears to be caused by browning enzymes not in the fruit itself, but from a fungus, *Glomerella cingulata,* which is present even in healthy pepper berries. White pepper is made by allowing the fruits to turn red, soaking them in water and rubbing the skin and thin pulp off, and then drying the seeds. The

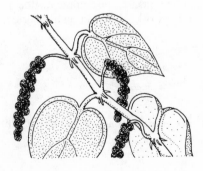

The tropical berries that are made into black and white pepper

milder white pepper is preferred in lightly colored sauces. The bite of black and white pepper is caused by an alkaloid, piperine.

Capsicum Peppers, also called chili or red peppers, are the dried fruit of certain New World relatives of the tomato (most are varieties of two species, *Capsicum annum* and the frequently smaller, hotter *C. frutescens*) which came back to Europe with the Spanish. Today, India and Mexico are among the largest producers. One variety has become well established in Hungary as paprika. Capsicum peppers can be a useful source of vitamins for heavy consumers; they contain from 6 to 9 times the amount of vitamin C found in tomatoes, by weight.

The pungency of these peppers derives from the alkaloid capsaicin. Recently, investigators have isolated five "capsaicinoid" components that have different effects on the mouth. Three give "rapid bite sensations" in the back of the palate and the throat, and the other two a long, low-intensity bite on the tongue and mid-palate. Variation in the proportions of these components may be responsible for the characteristic "burns" of different peppers. Capsaicin appears to accumulate in the fruit concurrently with the pigment during ripening, and is found primarily in the white placental tissues to which the seeds are attached. For every 100 parts of capsaicin in the placental tissue, there are 6 parts in the rest of the fruit tissue, and 4 in the seeds. If you are working with whole peppers, then, you can moderate their effect by removing the insides. Peppers must be handled with caution, however, because capsaicin is a general irritant and can "burn" skin that is already damaged by cuts or abrasions; it is a big mistake to rub your eye soon after handling hot peppers. Because of this property, capsaicin is used as the active component of antidog and antimugger sprays. Experimental work still underway indicates that capsaicin also has definite effects on the digestive system; it seems that the secretion of saliva and gastric juices, and peristaltic movements are increased by the ingestion of red peppers. And there is evidence that capsaicin has antibacterial effects as well.

Recently a psychologist, Paul Rozin, has suggested an interesting reason for the great popularity of the chili pepper, which has been adopted by many cuisines in the last 400 years and is now consumed in larger quantities by more people in the world than any other spice. After all, our bodies' natural response to capsaicin—pain, watery eyes, runny nose—is a defensive one meant to *prevent* us from eating such irritants, not encourage it. Rozin calls chili peppers an example of a "constrained risk." Like riding a roller coaster

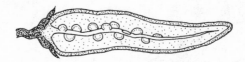

One variety of red pepper. Its "heat" is especially concentrated in the white tissue that bears the seeds.

or jumping into Lake Michigan in January, eating chemically "hot" food makes our body respond with warning signals. But because such situations are not genuinely dangerous, we can ignore the "meaning" of these sensations and savor the vertigo, shock, and burning for their own sake. It is also possible that the brain secretes endorphins, its own opiate substances, in response to a burning tongue, and that these contribute to the pleasurable "hangover" of a fiery meal.

Poppy seed comes from the same plant that gives us opium and the opium derivatives from its latex. The latex does not reach the seeds, however. They and the oil pressed from them have been used for thousands of years by the inhabitants of the plant's native region, Asia Minor. Today the tiny, kidney-shaped seeds, 900,000 of which make a pound, are used mostly in breads and sweets. Holland is now a major exporter.

Saffron consists of the stigmas—the flower parts that catch pollen for the ovary—of a certain crocus *(Crocus sativus)* native to the eastern Mediterranean. It is, at several hundred dollars a pound, the most expensive spice in the world; it is harvested by hand, and one ounce of saffron is made up of 13,000 stigmas, to which a single plant can contribute three. Saffron was

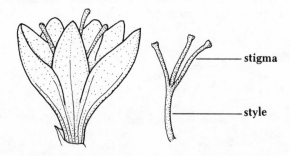

The saffron crocus. Pollen caught on the stigmas moves down the style to the ovary.

widely used many centuries before Christ as a spice, perfume, medicine, and dye. Arabs reintroduced saffron to medieval Europe (the English word comes from the Arabic for "yellow"), and it became very popular again, especially as a dye. In 15th century Germany, people were burned and buried alive as punishment for adulterating saffron. For a while, it was an important crop in England, where there is still a town named Saffron Walden. Today, saffron is often adulterated or replaced by turmeric or other, cheaper, more abundant plant parts. Saffron is the traditional flavoring in French *bouillabaise* and the Spanish *paella*.

Sesame seed comes from a herbaceous plant native to Indonesia and East Africa; the first known evidence of its cultivation is from the Middle East in 3000 B.C. Since then, it has become perhaps the Asian and Middle Eastern equivalent of the poppy seed, most frequently grown for its oil but also used as a garnish on sweets and breads.

Tarragon was first brought to Europe from Russia and western Asia by the Arabs, who seem to have been the first to use it as a flavoring in the 13th century. The name is a corruption of the Arabic for "little dragon." Today it is one of the French *fines herbes* and gives *sauce béarnaise* its characteristic flavor; it is also found in fish sauces and salad dressings and is sometimes stored in vinegar, both to preserve the herb and flavor the vinegar. Tarragon is the only commonly used herb or spice to come from a plant in the daisy family, the second largest family of flowering plants.

Vanilla comes from the pod fruit of a climbing vine in the orchid family, the largest family of flowering plants, and it is the only significant food plant from this family. Native to Central America, vanilla was enjoyed long before the arrival of European explorers. The characteristic flavor is developed from the initially tasteless, underripe pods (also called beans) by alternately "sweating" them in the sun for 10 or 20 days and then drying them slowly for several months more. In the process, the pod's enzymes free its principal flavor component, *vanillin*, from an interfering linkage to glucose, and browning reactions deepen its color and aroma.

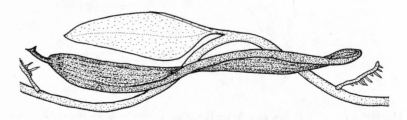

The vanilla vine and pod

Attempts to introduce the vine to other tropical areas met with little success because it would never set fruit. This problem was partly solved in the 19th century by the Belgian botanist Charles Morren, who introduced the practice of hand pollination. He discovered that in the vine's original habitat, one species of bee and one species of hummingbird were the only agents that would pollinate its flowers, and other sites simply didn't have these specialized pollinators. The bees were introduced to the new habitats, and soon vanilla was thriving there as well. Hand pollination remains the usual practice in commercial plantings, and is one reason for vanilla's steep price, second (though a distant second) only to saffron. Between 1875 and

1925, vanillin was chemically synthesized from the essential oil of cloves, and more recently it has been made from the lignin in wood wastes. Of course, artificial vanillin does not have the full, complex aroma of the actual pod and its many minor flavor components. By U.S. law, that which is labeled "vanilla extract" must derive from true vanilla.

TEA AND COFFEE

Among the foods we obtain from plants, tea leaves and coffee beans stand apart. We don't eat them intact as vegetables, nor do we normally use them to flavor other foods (coffee and green tea ice creams are delicious exceptions). Instead, we modify them chemically by various means, and then soak them in hot water. The resulting extracts are interesting enough—in both gustatory and pharmacological terms—that they can constitute "dishes" in themselves.

Tea

In prehistoric times the leaves of the tea plant, an evergreen bush in the Camellia family native to Southeast Asia, were probably eaten or boiled into a beverage. The earliest record of its cultivation comes from China in the 4th century A.D. By the 8th century, it was enjoying a vogue in China and becoming established in Japan. Over the next 600 or 700 years, the method of infusing the leaf in water slowly evolved. In Lu Yu's *The Classic of Tea*, written around A.D. 800, the tea is formed into cakes, which are then boiled. After a while, the preferred technique came to be powdering the leaf and whipping it into hot water the way we do cocoa. It was in the Ming Dynasty, from 1368 to 1644, that steeping individual sprigs of tea in hot water became standard. This was the period when the West discovered tea. The Venetian spice traders, together with the first Europeans to visit China, brought back word of the new drink, and some samples, beginning around 1560. Within a century the Dutch and English had established a regular sea trade, and the Russians sent overland caravans. During the 18th century tea came to replace ale at English breakfast tables, and soon the custom of afternoon tea caught on. In 1700, England imported 20,000 pounds of tea; in 1800, the figure was 20 *million* pounds. Not everyone was pleased with this huge shift in drinking habits. William Cobbett thought that good beer was infinitely preferable to tea. He wrote in *Cottage Economy* (1821) that

> The drink, which has come to supply the place of beer, has, in general, been *tea*. It is notorious, that tea has no *useful strength* in it; that it contains nothing *nutritious;* that it, besides being *good* for nothing, has *badness* in it, because it is well-known to produce want of sleep in many cases, and in all cases, to shake and weaken the nerves.

But such protests availed little. Today the British still consume more than twice as much tea per capita as the Japanese.

Because the Chinese and Japanese were secretive about the preparation of tea from the raw leaves, it wasn't until the early 19th century that plantings were extended by the Dutch to Indonesia and by the English to India and Ceylon. The most recent milestone in tea's history was the 1904 invention of the silk tea bag by Thomas Sullivan, a New York merchant who sent out samples in these containers. Today, half of the tea used in the American home is bagged.

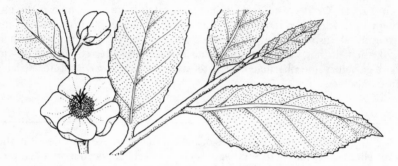

Tea. The choicest pluck consists of the terminal bud and two youngest leaves of each branch.

The tea bush grows in a wide range of climates, although it is commercially productive only in the tropics and subtropics. The principal producers today are China and Japan (which don't export much), Taiwan, Indonesia, India, and Sri Lanka (Ceylon). Plantings made at high altitudes and those partly shaded by trees yield fewer leaves, but these are finer in quality (the same rule holds for coffee beans). The best tea is made from the small young shoots and unopened leaf buds, which contain the highest concentrations of phenolic substances, enzymes, and caffeine and are more easily crushed than older leaves. The choice "pluck" is the terminal bud and the two adjacent leaves. The bush is pruned back regularly to maximize the number of new shoots. In colder climates, the bushes may be plucked 5 times in 7 months, while in the tropics, a harvest can be gathered every 10 days throughout the year. Once the leaves have been plucked, they are made into any of three kinds of tea, depending on the treatment. The fresh leaf alone brews only into a raw, thin liquid with no flavor or body. These qualities are developed by crushing and heating the leaves, which produces various aromatic molecules. Black and oolong teas are made by encouraging the leaf's enzymes to transform its chemical composition, while green tea is spared this "fermentation," as it is traditionally but inaccurately called.

Making Tea

There are four stages in the making of tea: withering, rolling, fermenting, and firing. The leaves are withered by drying them out to the point that they become wilted and structurally weak. This may take a day in the open air or a few hours in a drum dryer. Once withered, the leaves are rolled repeatedly for an hour or two. It used to be that this was done with the heel of the hand, a few leaves at a time, and was the most laborious part of the process, but machines have changed that. The most important effect of rolling is to crush the leaf cells and mix their many chemical components together. From this mixture will arise all the color, flavor, and astringency of the finished tea. Rolling also gives the leaves a twist that slows down the rate at which they give up their essence to the hot water.

The next stage, the fermentation, begins as soon as the leaves are rolled and enzymes, oxygen, and cell contents are mixed. During the 1 to 3 hours that the leaves are allowed to stand at about 80°F (27°C), they turn a coppery brown and develop both flavor and astringency. When the fermentation has proceeded as far as the maker desires, the leaves are "fired," or dried at temperatures slowly increasing to about 200°F (93°C), until they are left with a moisture content of about 5%. The tea is then ready for grading.

Fermentation does not involve yeasts or other microbes to any significant extent, and even has the effect of greatly reducing the number of microbes naturally present on the leaf surface. This happens because the major event during fermentation is the transformation of mostly colorless, flavorless phenolic substances into pigment molecules that are also astringent *tannins*. Tannins bind surface proteins in the mouth, where they produce a constricting feeling and the impression of a full-bodied liquid, and on microbes, which they thereby incapacitate. This transformation is caused by the same enzyme, polyphenoloxidase, that causes browning in many fruits and vegetables when their tissue is damaged by cutting or bruising. The action of the enzyme and oxygen produces a series of intermediate compounds: first, yellow, tannic complexes, then orange-reddish tannic ones, and finally condensed brownish complexes that are less tannic because they have already bound to each other. A good, well-colored *brisk* tea will have a ratio of 10 to 12 red or brown phenolics for every 1 yellow, while in a brown, flat tea the numbers would be more like 20 to 1. The fermentation supervisor, then, must make sure that oxidation of the phenolics doesn't go too far. About 3½ hours in ordinary air (much less in an oxygen-charged atmosphere) is the usual limit for fully fermented *black* tea. Its blackness is only a surface feature resulting from a layer of completely condensed tannins; underneath that layer, it is brownish. *Green* tea is not allowed to ferment at all: its enzymes are destroyed by steam before it is rolled and fired. It brews into a pale yellow liquid with a less complex, full aroma than the amber drink that black tea makes. *Oolong* tea is a compromise between black and green. It is fermented briefly, once before rolling and once after.

Tea Classification

The classification of teas is not a straightforward matter. Black teas, for example, are graded by the size of the pieces, not by quality. Consistency in size is important because small pieces brew much faster than large ones. Orange pekoe denotes long, thin, closely twisted leaves, pekoe more open leaves, and souchong large, coarse leaves. Orange pekoe and souchong tend to produce paler liquids than pekoe because it takes longer for the water to suffuse rolled or large leaves, but broken subgrades of these major types, with their larger surface area, will give a darker brew. The finest (smallest) grade is called dust, and is used in tea bags. Green teas are graded on the basis of the age of the leaves and the style of preparation. Gunpowder tea is small, tight balls of young and medium-aged leaves, imperial tea is a larger, looser, older version of gunpowder, and hyson types are loose, long, twisted leaves of various ages.

Generally, the best way to predict the quality of tea is simply to know which regions are renowned for producing the best. Here are a few of the more familiar names associated with good quality: black teas from Darjeeling, an eastern Indian district in the foothills of the Himalayas, and from Assam, in the same region but at a lower elevation; high-grown Ceylon teas from Sri Lanka, Oolongs and smoky Lapsang Souchongs from Taiwan; black tea from Keemun in northern China. Most commercial teas are blends of several, even (in the case of brand-name bags) 20 different teas, a practice that guarantees the consistency of the product even if one or two components suddenly become unavailable. English breakfast tea, traditionally Keemun, is no longer necessarily that; Constant Comment is a Ceylon blend flavored with orange peel, and Earl Grey is a blend perfumed with oil of bergamot (the bergamot is an inedible variety of orange).

Brewing Tea

Tea is brewed by infusing about a teaspoon of the dried leaf for every 6-ounce cup of water. The water should come cold from the tap—hot or reheated water contains less dissolved air, and is reputed to make flat-tasting tea—and should be poured just after reaching the boil to maximize its powers of penetrating the leaf and dissolving the components of color and flavor. The pot is preheated to prevent it from cooling the water too much. It has been said of Americans that they drink tea with the eye, tending to remove the leaves from the brew just as the desirable color is reached, which may be a matter of a minute or less. But connoisseurs claim that the flavor and caffeine in the leaf take longer to be extracted than the color does, and that a brewing period of 3 to 5 minutes is necessary to bring out the full quality of the tea. If this length of time results in too strong a flavor, then less tea should be used to begin with. And remember that fine pieces infuse faster than large ones. A longer brew, on the other hand, or reusing the leaves, will result in a harsh, bitter flavor. Generally, the browner or more condensed the pigments, the less astringent or brisk the tea. An orange-red liquid is the most desirable for a black tea.

If milk is added to the tea, the tannins will immediately bind to the milk proteins, and the taste will be much less astringent. Nearly sour milk may be curdled by the combination of high temperature and tannins. And squeezing lemon juice into a cup of tea lightens its color quite noticeably. This happens because the reddish pigments are also weak acids whose anionic (negatively charged) component is highly colored. When another acid is added to tea, the increased number of positive hydrogen ions neutralizes many of the pigment anions, and thereby reduces the intensity of the tea's hue. (Alkaline baking soda, conversely, will turn orange tea nearly blood red.)

Iced tea, an innovation popular only on this side of the Atlantic, presents a problem. Tea brewed in the normal fashion, and even made somewhat stronger than usual to compensate for the dilution caused by melting ice, tends to become cloudy when cooled down. Not a great fault, but somehow a sparklingly clear beverage seems more refreshing than an opaque one. It turns out that cloudiness is one of the factors used by professional tasters to judge the quality of tea. If a tea "creams down" with a good amount of fine, white precipitate, it is a good tea. If the precipitate is coarser and "muddy," the tea has probably been overfermented; and if scanty, underfermented. The precipitate is a combination of caffeine and pigment molecules; these materials crystallize as the temperature drops because they have only a limited solubility in water, and are much more soluble in hot water than in cool. Near-boiling water extracts so much of each component that they cannot remain dissolved in colder water, and, when the tea is poured over ice, it suddenly clouds up with tiny particles. The way to avoid clouding is simply not to saturate the water with more caffeine and pigment than it can hold at low temperatures. This is done by soaking the tea in lukewarm water— for several hours instead of several minutes, since the infusion will be much slower—which will not dissolve much more of these substances than the iced beverage can *keep* dissolved.

Coffee

The coffee tree is indigenous to tropical Africa, where its fruit was early put to good but rather different uses from the one that we know today. The flesh of the berries was made into a kind of wine, and the large inner seeds or "beans," like most seeds rich in protein and carbohydrates, were a useful food as well as a stimulant. It was not until around A.D. 1000 that Arabs in Ethiopia began making a hot drink from the coffee bean, and even then it was done as we now make cocoa: they pulverized the roasted beans and whipped the powder in hot water. The new drink caught on, and soon the first coffee houses were established in Mecca, Damascus, and Constantinople. The coffee tree was cultivated in Yemen, and Mocha, a Yemeni port, became synonymous with its principal commodity (the use of this name to mean a combination of coffee and chocolate is only a few decades old). Venice became acquainted with coffee through the spice trade in the 15th century, and in the 16th and early 17th centuries, English travelers discovered it. William Biddulph wrote in 1609 of his visit with the Turks:

> Their most common drink is coffa, which is a black kind of drink made of a kind of pulse like pease, called coava; which being ground in the mill, and boiled in water, they drink it as hot as they can suffer it; which they find to agree very well with them against their crudities and feeding on herbs and raw meats.

Another adventurer, George Sandys, wrote in 1601 that coffee was "black as soote, and tasting not much unlike it."

Despite this preliminary judgment, the new drink was a sensation across Europe. First brought to England around 1630, it became ensconced in London coffee houses in 1652, and Parisian cafés (named with the French word for coffee) followed about 8 years later. In order to circumvent the Arabs' monopoly, the English, Dutch, and French smuggled the beans to their colonies in India, Java, and the West Indies. South America was therefore ready to take the lead when the older Asian plantings were attacked with a fungal disease in the mid-19th century. In the meantime, official opposition to coffee met with little success. J. S. Bach's *Coffee Cantata* of 1732 comically plays German medical opinion against a young woman's cravings for the beverage. And 50 years earlier, a monarch had to back down on an attempt to ban not coffee, but coffee houses. On December 23, 1675, Charles II of England issued "A Proclamation for the Suppression of Coffee Houses."

> Whereas it is most apparent that the multitude of Coffee Houses of late years set up and kept within this Kingdom . . . and the great resort of idle and disaffected persons to them, have produced very evil and dangerous effects; as well for that many tradesmen and others, do herein misspend much of their time, which might and probably would be employed in and about their Lawful Calling and Affairs; but also for that in such houses . . . divers, false, malitious, and scandalous reports are devised and spread abroad to the Defamation of His Majesty's Government, and to the disturbance of the Peace and Quiet of the Realm; his Majesty hath thought it fit and necessary, that the said Coffee Houses be (for the Future) put down and suppressed . . .

The public outcry was so great that the proclamation was revoked on January 8. Both coffee and cafés have survived quite well.

Two Kinds of Coffee Bean

Today two species of coffee tree, which are kept to 5 to 12 feet by pruning, are commercially exploited. *Coffea arabica* is the original source of the Arabian beverage, while *Coffea canephora*, better known as *robusta*, has more recently come under cultivation for its better resistance to frost and disease, its faster fruiting, and its tolerance of warmer climates and lower elevations. The problem is that robusta beans are more neutral in flavor than the Arabicas, and therefore less interesting. As is true of tea leaves and wine grapes, long, cool, difficult growing conditions and less hardy species make for more complex, distinctive beverages. Plantings in the Americas are mostly Arabica, in Africa mostly robusta, and in India, about half and half. Jamaican, Java, mocha, and kona (Hawaiian) coffees are prized because these high-grown Arabicas are not prolific and are grown in relatively small regions: they are both very good and very rare. The same is true to a lesser extent of Andes-grown Colombian coffees. Most coffees that go by these famous names are blends that may contain only a small amount of the desirable beans. Today, about half of the world's coffee comes from Brazil, a quarter from the rest of Latin America, and one sixth from Africa.

Processing

Coffee beans undergo two stages of processing, one in their country of origin and one at their final destination. First, the seeds must be removed cleanly from the berries that enclose them. There are two ways of doing this, both involving a period of "fermentation," or enzymatic and microbial activity, during which the berry pulp softens and becomes more easily separated from the seeds. The seeds are then dried and shipped. They are not roasted until they have reached the manufacturer, and it is this treatment that develops their typical color and flavor, through browning reactions. In addition to these changes, the beans lose water and become more brittle and porous (and

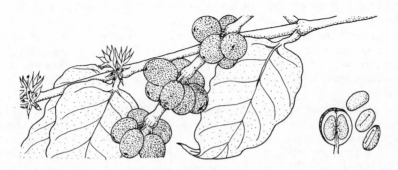

Coffee berries and seeds. Each berry contains two seeds.

so more easily ground and brewed); little caffeine is lost, however. The hotter and longer the beans are roasted, the darker and more strongly flavored they get. Lighter roasts are more delicate and individually distinctive in flavor, but also tend to be more acidic because fewer of the beans' acids are evaporated or decomposed by the heat. Roasting to taste can be done at home in a dry frying pan over medium heat; stir the green beans continuously to avoid burning. In the end, the roasted beans will contain at least 100 aromatic molecules that contribute flavor, together with protein, oil, starch, sugar, fiber, tannins, caffeine, and bitter phenolic substances.

Grinding and Brewing

Before the coffee can be brewed into a beverage it must be ground into smaller particles that will give up their contents to the water more rapidly than the whole bean can. Ideally, coffee beans should be kept whole and refrigerated or frozen in an airtight container until the water is put on the stove. Any contact with air causes the loss of some volatile flavor molecules and the oxidation of some oil: in other words, staleness and rancidity. (Oil allowed to collect inside a pot and to go rancid will taint subsequent brews.) Ground beans expose a much larger surface area to the air, and deteriorate much more rapidly than whole beans, so that it is best to delay grinding until the last minute. Similarly, the finer the beans are ground, the more rapidly and completely their soluble contents are extracted. Greek and Turkish coffees are as strong and bitter as they are because the coffee has been reduced nearly to a powder.

The same general rule holds for both tea and coffee: if they are brewed too long, their flavor deteriorates as astringent and bitter residues begin to dissolve and overpower the desirable flavor components. It has been figured that the optimum extraction of coffee is 20% of the solids in the grind. This takes about two minutes of contact with water at about 200°F (93°C). Underextraction results in a weak but sour flavor because the beans' acids are among the first substances to dissolve. An extraction rate of 30% is possible by steeping the grounds longer or re-passing the water through them, but the "strength" of overbrewed coffee is not simply a greater concentration of the desirable coffee flavor. To make *good* strong coffee, you must use more ground coffee per cup. The standard measure for ordinary coffee in this country is two tablespoons for every 6-ounce cup. To make flavorful but weaker coffee, brew it at normal strength, and then dilute it with hot water to taste.

The original method of brewing coffee was probably very much like our campfire style: the grounds were boiled in the pot, and the liquid decanted off. Because the grounds have been in contact with the water for a long time (especially for the second round), and the boiling drives off aroma molecules, this kind of coffee tends to be flat and harsh. The methods that are popular today—filtration and percolation—date from the early 19th century. In *The Physiology of Taste,* Brillat-Savarin attributes the technique of filtering through a perforated china plate to J. B. de Belloy, an Archbishop of Paris, who died in 1808. The French also developed the pumping percolator, which continually passes the brewing water over the grounds, in 1827. We should be very grateful for these innovations, especially the first. *How* grateful is evident when we read this recipe for brewing coffee from the 1844 *Kitchen Directory and American Housewife:*

> Use a tablespoonful ground to a pint of boiling water [less than a quarter of what we would use today]. Boil in tin pot twenty to twenty-five minutes. If boiled longer it will not taste fresh and lively. Let stand four or five minutes to settle, pour off grounds into a coffee pot or urn. Put fish skin or isinglass size of a ninepence in pot when put on to boil or else the white and shell of half an egg to a couple of quarts of coffee.

Not only boiled for 25 minutes, but boiled with fish skin or egg white—to retain any grounds or smaller particles that might cloud the coffee afterwards! This brew would have very little aroma left and plenty of harshness ("liveliness"?); the eggshell was probably valuable for neutralizing some of the no doubt excessive acidity. It is an excellent summary of how to go wrong.

Here, briefly, is the right approach. Use enough dry coffee to obtain the desired fullness of flavor without excessive bitterness. Use water just off the boil—it extracts the desired substances rapidly—but don't boil the water

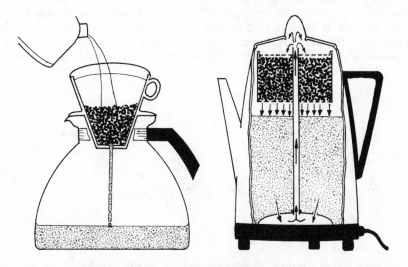

The drip and percolation methods of brewing coffee. The percolator uses steam pressure to force liquid continuously up the central tube and over the grounds.

while brewing, or the flavor will escape with the steam. And don't hold the brewed coffee at a high temperature for too long or reheat it too often. Again, much of the aroma will be lost, and certain complex molecules are broken down into acids, which give the coffee a sour edge. Because percolation involves boiling water—steam pressure forces the water up the central tube and over the grounds—and 10 to 15 minutes of extraction (because very little water is in contact with the grounds at a time), it generally does not produce as good a brew as the filtration or "drip" method.

Espresso, Instant, and Decaffeinated Coffee

Here are a few brief notes on common kinds of coffee not normally made in the home. Espresso, which is Italian for "pressed out," is a very strong coffee made very rapidly by forcing a combination of steam and water through about twice the normal quantity of grounds. It seems to have been developed in Italy just before World War II. "Instant" coffee is made by brewing the liquid and then removing the water by evaporation in a vacuum or by freeze-drying. Some experts say that instant is usually crystallized from overextracted, and so undesirably bitter, coffee. And decaffeinated coffee is made by steaming the unroasted beans and extracting the caffeine with a solvent, which must then be thoroughly washed out before drying and roasting the bean, or by steaming the beans and then abrading away the now caffeine-rich outer layers. The second process is preferred by people who are concerned about the safety of solvent residues, but very little of this Swiss water-processed coffee now makes it to this country. These extreme treatments, which were developed around 1900, do cause the loss of some flavor components. It happens that robusta coffees contain about twice as much caffeine as the Arabicas, so that as the use of robustas in blends has increased, the caffeine content of our coffee has gone up. This in turn may be increasing the demand for caffeine-free coffee.

The chemical structures of three closely related alkaloids: caffeine, theophylline (found in tea), and theobromine (found in chocolate). Caffeine and theophylline have by far the strongest effects on humans.

Caffeine

Finally, a word about caffeine, which is among the most commonly consumed drugs in the world, occurring as it does in coffee, tea, and cocoa, and added as it is to many soft drinks. Caffeine is an alkaloid that has several different effects on the human body. It stimulates the cortex of the brain, and in small doses can improve attention, concentration, and coordination. Caffeine acts on the kidneys to increase water elimination, and it stimulates gastric secretions. It stimulates the action of the heart, and causes blood vessels to widen everywhere but in the brain, where they are constricted. And by somehow altering the activity of calcium ions, caffeine increases the contracting power of skeletal muscles and makes them less susceptible to fatigue. A remarkable range of influences! It is important to remember, however, that an overdose of caffeine can cause just as broad a range of trouble: hyperesthesia, or overacute sensation, irregularities in the heartbeat, muscular trembling. Though tea leaves are 2% caffeine by weight, and coffee beans half that, a cup of coffee contains about twice the caffeine as a cup of tea—100 milligrams rather than 50—because the coffee is more completely extracted. Instant coffees have lost a quarter to a third of their caffeine. A 12-ounce can of cola contains about 50 milligrams of caffeine, a cup of cocoa about 15.

CHAPTER 5

Grains, Legumes, and Nuts

SEEDS AS FOOD

It would be hard to overestimate the importance of grains and legumes in the life of our species. They and the nuts are seeds: compact, desiccated packages that contain a plant's embryonic offspring, together with enough food for them to develop their roots and leaves and become self-sufficient. Because they are a concentrated source of protein and either carbohydrates or fats, and can easily be stored for long periods of time, the edible seeds have played a crucial role in human nutrition and cultural evolution. The turning point in the latter came about 10,000 years ago with the rise of agriculture and the selective propagation of food crops, and cereals and legumes were the first plants to be brought in some measure under human control. As the nomadic life of the hunter-gatherer gave way to settlements associated with large grain fields, centralization of the food supply and population brought the need for greater social organization, for planning and record keeping. The earliest known writing, alphabets, and arithmetical systems, dating from about 3000 B.C., are devoted to grain transactions. The culture of the fields made possible the culture of the mind.

Although it is less obvious to us than it would have been to our distant ancestors, cereals have continued to be the essential food of the human race. Today, they provide the bulk of the caloric intake for much of the world's population: around 70% for Egypt and India, and near 80% in China, or between 2 and 3 times the average for the developed West. The cereals and legumes put together account for more than two thirds of the world's dietary protein. Even the industrial countries are fed indirectly by the huge amounts of corn, wheat, and soybeans on which their cattle, hogs, and chickens are raised. When we learn that the cereals are members of the grass family, we find new significance in the Old Testament prophet Isaiah's admonition: "All flesh is grass."

Seeds are much more convenient than meat, milk, eggs, and other sources of protein that are quick to spoil. But they have an important drawback: unlike meat, milk, or eggs, any particular kind of seed is usually an incomplete protein source for animals, because it is deficient in one or more of the essential amino acids. Over the millennia, however, widely separated cultures have learned to combine different seeds in their diet so as to balance their protein intake. For example, the Asian diet of rice and soybeans and the Central American diet of corn and common beans have been traditional for many centuries. Today we know that the cereals are deficient mainly in lysine, the legumes in sulfur-containing amino acids; but when the two foods are blended together, these deficiencies are canceled out.

Some Definitions

The *grains*, or *cereals* (from *Ceres*, the Roman goddess of agriculture), are all plants in the grass family, the Gramineae. They produce many small, separate dry fruits, which we call the kernels or grains. We don't usually think of them as fruits because the layer between seed and outer skin, which is thick and succulent in culinary fruits, is thin and dry in the grains, for all practical purposes part of the tough bran coat. Wheat, barley, oats, and rye have been the most important grains in the Middle East and Europe; in Asia, rice; in the New World, maize, or corn; and in Africa sorghum and millets. The grains are of special culinary significance because they make possible beer and bread, both staples in the human diet at least since the early Egyptians (around 3000 B.C.). Since bread and beer are products of the complicated process of fermentation, they will be discussed in separate chapters. Here, we will examine the peculiar characteristics and histories of the major grains, together with their third significant manifestation in our daily lives, the breakfast cereals.

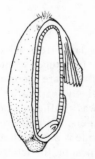

Drawing of an oat kernel, which is a complete fruit. The dry fruit and seed coats are shown enlarged at right.

The *legumes* (from the Latin *legere*, "to gather") constitute a family of plants that bear several seeds in a pod, a somewhat fleshy fruit that can be eaten as a vegetable before it dries out. Lentils, broad beans, peas, and chick peas are all native to the Fertile Crescent of the Near East, the soy and mung beans to Asia, and peanuts, lima beans, and common beans (navy, kidney, black, pinto, and many others) to the Americas. Legume and cereal crops have always been closely associated with each other: in Central America, for

Lentils. The surrounding pod is the characteristic fruit of the legumes.

example, bean vines were trained right onto corn stalks. This partnership is providential not only for the dietary protein balance it makes possible, but also because legumes have the unique property of forming a symbiotic relationship with certain soil bacteria that supply the plant with essential nitrogen compounds. Legumes, then, actually enrich the soil they grow in, while other plants, including the grains, deplete it. This is why various legumes, including clover and alfalfa, are often grown as rotation crops, a practice that goes back at least to Roman times.

The *nuts* comprise a culinary category that includes the edible seeds of fleshy fruits as well as true nuts, or one-seeded fruits with a hard, woody layer corresponding to the flesh of culinary fruits. Nuts are much less important in the human diet than grains and legumes, for two basic reasons. They have a very high fat content—on the order of 60% by weight—and this limits their usefulness as a protein source by limiting our capacity for them. Perhaps

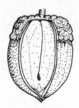

The acorn, a true nut, is a complete fruit. Its woody "shell" corresponds to the fleshy layer of culinary fruits and the thin bran of grains.

more important, nuts are generally the product of large trees, which do not begin to bear until years, sometimes decades, after planting, and cannot produce as much per acre as herbaceous plants. Nut trees are therefore inconvenient to cultivate, and their produce relatively expensive. Though undoubtedly a significant food source when man was a gatherer, and in more recent times where forests could supply local larders, the nuts are a minor dietary adjunct today. The one exception to this rule is the coconut, a major resource in many tropical countries. Because the peanut, though technically a legume, is almost universally thought of and treated as a nut, we will discuss it under this category.

The Green Revolution

The world's dependence on the seed crops has focused the attention of modern plant scientists on the problems of accelerating their improvement and increasing production enough to alleviate the hunger that is still all too common. This effort has gone hand in hand with the general interest in making modern agriculture more efficient and so more profitable. The result has been the so-called Green Revolution: the development of more productive varieties and culture techniques, and even the invention of a wholly new species called triticale, a hybrid of wheat *(Triticum)* and rye *(Secale)* that is hardier and more nutritious than either parent (it is used primarily as livestock feed). One measure of the revolution's success is the astonishing fact that, in the two decades of the fifties and sixties, world production of wheat increased by 100%.

This accelerated program of crop improvement had as its general goal uniformity: not only of high food yields per plant, but also of size, height, and time of ripening, which make mechanical harvesting possible, and of quality, which makes marketing and consumption more predictable. Uniformity of product is accomplished by assuring uniformity of genetic stock in the parent plant, and most growers today collectively raise only a handful of varieties, year after year.

GENETIC DIVERSITY OF U.S.

CROP PLANTINGS

Crop	Major Varieties	% Crop Acreage Planted in These Varieties (1970)
Wheat	9	50
Corn	6	71
Soybean	6	56
Potato	4	72

As a result, genetic diversity in the major food crops has declined sharply. One side effect of this trend has been the increased possibility of losing whole fields or even regions to insect pests or diseases, since resistance to them is conferred by genes (unlike animals, plants cannot develop immunity). In the corn blight epidemic of 1970, fully 15% of the American crop was destroyed. If all plants in the crop are identical, then their susceptibility to attack by other organisms is identical, and if one goes, all go. The disappearance of wild relatives and ancestral or older cultivated varieties is of great concern, because the genetic diversity that makes possible survival in changing conditions depends on these hereditary stocks. A National Seed Storage Laboratory has been established in Colorado in an attempt to maintain some reserves of scarce genetic material, but many scientists warn that much stronger conservation efforts are needed. Without them, the noble project of feeding the human race could conceivably end up starving it.

THE GRAINS

Of the approximately 8000 species in the grass family, only a handful play a significant role in the human diet. Aside from bamboo and sugar cane, these are the cereal plants. The cultivation of the major cereals—wheat, barley, rice, maize—is of great antiquity, while oats and rye are relatively recent domesticates. Originally weeds in Middle East plantings of wheat and barley, oats and rye supplanted their betters in cold, moist northern areas inhospitable to other grains. While the grains are very similar in structure and composition, their small differences have made for widely divergent culinary histories.

Structure and Composition

The edible portion of the cereal plant, commonly called the grain or kernel, is technically a complete fruit whose ovary-derived layer is very thin and dry. Three of the cereals—barley, oats, and rice—bear fruits that are covered by small, tough, leaflike structures that fuse to form the husk. Wheat, rye, and maize bear naked fruits and do not have to be husked before milling.

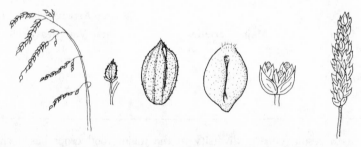

Rice *(left)*, like barley and oats, is borne in a protective husk that must be removed before the grain can be used as food. Wheat *(right)*, rye, and corn are borne as "naked" fruits, and are that much more convenient.

All the grains have the same basic structure. The fruit tissue consists of a layer of epidermis and several thin inner layers; altogether, it is only a few cells thick. Just underneath the seed coat is the *aleurone layer,* only one to four cells thick and yet containing oil, minerals, protein, and vitamins all out of proportion to its size. It in turn surrounds the *endosperm,* the organ that stores most of the carbohydrates and protein, and that takes up most of the

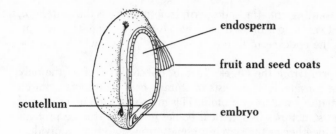

The anatomy of a wheat kernel. For a similar view through the scanning electron microscope, see photograph on page 286.

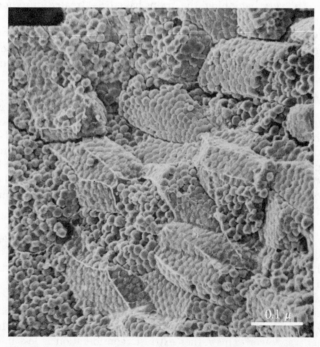

The endosperm of a high-starch corn kernel, viewed through the scanning electron microscope. Each cell is packed tightly with starch granules. (From Robutti et al., *Cer. Chem.* 51 (1974): 173–80. With permission of R. C. Hoseney.)

grain's volume. Abutting onto the endosperm from one side is the *scutellum*, a single modified cotyledon, or seed leaf, which absorbs, digests, and conducts food from the endosperm to the *embryo*, or "germ," which is at the base of the fruit.

The endosperm (from the Greek: "within the seed") is often the only part of the grain consumed. It consists of storage cells that contain starch granules embedded in a matrix of protein. This matrix is made up of normal cell contents, and sometimes of spherical bodies of special storage proteins which, squeezed together as the starch granules grow, lose their individual identity and become a continuous phase. There is generally less starch and more protein per cell as one moves from the center to the surface of the grain, and this gradient has consequences both for nutrient losses from milling, and for the behavior of grains during cooking, as we shall see when we get to popcorn.

Milling and Refining

From the very beginning, the grains have been treated to remove their tough protective layers and make them easier to cook and to chew. In prehistory, two rocks, and eventually a mortar and pestle, were the instruments used to crush the grain, after which the larger sheets of *bran* could be picked out of the fragmented endosperm. Milling was revolutionized around 800 B.C. with the invention of the rotary grinding machine, which could be powered by animals or waterwheels. This method prevailed until the mid-19th century, when roller milling was invented in Switzerland. Today, grooved rollers mesh with each other to create a combination of shearing, scraping, and crushing actions that open the grain up, scrape the endosperm from the bran, and grind it into a powder. The very different mechanical properties of endosperm, germ, and bran make this separation possible: the first is easily fragmented, the others oily and leathery respectively. Refined meal or flour is produced by sieving off the coarse particles, while in whole grain flours the germ and bran are added back at the end of the process.

The germ and the bran—which in practice includes the aleurone layer—together account for most of the fiber, oil, and B vitamins contained in the whole grain, as well as some 25% of its protein. Yet these parts of the grain are usually removed entirely from flours, cornmeal, and ordinary rice. Why this apparent waste? One reason is that white wheat flour has been a status symbol for centuries. It is "purer" than whole grain flour, and in the past was more expensive, since more refining was necessary and the yield per bushel was lower. This rather extrinsic consideration is less influential today, no doubt on account of the attention given to "natural" foods. In the last decade white bread's share of the American market has dropped by about a third.

But there are some more practical reasons for refining grains. Bran and germ do dilute the bread-making qualities of flour. Fiber can actually reduce

CEREALS: ORIGINS AND DATES OF
DOMESTICATION

Near East

Wheat	7000 B.C.
Barley	7000 B.C.
Rye	400 B.C. } domesticated in
Oats	A.D. 100? } Europe

Asia

Rice	4500 B.C.

Africa

Millet	4000 B.C.
Sorghum	4000 B.C.

Central America

Maize (corn)	4500 B.C.

WORLD GRAIN PRODUCTION (1980), COMPOSITION OF
WHOLE GRAIN (PERCENT)

	Metric Tons (Millions)	Water	Protein	Carbohydrate	Fat
Wheat	445	13	12	71	2
Rice	400	12	8	77	2
Maize	392	12	8	77	3
Barley	162	11	8	79	1
Millet, Sorghum	87	11	11	72	4
Oats	43	8	14	68	7
Rye	27	11	12	73	2
(Buckwheat)	2	11	12	73	2

From *FAO Production Yearbook*, 1980

the nutritional value of food by complexing with proteins (see pages 550–51), and wheat bran contains a substance, phytic acid, that similarly makes calcium unavailable to the body. During World War II, Dublin was put on whole grain bread, and the combination of bran and minimal calcium intake resulted in an epidemic of rickets; fully half the children of the city suffered from it. Finally, the most influential reason of all: the high lipid concentrations in the germ and aleurone layer shorten the shelf life of flour and whole grains substantially. The oils are susceptible to oxidation and develop rancid odors in a matter of weeks.

There are certainly drawbacks and waste involved in using refined flours and degerminated meals, but it should be clear that it is not simply a matter of industrial arrogance. People have been given what they want and have wanted for a long time: a pure-looking product that won't spoil quickly on the shelf. Today most refined cereals are fortified with B vitamins and iron in order to compensate for the nutrients lost with the bran.

Wheat

One of the two oldest cultivated food plants, wheat was the most important cereal in the ancient Mediterranean civilizations, and after a long hiatus from the Middle Ages to the 19th century, when hardier but less versatile cereals and potatoes became the principal staples, it has regained top status today. It was brought to America early in the 17th century and had reached the future Bread Basket, the Great Plains, by 1855. Compared to other temperate-zone cereals, wheat is a demanding crop. It is susceptible to disease in warm, humid regions and does best in a cool climate, but it cannot be grown as far north as can rye and oats.

Something on the order of 30,000 varieties of wheat are known, and they are classified into a few different types according to planting schedule and endosperm composition. Spring wheat is planted in the spring like most crops, but winter wheat is planted in the fall, lives through the winter as a small plant, and sends up its flowering stalks in the spring. Soft, hard, and durum (even harder) wheats are obviously classified according to the mechanical strength of the kernel, which in turn is a function of the protein-to-starch ratio in the endosperm. Hard wheats contain fewer and smaller starch grains, with the result that the protein matrix is more continuous and so stronger.

The protein content of wheat is not simply a matter of nutritional value. Wheat has long been the premier grain primarily because its storage protein has unique chemical properties. When ground up and mixed with water, this protein forms a complex, semisolid structure, called "gluten," which is both plastic and elastic: that is, it can stretch under pressure and yet tends to resist that pressure. It will expand to accommodate gases produced by yeast, and yet it contains the gases, rather than stretching to the point of bursting. Of the other grains, only rye has storage proteins with anything like these properties, and rye gluten is vastly inferior to wheat gluten. Without wheat, then, we would not have raised breads.

The different protein-to-starch ratios of the wheat varieties are best suited to different kinds of products. Durum semolina is used for pastas, hard flours for bread, and soft flours for pastries, biscuits, cookies, and cakes. New methods of separating flour particles in air streams have made it possible to get hard flour from soft wheat, and so kernel types may become less significant. Batters and doughs, which are remarkable systems in their own right, are covered in chapter 6.

Barley

Remains have been found of Stone Age cakes made of coarsely ground wheat and barley, and this is one of several pieces of evidence to suggest that the two grains are coeval. Barley has the advantages of a relatively short growing season and a hardy nature; it stands both drought and frost well, and is grown from the Arctic Circle to the tropical plains of northern India. It was as important as wheat in early Egypt and China, and in Athens was, according to Pliny, the special food of the gladiators, who were called *hordearii*, or "barley-eaters." By Roman times, though, wheat had displaced barley and become the preferred grain. The historian Polybius (2nd century B.C.) reported that reluctant soldiers were punished by confinement and barley rations, and Pliny said that "barley bread, which was extensively used by the ancients, has now fallen into universal disrepute" (Book 18). In the Middle Ages, and especially in northern Europe, barley and rye breads were the staple food of the peasantry, while wheat was reserved for the upper classes.

Today, barley is a staple only in the Middle East. In the West it is used primarily as animal feed (accounting for about 50% of consumption) and in the production of beers and distilled liquors (about 30%). Its most important culinary manifestation is *malt*, a powder made from barley grains that have been germinating for about a week. As it grows, the embryo accumulates an enzyme that it uses to digest its endosperm starch into sugar. Malt contains a great deal of this enzyme, and brewers and distillers use it to transform the starch in beer or liquor mashes into sugars that alcohol-producing yeast can feed on (the details of fermentation are discussed in full in chapter 9).

Rye

Rye seems to have originated in central Asia and, beginning about 4000 B.C., moved slowly westward as a weed contaminating the wheat and barley supplies of nomadic tribes. It reached the coast of the Baltic Sea around 2000 B.C., grew better than the other cereals in the typically poor soil and cool, moist climate, and was domesticated around 400 B.C. Rye was and remains little known along the Mediterranean. Up through the last century it was the predominant bread grain for the poor of northern Europe, and even today the taste for rye persists, especially in Scandinavia and eastern Europe. In West Germany, wheat production exceeded rye for the first time only in 1957, with East Germany following nine years later. In the United States today, 25% of a rather small crop is used for food, 25% for liquor production, and the rest for animal feed. Rye dough, on account of its poor gluten qualities, makes for very heavy bread (pumpernickel, for example), but one other characteristic makes it useful for weight-reducing diets. Rye flour contains an unusual amount of pentosans, or long chains of 5-carbon sugars. These

The ergot fungus growing in an ear of rye

compounds have a very high water-binding capacity, which means both that rye bread retains moisture better than wheat, and that dry rye crisps tend to swell in the stomach, thus giving the sensation of fullness. In addition, the pentosans are only very slowly broken down to sugar, and so take a long time to digest. This too reduces appetite.

Aside from matters culinary, rye is also remarkable for its pathological and pharmacological influences on human life. The poor soil and moist climate in which rye does well are favorable conditions for the growth of the *ergot* fungus, whose long, thin fruiting bodies grow out of the flowering stalk like grotesquely deformed kernels. From the 11th to the 16th centuries, ergot was responsible for frequent epidemics of what was called Holy Fire or Saint Anthony's Fire, a disease with two sets of symptoms: progressive gangrene, in which extremities hurt, went numb, turned black, shrank, and dropped off; and, more sporadically, mental derangement, twitching, and fits. Mild epidemics of ergot poisoning from contaminated flour have occurred as recently as 1927 in England, and 1951 in France.

Beginning in the Middle Ages, however, ergot was pressed into service as a drug. One important use was to hasten labor during childbirth, and in the 19th century this seems to have been standard procedure in difficult births. Early in the 20th century, chemists began to unravel the composition of ergot toxins, and discovered a handful of alkaloids with very different effects. One stimulates the uterine muscle; some are hallucinogens; and some constrict the blood vessels, an action which can cause gangrene, but also has useful applications to hypertension, migraine headaches, and the avoidance of postoperative shock. It turned out that all these alkaloids have a basic component in common—lysergic acid—and in 1943 the Swiss scientist Albert Hofmann, who was investigating the properties of synthetic ergotlike alkaloids, discovered the particular variant that would come to such prominence in the sixties: lysergic acid diethylamide (LSD).

Oats

The world produces more oats than rye today, but people eat more rye. Oats are fed mostly to animals, with about 5% of the world crop going to human consumption. Originating somewhere in Europe or Asia, the oat was domesticated in the early Christian era. In classical times it was considered a weed and used only for medicinal purposes. Theophrastus and Pliny believed that the oat was merely a diseased form of wheat. By 1600 it had become an important crop in northern Europe, in whose wet climate it does best; oats require more moisture than any other cereal but rice. Other countries, however, continued to disdain it. Dr. Johnson's *Dictionary* (1755) gives this definition for oats: "A grain, which in England is generally given to horses, but in Scotland supports the people." There are several reasons for the relatively minor status of oats. It is more difficult to process than other cereals because the husk which surrounds the kernel, or groat, is inedible and yet adherent. Oats contain from 2 to 5 times the fat that wheat does, together with large amounts of a fat-digesting enzyme, and the combination results in a tendency to become rancid. Steam treatment or careful removal of the bran, which contains the enzyme, is necessary to avoid rapid spoilage. And like barley, oats have no gluten-producing proteins, which means that no light breads can be made from oat flour. Porridges and heavy cakes are the best one can do.

On the other hand, the high fat content of oats has led to their use in face soaps. And the endosperm contains a natural antioxidant, similar in action to BHT. Fatty foods like potato chips, nuts, coffee, bacon, peanut butter, and margarine are sometimes preserved with a dusting or infusion of oat flour.

Rice

Rice is the principal food crop for about half of the world's population. It is a native of the Indian subcontinent (a minor species that bears red grain is native to Africa), and about 90% of the world crop is grown in monsoon Asia. We in the modern West unconsciously draw upon the ancient Eastern identification of rice with fertility when we pelt the bride and groom as they leave the church.

It is said that Alexander the Great introduced rice to Europe around 300 B.C., though it was not until the 8th century that the Moors first grew large quantities in Spain. Today, Italy is Europe's only major producer. South Carolina was the location of the first commercial American planting in 1685, though its output is now eclipsed by the lower Mississippi region, Texas, and California. Rice requires more water than any other cereal crop, a fact which the familiar image of the rice paddy would seem to bear out. Actually, the primary purpose of the standing water is to drown the rice seedlings' weedy competition, and some rice is grown "uphill," or in drained areas. Paddy rice is highly productive, and can give two harvests in a single season. But it is also a demanding crop, and historians have seen a connection between the necessary agreements on systems of irrigation and labor, and the strict social discipline characteristic of the East.

There are around 2500 different varieties of rice, some with red, blue, or purple coloration, but only one distinction is commercially significant. *Indian* rice is long grained and tends to be dry, flaky, and easily separated when cooked; *Japanese* is short grained, and moist, firm, and sticky when cooked. It turns out that Japanese rice has a higher proportion of waxy starch molecules, or amylopectin, in its starch granules than does Indian rice.

Milling Rice

Because rice is usually consumed as individual grains and not as meal or flour, rice milling is much more involved than the grind-and-sieve approach taken to wheat or corn. First the hull is removed, leaving what we call brown rice, or the intact kernel covered with the bran layers. Next, an abrasive process removes the bran and most of the germ—the beveled end of the grain sometimes retains part of the scutellum—and the result is milled, unpolished rice. Polishing in a wire-brush machine removes the aleurone layer which, with its high fat content, would otherwise limit the storage life of the grain. Sometimes sugar is added to give polished rice a brighter sheen. Today, polished rice is fortified with vitamins, which are applied in solution to the outside of the grain, coated with a protein powder, and dried. This exposed location of important nutrients is the reason that rice should not be rinsed before cooking, and should be cooked in as little water as possible. In some countries, including India, most rice is still cooked in the traditional way, with large amounts of excess water discarded at the end. About 15% of the kernel's protein is lost in milling and polishing, including much of the limiting amino acid, lysine. Rice ranks relatively low among the cereals in protein content, though it has a better-than-average balance of amino acids, even with the lysine lost in processing.

The development of mechanical milling and polishing techniques has certainly lowered spoilage losses and so helped feed more people, but at an unforeseen price. In the late 19th century the disease known as beriberi, in which the nerves, digestive system, and heart become inflamed and may degenerate, and which had been known but rare for many centuries, reached epidemic proportions in Asia. Dutch investigators discovered that it could be reversed by feeding rice bran to the victim, and later the vitamin thiamine was identified as the nutrient whose deficiency causes beriberi. The disease had been uncommon before not because milled rice was unpopular—on the contrary, it was valued because it was difficult to produce by hand, was scarce, and kept well (the parallel with white flour in Europe is very close)— but because very few could afford it, and those who could enjoyed a better diet than most. Techniques that brought the preferred rice to the masses also brought them unwanted suffering, which continues to this day in some regions.

Converted Rice

Beriberi had never been much of a problem in India and Pakistan, because rice there was parboiled before milling. Today, we call the product of this 2000-year-old technique "converted" rice. The whole grain is steeped in water, steamed, and then dried again before milling. This partial precooking makes milling easier by loosening the hull, and it improves the nutritional quality of the milled grain by causing the B vitamins in the bran and germ to diffuse into the endosperm, and by gelatinizing the aleurone layer, which then adheres to the grain rather than to the bran. Storage life is not compromised by this last effect because the steam heat inactivates the lipid-splitting enzymes that cause rancidity.

VITAMIN RETENTION IN REGULAR AND CONVERTED RICE

	Thiamine	Riboflavin	Niacin
regular rice	24%	44%	37%
converted rice	42%	66%	89%

Attempts to introduce converted rice to other cultures have been successful in Africa and the West Indies, but unsuccessful in the Philippines and Southeast Asia.

Converted rice takes as long or longer than ordinary rice to cook. Quick-cooking rice is manufactured by partially cooking and then fissuring the grain in order to speed the infiltration of hot water when the consumer cooks it. Freeze-drying, puffing, rolling, and microwave treatments are all used for the second step.

Wild Rice

This distant relative of common rice is a native not of Asia, but of the Great Lakes region of North America, where it grows in shallow lakes and marshes. Originally gathered by hand from its natural habitat by the Chippewa Indians, wild rice is no longer truly wild. A few decades ago, commercial producers began planting it in paddies, and mechanical harvesters are now used. In spite of this new systematic approach to wild rice production, the final product sells for several dollars a pound, many times the cost

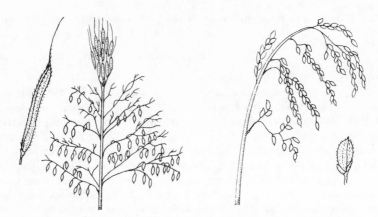

Wild rice *(left)*, a grass native to the northern U.S. and southern Canada, and common rice *(right)*, which comes from the tropics

of common rice, and accounts for less than 1% of the American rice market. It is processed somewhat more elaborately than its relatives: first fermented for a week or two to develop the nutty flavor and to ease hulling, and then heated to partly gelatinize the starch and to cause some browning. Wild rice contains more protein than ordinary rice, and is especially rich in lysine, the amino acid in which most of the grains are deficient.

Maize, or Corn

This grain is the New World's single most important contribution to the human diet, and today it is second only to wheat in acreage planted. It was under cultivation by at least 3500 B.C. in Central America, and because the large size of both plant and fruit made corn agriculture relatively easy, it quickly became the basic food plant of many other American cultures. The Incas of Peru, the Mayas and Aztecs of Mexico, Mississippi mound builders, the cliff dwellers of the American Southwest, and many seminomadic tribes in North and South America depended on corn as a dietary staple. It is thought that the crop required relatively little work of its growers—perhaps one day out of the week per person—and that the resulting spare time helped make possible the monumental temples and fortifications of the pre-Columbian civilizations. Corn remains the fundamental food plant of the United States, although three quarters of the crop now goes toward supplying us with eggs, milk, and meat. Columbus brought corn back with him to Europe, and within a generation it was being grown throughout southern Europe. It was originally known as "Indian corn" or "maize"; the latter word came from the term used by West Indian natives. *Corn* began as a generic term, denoting grain or grainlike objects (hence corned, or salted, beef), and in different parts of Britain will tend to be used for the most important grain locally. The use of *corn* to denote maize alone is peculiar to the United States.

Corn was one of the first crops to be the object of intensive efforts at improvement. Around the turn of the century there was a craze for raising the perfect ear of corn, with uniformity of both ear and kernel the standard. At this early stage, the interest was more aesthetic than anything else, reminiscent perhaps of the tulip mania in 17th-century Holland. Systematic efforts at hybridization followed, and in the thirties workers at the Connecticut Agricultural Station in New Haven first developed a commercially sound technique of producing hybrid seed (this had been a problem because the offspring of hybrid corn turn out to be less vigorous than their parents and cannot be used as seed in their turn; special seed corn has to be produced for every crop). By 1950, fully 75% of U.S. corn acreage was hybrid, and yields increased dramatically.

There are five different kinds of corn, each characterized by a different endosperm composition. Pop and flint corn have a relatively high protein content and hard rather than waxy starch. Dent corn, the variety most commonly grown for animal feed, has a localized deposit of soft, waxy starch at the crown of the kernel, which produces a depression, or dent, in the dried kernel. Flour corn, with little protein and mostly waxy starch, is grown only by native Americans for their own use. What we call Indian corn today are flour and flint varieties with variegated kernels. Finally, sweet corn, very popular as a vegetable when immature, stores more sugar than starch, and therefore has translucent kernels and loose, wrinkled skins (starch grains refract light and plump out the kernels in the other types). It appears that popcorn was the first kind of corn to be cultivated, but all five were known to native Americans long before the advent of Europeans.

The variegated kernels of decorative corn have a pigmented aleurone layer—that peripheral, nutrient-rich coat one or two cells thick—which generally accounts for hues other than white, yellow, and orange. It also con-

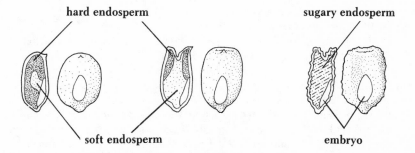

Left to right: kernels of pop, dent, and sweet corn. The composition of their endosperms determines their appearance and behavior.

tributes the turquoise flecks to tortillas made with "blue corn." The early American clergyman Cotton Mather recorded the first observation of sexuality in corn when he noticed that the windward side of a neighbor's corn planting, the side most exposed to pollen from other fields, developed multicolored cobs, while the rest of the planting was yellow. Yellow coloration resides not in the aleurone layer but in the endosperm, where both carotene and xanthophylls (found also in potatoes and onions) are deposited. The different locations of different pigments explain why it is that corn meal, which is ground from the endosperm, can be cream-colored or yellow, but never red.

Popcorn

Why does popcorn pop? No one seems to have investigated the problem in much detail, and so what follows is the current educated guess. Recall that starch grains are embedded in a protein matrix, and that this matrix is probably stronger in popcorn than in other types because its protein-to-starch ratio is higher. When the kernel is heated in hot oil, the small amount of moisture it contains partly gelatinizes the starch grains; this happens at around 150°F (66°C). Then, as the kernel temperature reaches the boiling point, the water vaporizes and expands rapidly in volume. The hard protein matrix holds until the pressure becomes too great, at which point the kernel bursts open and the endosperm expands in volume on account of the sudden pressure drop. At the same time, the already cooked starch granules are dried out as the water vapor escapes, and the endosperm texture becomes light and crisp. If, however, popcorn is cooked in a covered pan that offers no escape for the water vapor, the spongy endosperm will absorb it again, and chewy, tough popcorn is the result.

Popcorn pops well only when it falls within the narrow range of 11 to 14% moisture content, to which the better brands are adjusted before being packed in air-tight containers.

Corn Sickness: Pellagra

While corn has undoubtedly been a boon to a significant proportion of the human race, it has also been something of a bane. All the cereals are deficient in an important amino acid, lysine, but corn is more deficient than the others, and in addition is deficient in tryptophan. Beyond protein quality, 50 to 80% of corn's niacin is bound up with another molecule and unavailable to the human body. As a result, people who rely on corn as their primary food source risk both protein and niacin deficiencies. In areas like southern Europe and even the southern United States, where corn was introduced as a new staple crop, a new disease followed soon after: pellagra.

Once introduced to the Old World, corn quickly became popular with landowners for its comparatively high yields, and in several areas, notably Spain and Italy, it became a dietary staple for sharecroppers. In the 18th century, a peculiar syndrome that seemed to be associated with eating corn was described for the first time: red skin lesions, diarrhea, weakness, mental confusion, and in extreme cases a slow mental and physical degeneration. Often the symptoms would appear in the spring, after a long winter of living on corn, then gradually disappear during the summer as the diet improved, only to reappear intensified the following spring. An Italian monograph of 1771 studying the disease among sharecroppers who lived on polenta, or corn mush, first used the name *pellagra*, which means "rough skin." France recognized the occurrence of pellagra within its borders around 1820, and by the end of the century, Egypt and other parts of Africa were suffering epidemics. In the United States, few cases were reported before 1900. The first epidemic, at a black mental institution in Alabama, came in 1906, and by 1930 it was estimated that there were 200,000 pellagrins in the country, most of them southern cotton sharecroppers.

For 200 years it was thought that contaminated corn was the villain, something like ergotic rye, or, more cynically, that the victims were weak by nature and caught the disease from sheep. As late as 1915, a U.S. government commission decided that pellagra was an infectious disease transmitted from person to person. In the same year, a U.S. Public Health Service doctor, Joseph Goldberger, tested the received opinion carefully and suggested the likelihood of nutritional deficiency. In the late twenties he found that yeast extract and liver prevented a similar disease in dogs. In the late thirties, niacin was discovered and shown to be the decisive factor in preventing or reversing pellagra in humans. In 1941, Congress passed a law mandating the vitamin fortification of bread.

But pellagra had declined significantly in the United States between 1930 and 1933, during the depths of the Depression, and before proper treatment had been developed. The Depression had lowered the value of the cotton crop, and many sharecroppers planted more land in vegetables and fruits, thereby balancing their diet. Later, vitamin therapy and the jobs and income that the war effort brought to many of the poor all but eradicated the disease. Pellagra has slowly disappeared from Europe too. But it remains a serious problem in parts of South Africa, lower Egypt, and southwestern India, all areas where corn is a late introduction and where poverty is the social constant.

Traditional Alkaline Processing

One naturally wonders how the original cultivators and consumers of corn avoided pellagra. The answer is perhaps the most remarkable example of so-called primitive cultures developing rather sophisticated practices by long experience and anticipating the findings of science. In modern-day Mexico, corn flour, or masa, is made by boiling the corn in a 5% lime solution for about an hour, then washing, draining, and finally grinding the grain. It turns out that the use of an alkaline substance in corn processing is widespread across the Americas and goes back to early civilization: the Mayas and Aztecs used ashes or lime; North American tribes, ashes and naturally occurring soda; and a contemporary Mayan group burns mussel shells for the same purpose.

In the last thirty years or so, scientists have uncovered the fundamental purpose of alkaline processing. First, we recall from the effect of soda on vegetables that cell wall components are most soluble in alkaline conditions, which means that such treatment makes the removal of the corn "hull," or bran, easier. Second, alkalinity improves the amino acid balance in corn by *decreasing* the availability of the major storage protein, zein, to the human body. This protein is most deficient in lysine and tryptophan, so that reducing its contribution to the overall protein content reduces the *relative* deficiency of these amino acids. The relative availability of lysine increases 2.8 times, and that of tryptophan 1.3 times, when corn is made into masa. Of course, there is a net loss of usable protein overall, but what protein is left is better balanced in the essential amino acids. Finally, and most important for the avoidance of pellagra, alkaline conditions release corn's bound niacin, which can then be absorbed and used by the body.

The use of ashes and lime in corn preparation is a simple but remarkably effective technique for improving the nutritional value of this eccentric grain. The misfortune is that this culinary tradition was not introduced to the Mediterranean, to Africa, India, or to the American South, along with the crop itself.

Natives of the Americas had one other means of coping with corn's nutritional deficiencies: beans, which are well stocked with lysine, tryptophan, and niacin, and whose plants enrich the soil, were often grown as a companion crop. And there may be other stratagems of which we remain ignorant. Samuel de Champlain, while exploring just east of Lake Huron around 1616, observed what might be called a "fermenting" technique practiced by the Huron Indians. Here is a challenge for the anthropologist: is there a nutritional basis for this recipe, did it just involve the microbial conversion of starch to sugar, or is it the Huron equivalent of the "noble rot"?

> They have another way of eating Indian corn, to prepare which they take it in the ear and put it in water under the mud, leaving it two or three months in that state, until they judge that it is putrid; then they take it out and boil it with meat or fish and then eat it. They also roast it, and it is better this way than boiled, but I assure you that nothing smells so bad as this corn when it comes out of the water all covered with mud; yet the women and children take it and suck it like sugar cane, there being nothing they like better, as they plainly show.

The corn plant is used today to produce more different materials, culinary and otherwise, than any other cereal. Grits and corn meal, both coarsely ground endosperm, lose about half of their nutritional value during the hulling and degermination process, and so are usually fortified. Corn flakes are made by rolling and toasting grits. Corn mash is fermented to produce not only drinkable whiskey, but also alcohol and acetone for industrial use. Corn flour is used in fabric sizing and soaps, corn starch in plastics and dyes as well as sauces. Corn syrup is an ingredient in many candies and desserts, shoe polish, and rayon; corn oil is used in various foods, paint, and varnish. Corn stalks and leaves are turned into paper, and cobs are used as a cork substitute and for fuel. All this in addition to feeding our livestock. A versatile plant.

Minor and Pseudo-cereals

Millet and sorghum are two cereals known mostly as feed crops in the United States, but they are important in the human diet of many tropical countries. The millets, of which there are several distinct species, are native to Africa or Asia, and have been cultivated for 6000 years. The grain is remarkable for its high protein content, from 16 to 22%, but it has the disadvantage of being very small. The millets are especially important in arid lands because they have one of the lowest water requirements of any cereal, and will grow in poor soils. Sorghum originated in East Africa, and today is relied on both there and in Asia. In India and China its production is exceeded only by rice and wheat. Both millets and sorghum are made into porridge, unleavened bread, and beer.

Buckwheat is not a true cereal: it is not a grass, and its "kernels" are actually achenes, dry fruits analogous to the "seeds" of the strawberry. It is a native of central Asia, was cultivated in China, and introduced to Europe at the end of the Middle Ages. Today, buckwheat is a significant human food primarily in Russia. Elsewhere, it is used as animal fodder, as a cover crop, and an ingredient in special breads and cakes.

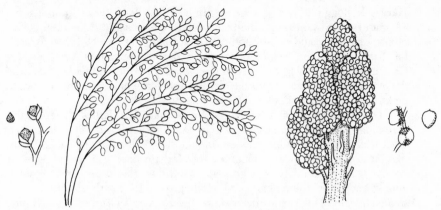

Millet and sorghum, two cereals of some importance in Africa and Asia

"Nuts May Save the Race": The Story of Breakfast Cereals

Apart from breads and pastries, the most common form in which we consume grain is probably the breakfast cereal, of which we can distinguish two basic types. The first, which requires cooking in water, has been known since the dawn of civilization; gruel and porridge are two of its more venerable names. Hot water softens the cell walls, gelatinizes the starch grains, and produces an edible, digestible, easily prepared, if somewhat tasteless mush. The only significant improvement brought by the machine age has been a reduction in cooking time, either by grinding the cereal finely enough that it is quickly cooked through, or by partly precooking it. Rolled oats are the most common example of this second technique. The grain is ground into a meal and steamed until some of its starch has been gelatinized (this step also inactivates troublesome enzymes). The meal is then rolled into flakes and dried back down to about 10% moisture by weight. The result is a convenient cereal that keeps well and can be cooked in a minute or two.

The more common breakfast cereal by far in the United States today is completely precooked and eaten with cold milk. Oddly enough, the industry that has recently come under such fire for giving children little more than empty calories, a sort of early-morning junk food, began as a self-consciously "pure" and "scientific" alternative to the destructive diet of turn-of-the-century America. Its story involves a uniquely American mix of eccentric health reformers, fringe religion, and commercial canniness.

Graham and Granula

In the middle third of the 19th century a vegetarian craze arose in opposition to the diet of salt beef and pork, hominy, condiments, and alkali-raised white bread that was prevalent at the time. A pure, plain diet for America was the object, and the issue was not only medical but moral. As Dr. John Harvey Kellogg put it somewhat later in his *Plain Facts for Old and Young,* "A man that lives on pork, fine-flour bread, rich pies and cakes, and condiments, drinks tea and coffee, and uses tobacco, might as well try to fly as to be chaste in thought." The movement's first chief spokesman was Sylvester Graham (1794–1851), a Presbyterian minister from Philadelphia who denounced white bread as pernicious and extolled whole grain flour, soon to be known as Graham flour, for its nutritiousness.

One doctor who was influenced by Graham was James C. Jackson, a Dansville, New York, advocate of the water cure who in 1863 developed what was probably the first modern breakfast cereal, which he called Granula. Jackson made a heavy dough out of Graham flour and water, baked it slowly in loaves until they were very dry, broke the loaves up into small chunks and baked them again, and finally ground the brittle chunks into yet smaller pieces. This is still the basic recipe for Post's Grape Nuts. Granula was served for breakfast after it had been soaked overnight in milk.

The Battle Creek Connection

In that same fateful year the future of Battle Creek, Michigan, as the breakfast cereal capital of the world was all but determined. Since 1855 Battle Creek had been the world headquarters of the Seventh Day Adventist Church, and for some time before that had been a midwestern outpost of the health movement, with mesmerists, phrenologists, water therapists, and other healers in residence. In 1863 the leader of the Church, Ellen White, was granted a revelation on the subject of human diet, and meatless, stimulantless, "pure" meals became the official Church policy. When a sanitarium run by followers of Graham in Battle Creek failed in 1866, the Church took it over and in 1876 brought in the son of a prominent Adventist family who was fresh out of medical school: Dr. John Harvey Kellogg.

Kellogg's interests were wide-ranging. He was familiar with current research in nutrition, gave a medical rationale for the vegetarian leanings of the time, and made diet the foundation of the Battle Creek Sanitarium's policy. As he put it in 1923, the purpose of this policy was to get the patient established in "a biologic mode of life," since "the average chronic invalid is suffering from the effects of a bad intestinal 'flora,'" and could do with much less protein and much more roughage in his diet.

There was one major problem with the sanitarium's meatless and spiceless menus: they were dull and unappetizing. For this reason, Kellogg was continually experimenting with novel preparations of nuts, legumes, and grains, the mainstays of the "biologic mode." In 1877 he came up with a breakfast food made of wheat, oats, and corn meal baked into biscuits and then ground up. Kellogg named the result Granula, and was promptly sued by Dr. Jackson. "Granola" was accepted as a compromise. But Kellogg then lost interest in cereals. He turned to yogurt and nut butters, and published a hopeful paper titled "Nuts May Save the Race."

Shredded Wheat and Wheat Flakes

In 1893 Kellogg met Henry D. Perky of Denver, who was producing little biscuits made of shredded, freshly steamed wheat grains. Kellogg was a devotee of dry heat treatment and made a strong case for baking the biscuits, at the same time offering to form a partnership with Perky. Perky took Kellogg's advice but not his offer. He went east, to Massachusetts and then Niagara Falls, and began producing shredded wheat as we know it.

This rebuff renewed Kellogg's interest in developing his own cereal, and he went to work to find something altogether original. In 1895 he came up with the first flake cereal by partly cooking whole wheat kernels, then rolling them flat and baking them until brown and dry. Wheat flakes, which he named "Granose," were a great novelty and spawned many imitations, but they weren't really very popular. In 1902 came the real brainstorm: corn flakes, flavored with barley malt. These were so successful that the sanitarium couldn't handle the business, and this led to the formation in 1906 of the

Battle Creek Toasted Corn Flake Company, managed by Kellogg's brother Will Keith, the precursor of the W. K. Kellogg Company.

C. W. Post

The Kelloggs were by no means without competition. Around the turn of the century, there were about 40 cereal companies trying to capitalize on the Battle Creek name, which was now synonymous with health and innovative foods. Only one, however, was much of a threat, and in fact beat the Kelloggs to the punch. Charles W. Post, an inventor and salesman from Springfield, Illinois, and then Texas, suffered from chronic constitutional illnesses, and came to the Battle Creek Sanitarium in 1891 to take the cure. After ten months under Dr. Kellogg's care with no change, Post left and was promptly "cured" by a local Christian Science practitioner who told him to go ahead and eat what he pleased. He then set up his own little retreat, La Vita Inn, where he studied dietetics, preached the influence of mind over body and the power of positive thinking, and manufactured his own brand of suspenders.

In 1895, Post brought out Postum, a coffee substitute made of wheat, bran, and molasses that was very similar to the caramel coffee served at the sanitarium. Unlike Kellogg, who was not much interested in reaching beyond local groceries and some Church missions, Post was a master of advertising and marketing, and built up an enterprise that the Kelloggs would overtake only after his death.

Post's first big cereal success was Grape Nuts, yet another version of Jackson's Granula but the only commercially successful one. The cereal made its debut in 1898, and was advertised as a "scientific food" that "steadies nerves" and "makes the blood red." Because added barley malt converts some of the cereal starch to glucose, or "grape sugar," Post called the preparation "pre-digested, and therefore easily assimilated."

Post's only big mistake in the business came in 1906, when he named his version of corn flakes "Elijah's Manna." The box showed a raven dropping manna into Elijah's outstretched hand. One of his fortunately unused sketches for a slogan ran, "Quote [sic] the Raven, never more will my tribe carry food for Prophets, modern day saints and sinners and citizens buy Elijah's Manna." Even without bringing Poe into it, Post was condemned from pulpits across the country and in England for exploiting the Bible. He withdrew the product and brought it out again in 1908 under the name by which we know it: Post Toasties.

Shredded Wheat, Kellogg's Corn Flakes, Post Grape Nuts: these were the foundations of the breakfast cereal industry, foundations that were strengthened by the contemporary scandals implicating much of the food industry with adulteration and dangerously unsanitary practices. Battle Creek cereals were professedly trustworthy, pure, and "scientific." Since then, thanks to massive infusions of sugar, vitamins, and marketing ingenuity into breakfast cereals, and despite the recent attacks on their nutritiousness, per capital consumption has steadily increased, while all other major grain products except pasta have shown just as steady a decline.

Puffed Cereals

Finally, a word about another kind of cereal that came along around 1900, the puffed cereals. The principle used here is analogous to the popping of popcorn. Either the grain or a composite dough can be used. The material is cooked to gelatinize the starch, and then partly dried. Then it is dropped into a pressure chamber, which builds up to a pressure of 100 to 200 pounds per square inch, and a temperature of 500° to 800°F (260° to 427°C). A release valve is triggered, and the material is ejected by the high pressure inside. The dough or endosperm is puffed out as its water vapor expands and escapes, and the result is a light, dry fragment. *Oven puffing* is a different process usually applied to rice. The grain is cooked and rolled flat into a thick flake, and then quickly toasted. Here, blistering lightens the texture.

The father of puffed cereals was Alexander P. Anderson, a native of Minnesota, who was interested in the nature of the starch granule. In 1901, working at Columbia University and the New York Botanical Gardens, he performed the experiment of filling test tubes with cereal starch, sealing them, and heating them. When the tubes were shattered, a porous, puffy mass of starch popped out. Anderson repeated the experiment using whole grains, and then, sensing the usefulness of his discoveries, returned to Minneapolis to find commercial backing. He soon joined Quaker Oats, originally a milling concern, in Chicago, where he developed the first steam-injected puffing guns. Puffed rice was first introduced to the public at the 1904 St. Louis World's Fair, and was marketed as a popcornlike snack. The following year it was transformed into a breakfast cereal, and quickly became popular.

LEGUMES

The legumes belong to the third largest family among the flowering plants (after the orchid and daisy families), and the second most important to the human diet, after the grasses. When it comes to culinary properties and applications, the legumes are a much more homogenous group than the grains. Their seeds are on the average twice as rich in protein as the grains, and especially well stocked in iron and the B vitamins. Some legumes are eaten green in the pod, when they contain more vitamins A and C but much less protein. A remarkable and as yet unexplained sign of their status in the ancient world is the fact that each of the four major legumes known to Rome lent its name to a prominent Roman family: Fabius comes from the faba bean, Lentulus from the lentil, Piso from the pea, and Cicero—most distinguished of them all—from the chick pea. No other food group has been so honored.

Some common legumes

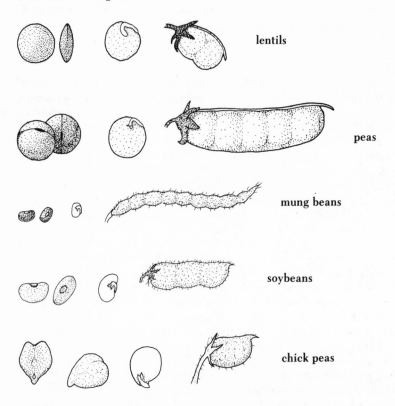

lentils

peas

mung beans

soybeans

chick peas

The *lentil* is probably the oldest cultivated legume, contemporaneous with wheat and barley and often associated physically with these crops. Its native ground is Southwest Asia, perhaps northern Syria, and it is now commonly eaten in India, the Middle East, and Eastern Europe. The Latin word for lentil, *lens,* gives us our word for a lentil-shaped, or doubly convex, piece of glass (the coinage dates from the 17th century).

The *pea* spread quite early from the Middle East to the Mediterranean, India, and China. It became especially important as an alternate protein source in Europe in the Middle Ages. The old children's rhyme, "Pease porridge hot,/Pease porridge cold,/Pease porridge in the pot/Nine days old" suggests how familiar a part of the diet it could be. The pea came to the New World with the earliest settlers. Today, two main varieties are cultivated: a starchy, smooth-coated one that gives us dried and split peas, and a wrinkly type with a higher sugar content, which is usually eaten when immature as a green vegetable. The pea occupies a special place in the history of science. In the 1860s, the Czech monk Gregor Mendel followed various traits, including seed color and shape, in successive generations of peas, and deduced from his observations two fundamental laws of genetic inheritance.

The so-called black-eyed pea is not really a pea, but a relative of the mung bean and other Asian legumes.

The *chick pea* was known in Rome as *cicer arietinum:* the second word means ramlike (the first names the legume itself), and refers to the seed's resemblance to a ram's head, complete with curling horns. Today, the chick pea, or garbanzo (the Spanish name, derived from the Greek), is a frequent ingredient in many Middle Eastern and Indian dishes. Hummus, for example, is a paste flavored with garlic, paprika, and lemon which is popular in many Mediterranean countries, with local variations. In India, the chick pea is planted on acreage fully three quarters of that devoted to wheat, and is

LEGUMES: ORIGINS AND
DATES OF DOMESTICATION

Near East

Lentil	7000 B.C.
Pea	6000 B.C.
Chick Pea	5000 B.C.
Broad Bean	3000 B.C.

Asia

Mung Bean	1000 B.C.(?)
Soybean	1000 B.C.(?)

Central and South America

Common Bean	6000 B.C.
Lima Bean	6000 B.C.

LEGUME COMPOSITION (PERCENT)

	Water	Protein	Carbohydrate	Fat
Broad Bean	12	25	58	1
Common Bean	11	22	61	2
Lima Bean	10	20	64	2
Mung Bean	11	24	60	1
sprout	89	4	7	0.2
Soybean	10	37	34	18
sprout	86	6	6	1
Lentil	11	25	60	1
Chick Pea	11	21	61	5
Pea	12	24	60	1

that country's most important legume; it is boiled, roasted, fried, sprouted, and made into soup *(dhal)* and flour.

The *broad bean,* also called fava or faba bean, was the only bean known to Europe until the discovery of the New World. The Egyptians, Greeks, and Romans cultivated it, though the philosopher Pythagoras forbade his follow-ers to eat it, on the grounds (it is thought) that beans contained the souls of the dead. This belief was also evident in Roman funeral banquets, at which beans were an important part of the offering made to dead relatives. Today this particular genus is little known in America, but it is popular in Europe, the Middle East, India, Mexico, and Brazil. People of Mediterranean back-ground may suffer from favism, a serious anemic condition that results from eating undercooked broad beans or inhaling the plant's pollen. The disease appears to involve a genetically determined sensitivity to the fava bean toxin vicine, which causes oxidative damage to the red blood cells of susceptible individuals.

The *mung bean,* known in the West primarily as material for sprouting, is native to India and spread early on to China and Asia generally. It is used in these areas for flour and for boiled, mashed dishes, as well as for sprouts.

The *soybean,* merely a curiosity in this country as late as 1900, is now the single largest cash crop in the United States, bringing more protein and oil into the economy than any other source. A native of northern China, its spread throughout Asia was probably contemporary with and hastened by the vegetarian doctrine of Buddhism. The soybean was not known in Europe until the 17th century, and was all but ignored in America until Commodore Matthew Perry's expedition to the Far East brought back two varieties in 1854. Commercial interest was aroused by the bean's high oil content—around 20%—and soy oil was first used industrially, in soaps, paints, and varnishes. It is highly unsaturated and very unstable, subject to oxidation and off-flavors, and had to await the invention of hydrogenation before it could be used in foods. During World War II, soy margarine replaced butter for most people; after the war margarine use continued, and the exceptionally high protein content of the soybean—around 40%, or nearly double the fig-ure for other legumes, five times the figure for corn—made soy meal a very attractive stock feed. Today, the United States produces about 75% of the world crop, primarily in the eastern part of the corn belt. Despite both its high protein content and that protein's *quality*—highest among the legumes, it approaches meat in its amino acid balance—the soybean is a significant human food only in the Far East, and though attempts have been made to introduce this valuable resource to India, Africa, and the Caribbean, they have had little success. The rather bland flavor of soybeans has perhaps encouraged the development of highly flavored fermented products, soy sauce being the most familiar in the West.

Bean curd is another form of processed soybeans that is becoming better known in this country. It was invented in China between A.D. 200 and 900.

TWO METHODS OF MAKING SOY SAUCE

I

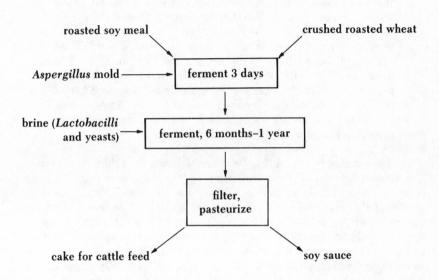

roasted soy meal   crushed roasted wheat

Aspergillus mold ⟶ ferment 3 days

brine (*Lactobacilli* and yeasts) ⟶ ferment, 6 months–1 year

filter, pasteurize

cake for cattle feed   soy sauce

II

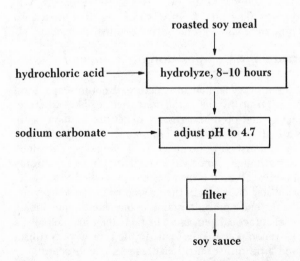

roasted soy meal

hydrochloric acid ⟶ hydrolyze, 8–10 hours

sodium carbonate ⟶ adjust pH to 4.7

filter

soy sauce

(many fewer flavor constituents)

One of the earliest European accounts of this novelty is Friar Domingo Navarrete's, which dates from the 17th century. He called it

> the most usual, common, and cheap sort of Food all China abounds in, and which all in that Empire eat, from the Emperor to the meanest Chinese; the Emperor and great Men as a Dainty, the common sort as necessary sustenance. It is called Teu Fu, that is Paste of Kidney Beans. I did not see how they made it. They drew the Milk out of the Kidney Beans, and turning it, make great Cakes of it like Cheeses, as big as a large Sive, and five or six fingers thick. All the Mass is as white as the very Snow, to look to nothing can be finer. . . . Alone it is insipid, but very good dress'd as I say and excellent fry'd in Butter.

The *common bean* and *lima bean* are both species in the Central American genus *Phaseolus*. The most important species, *Phaseolus vulgaris*, or the common bean, has been developed into literally hundreds of varieties, including the navy, field, kidney, pinto, and black beans. As many as 25 different varieties can be found for sale in some Mexican markets, each with a special culinary function: white types go with pork, black with tortillas, and so on. The ancestral plant was a native of southwestern Mexico. It first came under cultivation about 7000 years ago, and gradually diffused both north and south, reaching the major continents about 2000 years ago. Its use in Peru was predated, however, by the lima bean—the name derives from Peru's capital—which was native to Central America and domesticated somewhat later than the common bean. Both species were exported to Europe by Spanish explorers. The lima bean was introduced to Africa via the slave trade, and is now the main legume of that continent's tropics.

The Structure and Composition of Legumes

The legume seeds are all very similar in structure, if not in shape. Most of the seed's volume is taken up by the *cotyledons*, two leaflike structures modified to store the energy and protein necessary to get the seedling established. The cotyledons are attached to the *embryo*, a relatively small object consisting of a root, stem, and first pair of true leaves. The whole is contained by the *seed coat*, a thin but tough protective layer that is interrupted only at the hilum, the small bump where the seed has been attached to the pod, and where it will absorb water once it is in the ground or in the saucepan.

The cotyledons, like the endosperm in grains, determine the nutritional content of the seed for all practical purposes. In fact, they are actually a transformed endosperm. When pollen joins ovule in the process of fertilization, both an embryo and a primitive nutritional tissue, the endosperm, are formed. In the cereals the endosperm develops along with the embryo and remains the storage organ of the mature fruit. But in the legumes, it is absorbed by the embryo, which repackages the nutrients in the cotyledons.

MAKING BEAN CURD

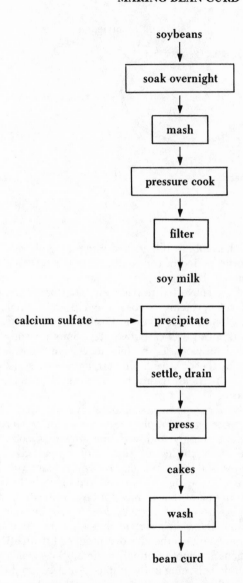

soybeans

soak overnight

mash

pressure cook

filter

soy milk

calcium sulfate ⟶ precipitate

settle, drain

press

cakes

wash

bean curd

88% water, 6% protein, 3.5% fat, 1.9% carbohydrate

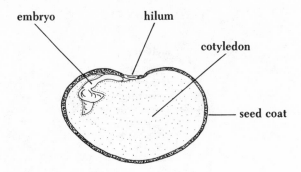

embryo   hilum

cotyledon

seed coat

The anatomy of a legume seed

The development of the embryo follows the same general pattern as fruit growth: a relatively brief period of cell multiplication is followed by a longer period of cell expansion. During the second stage, sugars and amino acids supplied by the plant are converted respectively into storage carbohydrates, mostly starch, and into proteins. Soybeans and peanuts, anomalies among the food legumes as the avocado is among fruits, store oil instead of starch (though they do contain cell wall carbohydrates). Starch is laid down in special cell bodies called amyloplasts, protein in cell vacuoles, which become "protein bodies." In the dry seed, this arrangement results in a matrix of protein and starch particles within the network of proteins that constituted the cell cytoplasm.

Legume seeds are also the repositories of defensive secondary compounds. As a general rule, those parts of a plant most important to its reproduction will be defended most strongly, and seeds and growing shoots are at the top of the list. Principal among the legume secondary compounds are the protease inhibitors, the lectins, and the cyanogens. As we have seen (in chapter 4), the first two interfere with the digestive process and so lower the nutritional value of the seeds to the point that it may not be worth eating them. Animals fed a diet of raw soybeans will actually *lose* weight and develop pancreatic problems, because the vain effort to digest the material takes more energy than it provides, and enzymes are overproduced in an attempt to compensate. The cyanogens are significant only in lima beans, which have caused serious cases of cyanide poisoning in the tropics. American and European breeders have developed low-cyanogen varieties, and most developed countries have laws that restrict commercial lima bean production to the safest types. These secondary compounds are all disabled or driven off by heat, and so are of no concern to most of us. If, however, legumes are eaten raw as sprouts, some care is advisable (see page 263).

The Problem of Legumes and Flatulence

It turns out that another chemical constituent of the legumes is responsible for a familiar, sometimes embarrassing consequence of their consumption. Flatulence has been the subject of various research efforts in recent decades, and it can be helpful to know why gas becomes a problem and how it can be minimized. Lest the subject be thought inappropriate for general discussion, however, we can point to ample and distinguished precedent. Saint Augustine saw involuntary bodily functions, flatulence among them, as unmistakable signs of man's fall from grace: once man failed to obey God, he became unable to obey even himself, and lost control over his physical nature. Another Church Father, Saint Jerome, recognized that beans were particularly offensive and forbade them to the nuns in his charge on the grounds that *"in partibus genitalibus titillationes producunt"*: "they tickle the genitals." This peculiar notion of the erotic persisted for some time; eleven centuries later, an English writer, Henry Buttes, wrote of beans' "flatulencie, whereby they provoke to lechery." Pleasure of yet another sort is probably at the root of the perennial children's rhyme that ends with the perverse wish, "Let's have beans for every meal!"

We are indebted to high-altitude aircraft flight and the space program for the recent spate of interest in flatulence. After World War II, it appeared that intestinal gas might prove to be a serious problem for test pilots in two ways: it was noticed that gravitational stress, vibration, and fatigue all affect the workings of the gastrointestinal tract, including gas production; and the volume of a given amount of gas increases as the pressure surrounding it decreases. This second fact means that a pilot's intestinal gas will expand as he flies higher into the atmosphere in an unpressurized cockpit. At 35,000 feet, for example, the volume will be 5.4 times what it would be at sea level. The resulting distension could cause substantial pain. When the first extended space flights were being planned, the confined atmosphere of the capsule gave rise to an additional worry. Several components of human gas, including hydrogen, methane, and hydrogen sulfide, are potentially toxic. There was the possibility that the astronauts might asphyxiate themselves. So the word went out across the land: study flatulence.

Work in the late sixties showed that normal individuals averaged an output of about a pint of gas a day, of which half is nitrogen and the result of swallowing air along with food and drink. Another 40% is carbon dioxide and the product of aerobic bacteria in the intestine. The remaining small fraction is a mixture of hydrogen, methane, and hydrogen sulfide, the products of anaerobic bacteria, and ammonia and the highly odorous indoles and skatoles, to which all the bacteria contribute. Fortunately, only the trace components are offensive to the smell; the others are merely "carriers." Some intestinal gas is also expired in the breath after diffusing through the intestinal wall into the blood, which carries it to the lungs as a waste product.

These are the few generalizations that can be made about human flatulence. The precise amount and composition of gas produced by a particular individual on a particular occasion depend on many different influences: how food is swallowed, the types of bacteria present in the intestine and their relative populations, the general activity of the gastrointestinal tract—all highly idiosyncratic traits—and on the food being eaten. With so many variables in play, it seemed to some space scientists that control could be exercised only at the level of personnel. In 1968, one wrote that "for extended space flights, it may prove advantageous to select astronauts who produce minimal amounts of methane and hydrogen and who do not normally produce very large quantities of flatus; and . . . selection criteria for astronauts might be established to eliminate those candidates who demonstrate marked or excessive gastroenterologic responses to stress. . . ." What a way to wash out!

In the decade before the space program, the offending ingredient in legumes had already been identified as certain *oligosaccharides*, or molecules consisting of three, four, and five sugar molecules linked together. Starch, of course, is a polysaccharide, made up of many thousands of glucose units. But the other sugars in oligosaccharides, like the glucoses in cellulose, are bonded together in ways that human digestive enzymes are not able to deal with. As a result, the oligosaccharides leave the upper intestine unchanged and enter the lower reaches in a form that the body cannot absorb. Here, the large bacterial population does the job we are unable to do, in the process giving off various gases, primarily carbon dioxide, as waste products. It is this sudden increase in bacterial metabolism, caused by the arrival of a food supply available to the bacteria alone and requiring some digestive action, that results in flatulence after eating legumes. The oligosaccharides are especially common in seeds because they are one form in which sugars can be stored for future use. Most other plant parts do not synthesize them in significant quantities, though for some reason grapes do. Different kinds of seeds contain very different mixtures of sugar-based molecules, depending on their own particular patterns of metabolism. Legumes happen to rank high in oligosaccharide content.

It has been found that oligosaccharides accumulate primarily in the final stages of seed development, and this finding is consistent with the general experience that green, immature beans are much less troublesome than dry beans. But it now also appears that there may be other factors involved in flatulence, since white, or navy, beans are known to have a lower oligosaccharide content than soybeans, but a higher gas activity. Some studies have implicated the fiber, or cellulose and hemicellulose content, with gas production, and it seems that some seeds contain an *anti*flatulence factor that has been tentatively identified as certain phenolic substances (similar to tannins) that tend to inhibit bacterial growth. Clearly, the field still awaits its Great Synthesis. Current knowledge can be boiled down to two statements. Navy and lima beans are generally the most offensive legumes, while peanuts, contrary to popular report, have little measurable activity. And both cooking in water and sprouting will decrease the absolute amount of oligosaccharides in individual seeds.

Sprouts

The sprouted seed is a culinary custom of ancient lineage in Asia, but a very recent arrival in the West. As late as 1903, the eminent Escoffier warned that "a prolonged soaking of dried vegetables may give rise to incipient germination, and this, by impairing the principles of the vegetables, depreciates the value of the food, and may even cause some harm to the diner." Chinese and Japanese restaurants probably gave sprouts their start in this country. Their current vogue has been accompanied by impressive and sweeping claims for the virtues of sprouted seeds. Their nutritional value "increases dramatically" during germination, the "natural" cookbooks say; lots of vitamins and minerals, few carbohydrates and fats. And anything is fair material. No need to stop with mung, soy, and alfalfa sprouts; on to fenugreek and garbanzos! All of this popular interest in sprouts has had its effect on food scientists, and several studies of the subject have been published in the last few years.

To date, though, the results of the research have been generally uncoordinated, unclear, sometimes contradictory, sometimes downright worthless. One paper claims that sprouts are stripped of their cotyledons before being served—news to most of us, who have no time for such tedium—and so the authors did the same to their experimental material. Others mix measurements in wet and dry weights, making it very difficult to compare the composition of a cooked dry bean with a sprout, the only datum of any practical interest. Another refers to earlier, basic research on germination in the common bean as background to its own report on the fate of oligosaccharides in the soybean, which, it is well known, behaves very differently. Still, it is possible to draw very broad generalizations about the nutritional value of seeds and sprouts.

Seed Germination

First it will help to know what is happening as a seed germinates. To begin with, the seed is a dry, dormant embryo surrounded by its cotyledons and a seed coat. When placed in a wet environment, the seed takes up water to a certain level—from 50 to 80% by weight, up from 5 to 10%—and this starts the biochemical machinery in motion. The embryonic stem and root begin to elongate, drawing on the proteins and starch stored in the cotyledons for building materials and energy. Since even sprouts grown in the light are able to photosynthesize very little at first, and most are grown in the dark to maintain a uniform pale color and delicate taste, the cotyledons and atmospheric oxygen remain the embryo's only nutritional sources. It is to be expected, then, that there will be a net loss of carbohydrates, or calories, as the seed sprouts, though there should be very little change in the total protein content. Amino acids, unlike starch, are not consumed to produce energy, but are rearranged to form new proteins. If the cotyledons are removed, however, most protein *and* carbohydrate will be lost.

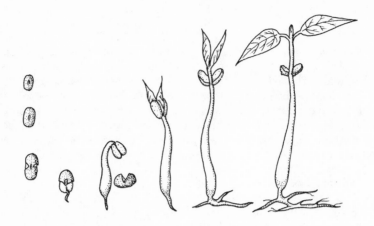

Several steps in the germination of a mung bean. The seed coat is shed as the leaves begin to emerge from between the cotyledons.

Perhaps the most significant change nutritionally is the first and most obvious one: the absorption of water. Since the seed now weighs two or three times as much as it did dry, one must eat twice as much by weight to get the same amounts of protein, carbohydrates, and so on. Of course, this will also be true of boiled legumes.

Now to the composition of sprouts—but with an initial qualification. Several investigators have reported large variations between different lots of the same legume, to say nothing of different legumes. The trends we summarize are just that—trends, and not absolute predictions that will hold true for every bean or pea.

As we would anticipate, carbohydrates and, in the case of the soybean, oils, decline during sprouting, while proteins mostly hold their own. Starch and oil are converted to carbon dioxide and water, while storage proteins are turned into other proteins. Of course, we have seen that the question of protein content is not as simple as this. The usefulness of protein to the human body depends on its structure and amino acid composition, and on antinutritional factors that interfere with digestion. The absolute amount of protein may not change from seed to sprout, but how about its availability?

Proteins and Vitamins

The data are scanty. A study of kidney beans showed that the digestibility of raw sprouts is equivalent to that of cooked beans, while cooked sprouts are even more digestible. The same pattern seems to hold for soybeans. It is also known that protease inhibitors actually *increase* in amount as common beans germinate, at least until the seventh day, or long after they would have been used as sprouts. The improvement after cooking can be attributed to the destruction of secondary compounds, while the improvement after sprouting appears to be unrelated to these factors. One possibility is a change in protein composition that makes the essential amino acids more available to the body. This is supported by a study of germinating corn, which showed a much better balance of amino acids after germination than before. Whether this is true of any or all legumes is just not known.

When we get to the vitamins, the picture is quite clear on at least one point. Germinating seeds are prodigious producers of ascorbic acid, or vitamin C. Even including the effect of dilution introduced by the large water uptake, vitamin C concentrations increase from 3 to 5 times their rather low initial levels in the seed. For reasons that remain largely unknown, ascorbic acid is especially concentrated in actively growing tissue like the seedling. Its accumulation is the single greatest nutritional advantage of sprouts over cooked seeds. But because the initial vitamin C content of the seed is so low, even a "dramatic increase" puts sprouts at the general level of green beans, and well behind cabbage and sweet peppers.

As for other vitamins: the concentrations of A and thiamine decline greatly; those of niacin and riboflavin may even increase slightly. Minerals decline due to the water uptake. Compared to most other vegetables, sprouts are ahead in iron and the B vitamins, but not by leaps and bounds.

We can summarize these results with what should be a fairly obvious statement. Legume sprouts have a nutritional value midway between that of the dry seed, which they have recently been, and the green vegetable, which they are on the way to becoming. Sprouts are higher in vitamin C and lower in calories than most seeds, higher in protein (5% versus about 2%) and in the B vitamins and iron than most vegetables. But they are hardly a wonder food.

It is quite reasonable to claim two other distinctions for sprouts: convenience and novelty. Anyone, anytime—even an apartment-dweller in Anchorage in February—can raise a good approximation to fresh vegetables with very little effort. And with their nutty flavor and crisp texture, sprouts are simply a nice change from the usual vegetables.

Finally, sprouts tend to be less troublesome than seeds as far as flatulence is concerned, for two reasons. They contain more water than cooked seeds, and so weight for weight less seed material is eaten in sprouts. And oligosaccharides, like other carbohydrates, are broken down to provide energy for the growing tissue. In soybeans, which have very little carbohydrate and a lot of oil, they are consumed quite rapidly—little is left by the third day—while in the common bean, which has plenty of starch to draw on, they decline only gradually. Again, sprouts occupy a middle ground between seeds and most vegetables.

Cooking Legumes

Boiling

Because mature legume seeds are very dry, boiling is far and away the most common means of making them palatable. The aim is to soften the seed anywhere from a firm texture to absolute mush, and this is accomplished by weakening the cotyledon cell walls and gelatinizing starch granules. As we saw in Chapter 4, the pH of the cooking water affects the texture of plant material. Cell wall hemicelluloses are more soluble in alkaline conditions, and seeds, like stems or leaves, will soften more readily for this reason than they would in acidic water. Veteran chili makers have probably noticed this effect when they put partially cooked beans into chili sauce: the beans simply do not get any softer, no matter how long the acidic sauce is simmered (the same thing happens with baked beans; in this case, calcium ions in the acidic molasses also complex with the cell wall materials and make them even less soluble). While this state of affairs is somewhat constraining—you can't save time by cooking the beans in the sauce, softening and flavoring them at the same time—it can also be put to good use. The texture of beans that are done just right but need to be kept warm, reheated, or cooked further in another dish, can be preserved if their new environment is made distinctly acid.

It is not advisable, however, to try for a world's record in soybean cooking time by adding tablespoons of baking soda to the water, since excessive alkalinity can affect both flavor and nutritional value adversely. The more cell-wall material that is lost, the more protein and vitamins will escape to the water, which is usually discarded. One study indicated that ⅛ teaspoon soda per cup of beans will hasten cooking without significant nutritional losses.

A better way to reduce cooking time by up to one half is to soak seeds for a few hours ahead of time in unheated water. The work of absorbing the water has thereby been accomplished, and only the heating remains to be done. It appears that there is little advantage to soaking beans for more than about four hours, though overnight periods may be most convenient. It has been reported that black beans do not cook any faster after soaking.

Then there is the matter of the water itself, and how much should be used. We have pointed out that in vegetable cookery, large volumes of water help minimize enzyme damage to vitamins and pigments by maintaining boiling temperatures even when the relatively cold vegetables are added. With legumes, whose cooking times are on the order of hours rather than minutes, different considerations predominate. For one thing, the amount of protein and carbohydrates lost to the water will increase as the water volume increases. And it turns out, contrary to what we would expect, that seeds will actually absorb *more* water in a smaller volume of water: the less cooking water, the fewer carbohydrates are leached out, and the carbohydrates will take up about 10 times their own weight in water. This means, then, that seeds will seem softer in a given time if cooked in a minimal amount of liquid. So give the seeds enough water both to soak up *and* to cook in (many a pan bottom has been charred because the cook forgot that beans imbibe), but don't drown them.

There is one conceivable reason for which one might want to contravene this advice, and lace huge volumes of water with baking soda. The noisome oligosaccharides are leached out in large quantities along with proteins, carbohydrates, and hemicelluloses, under these conditions. If nutrition and taste are less important considerations than comfort, then go right ahead.

Roasting

Soybeans and peanuts are the only legumes commonly roasted, probably because their high oil contents compensate for their dryness. In roasting seeds, changes in flavor are just as important as changes in texture, and perhaps more so. High temperatures and low water content make possible the complex reactions that result in the characteristic brown color and intense flavors of roasted foods. But because roasting is a dry heat technique, it will desiccate the already dry seeds and produce flavorful, brown jawbreakers. Unless, that is, they have been soaked, or even cooked to a firm texture, before roasting. A soaked-roasted bean will end up with about the same final moisture content as the initial dry bean, but with its starch partly gelatinized: it will be crisp rather than just hard. Like nuts, beans should be roasted at a fairly low temperature—around 250°F (121°C)—to prevent the surface from charring while the inside heats through.

Cooking and Legume Toxins

Protease inhibitors and lectins, both common components in seeds that lower their digestibility, are disabled by heat. The nutritional value of legumes, then is distinctly improved by cooking, and the same goes for sprouts. Bean curd, which is heated during production, does not improve further with cooking.

The lima bean is a special case, since it contains more than negligible amounts of cyanide precursors. These potentially toxic compounds are easily removed by boiling the beans in an uncovered pot: hydrogen cyanide gas forms and escapes with the steam. Lima beans that are raw and green or that have been sprouted, roasted, or boiled in a covered pot are probably not harmful if eaten in small quantities, but they are certainly not the safest forms of this particular food.

NUTS

From the first, the English word *nut* meant an edible kernel surrounded by a hard shell, and this remains the everyday meaning. But botanists have appropriated the word to refer specifically to one-seeded fruits with a tough, dry fruit layer rather than a fleshy, succulent one. Under this restricted definition, only acorns, hazelnuts, beechnuts, and sweet chestnuts—all members of one order of trees, the Fagales—qualify as true nuts among those that we commonly eat. Almonds and walnuts actually correspond to the stones of the peach or plum, and are surrounded by a fleshy layer while on the tree. The otherwise very different Brazil and pine nuts are both seeds, the first enclosed by the dozen in a woody pod, and the second borne without a covering (*Gymnosperm*, the class of plants that includes the pines, means "naked seed") on the pine cone.

We have seen that very few fruit, vegetable, or spice plants were known to more than one continent before the spice trade and discovery of the New World. The same is true of the cereals and legumes, but not of the nuts. Walnut, hazelnut, chestnut, oak, and pine trees all have both Old World and New World representatives. This is because the nut-bearing trees have simply been around a lot longer than the other food plants. Long enough, in fact, that they existed before North America and Europe had split apart, some 60 million years ago. The transcontinental distribution of many nut trees is, then, a mark of their great antiquity.

The Nutritional Value of Nuts

Most nuts are nutritionally similar to the cereals in two ways: they are good sources of B vitamins, and they have substantial reserves of proteins that are deficient in lysine. But their high fat content and relative scarcity rule them out as significant protein sources. Most nuts are never served as a separate dish in the course of a meal, and instead are used primarily as garnishes or snacks. The acorn and chestnut are two important exceptions to this rule; they are anomalously rich in starch, and have been used for centuries as a grain substitute by the poor, who grind them into coarse flour for porridge or flat breads.

Storing and Cooking Nuts

The high fat content of nuts renders them especially vulnerable to rancidity and the acquisition of other odors from their surroundings. Fat oxidation is favored by light, heat, moisture, and metal ions, and accordingly it is best to store nuts, cooked or raw, in dark, cool, dry areas, and in unreactive (glass or plastic) containers. Unshelled nuts also keep better than shelled. Peanuts, pecans, and walnuts are among the most sensitive (walnuts also darken with time), and almonds and cashews among the least.

Nuts are usually roasted or deep fried, the object in both cases being to crisp the tissue by drying it further, and to develop more intense flavors by way of browning reactions. Fairly low temperatures, around 275°F (135°C), are used to avoid scorching the outside before the inside is cooked through. Roasted nuts suffer losses in thiamine content, but the other vitamins are relatively unaffected.

COMPOSITION OF SOME COMMON NUTS
(PERCENT)

	Water	Protein	Fat	Carbohydrate
Acorn	14	8	5	68
Almond	5	19	54	20
Brazil	5	14	67	11
Cashew	5	17	46	29
Chestnut	52	3	2	42
Coconut (Meat)	51	4	35	9
Filbert, Hazel	6	13	62	17
Hickory	3	13	69	13
Macadamia	3	8	72	15
Peanut	6	26	48	19
Pecan	3	9	71	15
Pine	6	31	47	12
Pistachio	5	20	54	19
Sunflower	5	24	47	20
Walnut, Black	3	21	59	15
Walnut, English, Italian	4	15	64	16

Characteristics of Some Common Nuts

Acorn

The acorn is a true nut borne by the oak tree, and consists of two massive cotyledons, with the small embryo at the pointed end. Oaks are native to all continents but Australia, and of the 450-odd varieties known, white and live oaks are most frequently relied on for food. In Greek legend, the acorn was man's staple food during the Golden Age; since then, it has fallen somewhat in esteem and is more often fed to hogs. Its high carbohydrate content has led to its use as a pseudocereal by North American Indians and by Europeans during times of famine. The acorn has a very high tannin content, which can be reduced by soaking the ground nut in several changes of hot water.

Almond

The almond is the most popular common nut, though it is actually the seed of a plumlike fruit, or drupe (a fruit that has both succulent and stony layers surrounding the seed); the tree is a very close relative of the plum and

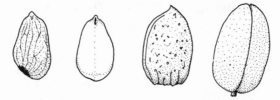

The almond, with its stony shell and fruit coat

peach. It is a native of western India, is mentioned frequently in the Old Testament, and was first cultivated in Europe by the Greeks. In the United States the almond is a major commercial crop in California, where it is grown on acreage second only to the grape's. Almonds are most often used in pastries and candies.

Brazil Nut

For once, the common name is a good guide: the Brazil nut is native to the Amazon, and attempts to cultivate the tree elsewhere have failed. Even in Brazil there are only a few commercial plantations; most trees remain in the wild. The tree is typically 150 feet tall and 6 feet in diameter, and bears 12 to 20 seeds in woody pods about 6 inches across. The pods are gathered only after they fall to the ground, but because they weigh about 5 pounds, they can be lethal missiles, and harvesters must carry shields to protect themselves. Brazil nuts are little consumed in Brazil itself; exports began around 1835.

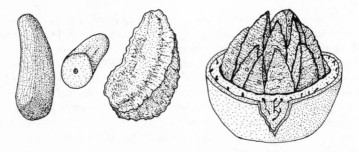

The Brazil nut in shell and pod

The edible portion of the seed is an immensely swollen embryonic stem. A pair of tiny cotyledons, more usually the storage organ in seeds, is visible at one end. Because of their size and high oil content, two Brazil nuts are the caloric equivalent of one egg. They are also especially rich in sulfur-containing amino acids.

Cashew

Like the Brazil nut, the cashew comes from the Amazon region, whose natives gave us its name. But it was successfully transplanted to India by the Portuguese in the 16th century, and today India and East Africa are the world's largest producers. The cashew is second only to the almond in world trade. Like the almond, the cashew is the seed of a drupe, but it is related not to the plum, but to poison ivy. This is why we never see cashews for sale in the shell. The shell contains an irritating oil that must be driven off by heating before the seed can be carefully extracted without contamination. The oil is used in paint, varnish, and rocket lubricants. In the producing countries, the seed-containing fruit is often discarded in favor of the swollen receptacle, or "false fruit," called the cashew apple. About 2 inches in diameter and 4 inches long, it is eaten raw or fermented into an alcoholic drink.

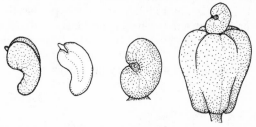

The cashew and its "apple," a false fruit prized for its own sake

Chestnut

In common with its distant relative, the acorn, the chestnut has an unusually high carbohydrate content and has been used as a pseudocereal. It is also occasionally steamed and eaten as a vegetable. It too is a native of several continents, and this is the basis for some hope in America's battle with the chestnut blight, which, since its explosion early in this century, has decimated the native population. Chinese and European varieties are now being tried out. The chestnut is not to be confused with the horse chestnut, which is not a true nut and belongs to an entirely different family. Horse chestnuts

The chestnut

are generally inedible on account of a high tannin content and the presence of other secondary compounds. Three known species, one of them Californian, are thought to be edible.

Because of its high initial moisture content, the chestnut is quite perishable, offering as it does a favorable climate for bacteria and molds. Chestnuts are best kept covered and refrigerated, and should be eaten fairly quickly. If freshly gathered, however, they should be cured at room temperature for a few days. This improves the flavor by permitting some starch to be converted into sugar before the cells' metabolism is slowed down.

Coconut

The coconut palm is one of the oldest food plants, its name recorded in Sanskrit, and today it is easily the most important nut commercially, growing as it does on poor, uncultivated soil and providing oil, fiber, lumber, and charcoal as well as food. About 20 billion nuts are produced each year. The coconut palm originated somewhere in the Malayan Archipelago, but has long been distributed throughout the tropics by both man and nature; the nuts have been known to survive flotation across entire oceans. Major producers today are the Philippines, India, and Indonesia.

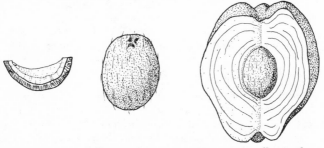

Though its meat is very different, the coconut is anatomically similar to the almond; both are the seeds of fruits called drupes. The coconut is shown here inside its fibrous fruit coat *(right)***.**

The coconut is the stone of a drupe fruit, and the meat and milk constitute the seed's endosperm. Coconut oil, one of the most useful of vegetable oils, is highly saturated and so very stable.

The word *coconut* comes from the Portuguese *coco*, which means goblin or monkey. The markings on the stem end of the nut can look uncannily like a face.

Filbert or Hazelnut

A true nut native to the temperate regions of the Northern Hemisphere, the filbert was known in China 5000 years ago and was gathered by the

The hazelnut, or filbert

Romans. It is guessed that the name comes from Saint Philibert, whose feast day, August 22, comes at the time of harvest in Europe. The filbert's close relative, the American hazel (in Europe, an older name for the filbert), is native to the Northeast and upper Midwest. Today, major plantings of the European varieties, which produce a larger nut, are located in the Northwest.

Hickory

Hickory trees are native to North America east of the Mississippi, and are little cultivated, with the exception of one species, the pecan. Hickories are members of the walnut family, and are the seeds of drupe fruits. The irregular nutmeat consists of the embryo's cotyledons.

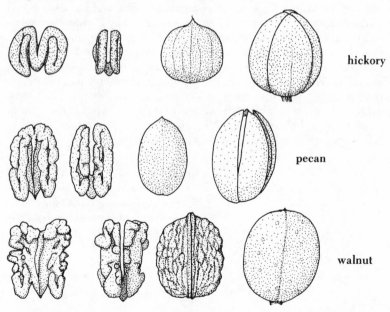

hickory

pecan

walnut

The hickory, pecan, and walnut are all members of the same family. The characteristic cotyledons are contained by a shell, and the shell by a tough, thin fruit coat.

Macadamia Nut

One of Australia's few contributions to the world's food plants is the macadamia nut, a native of that continent's northeastern coast. It is borne on a small evergreen tree that was named for John Macadam, a Scots-born chemist, in 1858. The macadamia was introduced to Hawaii in the 1890s, and became commercially significant there around 1930. The edible portion of the nut consists of the cotyledons.

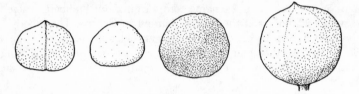

The macadamia nut, its shell, and fruit coat

Peanut

This most common of the nuts is not a nut, but the seed of a small leguminous bush that pushes its thin, woody fruit capsules below ground as they mature. The peanut is a native of South America, and has been found in Peruvian settlements dated at 800 B.C. It was spread by Portuguese explorers to East Africa, and by the Spanish to the Philippines; it came to North America via Africa and the slave trade. For a long time used primarily as livestock feed, the peanut became important as an oil and food source in World War II. Today, over half of the peanuts used in the United States go into peanut butter, which is eaten mostly by children. The world's largest producers are India and China; the largest exporter is Nigeria. The peanut consists of two cotyledons and a fairly large embryo. Peanuts are salted in the shell by being soaked in a brine solution before roasting.

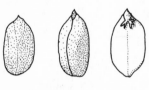 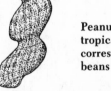

Peanuts are seeds of a tropical legume; the dry shell corresponds to the pods of beans and peas.

Pecan

A species of hickory native to the Mississippi River valley, the pecan was prized by the Indians (its name is Algonquian) and cultivated by Jefferson, who sent trees to Washington at Mount Vernon. The man who laid the foundation for serious improvement of pecan breeding, by working different grafts onto root stocks, was a slave named Antoine. In 1846 and 1847 he propagated a variety known as "Centennial" on 16 trees of Louisiana's Oak Alley plantation. Today, there are more than 300 named varieties grown mostly in the South and Southwest.

Pine Nut

The pine nut, or pignolia, is the naked seed of many different species of pine. It consists mostly of endosperm and several small cotyledons. A pine tree started from seed will take about 75 years to reach commercial levels of production. Italy and Spain are the leading exporters of pine nuts.

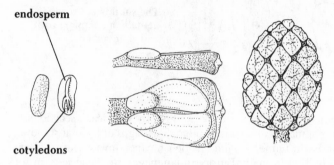

endosperm

cotyledons

The anatomy of a pine cone and its seeds

Pistachio

This native of central Asia has been cultivated for at least 3000 years, and was gathered from the wild long before that. The name derives from the Persian. In Genesis 43:11, Jacob sends gifts of almonds and pistachios to Egypt with his sons. A relative of the cashew, the pistachio is the seed of a drupe, and is unique for its greenish cotyledons. When the seed is ripe, the shell opens at one end. Pistachios first became available in the United States around 1880, but didn't become popular until the advent of vending machines in the thirties. The shell is sometimes dyed red. Pistachios are now grown in India, the Far East, Europe, North Africa, and the American Southwest; Iran and Turkey are the major exporters.

The pistachio

Sunflower Seed

This "seed" is actually a complete fruit, similar to the "seeds" on the exterior of the strawberry. It is the single North American native to become a significant world crop. Long cultivated by natives of the American Southwest, the sunflower was brought to Europe around 1510 as a decorative plant. The first large crops were grown in France and Bavaria in the 18th century as a source of vegetable oil. Today, the world's leading producer by far is, ironically, the Soviet Union. North Dakota, Minnesota, and California are tops in the United States. The seed is mainly storage cotyledons, and is also used for stock feed.

The sunflower "seed" is actually a complete fruit called an achene.

Walnuts

The walnut, a native of Asia, Europe, and North America, is second only to the almond in popularity and consumption. Two signs of its long-standing eminence are its Latin name, *Juglans*, which means "Jove's acorn" (*Jovis glans*), and the fact that in many European languages, the generic term for nut is also the word for walnut (the English word means "foreigner's nut"). The major American species, the black walnut, is native to the Appalachians; the English or Italian walnut, which came from Europe, is now grown mainly in California, and is preferred by producers for being easier to shell. The United States, France, and Italy are the major producers today. Like pecans and hickory nuts, its relatives, the walnut is the stone of a drupe, the edible portion being two irregularly shaped cotyledons.

Another species of walnut, the butternut, is little known, but esteemed by nut enthusiasts as among the best. It is remarkable for its high protein content—near 30%.

CHAPTER 6

Bread, Doughs, and Batters

THE ORIGINS AND HISTORY OF BREAD

It is sometimes said that domestication of the cereal plants made civilization materially possible. Perhaps a corollary to this idea is that bread represents the *culinary* domestication of grain, an achievement that made it possible to extract pleasure as well as nourishment from the hard, bland seeds. Before bread, grain was either simply parched on hot stones or boiled into a paste or gruel. Two discoveries ushered in a new age: first, that pastes too could be parched—hence, flat bread—and second, that paste set aside for a few days would ferment and become aerated—hence, raised bread. The original, simpler preparations may have satisfied the belly, but the new ones, with their yeasty flavor, variety of textures, and sculptural possibilities, satisfied the senses.

In rice-eating Southeast Asia and in those regions where starchy roots were the staple food, breads were little known. And peoples dependent on cereals other than wheat—for wheat alone produces gas-trapping gluten— would have developed only flat breads. But in the Middle East, the new techniques of bread making were quickly assimilated. The Bible indicates that in Moses' time bread in general was synonymous with nourishment. The familiar admonition of Jesus, "Man shall not live by bread alone," is a quotation from the Old Testament Book of Deuteronomy. And, after too much attention is paid to the miracle of the loaves, Jesus uses bread as a metaphor for *spiritual* nourishment, declaring, "I am the bread of life." That leavened bread was a common element of the Jewish diet is clear from Old Testament prohibitions against it during sacrifices and Passover, and from several New Testament passages that describe spiritual matters in everyday terms. Says

Jesus, emphasizing the covert but expansive, transforming power of yeast, "The kingdom of heaven is like unto leaven, which a woman took, and hid in three measures of meal, till the whole was leavened." Paul, on the other hand, likens yeast to spiritual adulteration: "Let us keep the feast, not with old leaven, neither with the leaven of malice and wickedness, but with the unleavened bread of sincerity and truth."

The biblical-sounding "Bread is the staff of life" is actually a 17th-century commonplace of the English pulpit and a sign of the remarkably persistent dominance of this particular food. In our own day, the slang senses of "bread" and "dough" play on the identification of two rather different necessities.

Bread has played a prominent role in Europe's sense of social relations and hierarchy. It has always been true that one's bread was determined by one's rank. From the Greeks on, white bread made from refined flour has been considered superior, and only the superior class enjoyed it. In Rome, the poor and criminals ate unleavened barley cakes or coarse dark breads; during the Middle Ages, they could also choose from oats and rye. Recall, too, the famous remark attributed to Marie Antoinette concerning the bread-less poor: "Let them eat cake"; literally, *"Qu'ils mangent de la brioche."* (Apparently, this was a favorite bon mot of the aristocracy; the wife of Louis XIV is reported to have asked, *"Qu'ils ne mangent de la croute de pâte?"*: "Why don't they eat pastry crust?") An interesting reversal has taken place recently in this country. Darker, whole grain bread is now thought to be more healthful than white and is now the loaf of preference for the educated and well-to-do, who can afford to pay premium prices for this onetime emblem of poverty.

The influence of bread extends to our very language of rank. The English word "lord" comes from the Anglo-Saxon *hlaford,* or "loaf-ward": the master is he who supplies food. "Lady" derives from *hlaefdige,* or "loaf-kneader"; she, or her retinue, produces what her husband distributes. And on the Latin side of our linguistic inheritance, the words "companion," "company," and variations thereon come from the late Latin *companio,* or "one who *shares* bread." Our social vocabulary has been shaped, in part, by this special food, which is the contemporary of civilization itself.

Bread has played many ritual and metaphorical roles throughout history. Diverse religions use bread in offerings and other ceremonies, the Jewish matzoh and the Christian communion wafer being two familiar examples, and the wedding cake a secularized one. In the British folk variation on Communion called "sin-eating," a willing soul is paid to sop up and consume away the sins of the newly dead by eating a loaf of bread over the coffin. Another superstition had it that a loaf containing a bit of mercury will indicate the location of a drowning victim by floating directly over the body (symbols of life were thought to be sensitive to life's recent loss). In a wonderful passage from *Huckleberry Finn,* Mark Twain has his hero convert this unlikely sonar device back into its original substance, the fine bread that he has seldom tasted. Several loaves have been launched to locate Huck after he fakes his own murder and hides out on an island in the Mississippi:

> A big double loaf come along, and I most got it with a long stick, but my foot slipped and she floated out further. . . . But by and by along comes another one, and this time I won. I took out the plug and shook out this little dab of quicksilver, and set my teeth in. It was "baker's bread"—what the quality eat; none of your low-down cornpone.

Many more examples could be cited, but it is clear enough from this brief sketch that bread, as substance and as idea, has long been a staple food of body and mind.

Early Developments

Ancient Times

The origins of bread, like those of most important culinary customs, are obscure. The best guess at present is that flat breads were a common feature of late Stone Age life; surviving versions include the tortilla, Indian johnny-cake, and Chinese pancake. About raised bread we can be somewhat more exact: according to archaeological evidence, it seems to have developed in Egypt around 4000 B.C. Raised bread had to await the selection of a new variety of wheat, one that could be easily husked. Wild wheats, with their characteristically adherent hulls, were parched in order to separate the grain, and heat denatures the gluten-forming proteins. Once a gruel of *raw* wheat could be made, the discovery of raised bread was only a matter of time.

The improvement of wheat was only one of several technical advances made in this most complicated of basic culinary operations. Grinding equipment progressed from the mortar and pestle to two flat stones and then, around 800 B.C. in Mesopotamia, to a circular motion that made feasible the eventual use of animal, water, and wind power. Fermentation, originally a matter of chance contamination by airborne yeasts, was promoted by the use of a piece of old dough—the method of choice to this day for the most prestigious bakers in Paris, as well as the makers of San Francisco sourdough—and then by the use of yeast-containing beer sediment. By Alexandrian times, around 300 B.C., yeast making was a specialized profession in Egypt. Finally, there were improvements in cooking equipment. The open fire was succeeded by the griddle stone, after which arose the primitive oven, which contained both coals and bread. The dough was stuck to an inside wall.

Leavened bread seems to have been a rather late arrival along the northern rim of the Mediterranean. The new wheat was not grown in Greece until about 400 B.C., and flat barley breads were probably the norm well

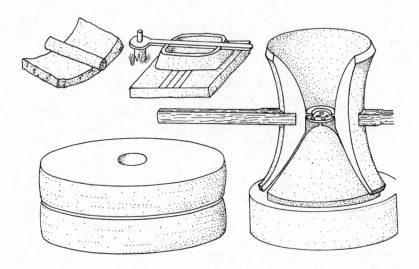

Four stages in the evolution of machines for grinding grain. *Clockwise from upper left:* the saddlestone and lever mill were limited by their need for discontinuous motion. The hourglass mill, which could be turned continuously in one direction by man or animal, was in widespread use by Roman times. Millstones finally made it possible to harness more elemental forces, and were put to use in water and wind mills.

after. We do know that the Greeks enjoyed breads and cakes flavored with honey, anise, sesame, or fruits. The Roman satirist Juvenal wrote that his countrymen were interested in only two things—"bread and circuses"—and huge amounts of wheat were imported from northern Africa and other parts of the empire to satisfy the public demand. During Pliny's lifetime, most bread was made in the home by women, though a corporation of bakers had been formed in 168 B.C.

Middle Ages

During the Middle Ages, Arabs brought the windmill to Europe from Persia, and the profession of bread making was established. Bakers' brotherhoods, and then guilds, arose in the 11th century. At first, bakers were specialists, producing either brown or white bread; it was not until the 17th century that improvements in milling and in per capita income led to the dissolution of the brown guild as a separate body. Bakeries were largely creatures of the city, while in the country, bread making continued to be a domestic task. In northern areas, rye, barley, and oats were more common than wheat and were made into coarse, heavy breads. One odd use of flat bread at this time was the "trencher," a dense, dry, thick slice that served as a plate at medieval meals. Once the meal was over, the trencher was either eaten or given away to the poor.

Adulteration and short weighing appear to have been fairly common problems during this period, and this led the civil authorities to control the quality and price of bread. In London, a baker whose loaf did not measure up would be dragged through the streets with the evidence hung around his neck.

Very few recipes for bread survive from the Middle Ages, probably because bread making was little practiced by the cooks who dictated the recipe collections. The first example quoted probably assumes some expertise in the reader, since no proportions are specified, and would probably have resulted in quite a variable product. "Berme" is the yeast produced by the fermentation of ale and includes some liquid.

Modern Times

While much of human life has been transformed in the last five or six centuries, domestic baking remains largely the same. Our ovens are much less troublesome, and our yeast packaged in foil, but as the 18th- and 19th-century recipes show, there is nothing new about the "sponge" method of mixing the water with a fraction of the flour, letting the yeast get a start, and then adding the rest of the flour to make up the dough. (One substantial change is the amount of bread baked at once: a bushel is the equivalent of 108 cups; a peck, 32 cups.)

ENGLISH BREAD RECIPES

Take fayre Flowre and the whyte of Eyroun and the yolke, a lytel. Than take Warme Berme, and putte al thes to-gederys and bete hem to-gederys with thin hond tyl it be schort and thikke y-now, and caste Sugre y-now ther-to, and thenne lat reste a whyle. An kaste in a fayre place in the oven and late bake y-now. And then with a knyf cutte yt round a-bove in maner of a crowne, and kepe the cruste that thou kyttest, and than caste ther-in clarifiyd Boter and Mille the cromes and botere to-gederes, and kevere it a-yen with the cruste that thou kyttest a-way. Than putte it in the ovyn ayen a lytel tyme and than take it out, and serve it forth.

—from a 14th-century manuscript

To make white bread, after the London way. You must take a bushel of the finest flour well dressed, put it in the kneading-trough at one end ready to mix, take a gallon of water (which we call liquor) and some yeast; stir it into the liquor till it looks of a good brown colour and begins to curdle, strain and mix it with your flour till it is about the thickness of a good seed-cake; then cover it with the lid of the trough, and let it stand three hours, and as soon as you see it begin to fall take a gallon more of liquor, and weigh three quarters of a pound of salt, and with your hands mix it well with the water: strain it, and with this liquor make your dough of a moderate thickness, fit to make up into loaves; then cover it with the lid, and let it stand three hours more. In the meantime, put the wood into the oven and heat it. It will take two hours heating. When your spunge has stood a proper time, clear the oven, and begin to make your bread. Set it in the oven, and close it up, and three hours will just bake it.

—from Hannah Glasse, 1774

Sift into a deep pan, or large wooden bowl, a peck of fine wheat flour, (adding a large tablespoonful of salt,) and mix the water with half a pint of strong fresh brewer's yeast, or near a whole pint if the yeast is home-made. Pour this into the hole, in the middle of the heap of flour. Mix in with a wooden spoon, a portion of the flour from the surrounding edges of the hole so as to make a thick batter, and having sprinkled dry flour on the top, let it rest for near an hour. This is called "*setting the sponge,*" or "*making the leaven.*" When it has swelled up to the surface, and burst through the coating of flour that covered the hole, pour in as much more lukewarm water as will suffice to mix the whole gradually into a dough. Knead it hard and thoroughly, leaving no lumps in it, and continue to knead till the dough leaves your hands. Throw over it a clean thick cloth, and set it in a warm place to rise again. When it is quite light, and cracked all over the surface, divide it into loaves, and give each loaf a little more kneading, and let it rest till it has risen as high as it will. Have the oven quite ready, and (having transferred the loaves to pans, sprinkled with flour,) bake them well.

—from Miss Leslie, 1857

English and American cookbooks from the 18th century on contain literally dozens of recipes for breads, cakes, and so on. French treatises, however, generally mention only sweet breads and pastry: everyday bread was either bought from the *boulangerie,* or making it was simply a task too well known to need explaining, or it was not an important part of refined dining. From the 17th through the 19th century, "French bread" was not the plain *baguette* it is today but what the French called *pain mollet,* or soft bread. It was made with butter, milk, and beer leavening, and despite learned opinion that it was "unhygienic," it was very popular.

Although domestic bread-making techniques have changed very little over the centuries, the proportion of bread baked in the home has fallen off drastically. In England around 1800, most bread was still being baked in domestic or communal village ovens. But as the Industrial Revolution spread and more of the population moved to crowded city quarters, the bakeries took over an ever-increasing share of bread production. This trend did not go unnoticed and was hotly challenged on economic, nutritional, and even moral grounds. The English political journalist William Cobbett wrote in *Cottage Economy* (1821), a tract addressed to the working class, that it is reasonable to buy bread only in cities where space and fuel are in short supply. Otherwise,

> How wasteful, then, and indeed, how shameful, for a labourer's
> wife to go to the baker's shop; and how negligent, how criminally
> careless of the welfare of his family, must the labourer be, who
> permits so scandalous an use of the proceeds of his labour!

Behind the somewhat shrill language is a public controversy over the adulteration of bakers' bread with whiteners (alum), fillers (chalk), and, according to some, human bones. There is also Cobbett's conviction that the traditional domestic life is intrinsically pure and honest:

> Give me, for a beautiful sight, a neat and smart woman, heating
> her oven and setting in her bread! And, if the bustle does make
> the sign of labour glisten on her brow, where is the man that
> would not kiss that off, rather than lick the plaster from the cheek
> of a duchess?

The moral note was sounded just as strongly by the Reverend Sylvester Graham, the Presbyterian minister and diet reformer after whom Graham crackers are named. In his 1837 *Treatise on Bread-Making,* he wrote long-windedly of the wife as her family's dietetic and spiritual savior, the only one who "justly appreciates the importance of good bread to their physical and moral welfare."

The effusions of Cobbett, Graham, and others failed, however, to reverse the trend. Bread making was one of the most time-consuming and laborious household tasks, a kiss on the sweaty forehead notwithstanding, and more and more of the work was delegated to the baker. Today, very little bread is made in the home. In 1900, 95% of all flour produced in this country was sold to families; in 1970, the figure was 15%. And as per capita income rises, per capita consumption of bread declines. If people can afford to get more of their calories from meat, they will, and they have. We lean less heavily than did our ancestors on the staff of life.

Leavening

Yeast

We have been eating raised breads for 6000 years, but it was not until the investigations of Pasteur a little over a century ago that the nature of the leavening process began to be understood. Today we know the secret to be the gas-producing metabolism of a particular kind of fungus, the yeasts. The word "yeast," however, is as old as the language and originally meant the froth or sediment of a fermenting liquid that could be used to leaven bread. ("Ferment" derives from the Latin *fervere*, to boil or seethe; the same root gives us "fervent.") In Egypt, beer froth was eventually used, but the first leaven was simply a piece of leftover dough in which yeast was already growing. The *first* raised doughs would have originated spontaneously, since yeast spores are ubiquitous and will readily infect a hospitable environment.

In Rome, the production of yeast was quite regularized, though beer froth was no longer part of the repertoire: Greece and Rome drank wine. But Pliny reports that bread made in barbarian, beer-drinking Spain and France "is lighter than that made elsewhere" because of the peculiarly bubbly leaven used (Book 18). Romans had several ways of making yeast. The simplest and oldest was to save a piece of old dough. During the grape harvest, however, millet or wheat bran was mixed with grape juice, allowed to become contaminated by the air, and then dried in the sun. The resulting cakes were soaked in water when needed. Barley cakes were sealed in clay pots until they became sour and yeasty, then dispersed in water. The Romans thought that raised breads were especially nutritious: Pliny says that they make the body strong.

Leap ahead 17 centuries, and very little has changed. According to A. Edlin's *Treatise on the Art of Breadmaking* (1805), the preferred leaven in England was beer froth, which could be preserved for later use by drying it slowly; when needed, it was mixed with water, flour, and sugar and allowed to ferment for a day. Said Edlin, "By means of this the finest, lightest bread is made." An old piece of dough was also still used, especially at sea, but was not favored because it produced "a disagreeable sourness" that had to be neutralized by adding potash. For some reason, sourness in bread had become a standard complaint by this time, though less than a century later

American gold miners made it into a mark of distinction. Around Edlin's time, the term "artificial yeast" was applied to what was probably the original way of making leavening: flour was mixed with water and allowed to ferment spontaneously. The idea first became popular on the continent; by mid-century, English bakers were producing yeast by fermenting malt and hops mixtures themselves rather than going to the brewer. They omitted flour from the formula on the grounds that it imparted a sourness to the yeast.

The American health movement of the mid-19th century reversed the Roman point of view and held that raised breads were likely to be harmful. There was probably ample if unspoken precedent for this distrust—it is thought that the prohibition against leavened bread in the sacrificial ceremonies of several religions has to do with the resemblance of leavening to spoilage and decay—and the reformers found various reasons to back it up. Sylvester Graham was fairly moderate, claiming that brewer's yeast is an "impure and poisonous substance" (probably a case of guilt by association) and that overfermentation, which produced the dreaded sour bread, could only be salvaged by the use of alkali, yet another adulterant. The scrupulous baker should use "sweet" domestic yeast and be very careful not to let the dough go sour.

The Boston Water Cure went further. In 1858, they published a pamphlet titled "Good Bread: How to Make It Light, Without Yeast, or Powders." Bread today, we read, is "rotted by fermentation or poisoned with acids and alkalis," to the point that " 'the staff of life' has well nigh become the staff of death." Their answer: to use such a hot oven that the sudden expansion of air and water vapor in the dough would do the job of puffing it up. The result must have been unacceptably heavy to most people's taste, since the suggestion sank without a trace.

Baking Powders

Pasteur's discovery in 1857 that yeast is composed of microscopic organisms whose activity caused fermentation did not make leavened bread any more popular. In his *Theory and Art of Breadmaking* (1861), Harvard professor Eben Horsford likened the yeast organisms to poisonous molds and quoted several physicians on the dangers of microbes. His suggestion was a mixture of sodium bicarbonate and "dry phosphate of lime," which could be extracted from bones. The supposed advantage of this particular formulation of baking powder was that the residues, lime and sodium phosphate, are both found in the human body and so are presumably harmless. If premixed with flour, Horsford noted, a handy "self-raising flour" would be the result, and loaves of dough could be ready for the oven in a matter of minutes, since no period of fermentation was necessary.

Of course, millennia of tradition were not overturned by these upstarts, and yeast has never lost its preeminence as a leavening agent, though today some commercial bread is mechanically aerated (yeast is added for flavor). But the chemical leavenings, or "powders," seen as a safe alternative to yeast by some and as yet another poison by others, have become important in their own sphere. The mixture of an acid with an alkali to produce carbon dioxide gas was developed around 1835 in England, and commercial baking powders first appeared around 1850. They are used almost exclusively in quickbreads, muffins, biscuits, and griddlecakes, whose batters are so thin that slow-acting yeast cannot aerate them properly. Even Graham allowed that bicarbonate and tartaric acid were handy for making pancakes, and the first successful self-rising flours (the idea preceded Horsford by about a decade) were marketed as pancake flours around 1905.

Pasteur's identification of yeast as a living organism has ultimately benefited the bread maker by making possible improved culturing techniques and purer, longer-lasting preparations. And on the question of health, contemporary opinion tends toward the Roman: culinary yeasts are not harmful; they are, in fact, a good source of B vitamins and lysine, the limiting amino acid in cereal proteins. Graham would have been shocked to see jars of brewer's yeast on health-food shelves.

White Bread versus Brown

As it turns out, we can also look back to the ancients for prescient advice on the relative merits of white and brown breads. Around 400 B.C., Hippocrates, the father of physicians, made this simple comparison in his *Regimen:* "Wheat is stronger and more nourishing than barley, but both it and its gruel are less laxative. Bread made of it without separating the bran dries and passes; when cleaned from the bran it nourishes more, but is less laxative." Or to make the phrase more pointed: brown bread is more laxative than white, white more nutritious than brown. This turns out to be not far from the truth.

Of course, the popular view today is that whole grain bread, because it contains the vitamin-rich germ and fiber-rich bran, is more nutritious and better for our general health than refined flour breads. This, in turn, is a relatively recent reaction against centuries, even millennia, of a rather unreflective preference for lighter breads. More labor was required to refine dark flour into light—not until the 19th century was it possible to remove *all* the bran—and the yield from a given measure of grain would be lower, so that light bread would have been scarcer and more expensive than dark, and therefore a status symbol. At least from the Greeks on, whiteness in bread was a mark of purity and distinction. Archestratus, a contemporary of Aristotle's and author of the *Gastronomia,* a sort of *Guide Michelin* of the ancient Mediterranean whose title gave us the word "gastronomy," accorded extravagant praise to a barley bread from Lesbos on just these grounds:

> And among the others [breads] the best can be found on Lesbos,
> there on the sailors' hill bathed by the billows, where the illus-
> trious Eresus is: bread so white that it outdoes the ethereal snow
> in purity. If the celestial gods eat barley bread, no doubt Hermes
> goes to Eresus to buy it for them.

We have already mentioned the decline of the brown bread guild in the late
Middle Ages, and the 18th-century scandals involving adulteration with
alum and chalk were probably encouraged by the public's color prejudice.

The first really influential partisan of whole grain bread was Sylvester
Graham, whose name became attached to whole wheat flour. Graham condemned the separation of bran in milling as "put[ting] asunder what God has
joined together"—the language of the wedding service coming naturally to
his pen—and claimed that coarse wheat was a remedy for various intestinal
complaints: "by its adaptation to the anatomical structure and to the physiological properties and functional powers of our organs, [it] serves to prevent
and remove the disorders and diseases of our bodies, only by preventing and
removing irritation and morbid action and condition. . . ." Not the clearest
of justifications, but Graham's influence extended into the 20th century. It
persuaded the breakfast cereal pioneer John Harvey Kellogg to advocate
enough roughage to induce at least three bowel movements a day and so
prevent "intestinal toxemia."

Fiber Rediscovered

Aside from a general association of roughage with "regularity," the issue
of refined foods seems to have faded from the turn of the century until the
mid-1970s, when it was revived and, with the help of news, health, and home
magazines, popularized to an unprecedented extent. A small group of British
surgeons published a report in the 1974 *Journal of the American Medical
Association* suggesting that through various mechanisms, dietary fiber—
indigestible components of the plant cell wall—could reduce the risk of diverticulosis, appendicitis, gallstones, varicose veins, hemorrhoids, colon and
rectal cancer, and atherosclerosis (hardening of the arteries). The rather
vague faith of Graham and Kellogg was now clarified and supported by population studies. And this time the country heeded the message, coinciding as
it did with the new interest in "natural," less processed foods. In the last two
decades, the percentage of commercial baking accounted for by white bread
has declined substantially.

As it turns out, fiber is not a panacea (see chapter 11 for details). As for
whole wheat in particular: it is true that whole grain flour contains more
protein, minerals, and vitamins than refined flour, including as it does the
nutritionally valuable germ and aleurone layer, as well as the mostly indigestible bran. But it is also true that some of these nutrients pass through the
digestive tract unabsorbed because the indigestible carbohydrates complex

with them and speed their passage out of the system. The nutrients in white bread do not suffer such losses. In normal diets, this drawback is negligible and probably leaves us ahead in protein and vitamins, but for people on marginal diets, whole grain bread can have disastrous consequences. We have already mentioned the epidemic of rickets that struck the children of Dublin after three years of wartime rations of dairy products and whole wheat bread. The combination of marginal supplies of calcium and vitamin D and the calcium-complexing activity of phytic acid, which is concentrated in the aleurone layer, was enough to tip the balance from health to serious disease. Similar problems with iron and zinc metabolism have been studied among the poor of Egypt and Iran. The moral: unrefined does not automatically mean healthful. Not all of us can afford to eat "naturally."

The irony is that following the Dublin outbreak and other evidence that mineral and vitamin deficiencies cause disease, the nutritional fortification of bread became mandatory in several countries, including the United States: but only white bread is affected, because whole grain breads are considered a specialty product. American consumers of brown bread are no longer the poor who cannot afford the price of refining, but rather a middle class interested in "pure" and "natural" products.

To summarize, brown bread contains more nutrients than white bread but makes it more difficult to absorb them. And while bran does ease bowel movements, there is no *direct* evidence that it thereby prevents disease. So we are left with a slight modification of Hippocrates' 2500-year-old observation: brown bread is more laxative, white bread more digestible. For people on an otherwise balanced and adequate diet, either is fully acceptable.

FLOUR

Types of Wheat

Three basic kinds of wheat are grown today, each with a characteristic kernel composition and particular culinary applications. The distinguishing factor is kernel "hardness," which is a measure of protein content. The higher the protein content, the fewer the starch granules, and so the more vitreous and translucent the endosperm. *Hard* wheat grains break up into large chunks of protein with relatively little free starch and form a strong gluten when the flour is mixed with water; hard wheats are therefore preferred for bread making and constitute about 75% of the American crop. *Durum* (a distinct species, *Triticum durum* rather than *T. aestivum*) is the hardest kind of wheat grown: too hard, in fact, for bread dough, which must have some give to it. It is usually milled into a coarse product called semolina, which is used to make the very stiff doughs necessary for dried pastas. Durum wheats account for about 5% of the crop. *Soft* wheats, with their high starch content, develop a weak gluten, and are made into cake flours and used in products that are meant to be tender and crumbly. Their flour has a distinctly soft, powdery texture compared to the gritty, coarse hard flours. "All-purpose," or "household," flour is a blend of hard and soft flours meant for use in a wide range of foods. Because it is a hybrid, it seldom gives the same results as commercially baked breads or cakes, which are made with specialized flours. Recent technical advances in particle separation have made it possible to extract a hard-flour fraction from soft wheat grains and vice versa, thus freeing the production of flour to some extent from dependence on the qualities of a particular crop.

FLOUR COMPOSITIONS
(PERCENT)

Flour	Protein	Carbohydrate
Whole wheat	13.3	71
Semolina	12.3	73.5
Straight hard	11.8	74.5
All-purpose	10.5	76.1
Straight soft	9.7	76.9
Cake	7.5	79.4
(12% moisture basis)		

Milling

Milling is the process of separating endosperm from bran and germ and then grinding down the endosperm into flour-sized particles. Though bran and germ are nutritionally valuable, the lipids they contain lower the shelf life of flour substantially, and neither plays any role in the physical structure of the dough, except to dilute the desirable properties of gluten and starch and to shear into the gas cells of raised breads. In fact, our word "flour" comes from the "flower," or best part of the meal: that is, the portion of the grain left after milling and then screening out these larger particles. ("Whole wheat flour" would have been an oxymoron for medieval Englishmen.) Wheat grain is prepared for milling by being cleaned, graded, and then "conditioned." This last step raises the moisture content of the kernel slightly, thereby toughening the bran, softening the endosperm, and so making it easier to separate the two. Grooved rollers shear open the grain, squeeze out the germ, and scrape the endosperm away to be ground, sieved, and reground until the particles reach the desired size.

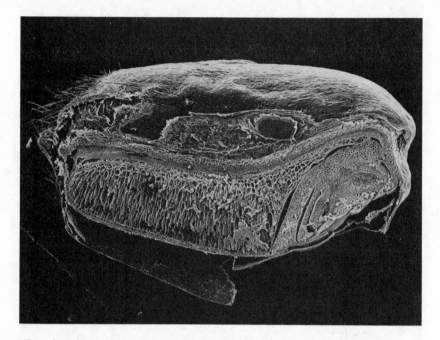

The wheat kernel before it is milled into flour. Its actual length is about 6 mm, or ¼ inch. The germ is at right, and the endosperm extends to the left; both are surrounded by the bran (see p. 231). (Scanning electron micrograph courtesy of Ann Hirsch.)

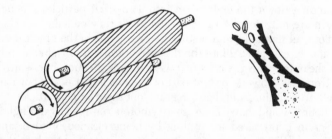

The first step in milling is to pass the grain between two grooved rollers, one rotating faster than the other. This kind of action separates the germ and bran from the endosperm, which is then ground further to make flour.

Durum semolina viewed through the scanning electron microscope. These chunks of endosperm are close to half a millimeter long. They consist of a dense, hard protein matrix surrounding starch granules. The absence of free starch granules makes for a tough dough that is well suited for pasta. (From Dexter et al., *Cer. Chem.* 55 (1978): 23–30. Courtesy B. L. Dronzek, Dept. of Plant Science, University of Manitoba, Winnipeg.)

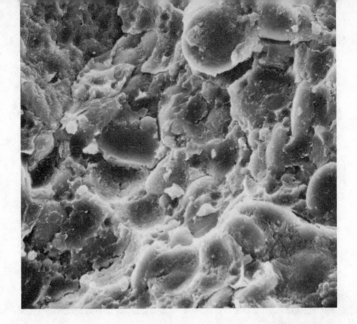

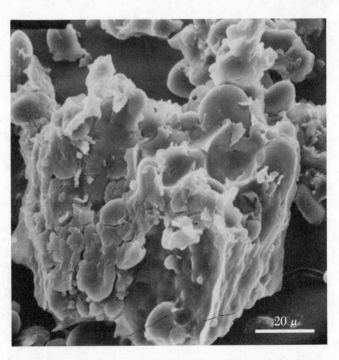

20 μ

Hard wheat endosperm *(top)* and flour *(bottom)*. The protein matrix is strong and solid enough to break off in chunks during milling, but compared to durum semolina there are many more starch granules, whole and damaged, projecting from the chunks and free in the flour. Hard flours make strong glutens and are preferred for bread making. (From Hoseney and Seib, *Bakers Dig.* 47 (1973): 26–28.)

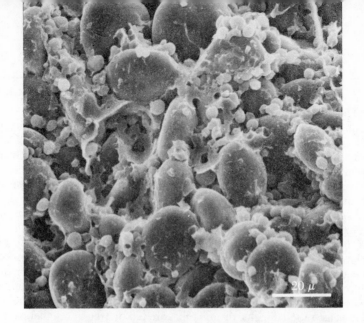

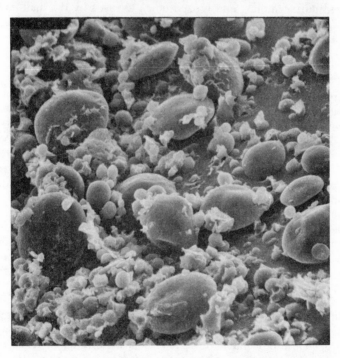

Soft wheat endosperm *(top)* and flour *(bottom)*. The protein in this kind of wheat comes in thin, weak sections interrupted by starch granules and air pockets. When milled, it produces very small particles, starch granules being among the more massive. Soft flour works into a weak gluten and is preferred for tender cakes. (From Hoseney and Seib, *Bakers Dig.* 47 (1973): 26–28.)

The final dimensions of endosperm material determine the protein content, and thereby the culinary behavior, of the flour. This is so because starch grains come in two basic sizes, with the larger—a few ten thousandths of an inch in diameter—accounting for most of the starch by weight. Flour particles in this range will tend to be mostly starch, while larger or smaller particles will contain more protein. Milling results in three classes of particles: individual starch grains, small chunks of the protein matrix, and larger chunks of protein that include one or more starch granules. The coarsest milling product, semolina, consists of large pieces of protein, a few hundredths of an inch across. It is especially well suited to dried pasta because more starch would result in a weaker dough and more easily broken noodles. Flour is generally composed of particles under 0.005 inch across and contains a mixture of all three particle classes. Free starch granules are a necessity in leavened doughs and batters, where they play an important structural role. And in yeast-raised doughs, the microbes are fed by free starch that has been damaged during milling and broken down to sugar by the malt enzymes. One cup of all-purpose flour contains about 100 billion particles of wheat endosperm.

Bleaching and Aging

After the flour has been ground and blended to the desired mix of particles, it is treated chemically to accomplish in a matter of minutes what otherwise takes weeks. Bleaching removes the light yellow color caused by xanthophylls, a variety of carotenoid pigment also found in potatoes and onions. The color has no practical or nutritional significance and is oxidized simply to obtain a uniform whiteness. Bleaching does, however, destroy the small amounts of vitamin E in flour, which probably accounts for its bad reputation in some circles. For historical reasons, yellow coloration is valued in pasta, and so semolina is never bleached.

Bleaching is often accomplished with the same gas, chlorine dioxide, that is used to age, or "improve," the flour. But even unbleached flour has been aged with potassium bromate or iodate. Aging has important practical results. It has long been known that flour allowed to sit for one to two months develops better baking qualities; hence the practice of letting flour "age" before use (during this period, it is also naturally bleached by oxygen in the air). But done in this way, aging is a time- and space-consuming, somewhat unpredictable procedure. Hence the use of chemicals both to accelerate and to control flour improvement. Aging affects the bonding characteristics of the gluten proteins in such a way that they form stronger, more elastic doughs. (This is discussed in more detail on page 296.)

The Components of Flour and the Behavior of Dough

Flour contains many different substances, including proteins, starch, lipids, sugars, and enzymes. Most of these need only a sentence or two of description. The one exception is the proteins, which, when mixed with water, form that remarkable material we call gluten. Without gluten, there would be no such thing as raised bread. Our understanding of gluten is, to put it mildly, imperfect; the subject is very complicated. Still, with a little work, we can make sense of many familiar aspects of bread making.

Wheat Proteins and Gluten

Wheat is preeminent among the grains because it is the only one whose endosperm proteins interact to form a gluten strong enough to produce raised breads (rye proteins form a very weak gluten). Gluten is both plastic and elastic; that is, it will both change its shape under pressure and tend to reassume its original shape when pressure is removed. Thanks to this delicate balance of properties, wheat dough can expand to incorporate the carbon dioxide produced by yeast and yet will put up enough resistance that it will not thin to the breaking point. If dough were wholly plastic, the gas would simply migrate to the surface and escape, while if it were merely elastic, the gas would accumulate in a few increasingly pressurized pockets, and the bread would come out heavy and coarse.

There are four general classes of storage protein—protein designed to feed the seedling until it has become self-supporting—in wheat endosperm, and each class includes up to 20 particular proteins. The water-soluble albumins and the globulins play a negligible role in bread making; they account for only 10 to 25% of the protein in flour. Gliadin and glutenin are the two insoluble classes that form *gluten,* the gumlike residue that remains after you have chewed on a piece of raw dough for a few minutes. Gliadin proteins are made up of several thousand atoms, and the glutenins may range from the thousands up to perhaps a million. By any measure, they are both long, sprawling molecules. It is obvious from these numbers that the interaction of the proteins with each other and with millions of water molecules in dough is horrendously complicated. Fortunately, it is possible to reduce the situation to a fairly simple scheme.

Proteins consist of an extended, occasionally coiled backbone of atoms, with various side groups of other atoms projecting out from this long chain (for more about protein structure, see chapter 13). Because the molecules are so long, it is possible for them to fold back upon themselves so that side groups on both adjacent and widely separated portions of one chain can end up next to each other and bond together. When such bonds do form, the molecule is fixed in this folded position. We can picture a single gluten molecule, then, as looking something like a fish line that is at the same time coiled up and folded or kinked onto itself. The gliadin molecules tend to form compact, ellipsoidal balls, while glutenin molecules are more extended. Mix them together in large quantities, and you have quite a tangle. Some side groups will react with others on other molecules, thereby crosslinking the proteins together. Gluten proteins do not dissolve in water (this is why they survive chewing, like gum) primarily because these bonds between dif-

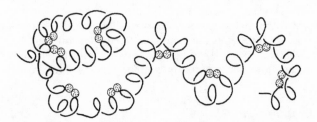

Schematic model of a gluten molecule. It is a very long, coiled chain of amino acids, some of whose side groups are attracted to each other. These bonds between different turns on the coil introduce kinks and folds into its shape.

ferent molecules are so numerous that water molecules cannot invade the protein complex and separate them from each other. Still, gluten proteins do absorb about twice their own weight in water, retaining it by means of hydrogen bonds. Dough gluten, then, is basically a complex network of intertwined proteins with water molecules filling the interstices.

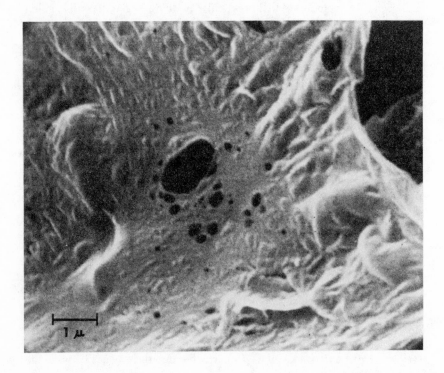

Plasticity and Elasticity

Knowing this much about the structure of gluten, we can make sense of the behavior of dough: its combination of plasticity and elasticity, the toughening that kneading brings about in it, and its tendency to break down if kneaded too long. When flour is first mixed with water, the proteins begin to unfold into a random network. Water molecules separate and lubricate the long chains to some extent by forming hydrogen bonds with them and so breaking some hydrogen bonds between proteins. The mixture behaves initially as a thick liquid. When it is stirred with a spoon, the protein chains are drawn together into visible filaments and form what has been vividly described as a "shaggy mass." The process of unfolding continues during kneading, an activity that both compresses and stretches the protein-water complex. The constant movement and stress have the effect of forcing the

Two instants in the process of gluten formation. When water is first added to flour, the gluten proteins are randomly oriented in a viscous liquid *(left)*. When this liquid is stirred, it quickly develops into a tangle of fibers *(right)* as the proteins unfold and gather in long bundles. (From Bernardin and Kasarda, *Cer. Chem.* 50 (1973): 735–45.)

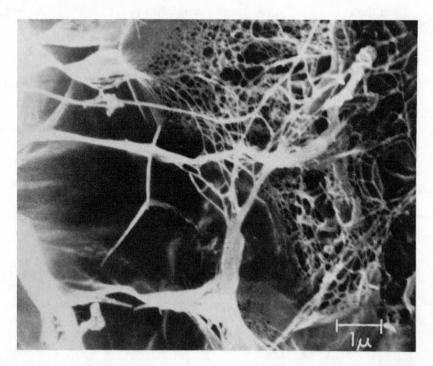

long molecules into a more orderly pattern, lining up local groups of them in roughly the same direction. (An analogy with this process might be shaking a box full of randomly oriented pencils.) As it develops, the side-by-side arrangement encourages the formation of more, and more regular, bonds between different molecules, and this crosslinking in turn makes the arrangement more stable and less easily deformed. The dough suddenly gets stiff, harder to manipulate, and takes on a smooth, shiny surface.

Kneading furthers the process of unfolding and aligning the long, initially tangled *(left)* gluten molecules. In the end, sheets of gluten associations *(right)* are interleaved in the dough, which takes on a smooth, fine texture.

At the same time that the dough gets stiff, it also becomes much more elastic: both more difficult to deform and, once deformed, tending to spring back to its original shape. The protein molecules have now been unfolded and elongated to a large extent, but many kinks in the chains remain, the product of bonds between nearby side groups on the same molecule. If stress is now put on the gluten complex by pulling on opposite ends, these kinking bonds will resist and then finally break. The individual molecules lengthen, and the mass as a whole therefore stretches. Once the deforming stress is removed, the attraction between side groups reasserts itself, and the molecules kink up again and shorten, pulling their crosslinked neighbors with them. The visible result: the dough creeps back to its original shape like a piece of elastic.

There are several important factors at work in the balance between plasticity and elasticity. One, the flour lipids, we will discuss shortly. Another is the ratio of gliadin and glutenin molecules; the former tend to remain compact and contribute fluidity to the dough, while the glutenin fraction does more of the stretching. And then there are the disulfide—sulfur-to-sulfur—bonds involved in some of the kinking and crosslinking among the gluten proteins. The reestablishment of these covalent bonds following stress, and even their resistance *during* stress, is compromised by the presence of another sulfur group called a thiol group—simply a combination of one sulfur and one hydrogen atom that can exist either on proteins or on much smaller molecules. If physical tension is making it difficult for two sulfur atoms to share their pair of electrons and a relatively mobile thiol group is

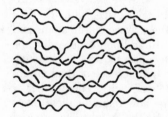

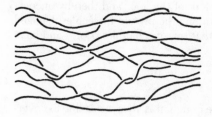

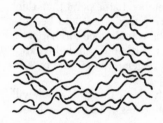

Dough is elastic because the aligned gluten molecules still have many kinks in them *(top)*. When a mass of gluten is pulled, the kinking bonds are broken, the proteins get longer, and the mass stretches *(middle)*. When it is released, the bonds re-form, and the dough shrinks back to its original shape *(bottom)*.

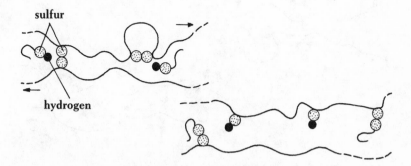

Tension being applied to two gluten molecules *(left)*, which are kinked and cross-linked by means of sulfide bonds. Sulfur-hydrogen groups on small, mobile molecules weaken gluten by blocking its sulfur atoms and preventing them from re-forming their bonds *(right)*.

nearby, then a stress-relieving interchange will take place, with the thiol group donating its hydrogen atom to one of the stressed sulfur atoms and forming a new disulfide bond with the other. In addition to lowering the proteins' resistance to stretching, this interchange also blocks the rekinking of the molecules once the stress is removed. So the presence of thiol groups contributes to a dough's plasticity and reduces its elasticity. The number of thiol groups in a particular flour will determine how many of these interchanges can occur; the more thiols, the less elastic the dough. The aging of flour, or the equivalent chemical treatment, is now understood to be an oxidizing process that lowers the thiol content of the flour and thereby encourages greater kinking and crosslinking among the protein molecules. Unaged flours produce relatively weak doughs that trap gas bubbles poorly and bake into flat, dense loaves.

Dough Breakdown

No matter what the strength of the dough, it is always possible to "overdevelop" it: to knead it so long that it breaks down into a thick fluid with no elasticity. In this case, the proteins have been stretched to the point that, thiol content notwithstanding, the disulfide bonds are pulled apart and become thiol groups by picking up hydrogen ions from the water. Once this happens, the suddenly increased thiol population makes further disulfide-breaking exchanges more likely, and the whole process of degeneration in the protein structure accelerates.

The bread maker's aim is to develop a highly elastic dough—which will mean maximal gas retention, loaf volume, and fineness of texture—without

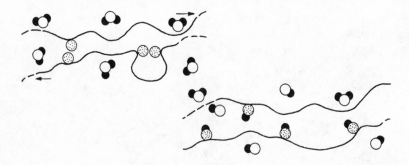

When dough is overworked, too much stress has been put on the proteins, and water molecules are sufficient to do the work of blocking sulfur atoms that thiol groups normally do (see previous drawing). The result of overworking is gluten with little ability to kink up or cross-link, and so a dough with little elasticity.

Farinograms are graphic representations of dough strength. The higher the pen moves, the more the dough is resisting the mixer blades. A strong flour produces a dough that becomes very stiff and stays that way under continued kneading *(left)*. A weak flour gives a dough that does not get as stiff, and that breaks down if it is kneaded after it develops *(right)*.

passing the invisible boundary that marks overdevelopment. It isn't easy to overwork bread dough by hand, but it is quite easy to do so in a food processor or a commercial dough developer. One important measure of flour quality is the susceptibility of its dough to breakdown. Strong flours made from high-protein wheat are generally stable, while weak flours from soft wheats are less so. This characteristic can be represented graphically by the "farinogram," which is produced by a pen coupled to the shaft of a mixing machine. The height of the pen's mark is proportional to the dough's resistance to the mixing blade. A strong, stable gluten will reach a fairly high plateau and decline only gradually as development continues, while a weak gluten will reach a lower peak of elasticity and then fall off suddenly. Very strong flours are especially necessary in commercial baking where high-speed developers work the dough a great deal in a matter of seconds. Commercial bread flour may be well suited to the food processor, but it requires substantially more hand kneading than does all-purpose flour. The resulting loaf, however, is unusually large and fine-grained.

Starch

Starch is by far the major component of flour, accounting for about 70% of its weight. It is present in the form of small granules, which are deposited in the wheat endosperm to provide the embryo with a compact energy source. The granule is composed of two kinds of starch molecules, amylose and amylopectin. Amylose usually accounts for about 25% of the total starch and is a long, linear molecule, while amylopectin is irregularly branched; individual molecules contain many tens of thousands of sugar units. In a granule, the two types of molecules are intermingled. In some areas, the oth-

erwise loosely organized molecules will happen to match up with each other
and become closely interlocked to form a "crystalline" region. A single long
molecule may pass through several crystalline regions and several amorphous
ones; on the average, about half of a given starch granule is crystalline. As
we shall see, this tendency of starch molecules to fall into very close associations is the single most important factor in bread staling—and it can be
reversed.

It was long thought that starch granules were contained by a thin membrane, perhaps made of cellulose. This idea was put forward to explain the
odd fact that the granules swell very little when put in water, until the temperature reaches a certain range, when they suddenly burst. Damaged granules, on the other hand, swell—that is, absorb water—even in cold water.
But no such protective coating has been found. It appears that the strongly
bound crystalline areas prevent water from entering the granule, until the
temperature is high enough to break up the intermolecular bonds. At that
temperature, around 140°F (60°C), the granule swells up all at once, and
the starch is "gelatinized," or forms a viscous complex with water. This is
why flour makes a good thickening agent in sauces and gravies, and why
these mixtures tend to thicken suddenly during cooking.

In batters and doughs, starch granules serve two important functions.
They help form the mechanical structure of the baked product by contributing a semisolid phase and by regulating the location of water in the cooking dough (more on this below). And, in yeast-raised breads, damaged starch
granules—from 10 to 35% of the total—are attacked by starch-breaking malt
enzymes, which convert them into sugars that the yeast cells can use as food.

Lipids

Flour contains on the order of 1% lipids, by weight—hardly a major
component numerically. But they are crucial to the development of a well-
raised bread. These fatty substances are concentrated in the gluten, and
when experimenters removed all the flour lipids from some dough before
baking, the dough was incapable of expansion to accommodate the yeast's
carbon dioxide. There is evidence that some lipids can form bonds with both
gliadin and glutenin molecules and may help bind these two populations
together into gluten, as well as bind gluten to the surface of starch granules.
And in order to explain the results of some X-ray studies of bread dough, a
model of gluten structure has been proposed that supplements the protein
picture outlined above. The gluten proteins are thought to form large, thin
sheets that are separated both by water and by very thin (two molecules thin)
layers of lipid material. This layering allows slippage on a large scale and,
together with thiol interchanges, would make an important contribution to
the phenomenon of plasticity in dough.

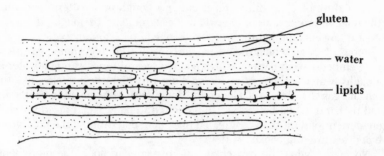

A model of dough structure which accounts for the observation that flour lipids are necessary to the proper expansion of dough during baking. Layers of wheat oils lubricate sheets of gluten, making it easier for these sheets to slide past each other and so accommodate the expansion of bubbles in the oven. (Redrawn from Grosskreutz, *Cer. Chem.* 38 (1961); 336–49.)

Minor Components: Carbohydrates and Enzymes

Starch is the principal but not the only carbohydrate to be found in flour. Endosperm cell walls contribute small amounts of cellulose and larger amounts of hemicelluloses and pentosans, long chains of five-carbon sugars. The latter two components, because they absorb water but do not dissolve, contribute somewhat to the structure of dough and the final moisture of the bread. Sugars, which constitute about 2% of flour weight, are very important in raised breads because they feed yeast. But since this amount is enough to do the job for only a short period of time, flour manufacturers have long supplemented the ground wheat not with additional sugar, but with malted wheat or barley: grains that have been allowed to sprout and develop the enzymes that break starch down to sugars. These enzymes begin to work when the flour is moistened, attacking damaged starch granules and producing a variety of sugars—glucose, maltose, oligosaccharides, and larger molecules—of which only glucose and maltose are metabolized by the yeast. Unsupplemented flour contains only small amounts of these starch enzymes, or amylases. All flours contain small amounts of proteases and lipases. By slowly digesting the gluten proteins into shorter molecules, the proteases can cause a dough to come out sticky and inelastic; fortunately, salt inhibits their activity. Flour lipases have no noticeable effects on bread making.

THE INGREDIENTS IN DOUGHS AND BATTERS

A mixture of flour and water is either a dough or a batter, depending on the relative proportions of the two major ingredients. In doughs, the water content is low enough that the water-protein complex, gluten, constitutes the continuous phase in which the other components (starch granules, gas pockets) are embedded. In batters, which contain several times as much water as do doughs, water is the continuous medium in which the proteins, like the starch and gas, are dispersed. Generally, doughs are stiff enough to be manipulated by hand, while batters are thin enough to pour. Gelatinized starch is often more important than coagulated protein in determining the final structure and texture of a cooked batter.

No matter what the particular ratio of water to flour, the ingredients we commonly add to this mixture—leavening, shortening, milk, eggs, sugar, salt—alter the balance of forces in batters and doughs in important ways.

Leavening

Yeast

The yeasts are a group of single-celled fungi; about 160 different species are known. Some cause human infections, some contribute to food spoilage, but one species in particular—*Saccharomyces cerevisiae*, "brewer's sugar fungus"—is put to good use in brewing and baking (see page 437 for a micrograph of yeast cells). In both processes, yeast gives the product a characteristic flavor; it leavens bread and converts grain or grape carbohydrates into alcohol.

Carbon dioxide and alcohol are normal products of yeast metabolism, and sugars are its raw material. The overall equation runs like this: $C_6H_{12}O_6 \rightarrow 2C_2H_5OH + 2CO_2$. The single-unit sugars glucose and fructose are the first to be consumed; then, after a time lag during which its metabolism shifts, the yeast feeds on the double-unit sugar maltose, which the malt enzymes produce from damaged starch granules. Added sugar will increase yeast activity until its concentration is high enough to have a deleterious osmotic effect on the cells. Because their metabolism declines sharply at this point, greater numbers of yeast cells must be used to compensate for the lowered gas output per cell. This is why extra yeast is often called for in recipes for sweet breads. Added salt has the same effect as excess sugar. Finally, yeast activity is dependent on temperature: the cells do best at about 95°F (35°C).

Compressed and dry yeast, the two most common forms, are different genetic strains of the same species and have significantly different traits. Compressed yeast is only partly dried, then is pressed into solid cakes, each of which contains on the order of 100 billion yeast cells (and a few 100 million bacteria). It easily survives freezing; in fact, it loses less activity during

DOUGHS AND BATTERS: REPRESENTATIVE COMPOSITIONS

	Flour	Total Water	Shortening	Milk Solids	Eggs	Sugar	Salt
Doughs							
Bread	100	65	3	3	0	5	2
Biscuit	100	70	15	6	0	3	2
Pastry	100	30	65	0	0	1	1
Cookie	100	20	40	3	6	45	1
Pasta	100	25	0	0	5	0	1
Batters							
Pancake, waffle	100	150	20	10	60	10	2
Crêpe, popover	100	230	0	15	60	0	2
Spongecake	100	75	0	0	100	100	1
Pound	100	80	50	4	50	100	2
Layer	100	130	40	7	50	130	3
Chiffon	100	150	40	0	140	130	2
Angel	100	220	0	0	250	45	3

Note: The numbers shown indicate parts measured by weight, with the amount of flour constant at 100. This chart is meant to give only a general idea of the relative proportions used in common baked foods; individual recipes will vary widely.

storage at −10°F (−23°C) than it does at 40°F (4°C). When added to water just before being mixed with the flour, it is fairly quiescent, showing little activity until it comes in contact with its sugar supply. Dry yeast (8% moisture, compared to 70% for compressed yeast) can be stored at room temperature, though it too does better in the freezer. It generally takes longer than cake yeast to shift over to maltose fermentation. But its most important peculiarity is its requirement of very warm water—the optimum is 105°–110°F (41°–43°C)—during rehydration. Lower or higher temperatures result in almost complete loss of fermenting power: that is, most of the yeast organisms are disabled. And, unlike compressed yeast, rehydrated dry yeast can sometimes be very active in water, bubbling away even before it is given sugar to feed on.

The behavior of dry yeast is not entirely understood. This is the current educated guess: when the yeast cells are dried out, some carbohydrate reserves are left concentrated inside. When water is added, some of these carbohydrates leave the cell to achieve an osmotic balance and then can constitute the "food" that enables the cells to metabolize so actively in the absence of flour. It is thought that dry yeast is sensitive to water temperature because reconstitution of the cell membranes must proceed rapidly enough that crucial cell contents are not lost to the solution. At temperatures lower than the optimum, reconstitution is too slow, while higher temperatures may begin to damage the cells and cause the loss of cell contents. An analysis of carbohydrates leaked from dry yeast into water at different temperatures supports this hypothesis.

Chemical Leavening: Baking Powder and Soda

Generally, yeast can be used as a leavening only in doughs, where gluten is the continuous phase. Yeast produces carbon dioxide relatively slowly, and the surrounding material must be elastic enough to contain it indefinitely: nothing is gained by an hour of fermentation if all but the last five minutes' production of gas escapes. Weak doughs and batters, then, must be raised with a faster-acting gas source: faster and yet not so fast that a cake, for example, will have time to collapse in the oven once the leavening period is over. This is the need filled by baking powders.

Chemical leavenings exploit the reaction between certain acidic and alkaline compounds, which results in the evolution of carbon dioxide, the same gas that yeast produces. The alkaline component is almost universally sodium bicarbonate (or sodium acid carbonate; the formula is $NaHCO_3$), commonly known as baking soda. It is ideal because it is cheap to produce, easily purified, nontoxic, and tasteless. Potassium bicarbonate is available for those on low-sodium diets, but this compound tends to absorb moisture and react prematurely, and it has a somewhat bitter flavor.

Baking soda can be the sole added leavening if the dough or batter is already acidic enough to react with it and evolve carbon dioxide. Yogurt and sour milk contain lactic acid and often are used instead of water or ordinary milk in such products. Sour milk can also be added along with the baking soda as a separate, "natural" component of the leavening. Two general rules of thumb: ½ teaspoon baking soda is neutralized by 1 cup of sour milk; and baking soda to be neutralized with sour milk provides the leavening action of four times its volume of baking powder, which contains both acid and alkaline materials. Sweet milk can be soured for use with baking soda by adding a tablespoon of lemon juice or vinegar, or 1¼ teaspoons cream of tartar, for every cup of milk.

Baking powder contains baking soda and an acid in the form of salt crystals that dissolve in water. Ground dry starch is also added to prevent premature reactions in humid air by absorbing moisture, and to dilute the powder. Most baking powders are "double-acting"; that is, they produce an initial set of gas bubbles upon mixing the powder into the batter and then a second set during the baking process. The first, and smaller, reaction is necessary to form many small gas cells in the batter; the second, to expand these cells to a size appropriate for a light final texture, but late enough in the cooking that the surrounding material has set enough to prevent the escape or coalescence of the bubbles and the ensuing collapse of the structure. This double action is accomplished by mixing two different acid salts into the powder, one of which reacts at room temperature, the other only at much higher temperatures. Cream of tartar ($KHC_4H_4O_6$), tartaric acid ($H_2C_4H_4O_6$), and monocalcium phosphate ($CaH_4(PO_4)_2$) are common fast-acting acid salts, and sodium aluminum sulfate ($Na_2SO_4 \cdot Al_2(SO_4)_3$) the most common high-temperature component. Different commercial brands of baking powder differ mainly in the proportions of acid salts. A home version of single-acting baking powder can be made by mixing ¼ teaspoon baking soda and ½ teaspoon cream of tartar; this is roughly the equivalent of 1 teaspoon baking powder.

Yeast not only leavens dough but gives it a characteristic flavor. Chemical leavenings too can affect flavor, but not beneficially. When acid and base are properly matched, neither is left behind in excess. But when too much soda is added, or when the batter is poorly mixed and the powder forms lumps—that is, when some material has not reacted—a bitter, soapy, or "chemical" flavor results. Excess alkalinity can alter and even destroy the flavor of chocolate and molasses. Colors will also be affected in alkaline conditions: browning reactions are enhanced, chocolate turns reddish, and blueberries turn green.

Shortening

This term has been used since the early 19th century to mean fats or oils added to baked goods that supposedly "shorten" or break up masses of gluten, thus weakening the structure and making the final product more tender. In fact, the role of added lipids in doughs and batters is not this straightforward. Here's what we know today.

The model of fat breaking up gluten sheets is appropriate in the case of pastry dough, in which large chunks of butter or lard are left intentionally unmixed with the flour. After repeated rollings and foldings, the dough becomes a mass of alternating gluten and fat layers. When baked, it results in a stack of separated flakes rather than an integrated, breadlike matrix. Solid fats work much better in pastry than liquid oils, which will seep throughout the flour phase. Ingredients and utensils are chilled (a cold marble slab is recommended for a working surface) before making pastry dough in order to minimize fat melting and the consequent loss of flakiness.

In cake batters, where gluten is not the continuous phase, the role of added lipids is quite different. In the first place, they serve as a sort of mechanical leavening agent. When butter is "creamed" with sugar, the sharp edges of the sugar crystals cut into the solid fat and create air cells, which can then be incorporated into the batter. Commercial shortening is a modern fat that is conveniently precreamed without sugar. Like other molecules, lipids form crystals just below their melting point, and the smaller the characteristic crystal size, the more air can be incorporated in the spaces between. (Liquid oils, of course, will not hold air any more than liquid water will.) Vegetable oils that have been partly hydrogenated to raise their melting point above room temperature turn out to have the optimal crystal type and so are most commonly used for shortening (lard has very large crystals and is much better suited to pastries). They are either charged with nitrogen or whipped to incorporate substantial amounts of air—about 10% of their volume is taken up by gas—and then tempered into a soft substance that requires much less creaming than butter does to attain a desired lightness. When the cake is baked, shortening tenderizes by separating starch granules from coagulated protein. And added fat will also make the cake moister and smoother in the mouth.

What fats do in bread dough is more of a mystery. We have seen that lipids naturally present in flour are necessary for bread dough to be extensible, and so capable of increasing in volume as the yeast does its work. Experiments have shown that added shortening up to the amount of 3 to 4.5% of the total dough weight will increase the final loaf volume by up to 20%, with most of this increase coming at very low shortening levels. And for some reason, the higher the melting point of the fat, the more it will increase loaf volume. Many models have been suggested to explain this dramatic effect, but none so far has seemed convincing. More understandable is the moistening and tenderizing effect of fats or oils on bread: lipids slow moisture loss by coating the starch granules.

Emulsifiers

A word is in order at this point about an ingredient known to the ordinary cook only as an additive in commercial shortening and in cake mixes. The emulsifiers—there are many—are molecules with a polar head and a nonpolar tail. By burying their tails in fatty droplets and so coating the droplet surfaces with their heads, the emulsifiers prevent the droplets from coalescing into larger and fewer drops and so separating from the water phase. Emulsifiers, then, preserve the emulsion, the fine suspension of lipids in water. This property of insulating lipid droplets is used in cake mixes to prevent the shortening from interfering with the protein-air foam and to keep the shortening itself in the smallest crystals with the greatest air capacity (this is their role in commercial shortening, too). Emulsifiers make possible sweet cakes of unprecedented lightness and volume; sugar tends to reduce cake volume, and very sweet cakes had previously been rather dense.

Bread bakers have been using emulsifiers for better than 40 years as an antistaling agent; they seem to prevent certain changes in the starch granule, but the exact mechanism is unknown.

Though the individual cook cannot buy emulsifiers off the shelf and use them at will, knowing what they are and what they do helps to explain *why* it is they are used commercially and why certain kinds of commercial products are difficult to match when made from scratch.

Sugar

As we have seen, sugar will, in moderate amounts, increase yeast fermentation by providing the cells with additional food, while in higher concentrations typical of raised sweet breads, it will inhibit fermentation by upsetting the cells' water balance. Sugar can also affect the development of gluten. It is hygroscopic and competes with the flour proteins for the available water; for this reason, high-sugar doughs take longer to form and develop. The same characteristic causes the final product to be moister, tenderer, and to stay that way longer, since moisture leaves the bread less readily when sugar is there to absorb it. Finally, added sugar enhances browning reactions and will make for a darker crust in a given period of baking. This, too, may make for a moister loaf if it causes the cook to take the bread out of the oven somewhat earlier than usual.

Salt

Small amounts of salt are incorporated into batters and doughs for the taste, but both yeast and gluten are also affected by its presence. Salt inhibits yeast activity and, if added to excess, will reduce the volume and lightness of the loaf. By forming strong ionic bonds with side chains on the flour proteins, salt tends to make them less mobile and so the gluten less extensible, more tough. Again, the end result is a denser loaf. But by inhibiting the activity of protein-digesting enzymes in the flour, salt prevents the gluten from being weakened into a sticky mass that might retain very little carbon dioxide.

Milk and Eggs

Both milk and eggs contribute three major ingredients: water, protein, and fat. The first must be taken into account when determining how much liquid to add to the flour. Milk and egg proteins coagulate into solid filaments when heated and so contribute to the structure of the baked product. Beaten egg whites incorporate large amounts of air and are often used with creamed butter to leaven cakes. And egg and milk fats will have the same effects as added shortening. Egg yolks will also contribute their characteristic color and flavor. Milk must be scalded—at 198°F (92°C) for 1 minute, or 185°F (85°C) for 7 minutes—before being used in doughs; this heat treatment apparently alters milk serum proteins that otherwise interact with flour proteins to produce a weak, "slack" dough. (The milk must be cooled before mixing to avoid damaging the yeast.)

BREAD

There are four basic phases in bread dough: gluten, starch granules, yeast, and carbon dioxide. The protein network forms a superstructure in which starch granules are embedded, and yeast activity divides it into thin, tender layers separated by pockets of gas. We will now follow the development and interaction of these phases, from mixing to staling.

Preparing the Dough

Mixing

The aim of the first step in making bread is obvious: flour, water, and yeast are mixed together, to get as homogenous a distribution of each ingredient as possible. As mixing occurs, several processes are initiated. The gluten proteins unfold and begin to form the water-protein complex. Damaged starch also absorbs some water, though the whole grains swell only slightly. Starch enzymes, both native to wheat and added in the form of malt, begin breaking the damaged starch down into sugars. And the yeast cells begin to feed on the sugars, multiplying and producing carbon dioxide and alcohol.

A variation on this mixing procedure that has its proponents today is the "sponge method." As we have seen, this method is at least two centuries old and has also been common practice in commercial bakeries. About half of the flour is added to all the water and yeast, and this "sponge" is allowed to sit for a while before the rest of the flour is mixed in. In effect, this interposes a period of fermentation in the middle of mixing (the name of the technique refers to the moist, bubbly texture that occurs at this point). The yeast is thereby given a head start in growth, and the fully mixed dough will be better aerated and better populated with yeast, compared to straight-mixed dough. The even moistening of flour may also be easier when accomplished in two steps.

Kneading

When the dough becomes too thick to stir, the cook abandons spoon and commits hands to kneading, an extended period of manipulation. Kneading improves the aeration of the dough and furthers the development of gluten. As the dough is compressed, folded over, compressed, and folded over many times, pockets of air are incorporated into the dough and squeezed under pressure into smaller, more numerous cells. Because the carbon dioxide produced by yeast does not create new pockets—the process is too gradual and diffuse—but instead leaks into and expands preexistent pockets, kneading will largely determine the final texture of the loaf. The fewer and larger the air cells, the coarser the bread. Commercial breads and homemade bread whose dough is kneaded in a food processor tend to have very fine, cakelike textures because machines are far more efficient at aerating the dough than are a wooden spoon and two hands.

Kneading also develops the gluten network. Repeated stretching and compressing unfolds the proteins further and encourages the development of crosslinking between the extended molecules. Oxygen in the air pockets oxidizes some thiol groups, thereby improving the dough's elasticity, while its plasticity is improved by the formation of gluten and lipid sheets. Kneading is continued as long as the dough continues to become stiffer. If the dough is worked long enough that many disulfide crosslinks are permanently disrupted, it breaks down and becomes sticky and inelastic. Overdevelopment is a real problem only when kneading is done mechanically; a food processor may need only a minute or two to ruin a dough, while humans are only too willing to stop work. The dough is well kneaded when it takes on a fine, satiny appearance. Because both added fats and sugar slow gluten development (fats by waterproofing the gluten and sugar by itself absorbing water), rich and sweet doughs must be kneaded longer than others.

Fermentation, or Rising

This is the stage during which the dough is set aside (covered to avoid moisture loss and contamination) for an hour or so in a warm place. Gluten development continues to some extent during fermentation because the expansion of gas pockets stretches the proteins. But the principal activity at this point is the multiplication of yeast cells and their production not only of carbon dioxide, but also of many minor compounds responsible for the characteristic yeast flavor. The carbon dioxide output expands the air pockets, and these, in turn, will influence the final texture of the bread. The optimum rising temperature is actually only 80°F (27°C), much lower than many cookbooks seem to suggest. The yeast do multiply and produce gas faster up to about 95°F (35°C), but they also secrete more noticeable quantities of sour and unpleasant-smelling by-products. And the warmer the dough, the stickier and more difficult to handle it is in subsequent steps.

The end of the fermentation period is signaled by the dough's volume—it approximately doubles—and by the condition of the gluten matrix. When poked with the finger, fully fermented dough will retain the impression and will not spring back: the gluten has been stretched to the limit of its elasticity. The dough is now punched back to relieve the stress on the gluten, squeeze out excess carbon dioxide and divide the gas pockets, redistribute the yeast and its food supply, and even out the temperature (fermentation generates heat) and moisture. While the surface of the expanding mass may have been dehydrated, the rest of the dough has become moister as the yeast converts sugar into water and carbon dioxide. Fermented dough will feel easier to work than newly kneaded dough for this reason.

Doughs made from hard bread flours are always put through a second complete rising to develop their tougher gluten fully; recipes for homemade breads may call for one or two risings. The second takes half as long as the first. Otherwise the dough is now shaped into loaves, placed in pans, and allowed to rise for a shorter period. Bakers call this step "proofing" (the same term is sometimes used for testing the activity of packaged yeast by starting it with some sugar in warm water). Subsequent fermentations proceed more rapidly than the first because more and more yeast is at work. The purpose of the proofing step is to set the structure of the final loaf in preparation for the "oven spring," the final expansion of the dough that occurs in the first few minutes of baking.

Taken together, the three stages of mixing, kneading, and fermentation require several hours of work and waiting from the bread maker. In commercial baking, where time and work are money, mechanical dough developers can produce a "ripe" dough, with good aeration and an optimum gluten, in 4 minutes. Yeast is added to such doughs only as flavoring.

Baking

Baking can be conveniently divided into three stages. The first begins when the partly risen loaf is put in a hot oven, and ends after about a quarter of the baking time has elapsed, when the interior of the loaf has reached about 140°F (60°C) and the yeast cells have been killed. The oven is usually set between 400° and 425°F (204°–218°C) for ordinary doughs, between 350° and 375°F (177°–191°C) for sweetened. (Sweet doughs must be cooked more gradually to prevent the surface from browning before the interior has set.) Baking temperature is determined by the necessity of coordinating two processes: the expansion of the gas cells on one hand, and the coagulation of gluten and gelatinization of starch on the other. If the oven temperature is too low, the dough will expand to its maximum long before the gluten and starch have set, and the loaf will collapse into a flat, dense mass. If the oven is too hot, protein and starch in the outer layers solidify too quickly, form a crust, and prevent the loaf as a whole from expanding. Again, texture suffers.

The yeast cells die when the temperature reaches about 140°F (60°C), but early in the baking they briefly become more active. This temporary increase in the production of carbon dioxide, and the expansion of the gas volume as it heats up, together cause the rapid rising called *oven spring*. This phenomenon must be anticipated, and the proofing loaves not allowed to rise to their limit before baking. Otherwise, gas cells will rupture before the gluten has solidified, and the loaf will collapse.

The second stage accounts for about half of the baking time. The center of the loaf now reaches its maximum temperature, just short of the boiling point, and as the starch and gluten phases are transformed, the semisolid foam of dough solidifies into bread. The two changes occur almost simulta-

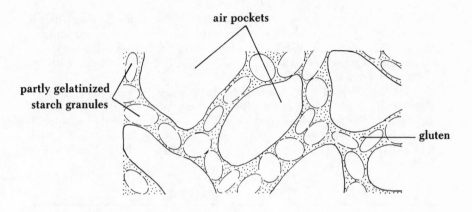

A schematic view of the structure of bread. Starch granules help reinforce the gluten walls that surround air pockets.

neously—starch gelatinizes at about 140°F (60°C), protein coagulates at 160°F (71°C)—and are a model of cooperation. The starch granules act as a sort of reinforcing material in the gluten: they furnish a solid surface for the gluten to adhere to, and they tend to line up along the walls of gas cells. Being discrete particles, they easily adjust position as the cells change shape. And as a phase that absorbs more water as the temperature rises, the starch relieves gluten of the water it *loses* as its proteins coagulate. The long protein chains align with each other and form rigid bonds, thereby squeezing out the water molecules that had occupied the interstices. Starch granules have a large water capacity—up to 10 times their own weight—and the gluten gives up only a fraction of this, so bread starch is never more than partly gelatinized. During gelatinization, the linear amylose molecules migrate out of the granule and form networks with water around the granule. This loss of starch disrupts the crystalline regions of the granule, and the amylopectin molecules left behind become surrounded with water. The granule expands and softens. The partly gelatinized starch phase will become more solid, as the gluten is now doing, when the bread cools.

In the last quarter of the baking period, stage three, the chemical and physical changes begun in stage two are completed, and the surface browning reactions, which improve both color and flavor, are initiated. Browning reactions between sugars and amino acids occur rapidly only above the boiling point and so follow the desiccation of the loaf surface. Though limited to the hot, dry crust, these reactions affect the flavor of the whole loaf because their products diffuse inward. A light-colored loaf will be noticeably less flavorful than a dark one.

A crust that is both brown and glossy can be obtained either by painting the surface with egg or by putting a pan of hot water in the oven or spraying the loaf surface with water. (Bagels are boiled briefly before baking.) In the second technique, the added moisture causes the surface starch to gelatinize into a thin, transparent coating. Since moisture prevents browning reactions, the treatment must be stopped early enough that the surface has time to dry out if a dark crust is desired.

Bread is judged to be done when its crust has browned and its inner structure has become fully set, the proteins coagulated, and the starch partly gelatinized. The second requirement can be verified indirectly by tapping on the bottom of the loaf. If the interior is still largely liquid, it will sound and feel heavy and dense. If it has cooked through, the loaf will sound hollow.

Cooling and Staling

Immediately after being removed from the oven, the loaf of bread is far from being a homogenous body. The outer layer is very dry and close to 400°F (204°C), the interior moist and around 200°F (93°C). During cooling, these differences slowly even themselves out. Moisture diffuses outward, and heat from the surface, both inward and out into the cooler air. As the temperature of the whole loaf declines, the starch solidifies but remains permeable to gases. If it didn't, the loaf would collapse despite its solid matrix. As it cools, the trapped gas contracts in volume and so becomes denser. As a result, the external air pressure becomes greater than the opposing internal gas pressure. If the solidifying matrix were impermeable, the loaf would be crushed by this imbalance. As it is, the starchy areas of the gas cell walls permit air to diffuse into them, thus equalizing the pressure and preventing collapse. This is just one more aspect of the intricate conspiracy that makes raised breads possible.

Starch and Staling

Most of us have only a vague idea, based on personal impressions, of what is involved in staling. It seems to arise during extended storage and to involve the loss of moisture: the crust gets leathery, the interior dry, hard, and crumbly. Our common sense turns out to be only partly right. The processes that cause staling actually begin during cooling, even before the starch has solidified enough that the loaf can be cut. And bread will stale even when there is no net loss of moisture from the loaf.

The landmark study of bread staling came as early as 1852, when the Frenchman Jean-Baptiste Boussingault, a pioneer in the study of nitrogen fixation (he demonstrated that certain plants increase the nitrogen content of soil and that soil alone—or, as we know today, certain soil bacteria—could do the same), showed that bread could be hermetically sealed to prevent it from losing water, and yet still go stale. He further established that staling could be reversed by reheating the bread to 140°F (60°C): the temperature,

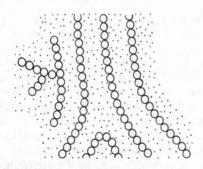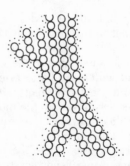

The recrystallization of starch after baking. The starch molecules (long chains of glucose units) slowly settle back into a tightly bonded mass *(right)*, squeezing out the water that separates them and producing a dry, crumbly texture in the bread.

we now know, at which starch gelatinizes. Subsequent research has shown that the starch phase is indeed the culprit, though gluten is involved in a minor way.

Picture the starch granule just after baking. It has taken up the water squeezed out of the gluten, the linear amylose molecules have formed a gel around the granule, and the branched amylopectin molecules left in the granule have become separated from each other by water. The once semi-crystalline granule has now become amorphous. But as cooling commences, some of these trends are reversed. Like a starch-based gravy, the cooling amylose gel gets stiffer, the long molecules more closely ordered, and some of the water held in the looser configuration is released. This amylose *retrogradation,* as it is called, is essentially complete once the bread has reached room temperature, and is responsible for the firming that improves its slicing quality.

The retrogradation of amylopectin is a slower process that continues over several days, but it is responsible for staling, as it is generally defined: the undesirable firming in texture *after* the bread has been cooled. Within the granule, the less regular amylopectin molecules reorder themselves only slowly, but in time crystalline regions do reappear, and again incorporated water is expelled. The same process occurs in the gluten phase, but most of its water has already migrated into the starch during baking. The cumulative result of all this molecular reordering: the starch and protein phases become more dense and more segregated from water, the texture firm and crumbly.

Staling is not, then, simply desiccation but a change in the location and distribution of water molecules. Where does the water go? Eventually, it goes to the crust, which begins much drier than the rest of the loaf. Here the excess water is absorbed by the gluten and starch, and crispness gives way to a tough, leathery kind of texture, the crust having very few pockets of gas to interrupt the solid matrix. Of course, if the bread is left uncovered, the water will eventually be lost to the air.

Because much of the water released by the starch remains in the loaf, bread staling can be reversed by heating it back to the gelatinization temperature of starch. Once more the intermolecular bonds are disrupted, water molecules move into the interstices, and the granule becomes tender again. One disadvantage of reheating is that the bread must be sealed to avoid evaporation, and this means that the crust will get even tougher, if anything. Reheating also noticeably revives the flavor of days-old bread. This may be because some flavor constituents become immobilized in the retrograded starch and are freed when the starch gelatinizes again or, more likely, because the volatile molecules responsible for flavor are *more* volatile, and so more available to the nose, at higher temperatures.

Certain emulsifying agents have been found to retard staling substantially and for this reason have been added to bread doughs for about 50 years by commercial bakers. It is thought that these substances complex with starch or in some other way interfere with water transport, thereby inhibiting recrystallization. For some reason, both the rate and the extent of staling are lower in lighter, less dense breads.

A last important point about staling: it is a temperature-dependent process. It proceeds most rapidly at temperatures just above freezing, and very

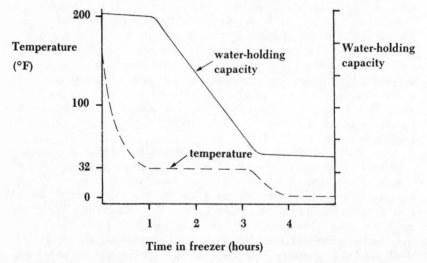

A graph showing the decline of starch's water-holding capacity as a cooked starch gel is cooled in a freezer. The decline is steepest when the gel is in the process of freezing solid at 32°F (0°C). (Redrawn from Chan and Toledo, *J. Food Sci.* 41 (1976): 301–03.)

slowly below freezing. In one experiment, bread stored at 46°F (7°C), a fairly typical refrigerator temperature, staled as much in one day as bread held at 86°F (30°C) did in six. The lesson is obvious. If you are going to freeze bread, freeze it as quickly as possible, so as to minimize the time it spends close to 32°F (0°C). If you are not going to freeze bread, store it wrapped tightly at room temperature. Some bakeries are now printing on their plastic bags the instructions "Store at room temperature or freeze." This is better advice than the still common "Store at room temperature or refrigerate."

Sour Breads

As we have seen, sourness in bread was long considered a serious defect, usually the result of poor yeast or letting the fermentation stage go on too long. In poorer areas of northern Europe, where rye bread was the rule, sourness was more acceptable.

Sourdough bread proper, the wheat bread associated with San Francisco, is another story, though it also begins with a kind of deprivation. Gold miners in California and Alaska did not have access to fresh supplies of yeast and had to revert to the ancient method of using a piece of leftover dough to start each new fermentation. Eventually, the "mother dough," or "starter," would come to be dominated by the microflora characteristic of the local area: contamination by airborne spores would dilute the initial yeast population. Breads with different mothers, then, would have slightly different compositions. This may be why it is reputedly difficult to keep sourdough starters from "deteriorating," or changing in microbial content, when they are brought to other parts of the country. Chicago's atmosphere is just not the same as San Francisco's.

A group of Bay Area scientists has analyzed the microflora of sourdough and come up with some interesting results. To begin with, the predominant yeast in sourdough is not the usual baker's yeast, *Saccharomyces cerevisiae*, but rather *Saccharomyces exiguus* (*exiguus* means small or scanty; *cerevisiae* means "of beer"), which thrives in very acidic environments and, unlike baker's yeast, cannot metabolize maltose at all. The acid responsible for the sour taste—about 75% is lactic acid, the rest acetic—is produced by a group of bacteria that has an absolute requirement for maltose and so would be unlikely to grow in a dough populated by baker's yeast. The several strains of bacteria involved are all very closely related to each other but not to any other known species, and so the investigators have dubbed them *Lactobacillus sanfrancisco*. They do best at a temperature of about 86°F (30°C) and a pH between 3.8 and 4.5. (Ordinary bread is more like 5.5.) Starters that have been maintained for decades appear to be very resistant to contamination, and it is thought that some sort of antibiotic action, analogous to that of the *Penicillium* molds in cheese, may be involved.

OTHER DOUGH PRODUCTS

Pastry

Pastry dough differs a great deal from bread dough, both in the proportions of standard ingredients, and in its final texture. It is unleavened, and contains about half the water of bread dough and many times the shortening. It bakes into a compact mass, with either a crumbly or a flaky texture. In the case of puff pastry, which is extremely flaky, the aim is to produce dozens, even hundreds of very thin, separate layers during baking. There are two keys to making good pastry dough. The first is to avoid gluten development as much as possible: in an unleavened dough, gluten would make for a very dense, tough texture. The low water content and minimal manipulation of the dough both reduce gluten development. The second key is to deploy the shortening so that it performs two tenderizing functions—isolating very small particles of starch and protein from each other, and isolating whole sheets of dough from each other.

While the best flour for bread making is a hard one that will develop a strong gluten, pastry requires a fairly soft flour. At the same time, too starchy a flour will not develop enough gluten to achieve a recognizably sheeted structure and will produce a very mealy pastry. All-purpose flour, which is a compromise between hard and soft, generally works quite well. In the case of puff pastry, which is manipulated more than plain pastry, very soft cake flour is usually added to household flour in the ratio of one part to every three or four in order to dilute the flour proteins.

The two ways in which shortening tenderizes pastry dough require very different mixtures of the two phases. In order to separate starch granules and gluten strands from each other, the lipids must be dispersed very finely throughout the dough. But only much coarser and less homogenous deposits will separate whole layers of dough from each other. Even dispersion is easiest with liquid oils, and the coarser dispersion, with solid fats or hydrogenated oils. If a crumbly, mealy crust is desired, oil is the best shortening, while recipes for flaky crusts will call for lard or solid vegetable shortening. Lard, the traditional favorite, produces superior flakiness because it forms crystals so large that they give the fat a noticeably grainy texture. However, lard tends to go rancid relatively quickly, and this fact, together with the bad reputation of cholesterol and saturated fats, limits its popularity today.

Technique for Plain Pastry

The mixing technique for basic pastry dough is dictated by the need to limit gluten development and to avoid liquefying all the shortening and so dispersing it too evenly. First, half of the shortening is mixed with the flour using knife blades or a special pastry cutter—these instruments mix with a minimum of kneading—until the two materials coat each other evenly and the mixture has a mealy texture. Then the rest of the shortening is cut in rather coarsely until it is broken into pea-sized pieces. A small amount of cold water (about 2 tablespoons per cup flour) is then added and the dough manipulated very briefly, just until the water is absorbed; in this way, little gluten is developed and the shortening is not melted. The dough is then put in the refrigerator to hydrate more fully and evenly. During this phase, some gluten does develop, but it is useful in forming a sheeted structure and has not been toughened by mechanical work. The dough is then rolled out and used to support or surround various sweet or savory mixtures. During baking, the larger particles of shortening, trapped air, and steam all help to separate the dough into layers and give it a flaky texture; the agitation as water vaporizes within the dough is quite visible after about 5 minutes. Because so little water is used, the starch granules do not undergo much gelatinization, and the flakes of pastry are dry and crisp. The bottom of a pie crust, which tends to come out soggy, can be improved by using as little liquid in the filling as possible and preheating the filling so that it takes less time to set in the oven.

Puff Pastry

Preparing this dough is much more elaborate and time consuming than preparing ordinary pastry. Here the flour (a combination of all-purpose and soft cake flours) is first mixed with the ice water, again avoiding excessive manipulation, and the resulting dough is patted into a large, thin rectangle, ideally on a prechilled slab of marble, which prevents the dough from warming to the work. Now the shortening, traditionally butter, and accounting for about a third of the dough's weight, is placed in small pieces or in a thin sheet onto two-thirds of the rectangle. The dough is then folded in thirds onto itself so that the butter forms a solid layer separating each fold of dough from its neighbor. This structure is then rolled out to the size of the original rectangle, folded in three once more, and then refrigerated for 30 minutes to an hour so that the butter has a chance to resolidify. The sequence of rolling, folding, and refrigerating is repeated two or three times more. The result is a dough made up of from 80 to 240 layers, give or take a few that stick together. When puff pastry is baked, usually in an initially very hot oven, the expanding air and water vapor puff the separate layers apart from each other and cause a volume increase of from 6 to 8 times. Puff pastry dough must be cut with a very sharp knife; a dull blade will press the many layers together at the edge and reduce their expansion during baking.

Pasta

Its History

We naturally think of pasta—thin pieces of stiff, unleavened dough that are molded into regular shapes and boiled, not baked—as a quintessentially Italian food. Both recipes and the vocabulary of shape, from the unfigurative macaroni, lasagna, and ravioli, to spaghetti (little strings), vermicelli (little worms), linguine (little tongues), and so on are of Italian origin, though the general term "noodle" comes from the German *Nudel*, whose own derivation is obscure. Yet the Italians were certainly not the only ones to invent pasta and were probably not the first.

According to a long-lived legend, noodles were first invented in China and then introduced to Italy by the traveler Marco Polo, who died in 1324. There is no doubt that the noodles were there for Polo to discover. It is thought today that the Chinese had noodles by the first century A.D., and special noodle shops became popular sometime during the Sung dynasty, from 960 to 1280. One early Chinese writer on the subject remarks that the common people invented noodles but learned the best ways to prepare them from foreigners, an indication that the food may have been invented independently by several peoples. In fact, there is good evidence that both India and the Middle East had known noodles at least by 1200 and probably long before.

The Italian side of the story is not at all clear, though the Marco Polo connection is quite improbable. Some partisans claim that the Etruscans had pasta, and there is some sense in the idea that this form of grain, which is so much easier to produce than bread—no fermentation or tricky baking is involved—would have been a "natural" discovery for primitive peoples to make. But the question is then raised as to how pasta is distinguished from other unleavened and boiled foods. For example, we know from the literature of Marco Polo's time that foods called lasagna, macaroni, and ravioli were already known in Italy. The poet Iacopone da Todi (died 1306) left us this instructive quatrain:

> *Chi guarda a maggioranza*
> *Spesse volte s'inganna.*
> *Granel di pepe vince*
> *Per virtù la lasagna.*

> He who looks at magnitude
> Is often mistaken.
> A grain of pepper conquers
> Lasagna with its strength.

And in the *Decamerone* of Boccaccio (died 1375), the story is told of an attempt to hoodwink a gluttonous peasant by describing the marvelous

> Berlinzone, the land of the Basques, in a country called Bengodi, where . . . there was a mountain made entirely of grated Parmesan cheese, on which lived people who did nothing but make macaroni and ravioli and cook them in capon broth. And then they threw them down, and the more of them you took, the more you had. And nearby ran a rivulet of white wine whose better was never drunk, and without a drop of water in it.
>
> [Day 8, Tale 3]

The problem is that these foods probably bore little resemblance to today's versions. It seems that they were more like dumplings, or gnocchi, and not thin noodles. In 1517, one Teofilo Folengo, under the pseudonym Merlinus Cocaius, published the *Liber Macaronices (Macaronic Book)*, which mixed classical and vernacular languages in a kind of literary burlesque: "pig Latin" would be a modern approximation. Folengo said that he took his title from macaroni, which he called "a certain food put together from flour, cheese, and butter: large, unrefined, and rustic." Two centuries after Marco Polo, then, macaroni was not necessarily the delicate hollow tube that we know today. A somewhat earlier recipe from Maestro Martino, however, sounds quite close to our macaroni. And the first English "macrows" were thin but flat noodles.

The beginnings of pasta's current popularity date from the 18th century, when mass production by machine was begun in Naples, and when English tourists developed a liking for it. "Macaroni" was used in England to describe a young man who had traveled on the Continent and now affected foreign tastes; it was a synonym for "fop" or "dandy." During the 19th century, homemade pasta slowly declined, especially in southern Italy, and the factory version became the basis of one or two daily meals in bour-

A notched rolling pin for cutting dough into noodles. From the *Works* of Bartolomeo Scappi (Venice, 1570.)

THREE EARLY NOODLE RECIPES

Rishta (Thread). Cut fat meat into middling pieces and put in the saucepan with a covering of water. Add cinnamon-bark, a little salt, a handful of peeled chick-peas, and half a handful of lentils. Boil until cooked: then add more water, and bring thoroughly to the boil. Now add threads (which are made by kneading flour and water well, then rolling out fine and cutting into thin threads four fingers long). Put over the fire and cook until set to a smooth consistency. When it has settled over a gentle fire for an hour, remove.

—from Baghdad, A.D. 1226

Macrows. Take and make a thynne foyle of douwh, and kerve it in pieces, and cast hem on boillying water, and seeth it wele. Take chese, and grate it, and butter, cast bynethen and above as losyns [lozenges], and serve forth.

—from the *Forme of Cury*, late 14th-century England

Roman macaroni. Take some good flour and mix with water and make a paste a little thicker than that of lasagna, and wrap it around a stick. Then throw away the stick and cut the paste the width of the little finger, and it will remain in the form of ribbons or lace. Cook it in fatty broth or in water according to the occasion.

—from Maestro Martino, mid-15th-century Italy

geois households. The peasantry took longer to accept what was thought to be a luxury (because "industrial") food. By the 1930s, pasta was so prominent a part of the Italian national diet that the Futurist writer F. T. Marinetti blamed it (in his *Cucina Futurista*) for "the weakness, pessimism, inactivity, nostalgia, and neutralism" of his country.

Today, the average Italian eats something like 65 pounds of pasta a year, more than twice the figure for any other nationality. The United States weighs in with about 8 pounds per person. But pasta is rising in popularity here, even as the consumption of bread declines.

Composition

There are two basic ingredients in pasta: water and either flour or semolina, the coarsest grade of milled wheat endosperm. Pasta dough is very stiff, and is only about 25% water by weight (bread dough is closer to 40% water). While ordinary or hard wheat flours are the usual choice for homemade pastas and commercial flat noodles (including fettucine and lasagna), durum wheat semolina is preferred for commercial spaghetti, macaroni, and other shaped pastas. Semolina contains very little free starch; it is almost entirely large chunks of endosperm protein with a few starch granules included (see the micrograph on page 287). Less water is required to make pasta dough from semolina—there is no damaged starch to compete with the protein in absorbing it—and this is an advantage if the dough is to be dried after shaping rather than cooked immediately. The gluten matrix made from semolina is also very strong and stands up best to the machine extrusion of spaghetti rods and the molding of various shaped pastas. And as makers of homemade pasta know, semolina-based noodles are much less brittle and easily broken than flour-based noodles.

Egg Noodles

Most homemade pastas and some commercial noodles are made with whole eggs or egg yolks, with a corresponding reduction in added water. Commercial egg noodles must be at least 5.5% egg solids by weight, and even at this low concentration, the cost of the ingredients is doubled. The point of using eggs is mainly their effect on flavor and color. In Italy, pasta has long been made from Mediterranean durum wheats that have a carotenoid pigment content about double that of most hard wheats, so that pasta is traditionally yellow. In order to maintain the color, semolina is never bleached as most flours are.

Drying and Cooking Pasta

Drying is the trickiest step in pasta manufacture. The problem is to find the best rate at which to reduce the moisture content from 25% to about 10%. If the material is dried too fast, the outer layers will shrink much faster than the inner, and the differential in movement results in "checking," or sheetlike cracks. If moisture is removed too slowly, airborne bacteria and molds will spoil the dough. Commercial producers today allow 15 to 36 hours for drying, beginning with a fast half hour, a slower hour or two, and then a very slow day or two during which the moisture is fully redistributed.

Pasta cooking is relatively simple. The noodles are boiled in enough water that they can absorb some and leave plenty behind to do the cooking. The cook tests them every few minutes until they are done *al dente*, or "to the tooth" (limp but slightly resistant to chewing). Tossing the drained noodles with some oil will coat the gluten and gelatinized starch on their surfaces and help prevent them from sticking to each other (oil added to the cooking water is partly wasted, since it floats to the top).

Biscuits

Biscuits today just aren't what they used to be, and for this we can be grateful. Our cookbooks give us directions for making moist, flaky, quickly baked morsels. But the name "biscuit" still carries vague overtones of per-durability and perhaps makes us think of the sailing expeditions or overland treks on which they were a daily feature. *Pain bis-cuit*, or "twice-cooked bread," was baked, sliced, and then baked again to make it as dry, light, and spoilage-resistant as possible; it was standard fare for armies, navies, and other inconveniently located groups a couple of centuries ago.

Biscuits are most often made today with baking powder. They contain about as much water as bread, but like pastry dough, biscuit dough is mixed only enough to incorporate all the ingredients without developing the gluten too much. Cold shortening is cut into the flour, and the liquid ingredients are added and kneaded into the dough for about one minute. The dough is then rolled out into a thick sheet and cut with a sharp blade; a blunt one will seal the edges and inhibit rising. Baking takes about 15 minutes.

Crackers and Pretzels

These products are not often made in the home, but it is interesting to know how they come to be. Both are made from relatively stiff doughs, though cracker and soft pretzel doughs are yeast-raised. A sheeting machine rolls cracker dough to a thickness of about 0.05 inch in a continuous ribbon that is folded onto itself to produce the characteristic layered texture. Baking takes only about 3 minutes. A tricky element in the process is the spacing of the puncture marks, which serve the important purpose of preventing the dough from blistering excessively. If too close together, the marks give a flat, degassed, tough cracker, while if they are too far apart, the upper layer detaches into a large blister and is easily shattered off.

Commercial pretzel dough is slightly stiffer than cracker dough. The novelty here is the characteristic hard, glossy, dark brown surface. Once it is shaped, the raw dough is sprayed with a 1% solution of sodium hydroxide (lye) or sodium carbonate that has been heated to about 200°F (93°C). The heat and moisture combine to gelatinize the surface starch. The dough is then salted and baked in a very hot oven for about 5 minutes. The starch gel now hardens to a shiny finish. Meanwhile, the alkaline conditions created by the lye are ideal for browning reactions, and dark pigments and intense fla-vor compounds rapidly accumulate. (The lye reacts with carbon dioxide in the oven to form a harmless carbonate.) The final step is a long, slow bake, from 20 to 25 minutes at 200°F (93°C), to dry the whole pretzel out. Soft and homemade pretzels are boiled briefly in a solution of baking soda and then baked for 10 or 15 minutes in a hot oven.

Cookies

Given the huge range of products covered by this term, everything from miniature cakes to a hardened sugar syrup to thin, crisp sheets of well-developed glutenous dough, it is impossible to explore cookies in much detail without writing a small book. Most cookies, however, fall into the category of doughs much higher in sugar and fat, and much lower in water, than bread dough. With this composition, the amount of water available to starch granules and gluten proteins in cookie dough is severely limited; sugar is hygroscopic and removes some of the already scarce moisture from circulation. Gluten development is minimized both by the water balance and by the mixing techniques, which avoid vigorous manipulation of the dough. A cakelike texture is achieved by beating the shortening, sugar, eggs, and liquid together and then gently folding in the flour and leavening; a stiffer, denser product is made by mixing all the ingredients together very slowly. During cooking, the starch granules can gelatinize only slightly. The outcome of all these trends: a tender, crumbly morsel.

BATTERS

Batters generally contain more fat, sugar, and water than doughs and bake into moister, softer, more dense and crumbly products. Gluten development, which has a binding, toughening effect, is minimized by using low-protein cake flour, by folding the flour into the other ingredients with the least agitation possible, and by making the continuous phase not the protein-water complex characteristic of bread dough, but water alone, with flour proteins greatly diluted. Gluten's role as a structural support is taken over by starch, which is almost completely gelatinized in the abundant water, and by egg or milk proteins that coagulate during cooking. Because batters are so watery and thin, they cannot retain the gas slowly evolved by yeast and must be leavened either chemically, or else mechanically, by beating air into the batter or its shortening or egg components. Both the fine particles characteristic of cake flours and finely granulated sugar (not powdered; sharp edges are important) are ideal for incorporating large numbers of very small air cells into batters.

Flat Cakes

These products have in common very thin batters used in fairly small quantities, although one of them, popovers, ends up anything but flat.

Crêpe batter is so dilute and spreads so quickly on the pan that hardly any air pockets, no matter how produced, survive. The delicacy of a crêpe is due not to its light texture but its exceeding thinness. The trick in making crêpes is precisely to spread a small amount of batter over the entire surface of the pan before it can set. Today, tilting the pan is the recommended method, while the late medieval *Ménagier de Paris* instructed its readers to "have a bowl pierced with a hole as large as your little finger; then put some of the batter in the bowl and, beginning in the middle, let it pour all around the pan."

Popover batter is almost identical to crêpe batter, yet popovers turn into a huge air pocket surrounded by a thin layer of pastry. The difference is a matter of cooking procedure. Popover batter is beaten to incorporate air and then is poured into small, deep cups in a preheated, liberally oiled pan and set in an oven. The surfaces of the batter set long before the trapped air in the center has expanded with the heat. When the individual air pockets do expand, they are forced by the rigid exterior to coalesce into one large chamber, and the batter balloons into an outsized blister.

Fritters, likewise, when dropped into hot oil have a coagulated surface within a matter of seconds. But because the hot oil cooks faster than an oven, there is little time for the air pockets to coalesce, and a multichambered morsel results.

Pancakes and waffles are made from a more floury, viscous batter than the others and so can retain gas cells for some time. They are leavened with baking powder or soda and then quickly fried on both sides. Pancakes are flipped just as bubbles begin to break on the upper surface, before the gas escapes.

Raised Cakes

The characteristics traditionally prized in raised cakes are lightness and tenderness, and for this reason the primary goal is to achieve the greatest possible volume. Cake batters are more complicated physical systems than bread doughs and are more easily thrown out of balance. Flour and egg components form a stabilizing network, while sugar, shortening, and additions such as cocoa tend to weaken this network. Because a change in one ingredient necessitates a compensatory change in others, it is very difficult to alter established recipes with success.

There are two basic kinds of raised cakes: shortened cakes, in which leavening is accomplished by a combination of creamed shortening and baking powder, and foam cakes, in which whole eggs or egg whites are beaten to produce air cells. (Actually, all doughs and batters are foams, but this is the standard terminology.) Vegetable shortenings with their high air capacity, chemical leavenings, and the electric mixer have all helped to turn cake making into a less onerous task than it once was. In 1857, Miss Leslie described a technique by which one could beat eggs "for an hour without fatigue" and then added: "But to stir butter and sugar is the hardest part of cake making. Have this done by a manservant."

Shortened Cakes

The standard shortened cake for many years was the pound cake, so named because it was made with one pound each of flour, butter, eggs, and sugar. Air was incorporated into the batter by creaming the butter and sugar together and whipping them to a very light texture. The eggs were folded in, and then, very gently, the flour, with as little disruption of the air-cell system as possible. The resulting cake was rather dense. Partly for this reason and partly because the formula is an expensive one, true pound cakes are rarely made today. Contemporary adaptations replace butter with vegetable shortening and some eggs with milk, and make use of baking powder. The most common shortened cake today is the layer cake and its many variations, which contain proportionally more sugar and liquid than pound cakes.

Vegetable shortening is generally preferred today in cakes for its superior creaming properties. The smaller the average crystal size in a given mixture of lipids, the more air it can incorporate because the more and smaller spaces there are between crystals. Liquid oils are not crystalline at all and cannot retain air, while animal fats tend to form large crystals that are useful in shortening pastry doughs but less so in leavening cakes. Large air pockets are buoyant enough that they will rise in thin batter and escape, while smaller ones remain evenly dispersed to serve as the reservoirs for carbon dioxide produced later by baking powder. Once liquid is added to the creamed shortening, it appears that the air cells leak into the liquid phase rather than remaining associated with the lipids.

In the conventional method of mixing shortened cakes, the solid shortening is first creamed, then the sugar added gradually, and the mixture is beaten until it reaches the fluffy consistency of whipped cream. The sugar does not dissolve but interrupts the fat structure and creates more air cells. Creaming is most effective at a warm room temperature, between 75° and 80°F (24°–27°C). The beaten eggs are now blended in. Flour, baking powder, and salt are sifted together and added to the leavened phase in four or five increments, each mixed only for about 10 seconds and each in alternation with portions of milk. When this procedure is finished, the batter is beaten for about a minute. If egg whites are being added separately to incorporate still more air, they are beaten until stiff and then folded into the batter gently.

Less elaborate protocols have been made possible by the even and very fast mixing of the electric mixer. Shortening, sugar, and eggs can be beaten together all at once and then the dry ingredients mixed in alternately with milk. In the "one-bowl method," the dry ingredients, shortening, and milk are beaten together, and then the eggs are beaten in, or all the ingredients are beaten together at once. This method requires higher proportions of sugar and liquid than normal recipes call for, and the shortening should contain emulsifiers. The extra water makes it easier to disperse all the ingredients evenly, and the emulsifiers help stabilize the air bubbles incorporated during beating. In the "dough-batter method," a "dough" is made by mixing together the flour, baking powder, salt, and fat. The batter is produced by beating half the milk and the sugar, and then the rest of the milk and the eggs, into the initial dough.

Foam Cakes

Under this heading fall three familiar cakes: angel and sponge, neither in its true form containing shortening or baking powder, and the chiffon cake, which is something of a hybrid. It is leavened in the same way as the other foam cakes, but it includes oil, which gives it a characteristic moistness and tenderness, and is usually helped along with baking powder.

Foam cakes are leavened by beating egg proteins into a light, air-filled foam. Just as is true of dough gluten, the key to whipped eggs is the manipulation of proteins. When air is incorporated into egg whites, the normally compact proteins unfold, partly coagulate, and so form a semisolid network (see chapter 2 for details about foamed egg whites). Cream of tartar, an acid salt, is usually added to stabilize the foam; it firms the texture and also lightens the color of angel cakes. On the other hand, lipids of any kind are extremely effective in preventing the formation of an air foam by surrounding the proteins and inhibiting coagulation. For this reason, whites are usually whipped separately from the yolks, which are added after the foam is formed, and all mixing utensils must be free of any residual oil or grease.

Angel cakes are made with egg whites only: hence their pristine color. The whites are beaten with salt and cream of tartar and most of the sugar; the rest of the sugar and the flour are then gently folded in. Sponge cakes are made with whole eggs and for this reason are more dense than angel cakes, richer, and tend to stale more easily. Shortening is sometimes substituted for the yolk to delay staling, and leavening may be added to lighten the texture. The whites are separated and beaten with sugar, the yolks beaten separately with lemon juice and sugar. The yolks are folded into the whites, and the flour and salt are then folded into this mixture. If an electric mixer is used, then the eggs need not be separated, though they must be beaten for a good 15 minutes with lemon juice before the sugar is beaten in and then the flour and salt folded in. The hybrid chiffon batter is made by beating the dry ingredients with the oil, egg yolks, and liquid until smooth. The whites are then beaten with cream of tartar until quite stiff, and then the batter is slowly folded into the whites. The use of oil gives chiffon cakes a very moist and tender texture similar to that of shortened cakes, but the use of foamed egg whites as a leavening makes them seem much lighter.

Baking Shortened and Foam Cakes

The baking process follows the same basic pattern in cakes as it does in bread. During the first stage, the batter structure is stabilized. As the batter temperature rises, the gas cells expand and the chemical leavening, if any, releases carbon dioxide. Because the batter is liquid and heating is uneven, there is some movement in the material until it has begun to set. Areas of batter heated rapidly at the bottom of the pan rise along the walls, flow along the surface, and plunge back into the middle. This can cause ringlike layering in the finished cake. It does not occur in cakes made from commercial mixes, apparently because added vegetable gums increase the batter viscosity and so prevent differential movement.

During the second stage of cake baking, the risen batter is set into its permanent shape by the oven heat. Flour, egg, and milk proteins coagulate, and starch gelatinizes. In the last stage, batter solidification is completed, and flavor-enhancing browning reactions take place in the now-desiccated surface.

Cakes are generally baked at temperatures of 365° to 375°F (185°–191°C). This range produces a finer, lighter texture than lower temperatures because the batter solidifies more quickly, before the gas cells can expand too far and begin to coalesce. Still higher temperatures cause a great deal of agitation in the batter and an unevenly shaped top.

Pan Size and Finish

By affecting the rate and distribution of heating, cake pans can have an important influence on their contents. The thickness of the pan does not matter very much, but its finish and its volume relative to the baked cake matter a great deal. Generally, the faster a cake is heated, the better; its gas cells expand more before the batter sets, and so its final volume and tenderness are maximized. The ideal pan size is that which matches the final volume of the cake. Too tall a pan, because its unused side area shields the batter from radiant energy, slows the heating rate and produces a drier cake with a less browned surface. And because it sets the sides of the cakes earlier than the center by a great deal, the center tends to rise longer and hump up. Too small a pan, because it cannot support the rising batter, will of course spill some, and what remains will be collapsed and dense. Tube pans work especially well because they speed the penetration of heat to the center of the batter, but the interest here is just as much in the novel shape as anything else. Finally, bright surfaces, because they reflect radiant heat, transmit heat poorly to the food they contain and slow the baking process. A rough, dull metal pan or a glass one (which also transmits radiant heat well) will require as much as 20% less baking time than a shiny-surfaced pan. Raising the baking temperature to compensate for the slower heating in shiny pans does not make up for the losses in moistness, cake volume, and fineness of grain.

Cooling Foam Cakes

Unlike shortened cakes, whose structure is determined almost entirely by gelatinized starch, the foam cakes are strengthened to a large extent by the egg proteins, and for this reason have a somewhat elastic texture. This is the reason that foam cakes are usually inverted in their tube pans over a bottle to cool; rather than slowly settling, the cake can remain stretched to its maximum volume before being served.

Cake Mixes

Commercial cake mixes often produce cakes whose lightness and tenderness are hard to match from scratch. This is due largely to two materials: chlorinated flour and fat emulsifiers. Both are available in some form to the home baker, but in corporate laboratories they can be carefully modified to the precise specifications of particular cake formulas.

Flour chlorination, or strong bleaching with chlorine gas, has several unique effects, none very well understood. It alters the surface properties of starch and flour lipids and inhibits the gluten proteins from associating (bread flours are bleached, but never with chlorine). Chlorine-treated flour can tolerate more of the structure-enfeebling ingredients, sugar and shortening, than normal flour and so can be made into sweeter, tenderer cakes.

Emulsifiers are used in most vegetable shortenings to stabilize the fat-air foam. In cakes, they prevent the shortening from reducing the volume of the air foam. Lipids finely dispersed in a water-air foam will interrupt the structure by coalescing into larger droplets, but emulsifiers help maintain an optimal droplet size. For a given level of shortening, then, the cake will rise to a greater volume with emulsifiers than it will without.

CHAPTER 7

Sauces

Of all culinary operations, sauce making surely seems one of the quirkiest, the least predictable, and often one of the most forbidding. Even the less elaborate preparations are fickle, suddenly separating into pools of oil and water, or thickening into clumps, or *never* thickening at all, or curdling. This chapter will not reveal any key to infallible sauce making, but it should provide both aid and comfort in the knowledge that sauces are by nature very tricky physical and chemical systems; that they are greatly influenced by the quality of the ingredients, the ways in which they are heated, and the particular action of the cook's arm; and that for these reasons, experience—failure as much as success—is the best teacher.

THE DEVELOPMENT OF SAUCES

Antique Sauces

Our first real knowledge of saucelike preparations comes from the late Latin recipe book attributed to Apicius. These mixtures were known as "juices"—the singular form is *ius*—and were heavily spiced and thickened with pieces of bread. A typical recipe is *ius candidum in elixam*, a white sauce for boiled foods: "pepper, liquamen [a fermented fish paste, probably similar in flavor to anchovy paste], wine, rue, onion, pine nuts, spiced wine, a few pieces of bread cut up to thicken, oil" (quantities were not specified). Nothing subtle about this, either in flavor or texture. The sauce seems designed not to complement the flavor of the food but to overpower it. It may be, however, that the ancient *ius* was more of a condiment than a sauce and used sparingly.

Five hundred years later, relatively little had changed. Spices, especially pepper and ginger, continued to be used in overwhelming amounts, and either bread or ground almonds was used as a thickener. But the terminology began to resemble our own. The French word *sauce* (in Italian or Spanish, *salsa*) replaced *ius*. It derives from the late Latin *saltus*, or "salted," and so its principal reference is to a flavor of a very simple, basic kind. As they have evolved, however, the one real constant of all sauces has been a particular kind of consistency, with flavor the infinitely variable component. The French also introduced a distinction between the *sauce* and the *grané*. The latter, whose name derives from the Latin *granatus*, "made with grains, grainy," was a kind of stew made with meat and meat juices, and not a separate mixture of spices and liquid.

It turns out that the English word "gravy" comes from the French *grané*. A *v* has replaced the *n* because, it is thought, at some point in the transfer of recipes from one manuscript to another, the transcriber mistook or miscopied that single letter. The evidence for this surmise is the simple fact that nearly identical recipes are called "grané," "grave," and "gravy" in 14th-century collections. Why the misspelled form is the one that survived in English remains a mystery. In any case, for some reason that country has always been partial to gravies, while the grané is now unknown in France. The modern gravy, a sauce made from the escaped juices of cooked meat, doesn't really have a precise equivalent in French; *jus*, or "juice," is the closest it gets.

GRANÉ AND GRAVY, CIRCA 1390

Grané de petis oiseaulx, or "gravy of small birds": *Grané* of fine or such grain as you wish, fry them [the grains] in clear lard; take white bread, dissolve it in some beef broth and strain; boil with your meat. Grind ginger and cinnamon, dissolve in verjus [green grape juice], boil everything together; it shouldn't be too thick.

—from Taillevent's *Viandier*

Oysters in gravey: Shell oysters and simmer them in wine and in their own broth, strain the broth through a cloth, take blanched almonds, grind them and mix with the same broth and anoint with flour of rice and put the oysters therein, and cast in powder of ginger, sugar, and mace.

—from the *Forme of Cury*

La Varenne

There seems to have been some gradual change in sauce making between 1400 and 1600. Spices were used more discriminatingly, and flour rather than coarse bread or almond powder was thrown into the sauces to thicken them. By most accounts, the great watershed in European culinary history came in 1553, when Catherine de Medici went to Paris to marry the future Henri II of France. She brought along her favorite cooks, who, it appears, were well ahead of the French in the simplicity, delicacy, and refinement of their art. Actually, direct evidence of their influence is scanty, though in his recent Italian cookbook, Giuliano Bugialli reproduces a 15th-century manuscript page describing a *roux*—flour browned in fat—which later became an indispensable ingredient in French sauce making. But a simple comparison of Taillevent's (Guillaume Tirel's) coarse and heavily spiced pre-Florentine sauces with the recipes of Pierre François de La Varenne a century after the fateful marriage (1651) shows that something radical had

---

RECIPES FROM LA VARENNE

Sauce Robert: Lard the loin [of pork], then roast, and baste with verjus and vinegar and a bouquet of sage. The fat having fallen, take some and fry an onion, then place this under the loin with the sauce you basted it with, and after letting the whole simmer a while, being careful that it doesn't get tough, serve.

Flour liaison: Melt your lard and remove the scrapings; throw your flour into your melted lard, but take care that it not stick to the pan; mix in some onion; it being cooked, add some good bouillon, mushrooms, and a dash of vinegar; then having boiled it with the seasoning, pass it through a sieve, and put it in a pot; when you would use some of it, hold it over hot ashes to thicken your sauces.

Fragrant sauce: . . . make a sauce with some good fresh butter, a little vinegar, salt, and nutmeg, and an egg yolk to bind the sauce; take care that it doesn't curdle.

—*Le Cuisinier François* (1651)

happened. Several of La Varenne's dishes are recognizably "modern." His "sauce Robert" demonstrates that seasonings are now less likely to overwhelm the taste of the food itself. His "flour liaison" shows that the *roux* (a word that actually appears elsewhere in the book) is now a standard element in the cook's repertoire, though it exists side by side with the older almond thickener and mushroom and truffle purées. And his "fragrant sauce" for asparagus appears to be very close to our hollandaise; in fact, it may be the first recorded recipe for an egg-based emulsified sauce (Taillevent used egg yolks to thicken soups, adding them in the last stage of cooking).

With La Varenne, French sauces were off and running. A century later, one cookbook would list 80 different sauce recipes. In the interim, many of the standard sauces of French cuisine were developed and named. Louis de Béchameil, one of Louis XIV's courtiers, gave his name to the basic white sauce, and two variants of that sauce, Soubise and Mornay, memorialize a commander of the French armies and a Huguenot family, respectively. Hollandaise is so called because it was developed by Huguenots exiled in Holland, and "mayonnaise" is thought to derive from a Minorcan port, Mahon, which was taken from the English by the Duc de Richelieu in 1756. It is very unlikely that the noblemen involved actually invented the sauces that bear their names; their cooks did the work, and the dishes became associated with their tables.

The French System

In 1789 came the French Revolution and with it changes in the social order that had supported cuisine in the high style. The great houses of France were much reduced, and cooks no longer had unlimited help and resources. With many fewer private families to employ them, chefs were forced to make a living in some other way: and so the first fine restaurants arose. The culinary impact of these upheavals was assessed by the first renowned French chef, Marie-Antoine Carême (1784–1833). Carême served English royalty, Talleyrand, and the Rothschilds and was a culinary scholar and copious writer to boot. In the "Preliminary Discourse" to his *Maître d'Hôtel français,* he notes that the "splendor of the old cuisine" was due to the lavish expenditures of the master on personnel and materials. With such liberality a thing of the past, cooks lucky enough to retain a position

> were thus obliged, for want of help, to simplify the work in order to be able to serve dinner, and then to do a great deal with very little. Necessity brought emulation; talent made up for everything, and experience, that mother of all perfection, brought important improvements to modern cuisine, making it at the same time both healthier and simpler.

THE FRENCH SAUCE FAMILIES

Brown, or Espagnole: brown stock, brown roux, tomatoes

Bordelaise ("from Bordeaux")	red wine, shallots
Diable ("devil")	white wine, shallots, cayenne
Lyonnaise ("from Lyon")	white wine, onion
Madeira	Madeira wine
Perigueux (village in Perigord)	Madeira, veal stock, truffles
Piquante	white wine, vinegar, gherkins, capers
Poivrade ("peppered")	vinegar, peppercorns
Red wine sauces	red wine (reduced)
Robert	white wine, onion, mustard

Velouté ("velvety"): white stock, yellow roux

White Bordelaise	white wine, shallots
Ravigote ("invigorated")	white wine, vinegar
Suprème	poultry stock, cream

Béchamel (a gourmand): milk, white roux

Crème	cream
Mornay (a family)	cheese, fish or poultry stock
Soubise (army commander)	onion puree

Hollandaise ("from Holland"): butter, eggs, lemon juice or vinegar

Mousseline (light cloth)	whipped cream
Béarnaise ("from Béarn")	white wine, vinegar, shallots, tarragon

Mayonnaise (Port Mahon, Minorca): vegetable oil, eggs, vinegar or lemon juice

Remoulade (twice ground)	gherkins, capers, mustard, anchovy paste

Restaurants too brought improvements; "in order to flatter the public taste," the commercial chefs had to come up with novel, ever more "elegant" and "exquisite" preparations. Social revolution had brought culinary progress.

Sauce Families

Carême's own conscious contribution to this progress involved the sauces. In his *Art of French Cooking in the 19th Century*, he celebrates the great advances in texture and flavor brought in his own lifetime by the use of concentrated stocks rather than plain broth, and *roux* rather than raw flour. Sauces were now suave, velvety, and succulent where they used to be thin and floury in flavor, coarse and uneven in texture. Carême's idea was to classify the sauces of the time into four families, each headed by a basic or leading sauce and each indefinitely extendible by playing variations on that sauce. This scheme fit the limits on postrevolutionary cuisine; the parent sauces could be prepared in advance, with the minor modifications and seasoning left to be done at the last minute on the day of the meal. As Raymond Sokolov puts it in his guide to the classic sauces, *The Saucier's Apprentice*, these sauces were conceived as "convenience foods at the highest level." Carême was proud of his conceptual breakthrough and the possibilities for invention that it opened up:

> We must consider as leading sauces the espagnole, velouté, allemande, and béchamel, because with these four sauces we can compose a very great number of small sauces, of which the seasoning differs infinitely. . . .
>
> It is in following this reasoning that I have, I dare say, created an infinite number of new things of which my books carry the indelible mark; and in order to obtain the same results, all that is necessary is a little good sense, and occupying oneself without relaxation in the progress of the science which one professes.

As one measure of the success of Carême's system, the great compilation of classic French cuisine, Auguste Escoffier's *Guide Culinaire* (1902), lists nearly 200 different sauces—not including those used in desserts. And Escoffier attributes the eminence of French cooking directly to its sauces. "The sauces represent the *partie capitale* of the cuisine. It is they which have created and maintained to this day the universal preponderance of French cuisine."

Innovation in sauce making did not end with Carême. Escoffier slightly redefined the family tree by naming espagnole, velouté, béchamel, tomate, and hollandaise as the five leading sauces, and he was not entirely happy with flour as the thickening agent. It had been learned by his time that flour contained extraneous substances—mainly proteins—that had to be skimmed off the sauce during long, slow cooking. Escoffier openly looked forward to the day when pure starch would replace flour as the thickener.

> The assistant of the stock, the roux, brings to the brown sauce only a flavor note of little importance, beyond its thickening principle, and it has the disadvantage of requiring, in order that the sauce be perfect, an almost absolute elimination of its components. Only the starchy principle remains in a sauce properly skimmed. Indeed, if this element is absolutely necessary to give

mellowness and velvetiness to the sauce, it is much simpler to give it pure, which permits one to bring it to the point in as little time as possible, and to avoid a too prolonged sojourn on the fire. It is therefore infinitely probable that before long starch, fecula, or arrowroot obtained in a state of absolute purity will replace flour in the roux.

In fact, Escoffier's prophecy has not been borne out, perhaps because pure starches are considered too modern or artificial an ingredient to be used in so noble a cause. In recent years, however, the proponents of "nouvelle cuisine" have moved away from the traditional roux thickener, preferring instead to concentrate their stocks to a pure velvetiness or to use heavy cream. And French domestic cookery, disinclined to the labor and expense of long-simmered stocks and sauces, has developed other alternatives: making a broth from the trimmings of a roast, deglazing the roasting pan and reducing this relatively small amount of liquid or binding it with cream or roux, and relying on the egg-based warm sauces—hollandaise, béarnaise, and so on.

Italy and England

While France has been and remains the country renowned for its sauces, Italy, which may have set France on the right track 400 years ago, has come to favor fresh vegetable purées—tomato and the basil *pesto,* for example—and simple mixtures of olive oil and wine, although the white sauce and mayonnaise are also commonly employed. Contemporary writers on Italian cuisine explain the lack of emphasis on sauces by saying that Italian cooks are more interested in bringing out the flavor of the meat or vegetable in question than improving or masking it by pouring something entirely different on top. In fact, this is a common argument against, or explanation of, the elaborate French system. In the 19th century, the English said that French meats and produce were generally inferior or unpredictable, and that this disadvantage gave rise to a certain ingenuity simply to make the food palatable.

As for England, an 18th-century bon mot attributed to Domenico Caracciolli put it this way, with implicit contrast to France: "England has sixty religions and one sauce." Its culinary standards did not filter down from the court but remained grounded in domestic habits and economies. Elegance and refinement were not much esteemed, and the French were often ridiculed for their pretensions. As one example of this basic divergence in attitude, compare the French gastronome Brillat-Savarin (1755–1826) with Hannah Glasse, author of a very popular 18th-century cookbook. Brillat-Savarin tells the story of the prince of Soubise being presented with a request from his chef for 50 hams, to be used at one supper party. Accused of thievery, the chef responds that all this meat is essential for the sauces to be made:

"Command me, and I can put these fifty hams which seem to bother you into a glass bottle no bigger than your thumb!" The prince is astonished, and won over, by this assertion of the cook's power to concentrate flavor. Glasse, on the other hand, gives several French sauce recipes that involve more meat than the meal they will accompany, and then makes a remark about "the Folly of these fine French Cooks" in running up such huge expenses for so little. Glasse's principal sauce is "gravy," made by browning some meat, carrots, onions, several herbs and spices, shaking in some flour, adding water, and stewing: a clear descendant of the medieval *grané* and a total stranger to the sauces being made just across the Channel. A few decades later, a contemporary of Carême's, William Kitchiner, does include a recipe for béchamel, but repeats a Glasse-like beef gravy and also presents "Wow Wow Sauce," which contains parsley, pickled cucumbers or walnuts, butter, flour, broth, vinegar, catsup, and mustard. On this and other evidence, there is probably some justice in the criticism of the English by Alberto Denti di Piranjo in his *Educated Gastronome* (Venice, 1950). The chapter on sauces begins with these pointed sentences:

> Doctor Johnson defined a sauce as something which is eaten with food, in order to improve its flavor.
> It would be difficult to believe that a man of the intelligence and culture of Dr. Johnson . . . had expressed himself in these terms, if we did not know that Dr. Johnson was English. Even today his compatriots, incapable of giving any flavor to their food, call on sauces to furnish to their dishes that which their dishes do not have. This explains the sauces, the jellies and prepared extracts, the bottled sauces, the *chutneys*, the *ketchups* which populate the tables of this unfortunate people.

The Science of Sauces

The nature of culinary sauces is still somewhat mysterious, as we shall see, but any real understanding had to await the development of colloidal chemistry late in the 19th century. A colloid is basically a suspension of very small particles in another substance. The word comes from the Greek for "glue" and was appropriated in 1861 by the English chemist Thomas Graham, who took gelatin, from which glue was made, to be the prototype of such a system. The adhesive connotation is no longer helpful, since smoke, clouds, and butter all qualify as colloids. And while it is true that glues are still made from milk proteins and cereal starch, and gravy and egg frequently do join cookbook pages together, our interest is in fully liquid colloids. In particular, we will see how it is that flour and egg yolks thicken up flavorful but initially runny fluids, binding them into cohesive, stately sauces. The French used the word *liaison*, meaning a close connection or bond, whether physical, political, or amorous, to name thickening agents, and when the English got around to borrowing the word in the 17th century, it was the culinary application that came first; military and illicit liaisons didn't arrive until the 19th century. As we shall see, the term is quite apt.

Rather than try to cover every kind of sauce, from the puréed vegetable to the chocolate syrup, we will concentrate on the two main divisions of the classic repertoire, starch-based and egg-based sauces. These concoctions are two different kinds of colloidal systems. Hollandaise, mayonnaise, and their relatives are *emulsions*, or a suspension of one liquid in another with which it cannot evenly mix: in particular, butter or oil in water. Flour-thickened sauces are examples of *sols*, or a dispersion of solid material in a liquid: here, starch in water. Sols differ from simple solutions mainly in the size of the dispersed molecules; a sugar syrup is a solution, but starch molecules are made up of thousands of sugar units and behave very differently. First, we will discuss the sols.

STARCH-THICKENED SAUCES

The Nature of Starch

The word "starch" comes from a Germanic root meaning "stiff" or "strong"; in German today, *Stärke* means both "starch" and "strength." This may be a relic of the ancient world's use of starch to stiffen, or "size," paper.

Chemically, starch is simply thousands of glucose sugar molecules linked up together into long molecules, or polymers, and is produced by plants from the sugar photosynthesized in their leaves. Unlike sugar crystals, starch granules are not soluble in water at ordinary temperatures, and so are a convenient form in which to store the plant's excess energy supplies. Roots

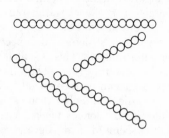 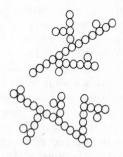

Starch molecules are made up of many thousands of glucose molecules (represented here by small circles) bonded together. They come in two forms: straight chains of amylose *(left)*, and branched molecules called amylopectin *(right)*.

and seeds are the organs in which starch is usually concentrated: tapioca and arrowroot are root starches; cornstarch and ordinary flour come from grains. The plant deposits starch in tiny granules, which may range in size from 2 to 50 microns, or in the neighborhood of a few ten-thousandths of an inch. The size, shape, location, and exact composition of the starch granules vary from species to species (see page 288 for micrographs that show starch granules in wheat flour).

There are two different kinds of starch molecules. One, called *amylose*, is simply a linear chain of glucose units. The other, called *amylopectin*, has a branched structure and resembles a bush more than a chain. The proportions of these two types in a granule and their average size—whether 10,000 glucose units or 50,000—depend on the plant from which they have been isolated and have important consequences for the behavior of the starch during cooking, as we shall see. Generally, amylose accounts for between 20 and 30% of the total. The amylose and amylopectin molecules are packed tightly into the solid granules; they form this strong association via the many hydrogen bonds that are possible along the sugar units. In some regions of the granule, the molecules are so regularly ordered that they form a kind of crystalline structure that is very strong. In other regions, they are more randomly arranged and have an amorphous structure that is relatively easily disrupted.

Gelatinization and Thickening

What makes starch so useful, whether in cooking or gluing, is its behavior in hot water. Mix some flour or cornstarch into cold water, and nothing much happens; the granules slowly sink to the bottom. But when the water gets hot enough, the energy of its molecules is sufficient to disrupt the weaker, amorphous regions of the granule, and so to permit hydrogen bonding between starch and water molecules. The granules then absorb water and swell up, thereby putting greater and greater stress on the crystalline regions. Within a certain range of temperatures characteristic of each starch source— for wheat flour, 140°–148°F (60°–65°C), for corn starch, 144°–158°F (62°– 70°C)—the granules suddenly lose all organized structure and become amorphous networks of starch and water intermingled. This is called the *gelatinization range*, because the granules become tiny *gels*, or liquid-containing meshworks of long molecules. This range can be recognized by the fact that the initially cloudy suspension of granules suddenly becomes more translucent. The individual starch molecules are no longer packed closely enough to deflect light rays, and so the mixture seems to clear. (Because flour also contains proteins, it will not clear as much as cornstarch.) Depending on the concentration of granules in the water, the mixture will begin to thicken, or become more viscous, either before, during, or after gelatinization, as they swell enough to get in each other's way. Most sauces are rather dilute and thicken after the mixture begins to clear.

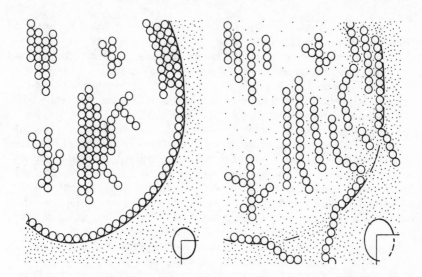

The swelling and gelatinization of a starch granule. To begin with, the starch molecules are tightly packed in the granule (*left;* only a few molecules shown), and water cannot penetrate it. When the water gets hot enough—that is, energetic enough—it succeeds in penetrating the granule and forcing individual starch molecules apart *(right).*

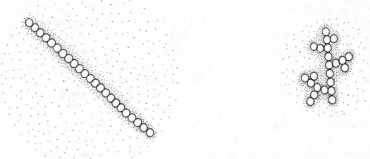

A long chain of amylose sweeps out a much greater volume of space than a branched amylopectin molecule containing the same number of glucose units. It is much more likely to tangle with other molecules, and so is a more effective thickener than amylopectin. Both molecules are shown here with their sheaths of immobilized water.

A mixture of starch and water will reach its greatest viscosity when it heats up to between 175° and 205°F (79°–96°C), or well beyond the gelatinization range. In this region, the swollen granules begin to leak amylose and amylopectin molecules into the surrounding liquid. It is the minority population that does the work of thickening. The straight amylose chains coil up into long helical structures when dissolved in water, but they retain their basically linear shape. The extended spirals immobilize a great many water molecules all around themselves by means of hydrogen bonding, and this reduces the amount of water that flows freely between amylose chains. The long, linear shape makes it very likely that one chain will knock into another or into a granule: each sweeps through a relatively large volume of liquid. By contrast, the branched shape of amylopectin makes for a very compact target and therefore a molecule less likely to collide with others. Even if it does collide, it is less likely to get tangled up and slow the motion of other molecules and granules in the vicinity. A small number of amylose molecules, then, will do the job of many more amylopectins. For this reason, so-called "waxy" varieties of corn and rice, which have high amylopectin contents, are relatively poor thickeners. A surprisingly small amount of ordinary flour, on the other hand, will have a useful effect. A thin sauce that calls for 1 tablespoon per cup of liquid contains less than 5% starch by weight.

In what sense is flour a *liaison*, an agent of bonding or connection? The starch molecules take some water out of circulation and slow the rest down by forming a tangled mesh. Chemical bonds—to be exact, hydrogen bonds—between starch and the other molecules are the means of thickening in both activities. In the more significant case of the mesh, the bonding agent is bonding to itself, and in coming between one area of liquid and another, the starch does not so much connect those areas as separate them and impede their movement. But we can save the idea of the liaison by observing that once the flour is added to the liquid, it becomes an intrinsic part of the sauce. And so even though the bonding is mostly confined to the starch molecules, they do indeed cause the sauce to cohere with itself, to hold together.

Once it reaches its thickest at around 200°F (93°C), the starch-water sol will slowly thin out again. There are, in fact, three different conditions that encourage thinning: heating for a long period of time after thickening occurs, heating up to the boiling point, and vigorous stirring. Each of these probably has the same immediate effect of shattering the swollen and fragile granules into very small fragments. While this does mean that even more amylose is released into the water, it also means that there are many fewer large bodies to get caught in the amylose tangle. In other words, the amount of netting increases, the mesh grows finer, but at the same time the whales become minnows. This effect is especially striking in the case of very thick pastes, made with 5 tablespoons of flour for every cup of liquid, but much less obvious in normal sauces, which call for only 1 or 2 tablespoons. If the granules are few and far between to begin with, their disintegration is less noticeable.

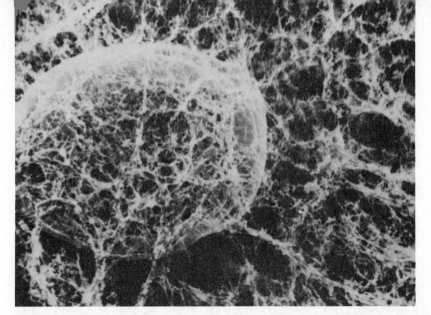

A swollen granule of potato starch caught in a meshwork of molecules freed from it and other granules. It is at this stage that a sauce will be at its thickest. (From Miller et al., *Cer. Chem.* 50 (1973): 271–80. Courtesy H. B. Trimbo.)

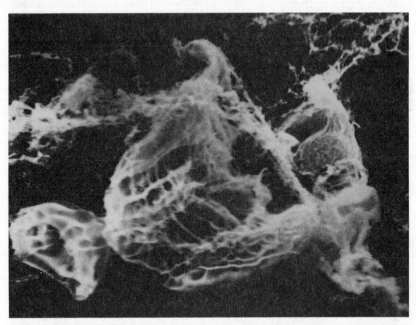

A granule of wheat starch that has lost nearly all of its substance to the surrounding liquid. As the granules disintegrate, they no longer offer resistance to the mesh of free starch, and the sauce thins out somewhat. (From Miller et al., *Cer. Chem.* 50 (1973): 271–80. Courtesy H. B. Trimbo.)

Cooling and Congealing

Another important aspect of the behavior of water-starch sols is the change they undergo when they cool down below 100°F (38°C) or so: if concentrated enough, they are transformed from sols to *gels*, or from a solid dispersed in a liquid to a liquid immobilized in the solid. As the mixture cools down, the water and starch molecules move with less and less energy, and at a certain point the force of hydrogen bonding begins to hold the molecules together longer than they are kept apart by random collisions. Gradually, the long amylose molecules form stable bonds among themselves, the kind of bonds that held them together in the granule initially. But because they have since been dispersed in water, the molecules now form a much looser network, a mesh with frequent and large gaps. And the water molecules settle in these pockets and along the unbonded sections of the amylose. The result: the liquid colloid gets progressively thicker and finally solidifies. This is the way in which pie fillings, puddings, and similar solid but moist concoctions are made; long gelatin and pectin molecules can also provide the solid phase. The strongest starch gels are produced from sols in which the granules have been completely dissolved—granule fragments can't contribute to the amylose network—and so necessitate thorough cooking. (When the liquid evaporates completely from a sol that has partly infiltrated the surface of a solid object, the molecular mesh hardens onto that surface. This is how glues work.)

The transition from sol to gel is of significance to the sauce maker for one reason. Sauces are created and monitored by the cook at rather high temperatures, around 200°F (93°C), but they are served at a much lower temperature, and when poured in a thin layer onto food, they cool even further. However thick a sauce is in the pan, it is going to be even thicker when it actually does the job of pleasing the diner, and may even congeal. So make your sauces lighter at the stove than you want them to be at the table. Experience is the best guide here, though one way to predict the final texture is to pour a spoonful into a cool dish and then sample it. Minimizing the amount of thickener has another advantage: it reduces the extent to which the sauce's flavor is diluted.

Flour and Cornstarch

Wheat flour is the thickener most often used in European and American cooking, and cornstarch is preferred in Oriental cuisines, though it sometimes substitutes for flour in Western recipes. Cornstarch is purified by soaking the whole grain, milling it coarsely to remove the germ and hull, and grinding, sieving, and centrifuging the remainder to separate the seed proteins. The resulting starch is washed, dried, and reground into a fine powder consisting of single granules or small aggregates.

Wheat starch will begin to thicken at a somewhat lower temperature than cornstarch. More important, flour contains a significant amount of pro-

When a sauce cools sufficiently to allow the starch molecules to reassociate with each other, it congeals. The original liquid dispersion of starch in water *(left)* becomes a solid network of starch with water dispersed in small pockets *(right)*. The sol becomes a gel.

tein, about 10% by weight, while cornstarch is practically pure and so thickens more efficiently. Cookbooks vary in their advice, with either 1½ or 2 tablespoons flour given as the equivalent of 1 tablespoon cornstarch. Perhaps the most immediately noticeable difference between these two thickeners is a visual one. A flour sol will remain mostly opaque, while a cornstarch-based sauce becomes translucent and even gives a glossy finish to the food it is poured on. Again, the key is flour proteins. Gluten proteins are not soluble and, when mixed in water, form clumps that will not break up with cooking. Light that may pass right through the starch-water mesh is scattered by the gluten bloblets, producing a milky, impenetrable appearance. Some authorities consider translucence and gloss to be absolute virtues. The classic French brown sauce is clarified by skimming the protein off the surface during hours of slow cooking, and Escoffier, as we have seen, looked forward to purified starches precisely to avoid this long procedure, Others, however, think of flour sauces as less flashy, more honest and "traditional" than cornstarch-based sauces. This is clearly a matter of taste. Do remember, though, that the two materials are not equal in thickening power, so that some juggling of proportions is necessary when substituting one for the other.

Root Starches

Both flour and cornstarch are made from seeds, and seed starches have two general characteristics: relatively short amylose chains, which means relatively low thickening power for the amount used, and a noticeable "cereal" taste. One of the reasons for making a roux, or cooking the flour in butter before adding it to the liquid, is to get rid of this raw flavor. The less com-

monly used root starches, on the other hand, have little taste at all and relatively long amylose chains. That is, they do not need to be precooked, and
will do an equivalent job of thickening in smaller quantities. Root starches
also gelatinize at lower temperatures. Potato starch was the first commercially important refined starch and is still quite popular in Europe. In the
United States, arrowroot, refined from a West Indian plant *(Maranta arun-
dinacea)*, is commonly available. The properties of root starches make them
especially useful for last-minute corrections, as we shall see.

Tapioca, Modified Starches

Tapioca, derived from the root of a tropical plant known as manioc or
cassava *(Manihot esculenta)*, is a third root starch used mostly in puddings.
It tends to form unpleasantly stringy associations in water and so is usually
made into large pregelatinized pearls (drops of the sol are shaken onto a hot
plate, where they gelatinize and then dry out into discrete blobs), which are
then cooked only long enough to be softened. The food industry also makes
and uses a variety of modified starches—acids, bases, and other chemical and
physical treatments alter the interaction of the starch molecules—with a
wide range of thickening and gelling properties.

Other Ingredients

Flavorings

Although sauces are not just mixtures of starch and water, these two
components are the basis for a sauce's structure, and most other ingredients
have only secondary effects on it. Salt and acid, for example, are frequently
added for their contributions to flavor. In large quantities, salt, which disso-
ciates into positive sodium and negative chloride ions, competes for water
molecules with the starch chains. As a result, the chains encounter each other
more often than they otherwise would and tend to clump together. Fortunately, this does not happen to any appreciable extent at palatable salt concentrations. Acids in the guise of wine or vinegar have the reverse effect.
They cause the fragmentation of starch chains into much shorter lengths, so
that starch granules gelatinize and disintegrate at lower temperatures, and
the final product is less viscous for a given amount of starch. Gentler cooking
will compensate to some extent for these changes, though again the effects
are minimal in all but the sourest of sauces. Other flavorings such as herbs
and spices are generally large, insoluble particles that have little influence on
the starch-water system; any essential oils that leak into the liquid do so in
physically negligible amounts.

Proteins, Fats

Two other materials are commonly found in sauces and have some influence on their texture. Flour is about 10% protein by weight, and much of this fraction is insoluble. Gluten aggregations probably get caught in the starch net and so slightly increase the viscosity of the solution, though the pure starches are generally more powerful thickeners overall. Sauces based on concentrated meat stocks also contain a good deal of *gelatin*. This protein is produced when collagen, a component of animal connective tissue, is heated in water. Collagen itself is not soluble in water, but gelatin is, and its long molecules make a water solution viscous just as starch does; it can immobilize up to 100 times its weight in water. Gelatin owes its remarkable properties to a peculiar amino acid makeup (large numbers of polar and either proline or hydroxyproline residues), which gives it a great affinity for water molecules, prevents its chains from coiling into more compact helical structures, and prevents the chains from coagulating when heated. Some sauces, especially the concentrated, starchless reductions favored in the "nouvelle cuisine," owe their texture entirely to their gelatin content.

Finally, fats are usually present in the form of butter, oil, or the drippings from a roast. They do not mix with water or water-soluble compounds and do not seem to have much of an effect on starch. Fat does have an important effect on what is called the "mouth feel" of a sauce—it contributes smoothness and moistness—and when used to precook the flour in a roux, it coats the flour particles, prevents them from clumping together in the water, and so safeguards against a lumpy sauce.

With this much understood about the behavior of starch in water, we can make some sense of the traditional procedures for making flour-based sauces.

The Classic Sauces

In the code formalized by Escoffier in 1902, there are three leading or mother sauces that are thickened with flour: the brown, or espagnole; the white, or velouté; and the béchamel. (Actually, Escoffier includes a tomato sauce, but it has no offspring in the system.) Each of these relies on a distinctive combination of roux and liquid. Brown sauce consists of a stock made from browned vegetables, meat, and bones, thickened with a roux that is cooked until the flour browns as well. White sauce uses a stock made from *un*browned meat, vegetables, and bones and is bound with a pale yellow roux. Béchamel combines milk with a roux that is not allowed to change color at all. From these three parent sauces, scores of offspring can be produced simply by varying the seasonings.

Escoffier said that a sauce should have three characteristics: a "decided" taste, a texture that is smooth and light without being runny, and a glossy appearance. The first is a matter of making fine stocks and being judicious in seasoning, while the second and third depend on how the thickening is accomplished. Generally, long and patient simmering is necessary, so that there will be little or no vestige of granular structure left to the starch and the insoluble gluten proteins will be caught up in the surface scum and so removed from the sauce.

The Stock

Béchamel aside, the first step in making the leading sauces is the preparation of the stock, which the French call the "foundation" *(fonds)* of cooking. The rich, assertive flavor of the brown sauce is created in large part by browning the meat, bones, and vegetables before adding the water, which brings a halt to the browning reactions (see chapter 14 for details about these reactions). The more delicate white sauce omits browning and so does not mask the particular flavors of the ingredients. In either case, the eventual liquid phase of the sauce is imbued with flavor—and with an important contributor to its final texture. The principal function of the bones, which are simmered in the water for hours before the meat and vegetables are added, is to provide gelatin. Veal bones, since they come from young animals, are especially rich in gelatin-producing collagen, while the largely mineralized bones of adult animals are less so. Even before the addition of flour, then, the stock already has the makings of a liaison. (Aspics and meat jellies are made by concentrating such stocks to the point that they will gel alone on cooling.) The sauce foundation is finished by putting the liquid through a strainer and removing the solid remains of meat and vegetables.

The Roux

The second step in classical sauce making is the preparation of the roux. *Roux* is the French word for "red" and at some point in the 18th century came to mean a flour that had been cooked long enough to change color. Roughly equal volumes of butter and flour are gently heated in a pan and stirred until the flour no longer has that characteristic "cereal" odor and the desired color has been attained. In the case of brown sauce, the flour is made to contribute to the color and flavor by being browned as the meat and vegetables have been. The more the flour is browned, however, the less power it has to thicken, for the same reason that caramelized sugar is less sweet than plain sugar. In each case, the carbohydrates are broken down and transformed in the browning reaction, so that there is a net loss of the original material and its particular properties. You will need more of a dark brown roux to thicken a given amount of liquid, then, than you will of a pale roux. In addition to improving the flavor and sometimes the color of sauces, the practice of precooking the flour improves the dispersion of both starch and fat in the liquid. Starch granules coated in oil are much less likely to clump together when added to a watery stock, and if the butter were dropped in as a lump, it would take a great deal more stirring than a roux does to be evenly dispersed. Finally, precooking inactivates the starch-digesting enzymes present in the flour that might otherwise break down some amylose molecules when liquid is added, and thus thin the sauce.

Once the roux has been added to the stock, the mixture is allowed to simmer for quite a while—an hour in the case of béchamel, 2 hours for velouté, and up to 10 in the case of brown sauce. During this time, the flavor is concentrated as water evaporates, and the starch granules dissolve and disperse among the gelatin molecules, with a very smooth texture the result. Brown sauce is cooked for the longest time because it is meant to be quite clear to the eye, and this requires that the gluten proteins coagulate and be carried to the surface, where they and the tomato solids can be skimmed off. This procedure was apparently onerous even for Escoffier, who said that espagnole could be reduced in *one* hour if pure starch were used. Today's proponents of the classic sauces, however, generally remain loyal to flour.

Finishing

All this may not sound much like a labor-saving procedure or the making of a "convenience food." It takes a day to prepare the stock for brown sauce and then another day to finish the sauce itself. The idea, though, is to make the leading sauces in quantity and store them in portions, to be used as needed over the course of several weeks. Some care is necessary in reheating them, since the solids can stick to the pan bottom and burn: the portion should be dissolved into a small amount of hot water. The leading sauce is then "finished," or enriched, usually with butter or egg yolk and seasonings, or is made into a "small" or "composed" sauce by adding other ingredients. Because the stocks have been skimmed of all fat and the roux contributes only about a tablespoon of butter per cup of sauce, the plain sauces are relatively lean; a final swirl of butter contributes a rich, succulent note. Egg yolk has the same effect, but because it contains some protein as well as fat, it has to be treated with caution. The protein will coagulate at high temperatures and so can curdle the sauce. Curdling is avoided by the simple precaution of dispersing the proteins so widely that they cannot form noticeable aggregations. A small amount of hot sauce is beaten into the raw yolk so as to warm it gradually and dilute it at the same time; this mixture is beaten quickly into the rest of the sauce, which can then be simmered without worry.

One need not be a culinary purist to profit from the classic saucemaking techniques. The stocks can be replaced by broths and bouillons and the long clarifications and reductions eliminated. But the use of the roux will improve the flavor, color, and consistency of any sauce. Because it leaves some starch granules intact, cooking the sauce only to the point of thickening will give it a slightly rough texture, while a long simmering will result in a somewhat thinner but smoother, "suave" result. And there is much to be said for learning to make a basic flour sauce or two and then developing your own special set of variations. This is a good way to broaden a culinary repertoire.

Gravy

We come now to the homely English cousin of French sauces, the gravy typically made to accompany a roast. This is a last-minute sauce that is put together just before serving. The drippings, both fat and browned solids, give the gravy its flavor and color. First the fat is poured off, and the roasting pan is "deglazed" with a small amount of water, wine, beer, or stock. The liquid dissolves the browning-reaction products that have stuck to the pan and so liberates their especially rich flavors for use. Now some flour is cooked in about an equal volume of the fat until it has lost its raw aroma, and the deglazing liquid is added, together with enough additional liquid to make up a sufficient volume. The mixture is cooked until it thickens, a matter of a few minutes. Because they are made at the last minute, gravies are not cooked long enough to cause the disintegration of the starch granules and therefore generally have a rather coarse texture, even when lump-free. This gives gravies a character very different from that of the suave sauce: hearty, one might say, or unpretentious, or, when they are extremely thick, almost bready.

If there is one problem or disadvantage connected with gravy, it is the temptation to use too much flour and thereby dilute the gravy's flavor, which is limited by the amount of drippings retrieved from the pan. The French prefer not to thicken the juices from a roast with flour and instead deglaze the pan, reduce the liquid by boiling it, and finish with a swirl of butter; the result is a very concentrated flavoring. Because we are often impatient for the meal to begin as we stand over the saucepan, we think that the original measure of flour wasn't enough to thicken the liquid and add some more to hasten the process. And because we forget that a sauce thickens once it leaves the stove, we aim for the final thickness during cooking and end up with a gravy that quickly congeals and that has a dilute, almost neutral taste. There is also an idea that a gravy has to be plumped up with flour in order to make enough for everyone. But increasing the volume only decreases the amount of flavor in a spoonful. It is preferable to have half the volume of gravy on one's food if it is twice as concentrated in its effect on the palate. Beware the blandishments of flour, then, if you want a liquid and flavorful gravy. If you have used between 1 and 2 tablespoons of flour for every cup of liquid, you have used plenty. Give it time to thicken.

If you like the help of flour in thickening the juices of a roast but would prefer a smoother texture, one approaching that of a long-simmered sauce, then there is a relatively quick way of breaking the starch granules into finer fragments, if not of dissolving them altogether. As we have seen, if a water-starch mixture is made thick enough to be pasty, then both stirring and boiling tend to thin that paste out dramatically. When the starch granules are crowded together like this, the main cause of increased viscosity is the swelling of the granules: the bigger they get, the harder it is for them to move past each other. And when they are this close, agitation will smash the granules into each other and break them up, thereby thinning the texture. Sauces and gravies don't thin out in this way because they are not populated heavily enough with granules and owe their viscosity primarily to the release of long amylose molecules into the solution. But we *can* break the granules up quickly by making a pregravy paste.

Here is the procedure. Brown the flour in fat as usual. Then make a paste by adding only a small amount of liquid to the roux, about one-fifth of the final volume. The mixture will thicken very quickly just as soon as the granules begin to swell. Now start beating the paste vigorously over moderate heat, and continue to do so for several minutes. At a certain point, you will notice that it begins to thin out; you are now breaking up the granules. The longer you beat the paste, the smoother the final gravy will be. Just be sure that you don't lose so much moisture that the starch begins to stick to the pan and burn. When you are through beating, dissolve a little of the paste in a little of the deglazing liquid or stock, and continue to mix small amounts together until all of the paste has been incorporated into the liquid. Then simmer as you would a normal gravy. Because the beating takes several minutes, you may want to do it in advance of the gravy making and then set the paste aside until you take the roast out and deglaze the pan.

Last-Minute Adjustments

If, once the sauce or gravy is made, you decide that the texture needs some fine adjustment, there are two obvious remedies. If the sauce is too thick, carefully dilute it with a little liquid; if too thin, add more starch. The second is trickier than the first—the effect of the starch is delayed until the granules gelatinize and start to leak amylose—and for this reason some authorities recommend an initial error on the thick side. If last-minute thickening is necessary, there are a couple of useful techniques to keep in mind. When dry starch is added to a hot liquid, it clumps. The first granules to make contact gelatinize almost immediately and form a barrier that keeps the other material dry. This problem can be avoided by mixing the flour with cold water, stirring it up, and adding a few drops at a time, or by using *beurre manié*, or kneaded butter, which is an uncooked version of a roux. Flour and soft butter are kneaded together in approximately equal volumes until the particles are evenly dispersed, and then small pieces of this mixture are stirred into the sauce. The fat keeps the flour from clumping, and because the butter must melt before all the starch can reach the liquid, the liaison is released gradually and evenly. Both flour and cornstarch have the disadvantage of their typical cereal flavor, which will not be cooked out in last-minute adjustments. Root starches are made to order for this situation because they are tasteless and thicken more efficiently than the seed starches, they cause less dilution of the sauce's native flavor; and because they gelatinize at a lower temperature, they thicken more rapidly.

EMULSIFIED SAUCES

Most culinary sauces are given body with starch or a mixture of starch and gelatin, but a few celebrated exceptions are thickened with fats, either butter or some kind of vegetable oil. Rather than forming a tangled mesh of long molecules, fats break up into tiny droplets that are packed into the water phase. Because the droplets are much more massive and slow-moving than individual molecules of water or fat, they impede the motion of the water and lend a thick, viscous, semisolid consistency to the mixture as a whole. Such a mixture is called an *emulsion,* a word that derives from the

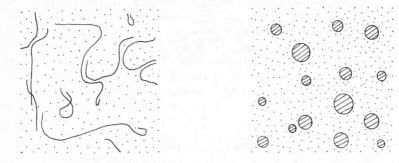

A starch-thickened sauce is a dispersion in water of large molecules *(left)*. A fat-thickened sauce is a dispersion in water of droplets, each containing many molecules *(right)*.

Latin for "to milk out" and that referred originally to the milky fluids that can be pressed from nuts and other plant tissues. Mayonnaise, hollandaise, and béarnaise are the three most celebrated emulsified sauces.

The Nature of Emulsions

An emulsion is a colloidal system in which one liquid is dispersed in the form of fine droplets throughout another liquid with which it cannot evenly mix. (The molecules of water and alcohol, for example, mix freely together and so cannot form an emulsion.) The most common, indeed ubiquitous emulsions in everyday life are made up of water and oil. Mayonnaise, cream,

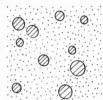 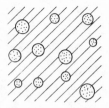

An oil-in-water emulsion *(left)* consists of oil droplets in water. A water-in-oil emulsion *(right)* consists of water droplets in oil.

and milk are all emulsions of fat dispersed in water (with about 70, 38, and 4% fat contents by weight, respectively), and butter is a water-in-oil emulsion, with water accounting for about 20% of the weight. (Because some culinary oils have relatively high melting points, one phase of these emulsions may frequently be solid rather than liquid, but they are all formed when both phases are liquid.) Cosmetic creams, floor and furniture waxes, some paints, asphalt, and even crude oil are all emulsions of water and oil.

An emulsion consists of two phases, one continuous and the other discontinuous, or dispersed. In the usual shorthand, an "oil-in-water" emulsion is one in which oil is dispersed in a continuous water phase; "water-in-oil" names the reverse situation. The dispersed liquid generally takes the form of tiny spheres, between 0.1 and 1.0 micron in diameter (a few millionths of an inch), separated from each other by the continuous phase. The droplets, like intact starch granules, are large enough to deflect light rays from their normal path through the surrounding liquid, and give emulsions their characteristically opaque, milky appearance and thereby their name. The more of these droplets there are concentrated in the continuous phase, the more they squeeze the continuous phase into a very thin film between them, and the more viscous the emulsion will be. If you introduce more of the continuous phase, on the other hand, you dilute the droplets, and the emulsion will thin out. Clearly, it is important to know which phase is which when working with emulsified sauces.

Fighting Surface Tension

Perhaps the most important characteristic of emulsions is that they require energy, and sometimes ingenuity, for their formation. Those who make mayonnaise or hollandaise by hand may find that a sore arm the next day is part of the price; shaking a bottle of oil and vinegar is a less extreme example. We all know from experience that when water and oil are poured into the same bowl, they separate into two layers: one does not spontaneously form tiny droplets and invade the other. This behavior can be described by saying that when liquids cannot mix for chemical reasons, they tend to

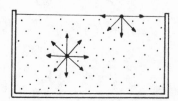

Surface tension works to minimize surface area. Molecules at the surface of a liquid feel an attractive force *(arrows)* from their neighbors alongside and below, but not from above. This force is what keeps a liquid from flying apart into a gas. When a small portion of liquid is isolated *(right)*, all the surface molecules are pulled inward, and the mass as a whole forms a sphere: the solid shape with the smallest surface area for a given volume.

arrange themselves in the way that exposes the least possible surface area to each other. The way they do this is to form a single large mass, which exposes less of itself to the other liquid than does the same total mass broken into pieces. As one relevant example, a cube of material measuring an inch along each edge, and thus a total surface area of 6 square inches, would have a surface area of several *acres* if divided into particles of colloidal size. This tendency of liquids to minimize their surface area is an expression of the force called *surface tension,* which arises from the attractive forces that hold the molecules of a liquid together. The surface of a liquid is the interface between that liquid and the air, or another liquid with which it cannot mix. At the surface, a molecule of the liquid feels the attractive force of molecules below the surface, but this force is not balanced by similar forces from the other direction. Surface molecules, then, are pulled continuously into the liquid, away from the air or the foreign liquid. If you isolate a small amount of the liquid and surround it by some other substance, all the surface molecules feel the same inward force, and a sphere—whether dewdrop or oil drop—is formed. If the liquid touches a solid or if there is enough of it to be distorted by its own weight, its shape may depart from the spherical, but the force of inward attraction will always work toward minimizing the surface area and maintaining a monolithic shape.

Emulsifiers

It is on account of surface tension, then, that the cook must supply energy to the liquid that is to be dispersed; its preferred monolithic arrangement must be shattered. And surface tension also necessitates a certain ingenuity for the purpose of *maintaining* an emulsion in the emulsified state.

For once you have dispersed one liquid in another and boosted its surface area by a factor of a million or so, its natural tendency will be simply to coalesce once more into a single large blob. In the time it takes to pour an emulsified sauce into a serving dish, then onto the food, and then to pick up a fork, the sauce should by all accounts have separated into puddles of butter and water. From the evidence of old cookbooks, it would seem that someone had discovered the egg yolk liaison by the 17th century. But it was not until the 20th that the egg's secret was understood. It turns out that many different substances can be used to stabilize emulsions and prevent them from separating into two layers. Certain proteins, plant resins and gums, other large carbohydrate molecules, including starch, and very fine particles, whether clay or graphite or ground mustard, can surround the droplets of the dispersed phase and interfere with their coalescence. Milk fat, for example, is naturally stabilized in the water solution that carries it by proteins, salts, and derivatives of fatty acids.

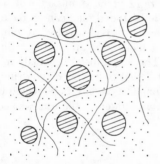 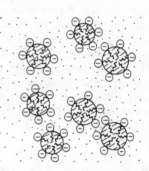

Long molecules like starch and proteins stabilize emulsions by getting in between droplets and interfering with coalescence (left). Soap-type emulsifiers (right) actually embed themselves partly in the fat droplets and leave their electrically charged heads projecting into the water phase. Coated in this way with a shell of negative charge, the droplets actually repel each other.

Such stabilizing substances are called *emulsifiers*. An especially important class of emulsifiers relies on the properties of fatty acids, which have a fat-soluble hydrocarbon tail attached to a water-soluble head. With one end immersed in the droplet phase and the other in the continuous phase, such molecules are extremely effective at preventing one droplet from recognizing another. Soaps belong to this group, and the versatile egg yolk contains several similar substances, most notably *lecithin*. Emulsifiers more obviously fit the culinary term "liaison" than starch does, though here, too, blockage and obstruction—one droplet from another—may suggest the opposite sort of activity. But by collecting at the interface between two different phases

and lowering the surface tensions, emulsifiers have a recognizably interme-
diary role and make possible the intimate coexistence of otherwise unmixable
liquids.

An emulsion is made, then, by a combination of work—breaking one
liquid into droplets—and the use of an appropriate emulsifier to stabilize the
system. Neither step is entirely straightforward. Some droplets sizes, mostly
very small ones, are more stable than others, and the properties of emulsifiers
vary greatly, depending on their exact structure and the particular liquids to
be emulsified. Some, for example, tend to favor water-in-oil emulsions, others
the reverse, and the same emulsifier may have very different effects on butter
and corn oil.

Emulsions have always been considered fickle concoctions. One of the
early colloid chemists, W. D. Bancroft, wrote in 1921 that contemporary
books on pharmacy were "filled with elaborate details as to the method of
making emulsions" and recorded two such details: "If one starts stirring to
the right, one must continue stirring to the right or no emulsion will be
formed. Some books go so far as to say that a left-handed man cannot make
an emulsion, but that seems a little absurd." The worry is always that at some
point the emulsion may break and separate into blobs of oil and water again.
When this happens, it happens in two stages. The first, called "flocculation,"
is the aggregation of several droplets without actual coalescence occurring.
Gentle agitation can usually reverse flocculation. The second stage, which is
irreversible without remaking the emulsion from scratch, is the coalescence
of the droplets.

The Breakdown of Emulsions

Several conditions can lead to the breakdown of an emulsion. One is a
too rapid addition of the dispersed phase to the continuous. If one portion
goes in before the previous one has been divided into sufficiently small, stable
droplets, all of that liquid is more likely to coalesce. Another error is to add
too much of the dispersed phase. If the droplets are crowded so close together
that they are in continuous contact, then they are almost certain to pool
together. It has been determined from simple geometry that a group of identical spheres just touching each other fills 74% of the total volume that contains them; the other 26% consists of space between the spheres and must be
filled by the continuous phase. It happens that actual emulsions contain a
wide range of sphere sizes, and this variety makes closer packing possible.
Mayonnaise sometimes exceeds 74% oil content, and stable emulsions have
been prepared that are 90% dispersed phase. But from about 70% on, trouble
is not very far away.

High and low temperatures can also break emulsions, and this fact limits
the abuse to which they can be subjected in the kitchen, whether in the making, serving, or in storage. At high temperatures, the molecules in a sauce are
moving very energetically, and the harder and more frequent the collisions

between droplets, the more likely they are to coalesce. At some temperature, which will depend on the particular emulsion involved, the odds shift in favor of coalescence, and the sauce separates. Temperatures above 140°F (60°C) can also cause the protein in egg yolk to coagulate into little lumps. Aside from the resulting deterioration in appearance and texture, this event has the unfortunate effect of removing large molecules from blocking duty and so making it easier for droplets to aggregate. And a sauce that is held before serving at a high temperature may lose enough water—the continuous phase—by evaporation that the dispersed oil droplets become overcrowded and pool together. So: emulsified sauces should be made and held at warm, rather than hot, temperatures and should not be spooned onto a piece of food still sizzling from the pan.

At the other end of the temperature scale, greatly reduced molecular movement has the effect of increasing the surface tension of liquids, since the less energetic the molecules at the surface of a droplet, the more they will feel the attractive forces of the interior molecules. And the greater the surface tension, the more likely two neighboring droplets are to overcome the emulsifier and coalesce. If the temperature drops low enough that one of the phases solidifies, as is the case when a sauce made from butter or olive oil is refrigerated, then the emulsion is likely to break when it is rewarmed. Few emulsified sauces take well to freezing.

The Investigations of Sauce Béarnaise

With this much understood about the general nature of emulsions, we are in a position to profit from the recent interest shown by chemists, mathematicians, doctors, physicists, and high school students in sauce béarnaise, an oil-in-water emulsion developed in France around 1830 and currently in vogue because the dishes it traditionally accompanies, grilled meats and fish, are popular. There are three basic steps to making the sauce. First, one prepares a flavorful water-based infusion of wine, vinegar, and herbs—tarragon and shallots—and boils it down to reduce the amount of liquid. Then egg yolks are added, and the mixture is slowly heated until it thickens somewhat. Finally, melted butter is gradually beaten in until the desired texture is obtained. Because the ingredients are heated—the butter must be liquid, and the yolk mixture hot enough to keep it that way—the possibility of *over*heating is always there. Overheating can ruin the sauce in two ways: by causing the two phases to separate, or by coagulating the egg proteins and forming small solid curds in an otherwise creamy liquid. The aim of the investigators has been to understand the balance of forces in sauce béarnaise, and with this understanding to make its preparation as foolproof as possible.

On December 15, 1977, there appeared in the prestigious scientific journal *Nature* an article entitled "Interparticle Forces in Multiphase Colloid Systems: The Resurrection of Coagulated Sauce Béarnaise." The European authors, including a chemist and a mathematician, claimed "the first successful attempt to resurrect a *sauce béarnaise* based on the theory of the stability of lyophobic colloid." Then, a little over a year later (April 5, 1979), in the "Occasional Notes" of the *New England Journal of Medicine,* an M.D. and a Rochester, N.Y., chef published "Doctor in the Kitchen: Experiments on Sauces Béarnaise" and threw some doubt on the European study. In December 1979, the proprietor of the "Amateur Scientist" department of *Scientific American* surveyed the controversy and offered yet another interpretation. This article in turn led a group of Los Angeles high school students to do some useful experiments. The debate has been both entertaining and illuminating.

Double Layers and Vinegar

C. M. Perram, C. Nicolau, and J. W. Perram, the authors of the initial paper in *Nature,* based their approach on the electrical interactions in colloidal systems. It happens that in hydrophobic colloids, including those made with oils, the dispersed droplets accumulate an electrical charge on their surfaces. (Robert Millikan was the first to measure precisely the unit charge of the electron in his famous "oil drop experiments"; he used radiation rather than sauce making to charge his droplets.) This surface charge attracts a layer of oppositely-charged ions from the surrounding solution, and this in turn attracts a second layer. The resulting "double layer" around individual droplets shields them from each other, and so prevents coalescence. Perram *et al.* reasoned that the stability of the béarnaise emulsion could be improved by increasing the number of ions available to form the double layers. Accordingly, when the sauce suffered "heavy coagulation" (which apparently means separation of the two phases) at the hands of the experimenters, they added a solution of acetic acid—vinegar—and stirred vigorously. The acetic acid will dissociate into positive hydrogen and negative acetate ions and, according to the theory, strengthen the double layers. Taste—or better, texture—tests by several observers seemed to confirm the efficacy of the technique.

The Perram report is more than slightly tongue in cheek (it was originally submitted to the *Journal of Irreproducible Results*). "The aim of this

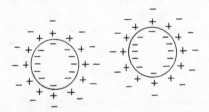

The double-layer interpretation of sauce béarnaise. Oil droplets accumulate a negative charge (an effect much like static electricity) which attracts a layer of positive ions from the solution, and this in turn attracts a layer of negative ions. So shielded, the droplets will not coalesce, and the sauce is stable. Further studies called the adequacy of this theory into serious question.

article is to indicate how an understanding of this interplay of forces can lead to more complete control of the stability of this extremely complex and important system." To judge from the return address, a chalet in Switzerland, and the fact that the authors are Danish and German, it would seem that the research was probably an interlude in a skiing vacation. And it leaves something to be desired experimentally: there was no control study, for example, to determine whether vigorous stirring *without* the addition of vinegar might also resurrect the sauce.

Vinegar Prevents Protein Coagulation

Dr. D. M. Small and Chef Michael Bernstein were more thorough. They tried several different sauce mixtures and carefully monitored the pH—the concentration of hydrogen ions—in each case. Free from the theoretical preconceptions of their predecessors, they discovered some useful facts. While the amount of vinegar had no influence on the stability of the emulsion itself—a point against the importance of the double layer in sauce making— it did make a significant difference in the tendency of egg proteins to coagulate, which is the one eventuality that cannot be repaired in any way once it occurs. In a sauce made with tap water at near neutral pH, the butter droplets flocculated at about 160°F (71°C), and the egg proteins curdled between 160° and 175°F (79°C). With enough vinegar to lower the pH to 4.5, flocculation occurred at the same temperature, but the proteins failed to coagulate even at 195°F (91°C), a temperature far above normal sauce-making conditions. So, as the authors put it, "the cook with a heavy hand on the burner might be advised to lower the pH of the sauce to prevent irreversible coagulation above 160°."

Small and Bernstein don't try to account for the advantages of an acidic sauce, but here's a guess, based on recent research into the behavior of the major egg protein, ovalbumin. These studies show that the kind of solid that this protein forms when heated is strongly dependent on pH. Two different effects are involved: one is the tendency for the normally globule-shaped molecules to unfold into long chains, and the other the tendency of the molecules to bond to each other, or aggregate. Both of these tendencies are encouraged by heat, which increases the general atomic motion, thereby disrupting the shapes of individual molecules and causing them to encounter each other more frequently. Depending on the pH of the surrounding liquid, unfolding may occur at a lower temperature than aggregation, and so precede it during cooking, or else aggregation may precede unfolding. If aggregation precedes unfolding, then dense, compact masses of protein are formed. But if unfolding comes first, then a more open, extended, and weak network is formed, with liquid continuing to occupy nooks and crannies within that network. In both very acidic and very alkaline conditions, ovalbumin will not coagulate at all, because the molecules accumulate a large amount of electrical charge and repel each other altogether.

Now the egg yolk is much more complicated, containing as it does sev-

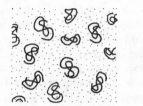

Two possible outcomes of heating egg proteins, which are folded into compact globules in their native state *(left)*. If conditions favor the disruption of their shape rather than aggregation, then they form a loose network of long chains *(center)* and will not disturb the texture of the sauce. If aggregation is favored, then the proteins come together in compact clumps *(right)* that are undesirably prominent in a smooth sauce.

eral different proteins, each with somewhat different properties. (Some of the protein that finds its way into an emulsified sauce derives from the white of the egg, a layer of which stubbornly adheres to the yolk.) Still, the same basic pattern in the ovalbumin studies is discernible in the experiments of Small and Bernstein. Near neutral pH, the egg proteins do not unfold much before aggregating, and so when they are heated enough, they form solid curds. At the more acidic pH of 4.5, they do unfold and then form a loose network widely dispersed throughout the sauce. This network is probably quite similar to the deployment of starch and gelatin molecules in flour- and stock-thickened sauces. It increases the viscosity of the béarnaise invisibly, while the denser clumps of unfolded protein have a more obvious, less pleasing effect on both the appearance and texture of the sauce. (It is easy to demonstrate to yourself the effect of acid on egg coagulation: first, stir an egg white and 3 tablespoons of water over low heat until it sets; then repeat, replacing the water with vinegar.) All told, it is fortunate that a pH of 4.5 is pleasantly piquant.

The contribution of Small and Bernstein was to show that vinegar is an important prophylactic against protein coagulation, while it seems to have little effect on the water-oil emulsion itself. It would appear, then, that the electrical forces on which the *Nature* paper was based are a negligible factor for this particular colloidal system. The doctor and chef, with their wider empirical background, point out that the two phases are readily separated by overheating, by cooling the sauce below 68°F (20°C) and then reheating, or by adding the butter oil too rapidly; but that this separation can be reversed by slowly beating the broken sauce into a pan containing a little water at 100° or 120°F (38°–49°C). From this report, temperature and technique appear to be the decisive factors.

The Emulsifiers in Egg Yolk

Now onto the field stepped Jearl Walker, whose "Amateur Scientist" column in *Scientific American* is spacious enough to permit him to summarize the recent research and then referee. Walker pointed out one weakness in the Small-Bernstein approach: namely, that the failure of added vinegar to change the temperature at which the oil droplets flocculate has no necessary bearing on the question of whether at any *given* temperature a more acid sauce is more stable. This question is much more difficult to answer experimentally. But Walker's most important contribution was to reintroduce to the debate a factor that had been recognized decades earlier in books on colloids and emulsions: the emulsifying action of the egg yolks and, in particular, of the phospholipid lecithin. He experimented with several different salts, thereby changing the distribution of charge in the sauce, and the results did not support the hypothesis of the electrical double layer. Emulsifiers appear to be a more likely source of stability.

If this is the case, then it seems obvious that the way to lessen the danger of a broken emulsion is to use plenty of egg yolk. Alas, it's not that easy. Yolk is about half water, so that adding some extra would dilute the sauce and require both more flavoring and more oil to produce a sauce of the same quality. And we all know that egg yolk contains a goodly amount of cholesterol. It turns out that cholesterol, like lecithin, is an emulsifier. Unlike lecithin, however, it stabilizes a water-in-oil emulsion, and so *de*stabilizes oil-in-water emulsions like sauce béarnaise. As we mentioned above, the properties of an emulsifier depend on its molecular structure. It is a general rule that whichever liquid a particular emulsifier is more soluble or dispersible in, that

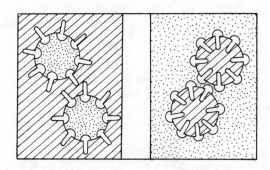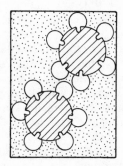

Different emulsifiers stabilize different kinds of emulsions. If a large part of the emulsifier molecule or particle extends into the continuous phase, then it will do an effective job of blocking the droplets from each other, and the emulsion will be stable *(left)*. But if most of the emulsifier is buried in the droplets, then their surface is poorly guarded *(center)*, and the emulsion is less stable than one in which a more appropriate emulsifier is used *(right)*.

liquid will constitute the continuous phase of the stable emulsion. The reason for this may be simply a matter of geometry. The more an emulsifying molecule is immersed in the phase outside the droplets, the more it blocks the droplets from each other. If most of the molecule is buried in the droplet, then not much of it is doing the desired work of guarding the droplet surface. Only one small corner of the cholesterol molecule can be ionized and so interact with water molecules; the bulk of it consists of nonpolar hydrocarbon rings. Lecithin, on the other hand, has a larger water-soluble head and so can stabilize oil-in-water systems (compare the structures of the two molecules on pages 532 and 607). Egg yolk, then, contains substances both favorable and unfavorable to the formation of emulsified sauces, and so is not an unqualified help.

Walker's suggestion that the yolk emulsifiers were more crucial to the cook's success than vinegar was convincingly verified by four of his readers, who were then students at the Westlake School for Girls in Los Angeles. Their results appeared in the "Amateur Scientist" for January 1981. Preparations of lecithin and cholesterol were obtained from a health-food store and a chemical supply house. In several different trials, the experimenters found that the addition of cholesterol to a previously stable sauce caused aggregation of the oil droplets, while lecithin could restore such separated sauces to smooth emulsions. And they confirmed one particular reason that only the freshest eggs should be used in the preparation of emulsified sauces. Walker had pointed out that as an egg ages, its lecithin content slowly declines, while its cholesterol content remains unchanged. The balance between the two, then, gets less favorable for oil-in-water emulsions as long as the egg is stored, and could conceivably reach the point at which such an emulsion would be impossible to maintain. Sure enough, when the students let a few fresh eggs age for a week after succeeding with other eggs from the same batch, the old eggs failed to emulsify the water and fat. When lecithin was beaten in, the sauce formed and stabilized.

While these experiments do demonstrate the importance of emulsifiers in the formation of stable sauces, the interpretation implied in the article— a straightforward battle between cholesterol and lecithin—is probably too simple. Several papers in such publications as *Poultry Science* and the *Journal of the American Oil Chemists' Society* have reported that when pure lecithin is added to the ingredients of mayonnaise, the sauce's stability is *reduced* rather than enhanced. But this does not mean that the students' experiments are invalidated. Their source of lecithin was a preparation sold in health-food stores, and such products are actually "soy phosphatides," or a whole range of phosphorus-fatty-acid complexes, lecithin among them, derived from soybeans. And it turns out that a variety of emulsifying molecules is more effective than just one. Says the *Encyclopedia of Chemical Technology,*

> The most stable emulsion systems usually consist of blends of two
> or more emulsifiers, one portion having lipophilic tendencies [an

attraction to fats], the other hydrophilic [an attraction to water]. . . . Only in relatively rare instances is a single emulsifier suitable. . . .

By this rule, even the cholesterol probably makes a contribution to the stability of the sauce. Adding pure lecithin or cholesterol upsets the delicate balance and breaks the emulsion, while introducing a mixture of soy phosphatides spreads the emulsifying duty over several other substances and maintains the emulsion. One report confirms that the purer the preparation of phosphatides—the less that lecithin is contaminated by other substances—the lower its emulsifying power.

Then there is the matter of the necessary ratio of lecithin to cholesterol. There is a general agreement in books and articles that the ratio in eggs, which is said to be around 7 to 1, is actually very close to the point at which an oil-in-water emulsion will break and "invert" into a water-in-oil system. Hence the trickiness of emulsifying with egg yolks and the necessity of fresh eggs. This argument is in fact based on a largely irrelevant experiment done in 1924 that involved not butter and egg fats but olive oil. When the two emulsifiers were initially present in the water phase, the inversion point came at a ratio of 8 to 1. But when the cholesterol was introduced in the oil phase—a closer approximation to the situation in the yolk—the figure was 2 to 1. So we really have no idea about the optimum conditions for the emulsified sauces. The question is probably pointless, since eggs, like soybeans, contain many potential emulsifiers, with cholesterol and lecithin only two of the more prominent. (For example, the yolk protein livetin is also known to be a very effective emulsifier.)

Finally, there is some wonderfully graphic evidence to take into account. In 1972, the team of C. M. Chang, W. D. Powrie, and O. Fennema published photographs, obtained with an electron microscope, of the oil droplets in mayonnaise. The pictures clearly show tiny particles clustered at the surface of the droplets, as well as very fine fibers dispersed between droplets. The food scientists interpreted the particles to be complex lipid-protein structures that have been freed from the egg yolk granules, and the fibers strands of the yolk protein phosvitin. Both the particles and the fibers appear to help separate the oil droplets from each other. So the job of emulsification in mayonnaise, and almost surely in béarnaise as well, is accomplished by a whole range of substances, from individual (and invisible) molecules of cholesterol, lecithin, and other emulsifiers to those comparatively massive molecular aggregations that do show up in the microscope.

What has the flurry of scientific attention to this *sauce célèbre* actually taught us? For one thing, that foods are very complicated chemical systems, and for another, that it is very easy to oversimplify such a system, to overlook important considerations and evidence and so jump to conclusions. But these lessons apply only to those engaged in advanced research. Amateur scientists may be encouraged by the fact that there is clearly room for their contri-

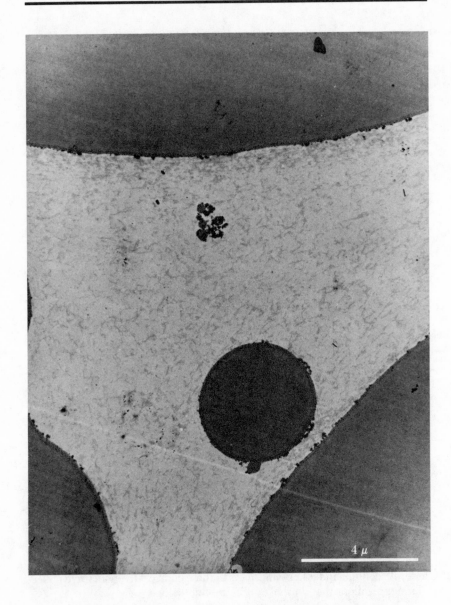

4 μ

A view through the electron microscope of oil droplets (the dark masses) in mayonnaise. The emulsion clearly owes its stability at least in part to the lipid-protein particles embedded in the droplet surfaces. The fibrous proteins floating between droplets may also help separate the droplets from each other. Both particles and proteins derive from egg yolk. (From Chang et al., *Can. Inst. Food Sci. Tech. J.* 5 (1972): 134–37. With permission of W. D. Powrie.)

butions. And for the cook, it comes down to a couple of strategies for avoiding total failure. The irreversible curdling of egg proteins, which accompanies overheating, can be minimized and even avoided by using enough vinegar and white wine to keep the yolk-liquid mixture distinctly acidic. So don't substitute water for vinegar, and minimize the amount of egg white that accompanies the yolk into the pan. Second, successful incorporation of the butter oil as droplets depends on the natural emulsifiers contained in the yolk. Use the freshest eggs possible to avoid the deterioration of some of these substances. And if success is more important than complete fidelity to tradition, keep a small container of soy phosphatides, a.k.a. "lecithin," hidden in the cabinet. A quick shake into a stubborn sauce may salvage it.

Sauce-making Techniques

Mayonnaise

The procedures for making the three principal emulsified sauces—hollandaise, béarnaise, mayonnaise—and their relatives are pretty much the same, with one exception. Mayonnaise is made from ingredients at room temperature, or even slightly cooler, while the other two sauces must be "cooked." The reason for this difference, of course, is that hollandaise and béarnaise are made with butter, which is a solid at room temperature. It and the ingredients to which it is added must be substantially warmer for an emulsion—a liquid dispersed in a liquid—to be formed. Because its vegetable oil is normally liquid, mayonnaise need not be heated as it is made. In fact, up to a point, the cooler the ingredients the better. Mayonnaise is by far the sauce most densely packed with oil droplets. In béarnaise, whose light body comes from some air incorporated during beating, the butter accounts for about a third of the sauce's volume, while in hollandaise the figure is closer to 60%; in mayonnaise, the oil takes up 75%. The cooler the ingredients, the more viscous the liquids, the fewer and slower the collisions between droplets, and so the less likely the droplets are to coalesce. It would be very difficult to crowd three parts of oil into one of water if the oil were butter and required elevated temperatures to be a liquid. At the same time, however, lower temperatures mean an increased surface tension in the two phases, and this fact makes it more difficult to divide the vegetable oil into droplets in the first place. It used to be that cookbooks would recommend chilling the ingredients and mixing bowl when making mayonnaise, but today room temperature is the near-unanimous choice.

Apart from heating the ingredients, the procedure for mayonnaise is representative of the preparation of the emulsified sauces. The egg yolks and water phase are always the first materials in the bowl, and the oil is added to them rather than the other way around. There are two very good reasons for this. The main job of the sauce making is to divide the oil into microscopic droplets, and this is clearly easier to do if one begins with one drop of oil in the water phase rather than the other way around. Similarly, the emulsifying agents will more quickly and evenly coat the oil droplets if initially there is more emulsifier present than there is oil. The first step, then, is to mix the yolks, vinegar or lemon juice, and seasonings together and beat them thoroughly until an even emulsion of yolk fats in water is formed. Now the real work begins. The oil must be dribbled in very slowly, and the mixture beaten continuously and vigorously, until about one quarter of the oil has been incorporated. After this point is reached, the cook's arm can rest for a moment and the oil is added in somewhat larger portions, a tablespoon at a time rather than drop by drop.

In the early mixing, when very little oil has been added, the problem is to produce small and widely separated droplets. As long as there is much more water than oil, it is possible for fairly large droplets to avoid the churning action of the whisk, and if they are large enough, these droplets will tend to rise to the surface of the mixture—oil is lighter than water—and there rejoin each other in a pool. Relentless beating and the addition of oil in very small amounts are necessary to produce and maintain small droplets. Once the volume of oil successfully incorporated is about the same as the initial volume of water and yolk, the droplets have become numerous and crowded enough that they effectively impede each other from rising, and the sauce is relatively stable. The sign that this point has been reached is a distinct thickness in consistency, like that of very heavy cream. The rest of the task is made easier by this initial care to keep the droplet size down. When additional oil is poured into an already well-emulsified system, the existing droplets work as a kind of mill, automatically breaking down the incoming oil into particles of their own dimensions. In the last stages of sauce making the cook's whisk need not break up the oil drops directly, but has the easier job of mixing the new oil with the sauce, distributing it evenly to all parts of the droplet "mill."

Because mayonnaise is so chock-full of oil—the continuous water phase accounts for only a quarter of the total volume—it is rather easily separated

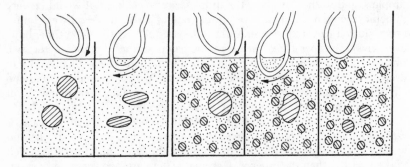

The droplet mill. When oil droplets are sparse *(left)*, they easily evade the whisk, and are hard to divide. When after much work they are smaller and more numerous *(right)*, the droplets themselves become obstacles to new drops of oil, and help break them up.

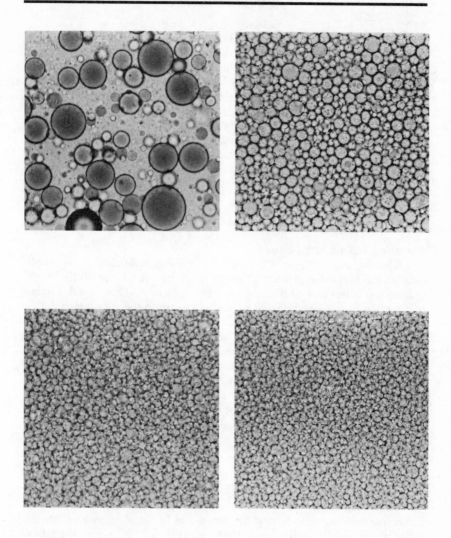

The making of mayonnaise as seen through a light microscope. *Clockwise from upper left:* one tablespoon of oil gives a sparse emulsion of coarse, unevenly sized droplets. With ¼ cup of oil beaten in, the number of droplets has greatly increased, and their size decreased. It is at this stage that the sauce thickens noticeably. The same trend continues as ⅜ cup and finally ½ cup have been incorporated into the egg yolk. Notice the distortion of some droplet spheres in the last two stages. The emulsifiers must be effective enough to withstand considerable physical pressure in order to prevent the oil droplets from coalescing into one big pool. (From P. C. Paul and H. H. Palmer, *Food Theory and Applications.* Copyright © 1972 John Wiley and Sons, Inc. Reprinted by permission.)

by extremes of cold, heat, and agitation. It will tend to leak oil in near-freezing refrigerators and on hot rather than warm food. And if it is beaten hard once the emulsion has formed, puddles of oil and water will result. This is one of the problems with making mayonnaise in a blender or food processor: the amount of energy poured into the mixture by the motor-driven blades is so high that the sauce can break down as quickly as it is formed. All of these problems are ameliorated to some extent by the addition of "stabilizers," usually long carbohydrate or protein molecules that fill the spaces between droplets. Blender mayonnaise, for example, is made with whole eggs; the albumen proteins weave themselves between droplets and prevent coalescence. The texture of such modified sauces, however, is noticeably different from the heavy, creamy original.

Rescuing a Separated Sauce

The rescue, or re-emulsification, of a broken mayonnaise is also applicable to the butter-based sauces, though the latter must be carefully heated to keep the butter melted. Once again, the problem is to get the oil droplets in a portion of the sauce down to the size and up to the concentration that they will almost automatically "mill" the remaining oil down. So a small amount of the sauce, including some of the yolk-and-vinegar phase, is put in a bowl with a little fresh yolk or water (and in the case of mayonnaise, some mustard; the seed particles are also excellent emulsifiers). These ingredients are beaten thoroughly to produce the initial emulsion, and then the rest of the sauce is very slowly dribbled in as the beating continues. All that this "rescue" amounts to is a repetition of the initial procedure, the only difference being that the water and oil phases are now partly mixed even before the beating begins.

Hollandaise and Béarnaise

Hollandaise and béarnaise are the prototypical butter-thickened sauces and differ only in seasoning and the amount of butter incorporated into the water phase. The thing that makes these sauces trickier to prepare than mayonnaise is the elevated temperature necessary to keep the butterfat liquid, a temperature high enough that irreversible coagulation of the egg proteins is a possibility. The standard precautions are to heat the yolk-water mixture very slowly and carefully, and to add the melted butter in small portions. A double boiler or a saucepan resting above a larger pan of simmering water will guarantee a gentle and even heat but will also retard the procedure perhaps more than is necessary; for this reason, some cooks prefer the comparatively risky direct heat of a burner. One useful trick is to have handy several pats of cold butter, which can be used to cool the mixture down quickly without interrupting the incorporation of fat into the yolk mixture. In any case, vigorous beating will help prevent the sauce at the bottom of the pan from overheating.

Here are the basic steps in the concoction of sauce hollandaise. First, the egg yolks are beaten alone to blend their various components, including the adhering white, evenly. Then water, lemon juice, and salt are beaten into the yolks, and this mixture is gradually heated and stirred until it becomes smooth, creamy, and noticeably thicker. The initial emulsion has now been formed. But if small white lumps appear, then the temperature has gotten too high, the egg proteins have coagulated, and the cook must start again from scratch. Otherwise, the butter is now incorporated by one of two techniques: one can either melt the butter in advance, then remove the yolk mixture from the heat and beat the butter into it; or, with the butter precut into small pieces, keep the pan over the heat and add the butter in solid form. In both cases, the butter must be added very slowly at first, then more rapidly once the emulsion has thickened. Yet another technique, notable for its simplicity, is to put all the ingredients fresh from the refrigerator—the yolks beaten together and the butter cut into small pieces—straight into an unheated frying pan, set it over medium-low heat, and start beating. After five or so minutes, the sauce is emulsified. Like mayonnaise, the butter-based sauces can also be made in a blender or food processor, but making them by hand is something like printing one's own photographs; both tasks hold the fascination of seeing the emulsion develop second by second. If for any reason the sauce separates, it can be rescued by keeping the whole mess warm, beginning with a tablespoon in a small amount of water or egg yolk, and carefully repeating the whole procedure.

Here, in summary, are some general rules for making emulsified sauces. Don't let any part of the sauce exceed 150°F (66°C); heat gently and beat continuously. Add the oil very slowly at first; better a tired arm than a broken emulsion. Don't try to incorporate more oil than the yolks can emulsify; according to Julia Child, the limits are ⅔ cup butter or ¾ cup oil per large yolk. If you want to thin a sauce slightly, do so with some of the continuous phase: water, not oil. Finally, make these sauces in stainless steel, enameled, or coated pans; metal oxides at the surface of aluminum or iron pans can discolor these pale concoctions.

Vinaigrette
Probably the most commonly made and consumed oil-water emulsion is the simple oil-and-vinegar salad dressing, or vinaigrette. The tangy flavor of this dressing is due, of course, to vinegar, whose name comes from the French *vin aigre*, or "sharp (sour) wine." It is in essence a water solution (normally 4 to 5%) of acetic acid, which gives it its characteristic pungent odor, though other substances make lesser contributions to its aroma. As its etymology suggests, vinegar is made by letting alcoholic liquids—wine, cider, fermented malt, distilled alcohol—"go sour," or be infected with bacteria (species of the genus *Acetobacter*) that metabolize alcohol and produce acetic acid. The various kinds of vinegar are valued for the particular aromas with which they accompany the fundamental acetic note.

Vinaigrette is the odd sauce out: very little of the preceding discussion applies to it. The usual proportions called for are three parts oil to one vinegar, or about the same ratio that goes into mayonnaise. But the preparation of vinaigrette is nowhere near as elaborate: usually we just put the oil, vinegar, and seasonings into a bottle and shake them up, or else beat them together for perhaps a minute. The resulting sauce does not thicken as much as the others do, and in fact it is not an oil-in-water emulsion but the reverse. Without the help of an emulsifier, one part of water simply cannot accommodate three parts of oil. By trying to form vinaigrette as if it were mayonnaise, one can actually observe the reversal of continuous and dispersed phases, or what is called "inversion." Put the vinegar—preferably red wine vinegar, for its color contrast with the oil—in a bowl, and slowly beat the oil in. An oil-in-water emulsion does form and gets noticeably thicker, until about equal volumes of oil and water have been mixed. All of a sudden the emulsion will thin out, and a close look will reveal droplets of the vinegar dispersed in oil. Let the sauce sit for a while, and a cloud of vinegar droplets slowly settles at the bottom of the bowl. Inversion can be prevented, and an oil-in-water emulsion formed, if two parts of oil are used rather than three and mustard and pulverized plant materials—for example, garlic or onion—are blended in as emulsifiers and stabilizers (the plant cell contents and wall materials help separate the oil droplets from each other).

Oil-and-vinegar emulsions are used almost exclusively as salad dressings, whose role is to coat with flavor the extensive surface area of lettuce leaves and cut vegetables. A thin, mobile sauce is more effective at this than a thick, creamy one. It makes sense to have oil be the continuous phase because it adheres to the vegetable surfaces better than the water-based vinegar, whose high surface tension causes it to bead up rather than leave a film. Because water and oil are antagonists, it is advisable to dry off the salad fixings before tossing them with vinaigrette; surfaces wet with water will repel the oil.

CHAPTER 8

Sugars, Chocolate, and Confectionery

THE APPEAL OF SWEETNESS

Sugar and candy are among the most popular and widely consumed foods today, even to the point of getting a bad reputation for displacing more nutritional foods from our diet. But they were once luxuries of the first order, and their continuing appeal is evidence of a deep-seated predilection. The source of this appeal is a handful of unassuming molecules, the simple carbohydrates called sugars. They are produced by the photosynthetic activity of plants, and are the standard currency of chemically stored energy for all plants and animals on earth. Many different kinds of sugars exist in nature, but only three are found in any quantity in the kitchen. Glucose and fructose both have the same chemical formula—$C_6H_{12}O_6$—but slightly different molecular structures (see page 586). Table sugar is nearly pure sucrose, each molecule of which is composed of one glucose and one fructose molecule joined together. Glucose and fructose are the major components of honey, and sucrose is extracted in quantity from sugar cane and beets.

The pure sugars are solid at room temperature, liquefy when heated, and at a certain point turn brown and develop a rich, characteristic "caramel" flavor. What Americans call candy (in Europe, the term is limited to hard confections) is basically pure sucrose mixed with small amounts of glucose, fructose, and other flavorings. Chocolate, which comes from the seed of a South American tree, has been married with sugar ever since its arrival in Europe 400 years ago, and today is preeminent among the nonsugar ingredients in candy. The cook can vary the texture of candy from rock hard to creamy simply by controlling the way in which hot sugar syrup cools and solidifies.

What makes sugars and candy so appealing, of course, is the fact that they taste sweet. We humans share an innate liking for sweetness—perhaps a relic of our early career as arboreal fruit-eaters (see chapter 4)—and as we shall see, this biological heritage has been a real force in human history. Equally notable is its prominence in our ways of thought. For one thing, our everyday language tends to identify sweetness with love and pleasure. "Honey," "sugar," and "sweetheart" are easily the most common terms of endearment in English (for some reason, cabbages and chickens are the French equivalents; "treasure," the German and Italian). The ancestor of "sweet" itself, the Latin *suavis*, comes from the verb meaning to persuade, to make pleasing to.

This painting found in the Spider Cave at Valencia, Spain, and dating back to about 8000 B.C. (greatly reduced from its original size) appears to show two people raiding a beehive. The leader *(enlarged at right)* may be carrying a basket for the honeycomb. Artificial hives, and so the domestication of bees, appear to date from 2500 B.C. in Egypt. (Redrawn from H. Ransome, *The Sacred Bee*, 1937.)

And consider the archaic evidence of myths from two very different cultures. First, the Hebrew account of the fall from Eden. Would forbidden nuts or vegetables be as convincing an embodiment of temptation as the forbidden fruit? Then there is the less familiar South American myth interpreted by the French anthropologist Claude Lévi-Strauss. A small native community in the Mato Grosso region of Brazil told the story that originally the wolf was the master of honey, but that the tortoise managed to procure it for the other animals. He made cuttings of honey, as if it were a plant, and gave one to every animal as a perpetual source. After a while, however, most of the animals had eaten their cuttings rather than cultivating them. Their leader then dispersed wild bees to form hives and said, "Now the forest is full of all kinds of honey. . . . All you have to do is go look for it." Lévi-Strauss finds the myth remarkable for its rare "antineolithic" outlook: when it comes to so desirable a food, simple gathering is preferable to the more advanced ways of civilization. "So powerful is its gastronomical appeal that, were it too easily obtained, men would partake of it too freely until the supply was exhausted." To this culture, then, wild hives are a sign of a falling back, a temporary regression caused by our insatiable hunger for sweetness.

As it turns out, the world supply of sugars has managed to keep up with demand, though at a certain cost to Europe's standing as a civilized power. And honey has been successfully "cultivated" all over the world.

HONEY

In the beginning, fruits must have been our principal source of sweetness. While some tropical fruits like the date can approach a sugar content of 60%, an even more concentrated source is honey, the stored food of certain species of bees. We know that badgers and bears will raid bee hives, and from a remarkable painting in the Spider Cave of Valencia, it is clear that humans too have gone out of their way to collect honey for at least 10,000 years. The "domestication" of bees—that is, housing them in artificial hives—is much more recent. The bee appears in Egyptian hieroglyphs around 4000 B.C., but the first representations of clay hives come some 1500 years later. Approximately the same chronology seems to apply in India.

However it was obtained, honey appears persistently as an emblem of the ideal in some of the earliest literature we know. A love poem inscribed 4000 years ago on a Sumerian clay tablet describes a bridegroom as "honey-sweet," the bride's caress as "more savory than honey," and their bedchamber as "honey-filled." In the Old Testament, of course, the promised land is pictured several times as a land flowing with milk and honey, a metaphor of delightful plenty that is itself used figuratively in the Song of Songs, where another bridegroom chants, "Thy lips, O my spouse, drop as the honeycomb: honey and milk are under thy tongue. . . ."

Honey was an important ingredient in both the cuisine and culture of classical Greece and Rome. Breads, cakes, and sauces were sweetened with it, and it was the base of a wide range of sweet wines, among them oxymel and hydromel. The late Latin recipe book ascribed to Apicius includes directions for preserving fruit and meat in honey and calls for its use in spiced wines and various dishes made from nuts, fruits, eggs, fresh cheese, and fried bread. For the Greeks, honey was also a food of talismanic significance. They offered it in ceremonies to the dead and the gods, and priestesses of the goddesses Demeter, Artemis, and Rhea were called *melissai:* the Greek *melissa,* like the Hebrew *deborah,* means "bee."

The status of honey was due in large part to its mysterious origins. Aristotle said, "One cannot well tell what is the substance [the bees] gather, nor the exact progress of their work" and suggested that they collect dew that has fallen from the sky. The belief that honey is a little bit of heaven fallen to earth was widespread. The Roman poet Vergil opens the fourth book of his *Georgics,* which is wholly devoted to bees and honey, with the line "Next I will discuss Heaven's gift, the honey from the skies." His contemporary, the natural historian Pliny, speculated in greater and somewhat unappetizing detail on honey's nature (Book 11).

> Honey comes out of the air. . . . At early dawn the leaves of trees are found bedewed with honey. . . . Whether this is the perspiration of the sky or a sort of saliva of the stars, or the moisture of the air purging itself, nevertheless it brings with it the great pleasure of its heavenly nature. It is always of the best quality when it is stored in the best flowers.

It was better than 1000 years before the true roles of flower and bee in the creation of honey were uncovered.

Honey was widely used in Europe until around 1500, at which point the increasing availability of easily stored cane sugar made it the less attractive alternative. Germany and the Slavic countries were leading producers in the meantime, and honey wine or "mead" (from the Sanskrit for honey) was a great favorite in both central Europe and Scandinavia. As late as the 17th century, honey-based alcoholic beverages were the object of some connoisseurship. The English courtier Sir Kenelm Digby collected his favorite recipes for various foods and drinks, and better than 100 of them, nearly half of the total, are formulas for mead, metheglin, hydromel, and so on. Sweet grape wines have turned these onetime staples into curiosities.

The Honey Bee

While the Western Hemisphere certainly knew and enjoyed honey before the arrival of European explorers, North America did not. The bees native to the New World, species of the genera *Melipona* and *Trigona,* are exclusively tropical. They also differ from the European honey bees in being

HONEY WINE

Hydromel as I Made It Weak For the Queen Mother

Take 18 quarts of spring-water, and one quart of honey; when the water is warm, put the honey into it. When it boileth up, skim it very well, and continue skimming it, as long as any scum will rise. Then put in one Race [root] of Ginger (sliced in thin slices), four Cloves, and a little sprig of green Rosemary. Let these boil in the Liquor so long, till in all it have boiled one hour. Then set it to cool, till it be blood-warm; and then put to it a spoonful of Ale-yest [yeast]. When it is worked up, put it into a vessel of a fit size; and after two or three days, bottle it up. You may drink it after six weeks, or two months.

Thus was the Hydromel made that I gave the Queen, which was exceedingly liked by everybody.

—*The Closet of Sir Kenelm Digby, Knight, Opened* (1699)

(Full-strength recipes call for 4–6 quarts of water and a spoonful of yeast for every quart of honey.)

stingless (although they do literally bite and can burn the skin with caustic chemicals) and in collecting fluids not just from flowers, but also from fruits, resins, and even carrion and excrement—sources that can make for unhealthy honeys as well as rich and strange flavors. European colonization brought a fundamental change to the New World by introducing, around 1625, the bee that produces practically all the honey in the world today, *Apis mellifera*.

We are lucky to have a near-contemporary and evocative description of the honey bee's movement across North America. In 1832, Washington Irving toured what is now the Oklahoma region and three years later published his observations in *A Tour on the Prairies*. The ninth chapter describes a "Bee-hunt," the practice of finding honey in the wild by following bees back to their hive, and begins with this paragraph.

The beautiful forest in which we were encamped abounded in bee-trees; that is to say, trees in the decayed trunks of which wild bees had established their hives. It is surprising in what countless swarms the bees have overspread the Far West within but a moderate number of years. The Indians consider them the harbinger of the white man, as the buffalo is of the red man; and say that, in proportion as the bee advances, the Indian and buffalo retire. We are always accustomed to associate the hum of the bee-hive with the farm-house and flower-garden, and to consider those industrious little animals as connected with the busy haunts of man, and I am told that the wild bee is seldom to be met with at any great distance from the frontier. They have been the heralds of civilization, steadfastly preceding it as it advanced from the Atlantic borders, and some of the ancient settlers of the West pretend to give the very year when the honey-bee first crossed the Mississippi. The Indians with surprise found the mouldering trees of their forests suddenly teeming with ambrosial sweets, and nothing, I am told, can exceed the greedy relish with which they banquet for the first time upon this unbought luxury of the wilderness.

For those of us who buy our luxury in jars, this initial sense of wonder is worth reimagining.

Bees are social insects that have evolved along with nectar-producing flowering plants. The two organisms help each other out: the plants provide the insects with food, and the insects carry pollen from one flower to another. Honey is the form in which flower nectar is stored in the hive. It appears from the fossil record that bees have been around for some 50 million years, though their social organization is only 10 or 20 million years old. *Apis*, the principal honey-producing genus, originated in western Asia and branched out into four major species. Two of these nest only in the open, and so have remained confined to tropical or semi-tropical Asia. The others nest in sheltered places, can therefore survive in harsher climates, and so have spread. *Apis mellifera*, the honey bee proper, now inhabits the whole of the northern hemisphere up to the Arctic Circle. (The infamous "killer bees" of Brazil are a cross between European and African races of *Apis;* they have been successful in South America largely because of the latter's tropical origin and innate aggressiveness.)

How Honey is Made

Nectar and Honeydew

The principal raw material of honey is the nectar collected from flowers. Secondary sources include nectaries elsewhere on the plant and honeydew, the secretions of a particular group of bugs. The plant glands that release nectar, whether in the flowers or along the stem and leaves, may have originated as sugar valves. By venting sugary sap to the outside of the plant, they can help regulate the osmotic pressure of the circulating plant fluids, in somewhat the same way that our kidneys control our sodium levels. Floral

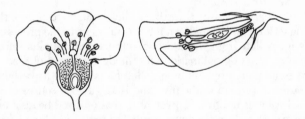

The nectar-secreting tissues (shaded) of a flower in the rose family *(left)* and of 3-leaved clover *(right, much enlarged)*. Their proximity to the reproductive organs helps to enlist bees in the job of cross-pollination.

nectaries now have the important role of attracting pollinators, whether bees, butterflies, or birds. The chemical composition of nectar varies widely. It often includes B vitamins, vitamin C, and some minerals, but its major ingredient by far, accounting for 10 to 80% of its weight, is sugar. Some nectars are mostly sucrose, some are evenly divided among sucrose, glucose, and fructose, and some are mostly fructose. A few nectars, while harmless to bees, are poisonous to humans, and so generate toxic honeys. Honey from the Pontic region of eastern Turkey was notorious in classical times; a local species of rhododendron was probably the culprit. Today a handful of plants, including some azaleas, is known to generate poisonous honey, but none is very common in North America.

The most important sources of nectar are leguminous plants, especially clover; linden trees; and members of the family Compositae, a large group that includes the sunflower, dandelion, and thistles. Sage, thyme, tupelo, buckwheat, and eucalyptus honeys, with their distinctive tastes, are especially valued.

Unlike nectar, honeydew is already partly processed before the bee gets to it. Insects of the order Homoptera, or the "true bugs," which include both bedbugs and aphids, have specialized mouth parts—two grooved bristles that, when drawn together, form a rigid tube—that allow them to penetrate the surface layers of other organisms and draw out nourishing fluids. Plant suckers tap into the phloem tubes, which carry sugars from the leaves to the rest of the plant, and remove the same sap that nectar glands naturally release. Honeydew is formed when these bugs excrete droplets of sap soon after ingesting it. Because the sap is so dilute, its consumers have developed a digestive tract that includes a filtering chamber just before the intestine. Here the nutrients are concentrated, and some excess moisture, together with small amounts of sugar and digestive enzymes, is sent directly out the other end, bypassing the intestine. When it reaches the air, water quickly evaporates from the honeydew droplets, and they become even more concentrated than the original sap. Bees and ants, among other insects, take advantage of this service by collecting the honeydew for their own use. Honeydew is especially important in forests dominated by conifers, which do not bear flowers.

Gathering Nectar

The bee gathers nectar by thrusting its head into the center of the flower and inserting its long proboscis down into the nectary. In the process, its hairy body picks up pollen from the flower's anthers and deposits some on the stigma, thereby initiating pollination and the development of seed. The bee also collects some pollen by combing its leg hairs against its body. Pollen is the main source of protein in the hive and is also a good source of vitamins; its nutritional value is roughly equivalent to that of dried beans and peas. It is stored in separate cells in the honeycomb, fed to the developing brood, and consumed by workers for the first half of their life. After that point, their sole food is honey.

Once the bee draws nectar up through its proboscis, the liquid passes through the esophagus into the honey sac, a chamber immediately preceding the intestine. The honey sac is mainly a storage tank that holds the nectar until the bee returns to its hive. But certain glands do secrete enzymes into the sac, and these work to break down starch into smaller chains of sugars and sucrose into its constituent glucose and fructose molecules. Any pollen ingested along with the nectar passes on into the gut, so that after 15 minutes, the pollen count is reduced to about one third its initial level.

A few remarkable figures are worth quoting. A strong hive contains one mature queen, a few hundred male drones, and some 20,000 female workers. For every pound of honey taken to market, 8 pounds are used by the hive in its everyday activities. The total flight path required for a bee to gather enough nectar for this pound of surplus honey has been estimated at 3 orbits around the earth. The average bee forages within one mile of the hive, makes up to 25 round trips each day, and carries a load of around 0.002 of an ounce—approximately half its weight. With its light chassis, a bee gets about 7 million miles to a gallon of honey. In a lifetime of gathering, a bee will contribute only a small fraction of an ounce of nectar to the hive.

By the time the foraging bee reaches the hive, the nectar has less sucrose and more glucose and fructose than it did originally, and is more dilute because the bee's saliva, which enables it to exploit nearly dry sources of sugar, has watered it down. The principal task in the hive is to concentrate the nectar to the point that it will resist bacteria and molds and so keep until it is needed. Many microbes normally feed on sugars, but they are killed if the sugar concentration is high enough for the osmotic pressure to draw moisture out of their cells.

In the Hive

The first stage in processing nectar is accomplished by "house bees" to which the forager passes its load. These workers—sometimes only one, sometimes a long chain—pump the nectar in and out of themselves for 15 or 20 minutes, repeatedly forming a thin droplet under their proboscises, until the carbohydrate content has reached 50 or 60%. The final bee in the chain then deposits the concentrated nectar in a thin film on the honeycomb, which is

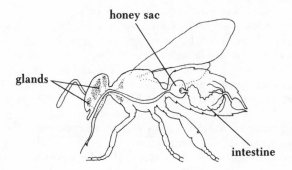

The anatomy of the worker bee. Nectar is held in the honey sac, together with enzymes from various glands, until the worker can return to the hive.

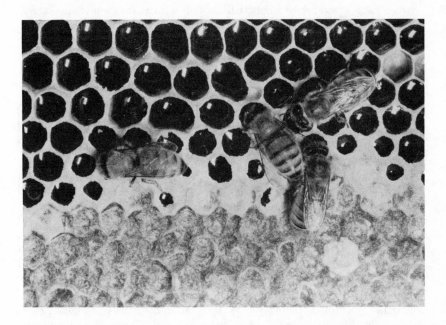

Worker bees sealing cells of ripe honey in the hive. (Photograph by Kenneth Lorenzen.)

a waxy network of hexagonal cylinders about ⅕ inch across, built up from the secretions of the wax glands of young workers. Here, the nectar evaporates further, until it is only 20% water. This process, known as "ripening," takes about 3 weeks, depending on the temperature and humidity, and is not entirely passive. The bees fill fresh nectar cells to only one-third capacity in order to leave a great deal of surface exposed to the air. Nearly ripe honey is transferred to cells that are three-quarters full, and cells of fully ripe honey are filled to capacity and then capped with a layer of wax. All the while, preforaging workers keep the hive air in continuous motion by fanning their wings.

The ripening of honey involves both evaporation and the continuing work of bee enzymes. For example, the disaccharide sucrose is converted almost entirely to glucose and fructose. Why? Because a mixture of single-unit sugars is more soluble than the equivalent amount of sucrose. That is, more fructose and glucose can be dissolved in a given amount of water than sucrose, and so can produce a more concentrated solution. Higher concentrations mean both a more compact supply of energy and a more effective defense against spoilage. Antimicrobial action is also the function of an enzyme that oxidizes glucose to form gluconic acid and peroxides. Gluconic acid lowers the honey's pH, and the peroxides act as an antiseptic. More mysterious are several enzymes that actually synthesize long-chain sugars, some of them very rare, in small quantities. In addition to all this sugar chemistry, ripening honey also undergoes complex changes in color and flavor. So far, upwards of 200 different substances have been identified in honey, and there are certainly others yet to be discovered. The following table summarizes the average composition of honeys produced in the United States.

COMPOSITION OF HONEY
(PERCENT)

Water	17	Higher sugars	1.5
Fructose	38	Acids	0.6
Glucose	31	Minerals	0.2
Sucrose	1.5		
Other disaccharides	7	pH 3.9	

It is worth noting briefly that despite the claims of some "natural" food advocates, honey is no wonder food. Its vitamin content is negligible; bees get most of theirs from pollen. Also, because the B vitamins are part of the machinery that liberates energy from carbohydrates and fats, a given amount of honey actually uses up more B vitamins than it supplies. It contains only about 3% of the thiamine and 6% of the niacin necessary to convert its sugars into energy. Honey has also been reputed to have antibacterial properties, and early physicians used it to dress wounds. In the late thirties, the existence of a natural antibiotic in honey, dubbed "inhibine," was postulated. In 1963, the identity of inhibine was discovered to be hydrogen peroxide, one of the products of a glucose-oxidizing enzyme and a substance well known and long employed in medicine. There are plenty of reasons for liking honey, but nutrition and medicinal value are not two of them. And although pollen and royal jelly do wonders for bees, there is no evidence that they do anything out of the ordinary for humans, even at the premium prices they fetch in health-food stores. One of the most entertaining, if not wholly reliable, accounts of the virtues of bee food is Roald Dahl's story "Royal Jelly" (included in the 1959 collection *Kiss Kiss*), in which that substance helps a sickly infant grow strong—but into something less than human.

From Hive to Table

Improving the Hive

Up until about 1500, harvesting honey was a rather crude affair. Many hives were simply tree trunks—the German for "hive" is *Bienenstock*, literally "bee trunk"—and even those made from straw or clay were simple, irregularly shaped containers. Beekeepers would break open all the hives but one and remove the combs, thereby ruining the colonies; the exception would be allowed to overwinter. In time, there evolved box hives, which took advantage of the fact that the hive is organized vertically, with the breeding and rearing areas below, pollen stores above them, and honey stores on top. Two-part hives allowed the lower, active section to remain undisturbed while the honeycomb was harvested. But even this system remained cumbersome, and the yield per hive rather low, until the arrival of the father of modern beekeeping, Lorenzo Lorraine Langstroth.

This Yale graduate, pastor, and Andover, Massachusetts, school principal retired in 1852 to Oxford, Ohio, to devote himself to his bees. His great discovery was the "bee space," a gap of 5⁄16 inch that, when left around thin wooden frames slipped vertically into the hive, prevents them from getting stuck. Before Langstroth, attempts to create removable frames that could be pulled from the hive filled with honeycomb, emptied, and replaced, had always failed. The bees would either fill the spaces with wax, or, if the spaces were large, they would build honeycombs in them. In either case, the frame would be immobilized. The bee space is the particular dimension that remains unplugged, apparently because it is very close to the diameter of a bee and can serve as a passageway. Langstroth patented his hive in 1852 but saw little reward for his innovation; all that had to be known to infringe on the patent was a single dimension. But it is Langstroth's pioneering work in hive construction and management that made modern, efficient honey production possible. Today, several "supers," or upper stories, will be filled with honey by a single hive; the simple expedient of a "queen excluder," a wire screen with a mesh size too small to allow the much larger queen to pass

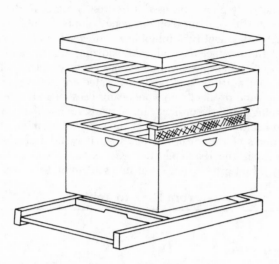

The man-made hive. Bees live, reproduce, and store pollen and honey in the lower section, and fill the upper section with surplus honey.

through, prevents their contamination by brood or pollen cells. World production of honey now runs around 600,000 tons a year, with 40% of this total coming from the New World.

Processing Honey

Once the honeycomb is removed from the hive, the individual cells are mechanically uncapped and the honey extracted in a centrifuge. It is first heated to around 155°F (68°C) to destroy sugar-fermenting yeasts, then strained to remove pieces of wax and debris, blended with other honeys, and finally filtered under pressure to remove pollen grains and very small air bubbles that would cloud the liquid. In the United States and Australia, honey is sold mostly in liquid form and so is bottled at this stage. Airtight storage is important because the sugars in honey are very hygroscopic, or moisture attracting, and will absorb water from the air if the relative humidity is over 60%. And if the water content of honey rises much above 17%, yeasts will grow and ferment the sugars to alcohol and carbon dioxide.

Spoilage also occurs when honey granulates, or crystallizes, something that will eventually happen, especially if the temperature stays around 60°F (16°C). It turns out that the glucose concentration in honey is close to that sugar's saturation point, so it tends to leave solution and form crystals fairly readily. (Tupelo and sage honeys, with much more fructose than glucose, resist granulation.) And when some sugar precipitates out, the remaining solution is less concentrated and so more hospitable to microbes. Because all sugars become increasingly soluble as the temperature rises, granulating honey can be reliquefied by putting the jar in a pan of hot water. Once it cools, of course, the whole process starts again.

In countries other than the United States and Australia, most honey is sold in the thick, opaque, pregranulated form. In the decades following World War I, "sweet clover," actually a kind of alfalfa and a prodigious producer of nectar, gained wide acceptance in North America as a forage and rotation crop. Honey production boomed as a consequence, and some way of preserving it during storage and export was needed. The answer was controlled crystallization: yeasts cannot grow on glucose crystals, and if the remaining solution is dispersed very finely so that there are no pockets large enough to support many yeast cells, noticeable fermentation will not occur. Fine dispersal of the solution is possible only if the glucose crystals are also very fine. This is a desirable characteristic for reasons of texture as well; large crystals feel grainy to the tongue and do not spread well on other foods. The problem, then, was to grow very small crystals. Around 1930, E. J. Dyce of Cornell University developed the process that still bears his name.

As we shall see in more detail when we get to candy, the secret of producing fine crystals is to produce many "seeds"—solid nuclei, whether other crystals, pollen, or even very small air bubbles, onto which glucose molecules will attach and build up a solid mass—all at the same time, and to make the process go quickly. Under these conditions, each seed crystal will draw sugar only from its immediate neighborhood, and because of the competition from all the others, no crystal will be able to grow very large. In the Dyce process, starter nuclei are provided to the cooling honey in the form of granulated honey that has been chilled and ground into a fine powder. In 3 days, the

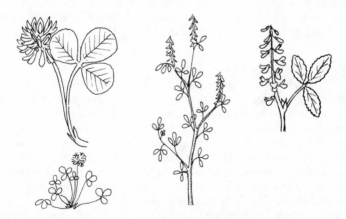

When sweet clover *(right)* was introduced to this continent as a forage crop, it caused a great increase in honey production. The more familiar 3-leaved clover with its ball of flowers *(left)* is a different plant, belonging not to the genus *Melilotus*, but to *Trifolium*.

seeded honey is firm, and in 6 it is ready for market, having reached a creamy consistency. Though it seems more like a solid than anything else, 85% of the honey remains in liquid form, dispersed around the 15% that has solidified into tiny crystals of glucose.

Culinary Uses

Unlike sugar, which is often a hidden ingredient in processed foods, honey is a very visible sweetener; most of it is added to foods by individual consumers. It is the characteristic sweetener in such traditional national foods as baklava and lebkuchen, nougat, torrone, halvah, and pasteli, and in such beverages as Benedictine, Drambuie, and Irish Mist. Although mead has all but disappeared, honey beer is popular in Africa. Americans use honey in many baked goods not only for its distinctive flavor but also for its improvement of keeping quality. It can be substituted for sugar—1 measure of honey is considered the equivalent of 1¼ of sugar, although the amount of added liquid must be decreased because honey does contain some water. Because it is more hygroscopic, or water attracting, than table sugar (fructose is more hygroscopic than sucrose), honey will keep breads and cakes moister than sugar will, losing water to the air more slowly, and even absorbing it on humid days.

MAPLE SYRUP

If we put aside the many fascinating details, bees perform one basic task when they make honey: they remove a very dilute solution of sugar from plants and concentrate it. What the bees have come to do instinctively, humans have learned to do. We make maple, cane, and beet sugars by extracting juices from these plants and boiling off the excess water.

The North American Sugar

Of these three man-made sweets, maple sugar is most like honey in that it retains nearly all the original contents of the sap and is not refined to the extent that cane and beet sugar are. Until the arrival of the honey bee, maple sugar was the only source of concentrated sweetness, outside of dried berries, that the North American continent provided its inhabitants. It is not known whether maple sugar is as old as cane sugar, but several Indian tribes, notably the Algonquins, Iroquois, and Ojibways, had well-established myths about and terminologies for maple sugaring by the time that European explorers encountered them. The Indians taught the colonists their techniques, and thanks to a remarkable document, we have some idea of what they were. In 1755, a young colonist named James Smith was captured and "adopted" by a small group of natives in the region that is now Ohio. He lived with them until 1759, when he escaped to become an Indian fighter and then a colonel in the Revolutionary Army. In 1799 he published *An Account of the*

Remarkable Occurrences in the Life and Travels of Col. James Smith, which contains two descriptions of sugaring. Here is the first:

> In this month [February] we began to make sugar. As some of the elm bark will strip at this season, the squaws after finding a tree that would do, cut it down, and with a crooked stick broad and sharp at the end, took the bark off the tree, and of this bark, made vessels in a curious manner, that would hold about two gallons each: they made above one hundred of these kind of vessels. In the sugar-tree they cut a notch, stooping down, and at the end of the notch, stuck in a tomahawk; in the place where they stuck the tomahawk, they drove a long chip, in order to carry the water out from the tree, and under this they set their vessel, to receive it. As sugar trees were plenty and large here, they seldom or never notched a tree that was not two or three feet over. They also made bark vessels for carrying the water, that would hold about four gallons each. They had two brass kettles, that held about fifteen gallons each, and other smaller kettles in which they boiled the water. But as they could not at all times boil the water away as fast as it was collected, they made vessels of bark, that would hold about one hundred gallons each, for retaining the water; and tho' the sugar trees did not run every day, they had always a sufficient quantity of water to keep them boiling during the whole sugar season.

This particular operation was rather advanced, and it would seem that the greatest problem was the production of suitable containers. North Americans did not have any knowledge of metalworking until the European migration. Before the arrival of brass kettles, the boiling took place in more fragile and clumsy clay pots and would have been a tricky business. The alternative method that Smith observed the next spring is more primitive and likely to be more authentic.

> Some time in February, we scaffolded up our fur and skins, and moved about ten miles in search of a sugar camp or a suitable place to make sugar. . . . We had no large kettles with us this year, and the squaws made the frost, in some measure, supply the place of fire, in making sugar. Their large bark vessels, for holding the stock-water, they made broad and shallow; and as the weather is very cold here, it frequently freezes at night in sugar time; and the ice they break and cast out of the vessels. I asked them if they were not throwing away the sugar? they said no; it was water they were casting away, sugar did not freeze and there was scarcely any in that ice. They said I might try the experiment, and boil some of it, and see what I would get. I never did try it; but I observed that after several times freezing, the

water that remained in the vessel, changed its colour and became brown and very sweet.

Despite the last sentence, this early version of freeze-drying would not of itself have turned the sap brown and developed the rich flavors we associate with maple syrup; these are the result of browning reactions that require high temperatures. In all likelihood, the nightly frosts were used to reduce the sap to volumes more easily handled in clay pots. According to Benjamin Rush, a physician, signer of the Declaration of Independence, and social reformer, some colonists also used the freezing technique, along with simple evaporation, but the unreliability of the weather led most people to boil their syrup down.

Maple sugar was an important part of the native Americans' diet. Smith reports that they worked it into bear fat and used this mixture as a dip for other foods. Rush said that they mixed it with corn meal to make a light, compact provision for journeys. For the colonists, maple sugar was much cheaper and more available than the heavily taxed refined sugar from the West Indies. Even after the Revolution, Americans found a moral reason for preferring maple sugar to cane; cane sugar was produced largely with slave labor. Almanacs reminded farmers of this fact in their advice for the early spring, and Benjamin Rush wrote to Thomas Jefferson in 1791 that the maple was an instrument of providence.

> In contemplating the present opening prospects in human affairs, I am led to expect that a material part of the general happiness which Heaven seems to have prepared for mankind will be derived from the manufactory and general use of maple sugar. . . . I cannot help contemplating a sugar maple tree with a species of affection and even veneration, for I have persuaded myself to behold in it the happy means of rendering the commerce and slavery of our African brethren in the sugar islands as unnecessary as it has always been inhuman and unjust.

The development of the sugar beet and competition in the cane sugar trade did conspire to weaken slavery in the Indies, but toward the end of the nineteenth century, cane and beet sugar became so much cheaper that maple sugar production declined steeply, and current production is running at about one fifth of 1900 levels. Maple syrup today is used in a very small number of foods, and its production remains very much a cottage industry, with several thousand small landowners in the northeast making it as a supplemental source of income.

The Mysterious Sap Run

Because not much is known about exactly why the trees do what they do in early spring, maple sugaring continues to be as much an art as a science. The maple family originated in China or Japan and numbers some 100 species which are spread throughout the Northern Hemisphere, though they are relatively rare in Europe. Of the four North American species good for sugaring, the hard or rock maple, *Acer saccharum*, produces sap of greater quality and in greater quantity than the others and accounts for most of the syrup produced today. Growing conditions that encourage the greatest possible photosynthesis—the leaves' conversion of sunlight, water, and carbon dioxide into sugar—will maximize the amount of sugar in the following spring's sap. A large crown of leaves, a sunny summer and fall, and a late frost are ideal. In the spring, sap is collected from the first major thaw until the leaf buds burst, at which point the tree fluids begin to carry substances that give the syrup a harsh flavor. The run itself is improved by four conditions: a severe winter that freezes the roots, snow cover that keeps the roots cold in the spring, extreme variations in temperature from day to night that are typical of mountain climates, and good exposure to the sun. The northeastern states meet these needs most consistently.

There is as yet no full explanation of the maple sap run, how or why it occurs. It appears to be an almost unique, anomalous process. Sap does run in other trees in early spring, and some of them—birch and elm, for example—have been tapped for sugar. But maples produce more and sweeter sap than any other tree. What is more peculiar is that sap, unlike nectar, does not come from the leaves (they have not been deployed yet) or from the roots, where sugar is often stored and where water enters the tree. Rather, it appears to flow from all directions—above, below, sideways—into the tap hole. And the logic of the optimal conditions is far from clear.

The current theory goes something like this. Sugar from the previous season is stored all along the trunk, particularly in specialized cells in the *xylem* tissue. A mixture of live cells and transport tubes formed by dead cells, the xylem normally carries water and nutrients from the roots to the rest of the tree. (*Phloem* tissue, from which honeydew and nectar derive, carries

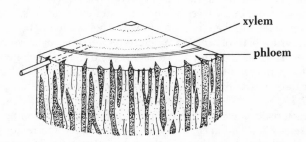

Maple sap flow occurs in the xylem, the tissue in the trunk that transports food to the leaves.

sugars from the leaves downward.) When the maple trunk thaws out in early spring, it appears that the specialized xylem cells actively pump their sugar into the xylem vessels, the transport tubes formed by now-dead cells. This work has two effects. One is that the increase in sugar content of the xylem vessels creates an osmotic pressure that draws water into the vessels, thereby increasing the water pressure. And the metabolic activity of the pumping cells, like all organic activity, releases waste products, among them the gas carbon dioxide. It happens that the solubility of carbon dioxide in water decreases as the temperature rises above freezing, so that as the trunk warms up on a sunny spring afternoon, gas pressure builds up in the xylem tissue along with the water pressure. The result: a substantial flow outward when the sap seeker taps into the xylem.

And what is the advantage to the tree of this intricate mechanism? It seems that the sugary sap is pushed via horizontal structures called rays from the xylem into the cambium, which is the area of active growth in the tree. It may be, then, that the maple sap run is a happy consequence of early preparations for the new year's growth.

Syrup Production

The Indian technique of making a wide gash in the tree trunk was fine if the same tree was not likely to be exploited year after year. The colonists introduced the practice of punching a small hole with an auger and fitting in a wooden or metal spout from which a bucket was hung. The collected sap was then boiled in shallow pans and the scum of impurities periodically removed, until the syrup's taste and viscosity dictated a halt. The final concentration was generally high enough that the syrup would solidify into sugar crystals on cooling; such terms as "sugarbush" for a stand of maples, "sugar house" for the shed in which the boiling was done, and "sugaring" for the whole process all have reference to this final point rather than to the syrup, which is more common today.

Twentieth-century technology has made some inroads into sugaring. Power drills are used to tap the trees, and on hills plastic tubing sometimes carries the sap from many trees down to a central holding tank. The concentrating process is controlled by monitoring the boiling temperature and density of the syrup, both of which are determined by the sugar concentration. The original sap averages around 3% sucrose, and a single tree may yield 12 gallons of sap in one season. It takes between 30 and 40 gallons of sap to make a gallon of syrup. The final composition of maple syrup is approximately 62% sucrose, 35% water, 1% glucose and fructose, and 1% malic acid. The characteristic flavor of the syrup is a product of browning reactions between sugars and amino acids in the sap; the longer and hotter the syrup is boiled, the darker the color and the heavier the taste. Maple syrups are graded according to color, flavor, and sugar content, with the higher grades (AA and A) assigned to the lighter, delicately flavored, and slightly less concentrated syrups.

Maple Sugar

Maple sugar is made by concentrating the sucrose to the point that it will crystallize out of solution when the syrup cools. This point is marked by a boiling temperature of about 25°F above the boiling point of water, or 237°F (114°C) at sea level. As we shall see in the discussion of confectionery, the way in which a sugar syrup cools determines the final texture of the solid. If cooling is slow and the syrup is stirred only occasionally, then large-grained crystals form, and the sugar is coarse. Maple cream, a malleable mixture of very fine crystals in a small amount of dispersed syrup, is made by cooling the syrup very rapidly to about 70°F (21°C) by immersing the pan in baths of iced water, and then beating it continuously until it becomes very stiff. This mass is then rewarmed in a double boiler and becomes smooth and semisoft.

A warning to those who wish to make their own maple sugar: most syrups sold in stores are only "maple flavored"; check the label carefully. Pure maple syrup is much more expensive than the imitations. And boiling syrup rises very high in the pan, so be sure you use one with plenty of extra capacity.

SUGAR

Its History

Ordinary table sugar, which is 99% pure sucrose and by far more commonly used today than either honey or maple sugar, was barely known to Europe nine centuries ago and was a luxury reserved for the fortunate few until 1700. Until the late 19th century, the principal source of sucrose was the sugar cane, *Saccharum officinarum*, a 20-foot-tall member of the grass family with an unusually high sucrose content—about 13%—in its fluids. The stalk of another grass, the corn or maize plant, was pressed for its sweet juice by the natives of Central America, according to some of the first European explorers. The sugar cane originated somewhere in the South Pacific and gradually spread into Asia, perhaps carried by early human migrations. The technology of making "raw" sugar by pressing out the cane juice and boiling it down into dark crystals was developed in India around 500 B.C.

Both sugar cane and the technique for extracting sugar were carried westward from India first by the Persians, who established them in the Tigris-Euphrates valley in the 6th century A.D., and then by the Arabs, who conquered Persia for Islam around 640 and introduced the cane to northern Africa, Syria, and Spain. Christian Europe first encountered sugar during the Crusades to the Holy Land in the 12th century, and shortly thereafter Venice was established as the hub of the sugar trade to the West. The first large shipment to England that we know of came in 1319. Our language still bears the traces of this history. "Sugar" comes from the Arabic imitation of the

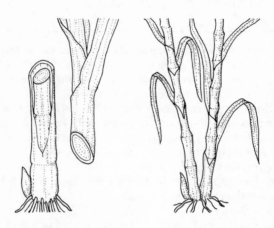

The sugar cane, *Saccharum officinarum*

Sanskrit *karkara,* meaning gravel or small chunks of material; "candy" from
the Arabic version of the Sanskrit for sugar itself, *khandakah.*

A Spice and a Medicine

At first, Europeans treated sugar the way they treated pepper, ginger,
and other exotic imports, as a flavoring and a medicine. Medieval recipes
from the French and English courts call for sugar to be added to fish and
fowl sauces, to ham, and to various fruit and cream-egg desserts. Chaucer's
Tale of Sir Topas, a 14th-century parody of the chivalric romance, indicates
the company to which sugar properly belonged at that time:

> They fette hym first the sweete wyn,   (fetched)
> And mede eek in a mazelyn,   (mead/also/goblet)
>   And royal spicerye
> Of gyngebread that was ful fyn,
> And lycorys, and eek comyn,   (cumin)
>   With sugre that is trye.   (excellent)

Sweets, or candy, began not as little treats but as "confections" (from the
Latin *conficere,* "to put together," "to prepare") produced by the apothecaries, or druggists. Sugar was used both as a drug in itself and as a material
with which to bind up all the other ingredients and to disguise their taste.
Walter Ryff begins his *Confectbuch und Haussapoteck,* or *Confection-book
and Home Druggist* (Frankfurt, 1571) with a discourse on this all-purpose
material. "The common proverb that honey and sugar are the druggist's
most valuable stock is true: for honey and sugar in particular are used in all

electuaries, confections, conserves, preserves, syrups, juleps, and other excellent preparations." Electuaries, for example, were probably solid pills of sugar and drugs. The word comes from the Greek for "to lick out" and so suggests something to be sucked on, like our cough drops or throat lozenges.

It is difficult to pinpoint sugar's transition from remedy to luxury, although it is said that the first nonmedical confection in Europe was invented around 1200 by a French druggist who coated almonds with sugar. And the medicinal origins of confection live on in contemporary language. While "honey" is almost invariably a term of praise, "sugar" is often ambivalent. Sugary words, a sugary personality, suggest a certain calculation and artificiality. And the idea of "sugaring over" something, the deception of hiding something distasteful in a sweet shell, would seem to be taken directly from the druggist's confections. As early as 1400, the phrase "Gall in his breast and sugar in his face" was used, and Shakespeare has Hamlet say to Ophelia,

> 'Tis too much prov'd, that with devotion's visage
> And pious action we do sugar o'er
> The devil himself.   (III.i)

The Rise of the Sugar Industry

By the 15th century, sugar had come into more general use, though it was still expensive. In the first printed book to contain recipes, *De Honesta Voluptate et Valitudine Vulgare (On Honest Pleasure and Physical Health)*, the Vatican librarian Bartolomeo Sacchi writes that sugar is now (around 1475) being produced in Crete and Sicily as well as India and Arabia, and adds, "The ancients used sugar only in medicines, and for this reason make no mention of sugar in their foods. They certainly missed out on a great delight, since nothing that is given to us to eat can be so tasty." He goes on to recommend various nuts and seeds coated in sugar syrup, together with marzipan, an almond-sugar paste, and various sweet cakes.

In the 18th century, we find the first cookbooks devoted entirely to confectionery; sweets were divided into 8 or 10 varieties, and the middle classes began to consume sugar by the pound. In 1825, the French gastronome Brillat-Savarin listed better than a dozen uses for this "universal flavoring" and observed that sugar "has become a staple food of the first necessity; there is not a woman, especially if she be well-to-do, who does not spend more for her sugar than she does for her bread." Sugar production has risen fairly steadily ever since. In many affluent countries today, per capita consumption exceeds 100 pounds a year. U.S. consumption has recently declined to about 80 pounds per person, apparently on account of the greater industrial reliance on corn-derived syrups (see pages 394–96) and the popularity of low-calorie foods.

The early stages of sugar's history were relatively benign, but the 18th-century explosion in consumption was made possible by European colonial rule in the West Indies. And millions of Africans were enslaved to satisfy the European craving for sweets. Early explorers returned with news of undefended tropical lands well suited for cultivating sugar cane, and the great sea powers lost little time in trying to free themselves from the Middle-Eastern monopoly. Columbus, whose father-in-law may have had a plantation in the Madeira Islands, carried the cane to Hispaniola (now Haiti and the Dominican Republic) on his second voyage in 1493. By about 1550, the coasts of western Africa, Brazil, Mexico, and many Caribbean islands had been colonized by the Spanish and Portuguese and were producing sugar in significant quantities, and the English, French, and Dutch followed in the next century. Already at this point, some 10,000 Africans were being traded via the Portuguese colony São Thome to the Americas every year. The sugar industry was not the only force behind the great expansion of slavery from 1550 on, but it probably was the major force and helped ease its introduction into the southern American colonies and the cotton plantations. According to one estimate, fully two thirds of the 20 million Africans enslaved worked on sugar plantations.

Thus was confectionery liberated from the druggists to find a place in middle-class cookbooks. The estimated per capita consumption of sugar in England rose from 4 pounds a year in 1700 to 12 pounds in 1780, a good fraction of which probably went into tea, coffee, and chocolate, all newly popular beverages. Rum, a new alcoholic drink made from by-products of sugar refining, became the official liquor of the British navy. Beyond merely culinary innovations, the intricate trade in sugar, slaves, rum, and manufactured goods made major ports out of the hitherto minor cities of Bristol and Liverpool in England, and Newport, Rhode Island. And the huge fortunes made by plantation owners helped finance the opening stages of the Industrial Revolution. All in all, this single plant species, which makes no particular contribution to human health or well-being, has had a remarkably wide-ranging influence on Western history.

In the 18th century, just when it seemed at its strongest, the West Indian sugar industry began a rapid decline. The horrors of slavery gave rise to abolition movements, especially in Britain, and some merchants refused to deal in anything but East Indian sugar. Slaves staged revolts, and received some support from the very countries that had carried them to the plantations. One by one, through the mid-19th century, European countries outlawed slavery in the colonies.

But the crowning blow to West Indian sugar was the development of an alternative to the sugar cane. In 1747, a Prussian chemist, Andreas Marggraf, showed that by extracting with alcohol (he used brandy) the juice of the white beet (*Beta vulgaris*, var. *altissima*), a common European vegetable, one could isolate crystals that were identical to those purified from sugar cane. Just as important, the yield of sugar was comparable (today, a 2-lb beet will contain 14 teaspoons of sugar). Marggraf foresaw a kind of cottage industry by which individual farmers could satisfy their own needs for sugar, but this never came about, and many years passed before the idea escaped

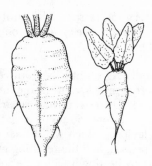

The sugar beet *(close-up on left)*, **a variety of the common beet** *Beta vulgaris*

the laboratory. Several pioneer ventures were set up in Prussia, Russia, and Austria around 1800, but they made very slow progress. Politics finally provided the necessary impetus. In 1811, the Emperor Napoleon officially set the goal of freeing France from dependence on the English colonies for various commodities, and in 1812 personally awarded a medal to Benjamin Delessert, who had developed a working sugar-beet factory. In the next year, 300 such factories sprang up. A treaty resuming trade between France and England was signed in 1814, making West Indian sugar available once again, and the fledgling industry crashed as suddenly as it had begun; but it rose again in the 1840s and has flourished ever since.

At present, beet sugar accounts for about 40% of the sucrose produced in the world. Russia, Germany, and the United States are the major beet growers, with California, Colorado, and Utah the leading states (the California cities Spreckels and Oxnard are named after sugar entrepreneurs). The Caribbean is now a minor source of cane sugar, its role having been assumed by India and Brazil. Spurred by the demand of an increasingly populous and affluent West, world sugar production increased sevenfold between 1900 and 1964, a rate matched by no other major crop in history.

How Sugar Is Made

The processing of cane and beet sugar is more involved than the production of honey and maple syrup for one basic reason: bees and maple sugarers begin with an isolated plant fluid that contains little else besides water and sugar, but the raw material for table sugar is the crushed whole stem of the cane or the whole root of the beet. Cane and beet juices are full of many compounds—proteins, complex carbohydrates, tannins, pigments, and so on—which not only interfere with the sweet taste by themselves, but also decompose at the high temperatures necessary for the concentration process and produce even less palatable chemicals. (This is why the intrusion of such compounds into maple sap when the tree begins to leaf out and grow signals the end of the sugaring season.) The purity that is the result of an initial selectivity in honey and maple syrup can only be obtained by refining in the case of cane and beet sugar.

From the late Middle Ages until the 19th century, when machinery changed nearly every sort of manufacturing, the treatment of sugar followed the same basic procedure. There were four separate stages: clarifying the cane juice, boiling it down to concentrate the sugar, separating the molasses, and a process called "claying," which was a final washing. The cane stalks were first crushed and pressed, and the resulting green juice was cleared of many organic impurities by heating it with lime and a substance such as egg white or animal blood, which would coagulate and take impurities with it into a scum that could be skimmed off. The remaining liquid was then boiled down in a series of shallow pans until it had lost nearly all of its water, and poured into cone-shaped clay molds a foot or two long with a capacity of 5 to 30 pounds. There it was cooled, stirred, and allowed to crystallize into "raw sugar," a dense mass of sucrose crystals surrounded by a film of water, some uncrystallized sucrose and other sugars, minerals, and other impurities. The clay cones were left to stand inverted for a few days, during which time the non-sucrose film, or *molasses,* would run off through a small hole in the tip. In the final phase, a fine wet clay was packed over the wide end of the cone, and its moisture allowed to percolate through the solid block of sugar crystals for 8 to 10 days. This washing, which could be repeated several times, would remove most of the remaining impurities, though the resulting sugar was generally yellowish and would be wrapped in blue paper to make it look whiter.

Until the late 19th century, sugar was sold in these conical masses, which were called loaves: hence the various topographical features that have been named "Sugarloaf" for their supposed resemblance. In 1872, a one-time grocer's assistant named Henry Tate, who had worked his way to the top of a Liverpool sugar refinery, was shown an invention that cut up sugar loaves into small pieces for household use. Tate patented the device, went into production, and in a short time made a fortune with "Tate's Cube Sugar." He became a philanthropist, was knighted, and just before he died, built the National Gallery of British Art, better known as the Tate Gallery, which he filled with his own collection.

Today, sugar is produced by somewhat different means. The initial clarification is accomplished without animal protein; only heat and lime are used. Rather than waiting for gravity to draw off the molasses, refiners use centrifuges, which spin the raw sugar and, much like a clothes dryer, force the liquid off the crystals. The sucrose is made absolutely colorless by the technique of decolorization, in which granular carbon—a material like activated charcoal that can absorb undesirable molecules on its large surface area—is added to the centrifuged, redissolved sugar. After it removes the last remaining impurities, it is filtered out. The final crystallization process is

Sugar making in the 16th century as depicted by Jan van der Straet (Stradanus) in his book *Nova Reperta (New Inventions)*. *Clockwise from foreground:* the cane is chopped into pieces, crushed in a mill, and pressed. The extracted juice is then boiled down into a concentrated syrup, and poured into molds to crystallize. The result: large, heavy "loaves" of sugar. (Courtesy of the Burndy Library, Norwalk, Connecticut.)

carefully controlled to give individual sugar crystals of uniform size; the older loaf form was simply broken into irregular chunks when some sugar was needed. Our table sugar is an astonishingly pure 99.8% sucrose.

Molasses

Molasses, a word derived from the late Latin for "honeylike" *(mella-ceus)*, is generally defined as the syrup left over after the available sucrose has been crystallized from the juice of the sugar cane. There is such a thing as beet molasses, but it has an unpleasant odor that prevents its use in food, at least for humans. In order to extract as much sucrose as possible from cane juice, crystallization is performed in several different steps, each of which results in a different grade of molasses. "First" molasses is the product of centrifuging off the raw sugar crystals and still contains some removable sucrose. It is then mixed with some uncrystallized sugar syrup, crystallized, and centrifuged. The resulting "second" molasses is even more concentrated in impurities than the first. Repeating this process once more yields "third," or final, or "blackstrap" (from the Dutch *stroop* for "syrup") molasses. The very dark color of final molasses is due to the extreme caramelization of the remaining sugars and to chemical reactions induced by the high temperatures repeatedly reached during boiling. One such reaction is the division of some sucrose into its constituent glucose and fructose molecules. Other reactions, together with the high concentration of minerals, give final molasses a very harsh flavor that makes it generally unfit for human consumption, although it is sometimes made edible by blending it with corn syrup; a small amount is also used in tobacco curing. Most final molasses is used as cattle feed—its high carbohydrate content makes it a good fattener—and as the raw material for yeast fermentation in making rum and industrial alcohol.

First and second molasses have been used in foods for many years, and for a long time were the only form of sugar available to the slaves and then to the poor of the rural South. Until fairly recently, most of the edible molasses produced in Louisiana had a heavy sulfurous taste because sulfur dioxide was used to clarify and lighten the color of cane juice. The current trend, however, is toward very mild flavors. Today, premium edible molasses is made by blending clarified cane syrups with first molasses in order to maintain better control over quality.

In the last few decades, great claims have been made for the nutritional excellence of molasses. Though it is true that, compared to 99.8% pure sucrose, molasses contains large amounts of minerals and B vitamins, it is also true that in absolute terms, these amounts are minor. A tablespoon of blackstrap molasses—which, given its flavor, is a great deal—contains less than a thirtieth of the recommended daily dose of B vitamins and about a sixth of the iron and calcium allowances. Premium molasses carries about half these amounts.

Brown Sugar

Similar claims of nutritional value have been made for brown sugar, which is essentially a mixture of white sugar and molasses, and so an even less likely candidate than molasses. It is also thought that brown sugar is somehow less "processed" than white and therefore more healthful. The sugar industry itself may have given unwitting credence to this myth by its early propaganda *against* brown sugar. One history of American food quotes from an 1898 advertisement warning against infestations of "a formidably organized, exceedingly lively, and decidedly ugly little animal" in all but white sugars. "The number of these creatures found in raw sugar is exceedingly great and in no instance is raw sugar quite free from either the insects or their eggs. Brown sugar should never be used. . . . It is fortunate to note, however, that these terrible creatures do not occur in refined sugar of any quality."

Brown sugar *is* refined. It is made by adding special syrups that have undergone the ideal amount of browning to refined, redissolved sucrose. This mixture is then crystallized, and the fine molasses film, instead of being

CANE SUGAR PRODUCTION

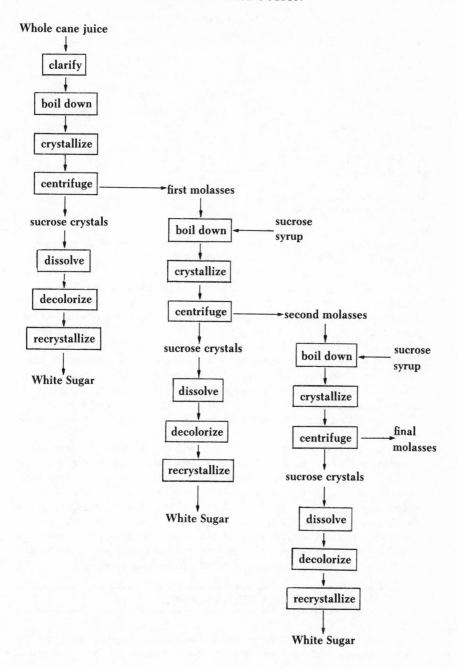

	Sucrose	Water	Glucose, Fructose	Other Carbohy-drates	Minerals	Nitro-gen°
COMPOSITION OF SUGARS **(PERCENT)**						
White sugar	99.8	0.1	0.05		0.02	
Brown sugar	92	3.5	4.0		0.5	
Final molasses	35	20	20	6	12	5

°Includes proteins, amino acids, etc.

washed off, is left on the crystals. The result is pure sucrose crystals with a thin coating that gives them a characteristic flavor, color, and moistness. Truly "raw," unrefined sugar contains soil, microbes, and other contaminants, and the FDA classifies it as unfit for direct use in food. "Turbinado" sugar, which is edible, is partly refined by washing the initially crystallized sucrose with steam in the centrifuge, but it is not redissolved or treated further. Turbinado has much the same composition as brown sugar. Brown sugar is soft and clingy because its molasses film—whose glucose and fructose are more hygroscopic than sucrose—contains about 35 times as much water as ordinary white sugar. Of course, if brown sugar is left exposed to dry air, it will lose its moisture through evaporation and become hard and lumpy. It can be kept moist by storing it in an airtight container or resoftened by closing it up with a damp towel or piece of apple from which it can absorb moisture.

The several different kinds of sugar sold today have somewhat different applications. Brown sugar can replace white in any recipe whose flavor will not be harmed by a note of molasses. Because brown sugar tends to trap air pockets between groups of adhering crystals, it should be packed down before its volume is measured. Powdered, or confectioners', sugar is used to sweeten uncooked foods without making them grainy. It is typically used in icings and frostings, and on sugar-dusted candies. Because powdered sugar absorbs moisture and clumps up quite readily, it is usually packed with small amounts of cornstarch. It should be sifted before measuring.

CORN SYRUP AND FRUCTOSE

We come now to a source of sugar that is relatively minor compared to cane and beets, but one with unique properties that are becoming more important all the time. In 1811, a Russian chemist, K. S. Kirchhof, found that if he heated potato starch in the presence of sulfuric acid, sweet crystals and a viscous syrup would result. A few years later, he discovered that malted barley had the same effect as the acid, and thereby laid the foundations for a scientific understanding of the brewing process. We now know that starch consists of long chains of glucose molecules, and that both acid and certain plant, animal, and microbial enzymes will break this chain down into smaller pieces and eventually into individual glucose molecules. These molecules can be crystallized, as Kirchhof found, and the longer glucose chains give the solution a thick, soupy texture. Corn syrup—rather than potato syrup—has been produced in the United States since the middle of the 19th century. Starch granules are extracted from the kernels and then treated with acid or with fungal, bacterial, or malt enzymes to develop a sweet syrup. Nowadays, enzymes from the easily cultured molds *Aspergillus oryzae* (also used in Japan to break rice starch down into fermentable sugars for *sake*) and *A. niger* are used almost exclusively.

Corn syrup is especially valued because its sweetness and physical properties can be varied widely. The proportion of the single- and double-unit sugars glucose and maltose—the two structures that the tongue registers as sweet—can be set at from 30 to 70% of the total carbohydrates simply by controlling the thoroughness of the enzymatic digestion of starch. At the other end of the scale, the number of chains with more than six sugar units might run between 10 and 50% of the total sugars present. The more glucose and maltose produced (the more complete the digestion), the sweeter the syrup. The more long chains there are (the less complete the digestion), the more viscous the syrup will be, because large molecules get tangled up in a random way and slow down the motion of all molecules in the solution. It is largely this physical effect that has made corn syrup increasingly important in confectionery and baking. Because it interferes with molecular motion, it

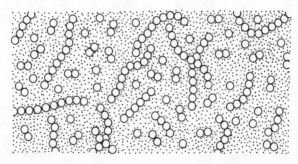

Corn syrup is a water solution of various glucose chains. One- and two-unit sugars taste sweet, while longer chains, which do not register on our taste buds, make the syrup viscous. By controlling the relative populations of different-sized chains, the manufacturer can tailor the taste and consistency of corn syrup to the particular requirements of a given food.

has the often valuable effect of preventing other sugars in candy from crystallizing and producing a grainy texture: everything is moving very slowly, and the crystal faces keep getting covered with 10-sugar chains that cannot ever become part of the sucrose crystal. Another consequence of corn syrup's viscosity is that it imparts a thick, chewy texture to foods. And because it consists of many sugar chains that are hygroscopic but do not actually taste sweet, corn syrup helps prevent moisture loss and increases the storage life of various foods without the cloying, sweet taste honey or sucrose syrup imparts.

Fructose Syrups and Crystals

When the corn starch is converted entirely into individual glucose molecules, the product is known as "dextrose," which is simply a synonym for glucose. It means "right-handed sugar" and refers to the direction in which the molecules rotate a beam of polarized light (such behavior is a clue to molecular structure). The glucose units can, in turn, be rearranged by an enzyme derived from various species of the bacterial genus *Streptomyces* to form fructose, a sugar with the identical formula—$C_6H_{12}O_6$—but a different structure and a sweetness exceeding that of sucrose. Fructose syrups of varying strengths are being used more and more as a replacement for sucrose in soft drinks, mostly because they are significantly cheaper and their source—corn—is a more predictable crop than beets and cane. Its liquid form is also the most convenient for the soft-drink industry, which accounts for 23% of the annual U.S. sugar consumption, more than the 14% used for cereals and baking, and the mere 10% used for confectionery. Less than 25% is packaged for home use.

Perhaps because its name invokes fruit, fructose has lately been praised as more "natural" than table sugar, more healthful, and less conducive to weight gain. However, any fructose sold in crystalline form as a replacement for table sugar is likely to be even more refined and even less "natural" than sucrose, since it is usually obtained by dividing pure sucrose into its glucose and fructose halves and then isolating the latter. It is true that because fructose tastes sweeter than sucrose but contains only half the calories—it being only half of a sucrose molecule—fructose supplies more sweetness per calorie consumed. But we do not eat simply to satisfy our sweet sensors; if we eat half as many calories, we simply get hungry earlier. Nor does fructose have any special health value, despite its pastoral-sounding name. "Fructose" does mean "fruit sugar," and it is most often found in fruit, but glucose and sucrose are also common in the plant kingdom. Perhaps fructose would be less appealing to faddists if it were called by its alternative name, levulose ("left-handed sugar").

CHOCOLATE

The Exotic Beverage

The story of chocolate begins in the New World with the accounts of early explorers and missionaries. Of pre-Columbian times we have only uncertain evidence. The cocoa tree originated in the river valleys of South America and was carried north into Mexico by the Mayas before the 7th century A.D. It was cultivated and the cocoa bean enjoyed by many peoples, including the Mayas, Aztecs, and Toltecs. The seed of the cocoa tree contains a good deal of fat and some starch and protein, and eaten in quantity would have been an important food. American natives valued it enough that cocoa beans were used as a form of currency. The first Europeans to see the cocoa bean were probably the crew of Columbus's fourth voyage in 1502, who brought some back to Spain. But it was not until the arrival of the Spanish *conquistadores* under Cortez that the use of cocoa was understood. In 1519 one of Cortez's lieutenants, Bernal Diaz del Castillo, saw the Aztec emperor Montezuma at table and later described the scene:

> Fruit of all the kinds that the country produced were laid before him; he ate very little, but from time to time a liquor prepared from cocoa, and of an aphrodisiac nature, as we were told, was presented to him in golden cups. . . . I observed a number of jars, above fifty, brought in, filled with foaming chocolate, of which he took some. . . .

The words "chocolate" and "cocoa" (an 18th-century corruption of the original "cacao") both derive from the Aztec, the latter being the name of the tree and the former meaning either "cocoa water" or "bitter water."

Several accounts of this novel preparation have survived from the 16th and 17th centuries. One of the first is to be found in the rather critical *History of the New World* (1565) by the Milanese Girolamo Benzoni, who traveled in Central America for 15 years beginning in 1541. He remarked that the region had made two unique contributions to the world: "Indian fowls," or turkeys, and "cavacate," or the cocoa bean,

> which they use as money, and is produced on a moderately sized tree that flourishes only in very warm and shady localities. . . . The fruit is like almonds, lying in a shell resembling a gourd in size. It ripens in a year, and being plucked when the season has arrived, they pick out the kernels and lay them on the mats to dry; then when they wish for the beverage, they roast them in an earthen pan over the fire, and grind them with the stones

which they use for preparing bread [*metate*]. Finally, they put
the paste into cups . . . and mixing it gradually with water, sometimes adding a little of their spice, they drink it, though it seems
more suited for pigs than men.

I was upwards of a year in that country without ever being
induced to taste this beverage; and when I passed through a tribe,
if an Indian wished occasionally to give me some, he was very
much surprised to see me refuse it, and went away laughing. But
subsequently, wine failing, and unwilling to drink nothing but
water, I did as others did. The flavor is somewhat bitter, but it
satisfies and refreshes the body without intoxicating: the Indians
esteem it above everything, wherever they are accustomed to it.

It seems that both the bitterness and frothy texture of this version of hot
chocolate—a mixture of roasted cocoa beans, red pepper, vanilla, and
water—put the first visitors off. But 60 or 70 years later the Spanish had
introduced some modifications and, as the Spanish Jesuit Joseph de Acosta
put it, would die for their chocolate. In 1648 the English Jesuit Thomas Gage
wrote that in Mexico only "the meaner sort of people"—that is, the natives—
took their chocolate with nothing but cocoa, achiote (or "annatto," a brick-
red tropical seed still used as a food coloring), maize flour, and chili pepper.
The Europeans now added a variety of flavorings, among them sugar, cinnamon, cloves, anise, almonds, hazelnuts, vanilla, orange-flower water, and
musk. The cocoa beans and spices were dried, ground up and mixed
together, and heated to form a paste. The paste was then scraped onto a
plantain leaf or piece of paper, allowed to solidify, and then peeled off as a
large tablet. According to Gage, there were several ways of preparing chocolate, both hot and cold.

> The one most used in Mexico is to take it hot with *atole* [a maize
> gruel], dissolving a tablet in hot water, and then stirring and beat-
> ing it in the cup with a molinet, and when it is well stirred to a
> scum or froth, then to fill the cup with hot *atole*, and so drink it
> sup by sup.

Unlike Benzoni, Gage liked chocolate, at least with its European additives,
and noticed the effects of its caffeine content:

> For myself I must say I used it twelve years constantly, drinking
> one cup in the morning, another yet before dinner between nine
> or ten of the clock, another within an hour or after dinner, and
> another between four and five in the afternoon, and when I was
> purposed to sit up late to study, I would take another cup about
> seven or eight at night, which would keep me waking till about
> midnight.

Not only would the Spanish die for their chocolate: they would also kill for it, if Gage is to be believed. He wrote that in the city of Chiapa Real, now San Cristobal de las Casas, the women insisted on having their maids bring them a cup during Mass. The bishop disapproved of the disruption and proclaimed that anyone who ate or drank in church would be excommunicated. Some women openly challenged him, and "this caused one day such an uproar in the Cathedral that many swords were drawn against the priests and prebends, who attempted to take away from the maids the cups of chocolate which they had brought unto their mistresses." The women then stayed away from the cathedral altogether—until the bishop suddenly took ill and died. He and his physician were sure that he had been poisoned; rumor blamed a carefully prepared cup of chocolate. "And it became afterwards a proverb in that country, Beware of the chocolate of Chiapa."

The church also waged an intermittent campaign against chocolate in Europe, largely on the grounds that it was tainted by the character of its inventors. Said one French cleric around 1620, chocolate is "the damnable agent of necromancers and sorcerers. It is well to abstain from chocolate in order to avoid the familiarity and company of a nation so suspected of sorcery." But such arguments availed little. The first European "factories," where the beans could be made into a paste fit for mixing with water, were built in Spain around 1580, and within 70 years, despite Spanish efforts to keep the secret, chocolate had found its way into Italy, France, and England. With the exceptions of sugar and vanilla, these countries purged the drink of all the added flavorings. At first, vendors of lemonade sold it in Paris; coffeehouses—themselves an innovation—served it in London. But soon chocolate houses were thriving in London as a kind of specialty, a fancy version of the coffeehouse. There were two of some fame: the Cocoa Tree, which catered to members of the Tory party, and White's, which became a favorite of Whig aristocrats, literary people, and gamblers. A couple of scenes from William Hogarth's famous series of paintings *The Rake's Progress* are situated in and just outside White's. The idea of making chocolate from milk and even eggs seems to have arisen in these clubs, though some writers objected to the scum or hardened egg white that resulted.

Milk Chocolate

The first great booster of "milk chocolate," as the drink was called, was Sir Hans Sloane, a naturalist and physician to Queen Anne who lived in Jamaica from 1687 to 1689 and there saw chocolate, whose fat he had considered a great tax on the stomach, fed even to the very young. Twice in his book *A Voyage to the Islands* (1707) Sloane remarked that "chocolate is given to young children here almost the first Meat they take except the Mother's Milk, and is found to agree with them as well as Milk-Meats in England." Apparently he thought that blending these two foods together would be especially healthful, as an advertisement from the period indicates.

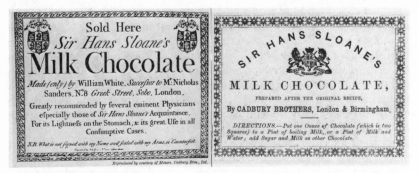

An 18th- and a 19th-century advertisement for the original milk chocolate, a hot drink which the respected physician Sir Hans Sloane thought beneficial to the health. (Courtesy Cadbury Schweppes Ltd.)

Coffee and Chocolate

An even more eminent name is associated with the sublime combination of chocolate and coffee. In 1750, Henri Lekain, then a young actor and later a director, met the aging Voltaire and was invited to "take his share of a dozen cups of chocolate mixed with some coffee." Lekain noted in his *Mémoires* (1801) that "this was the only nourishment of M. de Voltaire from 5 in the morning until 3 in the afternoon." It is highly unlikely that Voltaire was the first to enjoy these tastes together, especially since chocolate had been introduced in coffeehouses, but his great devotion to the combination makes him a worthy patron figure.

The Rise of Chocolate Candy

For the first couple of centuries, Europe knew chocolate almost exclusively as the beverage. The use of the cocoa bean in confectionery was quite limited. An English controversialist named Henry Stubbe noted in his treatise on chocolate, *The Indian Nectar* (1662), that in Spain and the Spanish colonies "there is another way of taking it made into *Lozenges,* or shaped into *Almonds,*" though the Spanish themselves claim that "to eat Chocolata in Cakes, or otherwise by bits, begets insuperable Obstructions." But Stubbe also found evidence of a certain advantage to taking chocolate in this concentrated form. He quotes one Roblez, a physician in Peru, who like Thomas Gage had noticed chocolate's stimulative effects: "The Cacao-nut being made into Confects, being eaten at night, makes Men to wake all night-long: and is therefore good for Souldiers, that are upon the Guard." Cookbooks of the 18th century would include perhaps two or three recipes for chocolate candy, with the closest approach to our idea of chocolates being a mixture of grated chocolate paste, fine sugar, and a preparation of plant gums to hold the two together. An interesting glimpse of the general attitude toward solid chocolate is provided by the great French *Encyclopédie* compiled around midcentury by Diderot and others. Under *"Cacao"* it lists two kinds of confection: one the flavoring of whole, unripe beans with citron and cinnamon, the other whole ripe beans candied in sugar syrup. Under the entry for *"Chocolat,"* we find that it was commonly sold as a half-cocoa, half-sugar cake flavored with some vanilla and cinnamon. And this mixture is thought of not as a sweet but as an emergency meal, perhaps the first instant breakfast.

> When one is in a hurry to leave one's lodgings, or when during travel one does not have the time to make it into a drink, one can eat a tablet of one ounce and drink a cup [of water] on top of that, and let the stomach churn to dissolve this impromptu breakfast.

Even in the middle of the 19th century, the English compendium *Gunter's Modern Confectioner* devoted only 4 pages out of 220 to chocolate recipes.

Dutch Innovation

The principal reason for this notable lack of interest in chocolate as a flavoring in candy was probably the texture of the chocolate paste, which would have been crumbly, coarse, and limited in its capacity to incorporate sugar. The suave confections that are so popular today were made possible by a single innovation in 1828. And the idea was not all that original. Conrad van Houten, whose family ran a chocolate business in Amsterdam, was trying to find a way to make chocolate less oily so that the drink would be less heavy and filling. The bean is better than half cocoa butter by weight, and the use of sticks called *moulinets* to whip the drink into a froth was in part an attempt to disperse the fat evenly throughout the liquid. Van Houten developed a screw press that would remove most of the butter from the bean. In fact, a picture of just such a press appears in a 17th-century French treatise by Nicolas de Blegny on the new beverages, *Le Bon Usage du thé, du caffé, et du chocolat* (1687). According to de Blegny, the "excessive fat" of chocolate makes it "unhealthy," and he gave two methods for reducing it: pressing the ground beans between warm metal plates lined with absorbent paper, and burning it off by mixing the upper, oily layer of liquid chocolate with alcohol and igniting it. But van Houten made something new with his press: cocoa powder. Ever since 1828, "hot chocolate" has been a very different drink from the "chocolate" of 1600 or 1800, which was something like chunks of unsweetened baker's chocolate dissolved in hot water or milk.

Van Houten's screw press made modern chocolate candy possible by accumulating excess butter from the production of cocoa powder: butter which could be *added* to ordinary ground cocoa beans to make the paste more malleable, smoother, and more tolerant of added sugar. The first "eating chocolate" was introduced by the English firm of Fry and Sons in 1847, and within 30 years it had caught on in a big way. The anonymous author of *The Candy-Maker* (New York, 1878) gave this summary of recent fashion:

> Ten years ago taffy cut up into various shapes, and variously flavored, was the favorite. Then gum drops couldn't be made fast enough to meet the call. Dealers began putting brandy and cordials into them, and with that the demand fell off, and the gum drop *furore* was killed. At one time New York women would scarcely eat any confectionery but cream-stuffed dates. Then fig paste had a run of about two years. Chocolate creams and chocolate caramels have had a long run, and promise to have an enduring demand, but new fashions may start up at any time.

New fashions there were, but most of them involved the new eating chocolate.

Milk and Bar Chocolates

By 1917, Alice Bradley's *Candy Cook Book* devoted an entire chapter to "Assorted Chocolates," whose list of some 60 recipes for centers "may be extended almost indefinitely." And Bradley noted that "more than one hundred different chocolates may be found in the price lists of some manufacturers." The South American bean had come of age as a major ingredient in candy. Two technical developments had helped: in 1876, a Swiss confectioner named Daniel Peter invented solid milk chocolate with the help of Henri Nestlé's new dried milk, and in 1913 another Swiss, Jules Séchaud, perfected the filled chocolate shell. The chocolate bar was introduced around 1910 but got its biggest boost during World War II, when it was issued to the American armed forces. In order to explain the limited quantities of chocolate in the stores, Nestlé's printed a full-page magazine ad that showed Ernie Pyle typing a news story: "When our infantry goes into a big push each man gets three bars of D-ration chocolate, enough to last one day. He takes no other food. . . ." The ad continued: "Yes, chocolate is a fighting food, it supplies the greatest amount of nourishment in the smallest possible bulk." As we shall see, chocolate is indeed a concentrated source of energy.

The Making of Chocolate

Chocolate Liquor

The cocoa tree, named *Theobroma cacao* by Linnaeus—*theobroma* is Greek for "food of the gods"—is an evergreen that thrives up to 20 degrees north and south of the equator, grows about 20 feet tall, and produces pods from 6 to 10 inches long, 3 or 4 inches in diameter, and containing 20 to 40 beans, each about an inch long. A handful of different varieties makes up

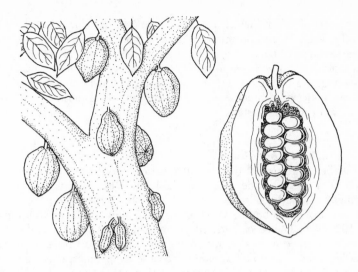

Cocoa pods grow along the trunk and branches of the tropical tree *Theobroma cacao.*

the bulk of the world crop, which stood at about 75,000 tons in 1900, and is close to 1.5 million today. West Africa now accounts for nearly three quarters of this figure, with Brazil the largest producer in the Americas.

Cocoa pods are harvested year-round, though most heavily from May to December. The first step in processing, sometimes omitted, is a brief fermentation of the beans and pulp, which are removed from the pod, piled on the ground, and allowed to sit in the sun for a few days. Various microbes multiply in the juicy pulp and raise the temperature, thereby killing the seeds' embryos and causing some biochemical changes: cell walls in the beans are broken down, various substances are mixed together, and the bitter, astringent phenolic compounds bind to each other and become less bothersome. The beans are then cleaned of pulp, dried, and shipped to consuming countries. There, the first step is to roast the beans for about an hour at about 250°F (121°C) to develop the rich, characteristic flavor of chocolate, which involves some 300 different chemicals, by way of "browning reactions." (See chapter 14.) At this point, the beans are cracked open and the kernels, called "nibs," are separated from the shells, which are then used as animal feed and fertilizer. From here on, the rather involved manipulations of the browned nibs are all aimed at achieving the desired consistency of the final product.

The nibs are ground up and, because they are more than half cocoa butter, form a thick liquid called *chocolate liquor*, which consists of small particles of solid nib—protein, carbohydrates, and so on—suspended in the oil. After this initial grinding comes refining, a second grinding done between sets of rollers, which brings the particle size down to the desired range, between 25 and 50 microns (around 0.001 inch). A relatively large particle size results in a coarse, grainy chocolate or, in the case of cocoa powder, a material that will sink to the bottom of the cup, while an overly fine grind will give a very pasty, gummy product. Most Swiss and German chocolates are noticeably smoother than the English and American, apparently for reasons of national taste.

Cocoa and Hot Chocolate

Further treatment of the chocolate varies according to the desired end product. For cocoa powder, the material from which chocolate drinks and flavorings for baked goods are made, the next step is the removal of substantial amounts of cocoa butter, with the final fat content of the powder ranging from 10 to 35%. The cocoa butter is pressed out of the liquor, and the resulting paste is formed into cakes and then ground up one last time. Some cocoa is also "dutched," a process so named because its inventor, once again, was Conrad van Houten. Dutching means treating either the whole nibs or the chocolate liquor with an alkaline solution, usually potassium carbonate, to raise its pH from 5.5 to 7 or 8. This has several consequences: it darkens the color, makes the flavor milder, and somehow improves the dispersion of cocoa particles in the liquid: they clump together less easily.

The recommended technique for confecting hot chocolate is first to make a paste with some scalded milk or boiled water and the cocoa powder, and then to mix it with the rest of the liquid and whisk vigorously. Because cocoa powder does not disperse well in cold water, so-called instant cocoas have been developed by adding lecithin, an emulsifier that helps separate the particles. Sugar is frequently added to instant cocoa mix and may account for up to 70% of its weight.

Eating Chocolate

Chocolate destined for candy or bars is treated very differently from cocoa powder. Eating chocolate is not defatted; rather, it is further enriched with cocoa butter. If bittersweet or sweet chocolate is to be made, then sugar is added; if milk chocolate, then sugar and milk solids. The mixture is then subjected to a process called "conching" (after the shell-shaped machines first used). The material is poured into a container in which a very heavy roller moves back and forth continuously, grinding, mixing, and slightly heating the ingredients. This step, invented around the turn of the century, serves the primary purpose of evaporating moisture and volatile acids and mellowing the flavor. It may also smooth the edges of sugar crystals, making the texture less grainy, and thoroughly mixes the milk, sugar, and chocolate particles in the cocoa butter medium. Sometimes the emulsifier lecithin is added at this point to improve the dispersion of solids; if so, about 5% less of the expensive cocoa butter is needed. Conching can go on for up to several days.

CHOCOLATE MANUFACTURE

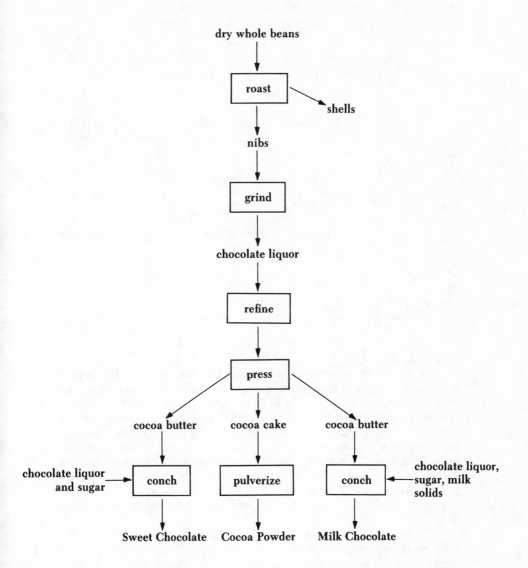

dry whole beans

roast → shells

nibs

grind

chocolate liquor

refine

press

cocoa butter   cocoa cake   cocoa butter

chocolate liquor and sugar → conch   pulverize   conch ← chocolate liquor, sugar, milk solids

Sweet Chocolate   Cocoa Powder   Milk Chocolate

Once this stage is completed, the chocolate mixture is shaped by molding or pouring and "tempered," or cooled slowly and carefully, a process made necessary by the special nature of the fat stored in cocoa beans.

COMPOSITION OF CHOCOLATE PRODUCTS
(PERCENT)

	Chocolate liquor	Added Cocoa Butter	Sugar	Milk Solids
Bitter	95	5		
Bittersweet	35–50	15	50–35	
Sweet	15	15	70	
Milk	10	20	50	15

	Fat	Carbohydrates	Protein	Minerals
Bitter	55	30	10	3
Sweet	35	60	4	1
Milk	30	60	5	1
Cocoa	15	55	18	5

Cocoa Butter

Very early in its European career, cocoa butter was noted for its resistance to rancidity and was prized as a cosmetic oil. It is remarkable among natural fats in being very regular in composition—most are mixtures of several different triglycerides—and this regularity gives it two properties that are important in culinary matters: a sharp melting point, and a tendency to form large crystals. The former means that chocolate is generally either solid or liquid; it does not soften gradually as butter does, with its mixture of triglycerides. Cocoa butter gives the impression of cooling the mouth as it melts because its melting point is just below body temperature, and the phase change from solid to liquid absorbs energy without raising the temperature of the fat.

Crystal formation, however, is a distinct drawback, since it means that if the conditions are favorable, the fat in chocolate will leave its initially even dispersion and congregate in fewer, larger masses. This is exactly why chocolate that has been stored for a long time will become white on the outside. The coating, called "fat bloom," is cocoa butter that has migrated toward the surface and crystallized out of the mixture. This process, which occurs most readily at temperatures between 70° and 75°F (21°–24°C)—certain crystal structures are unstable in this range, even though it is well below the fat's normal melting point—can be slowed down by adding to the chocolate some clarified butter, which makes the mix of fats more random and so retards the formation of crystals. Tempering prevents the initial formation of crystal nuclei and so improves the shelf life of the candy.

A similar kind of discoloration can be caused by "sugar bloom," which occurs when a combination of loose wrapping and refrigerator temperatures causes water to condense on the chocolate surface. Sugar dissolves in the moisture, and when the moisture evaporates, a white crust of sugar crystals is left behind.

Both of these problems are invisible in the case of white chocolate. This is made from no rare albino roast, but rather is chocolate without chocolate: it contains no cocoa solids whatsoever. White chocolate is simply a mixture of cocoa butter, milk solids, and sugar. It does not keep very well; something in the cocoa solids apparently helps prevent the milk fat from going rancid in ordinary milk chocolate.

Liquid-Center Chocolates

Many kinds of chocolate candy are made today, and many clever techniques have made them possible. One of these stands out and deserves a brief description. Liquid-center chocolates can be made by brute force by simply molding a bottom, filling it with flavored syrup, and then fitting a top to it. But a much more elegant method was worked out in 1924 by a government chemist, H. S. Paine. Remember that cane and beet sugar, or sucrose, consists of one glucose and one fructose unit joined together, and that a mixture of these two sugars is more soluble than the equivalent amount of sucrose; sucrose will crystallize from a solution at concentrations where its two components will not. And while living organisms, including humans, can use sucrose as a fuel, they first must break it down into the two simple sugars. Paine's idea was to isolate an enzyme that does exactly that—today it is taken from yeast cultures—and add it to a sugar center that has the consistency of fudge. The center does contain some water, but being solid, it is easily coated with melted chocolate. Then the enzyme goes to work, slowly breaking down the sucrose into its more soluble units, and the solid center liquefies. Because the reaction also removes water from the mixture,

$$C_{12}H_{22}O_{11} + H_2O \rightarrow C_6H_{12}O_6 + C_6H_{12}O_6$$
$$\text{sucrose} \qquad\qquad \text{glucose} \qquad \text{fructose}$$

it produces not a soupy syrup but a creamy, thick one. The yeast enzyme takes a couple of months at 65°F (18°C) to complete its work.

Chocolate as Food and Drug

Cocoa beans, like all seeds, are rich in nutrients that support the embryo until it develops leaves and roots. With its very high fat content, chocolate was probably an important source of energy for the inhabitants of Central

America. The German explorer Alexander von Humboldt, traveling in equatorial America around 1800, was impressed with chocolate's "salutary properties" and excellence as a provision:

> Alike easy to convey, and employ as an aliment, it contains a large quantity of nutritive and stimulating particles in a small volume. . . . In the New World, chocolate and the flower of maize have rendered accessible to man the tablelands of the Andes, and vast uninhabited forests.

An ounce of eating chocolate contains about 150 calories and 2 or 3 grams of protein. The original bean, like many seeds, does harbor significant amounts of B vitamins and vitamin E, but they are so diluted by fat and sugar in eating chocolate as to be negligible.

As for Humboldt's "stimulating particles," whose effects had been noticed by some of the earliest visitors to the New World, chocolate contains two related alkaloids, theobromine and caffeine, in the ratio of about 10 to 1. Theobromine does not stimulate the nervous system as caffeine does; its main effect seems to be a diuretic one. Commercial chocolate products are about 0.1% caffeine and have the power of a fraction of a cup of coffee. Straight chocolate liquor, which is available as unsweetened baking chocolate, is a more concentrated source of caffeine.

While chocolate may once have been used as a stimulant and source of nutrition, in its currently popular manifestations it is primarily a source of calories. And, of course, pleasure.

The Chocolate Binge as Medication?

Two New York psychiatrists have recently suggested that the tendency of some people, especially women, to go on chocolate binges after emotionally upsetting incidents may be a form of self-medication for an imbalance in the chemicals that control mood. Donald F. Klein and Michael R. Liebowitz argue that some people who suffer from wide fluctuations in mood may have a faulty mechanism for controlling the body levels of phenylethylamine, a naturally occurring substance that has amphetaminelike effects. The evidence for this is that certain drugs called monoamine oxidase inhibitors, which inhibit the enzymes that break down phenylethylamine, lift the spirits of people for whom the standard mood-elevating drugs don't work. According to the doctors, chocolate is rich in phenylethylamine. So the self-professed chocolate "addicts" who turn to the candy for solace after falling into a deep depression may be doing so in an unconscious effort to stabilize their body chemistry. Experiments so far have failed to verify this theory.

A Declining Part of the Diet

For some reason, the American hunger for this beneficent substance has recently been on the decline. We continue to consume the largest fraction of the world's production, but this is due mostly to our large population. The Swiss are the champion chocolate eaters, with an annual per capita consumption of around 22 pounds, and England, Germany, and Belgium are not far behind. In the midsixties the United States was up with the pack, but for some reason in the last decade we have fallen way back, to around 14 pounds per person.

CONFECTIONERY

An Obscure History

Much of the history of confectionery, and especially the origin of particular candies, remains obscure, although there are plenty of apocryphal and contradictory stories. The first foods that we would recognize as manufactured sweets were probably thick pastes of fruit and honey. When sugar reached the Near East, small lozenges and mixtures of gum arabic and sugar were soon produced. By the Middle Ages, druggists were preserving various herbs in sugar and may have made the first *dragée*—a candy dredged in powdered sugar—from almonds moistened with honey. Italy knew marzipan, a paste of ground almonds and fine sugar, by about 1300, probably through commerce with Arabs; this was the first particular candy to gain wide popularity, and it remains popular 700 years later. In the 17th century, hard sugar candies had become common, and around 1650 the praline, a hard paste made by mixing ground nuts with boiling syrup, was invented and named in honor of the duke of Plessis-Praslin, commander of the French armies under Louis XIII and Louis XIV.

Documentation improves in the 18th century as sugar became more widely available. Whole cookbooks were devoted to confectionery, a term that apparently covered everything from candies and preserves to pastries, wines, and liqueurs. Systems of marking different syrup concentrations— ancestors of today's thread-ball-crack scale—were developed, with from 5 to 12 different stages, or "degrees of refining." And in the *Encyclopédie* of Diderot and others (1751–76), we are given a brief but pointed analysis of this branch of cuisine. *Confiserie* is defined as

> the art of making preserves of all kinds, and many other works in sugar, such as biscuits, marzipans, macaroons, etc. It seems that this art was invented solely to flatter the taste in as many ways as it produces different works. There are no fruits, flowers, plants, however good they may naturally be, to which it cannot give a more complimentary, more agreeable taste. It sweetens the bitterness of the harshest fruits, and makes of them a delicious dish. It furnishes the tables of great noblemen with their prettiest ornament. Confectionery can execute in sugar all sorts of designs, schemes, figures, and even considerable pieces of architecture.

The origins of confectionery in medicine have been entirely forgotten by this time, and the whole point is now pleasure: the pleasures of novelty, visual and gustatory decoration, and just plain sweetness. But the standard repertoire of the time bore little resemblance to what we consider candy today. Aside from marzipan and hard candy bonbons, or sugarplums, the *Encyclopédie* (borrowing directly from Ephraim Chambers' *Cyclopaedia*, the first English encyclopedia) lists eight genres of *"Confiture."* These are liquid preserves (whole fruit or pieces in transparent syrup), marmalades (semiliquid preparations of fruit pulp), jellies (from fruit juice and sugar), pastes (marmalade dried in an oven), dry preserves (fruit boiled in syrup, drained, and dried in an oven), conserves (a dry preserve made with fruit pulp), candies (whole fruits boiled in syrup and covered with a hard sugar shell), and *dragées*. Only a few of these are still with us, and those only at the breakfast table.

It is in the 19th century that recognizably modern candy comes into being; eating chocolate is invented, and control of crystallization is refined. *Taffy* or *toffee*, from the Creole for a mixture of sugar and molasses, and *nougat*, from the vulgar Latin for "nut cake," enter the language early in the century; *fondant*, from the French for "melting," the basic material of fudge and all semisoft or creamy centers, is developed around 1850. Most candy today is a variation of some kind on bonbons, taffy, and fondant.

The Behavior of Sugar Syrups

Generally speaking, candy is made by cooking up a very hot, very concentrated sugar solution and then carefully controlling its transition to room temperature. In order to understand how we can measure sugar concentrations in the kitchen and how we can manipulate the transformation of syrup into candy with little more than a wooden spoon, we must look at the behavior of sugar in water.

Boiling Point and Sugar Concentration

As is explained in chapter 13, the addition of dissolved molecules to pure water raises the boiling point of the solution above that of water. The magnitude of the change depends on the amount of material dissolved; the more foreign molecules, the higher the boiling point. So the boiling point of a solution is an indirect sign of the concentration of dissolved material. The relationship between concentration and boiling point has been plotted for many combinations of solvent and solute, including water and sugar. The graph on the next page shows, for example, that a syrup that boils at 235°F (113°C) is about 85% sugar (measured as the volume of its granulated form).

As a liquid boils, it loses molecules from the liquid phase to the gaseous phase, while the solids stay behind. In a sugar syrup, this means that the sugar molecules account for a larger and larger proportion of all the molecules in the solution. In other words, the syrup gets more and more concen-

Boiling point (°F)

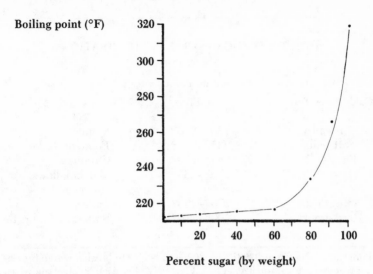

Percent sugar (by weight)

The boiling point of a sugar solution increases as the concentration of sugar increases.

trated as the water boils off. All the candy maker has to do is mix 2 or 3 cups of sugar in a cup of water and then heat the mixture. At a certain point, all the sugar will dissolve, and soon the solution begins to boil. As it does, it gets more concentrated, and so its boiling point rises. (A pure liquid cannot be heated above its boiling point; liquid water will never exceed 212°F (100°C), at least at sea level.) The process continues, with both the boiling point and concentration of the syrup increasing. Now experience has shown that certain syrup concentrations are best for making certain kinds of candy. Generally, the more water the syrup contains, the softer the final product will be. At 235°F (113°C), or about 85% sugar, the cook can stop the concentration process and make fudge; at 270°F (132°C), or 90%, taffy; at 300°F (149°C), nearing 100%, brittles and hard candy. A simple number on a candy thermometer, then, is an excellent indicator of the syrup's culinary possibilities.

Although it was invented 400 years ago by Sanctorius, the thermometer has been a common household appliance for only a few decades. Beginning in the 17th century and continuing to this day, confectioners have used a more direct means of sampling the syrup's fitness for different candies: spooning out a small amount, cooling it quickly, and noting its behavior. This technique is still useful when a candy thermometer is not available or when the amount of syrup being made is so small that an accurate reading is difficult to obtain. Today's "cold water test," or thread-ball-crack test, is not as picturesque as the early methods, but it is well calibrated. Thin syrups will simply form a thread in the air. Somewhat more concentrated syrups form a ball when dropped into cold water, and the ball will be soft and malleable

STAGES IN SUGAR SYRUP CONCENTRATION

Cold Water Test	Boiling Point (°F)	Candy Type
Thread	230°–235°	Syrup
Soft ball	235°–240°	Fondant, fudge
Firm ball	245°–250°	Caramels
Hard ball	250°–265°	Marshmallows, nougat
Soft crack	270°–290°	Taffies
Hard crack	300°–310°	Butterscotch, brittle

Pure sugar liquefies at 320°F (160°C), and begins to break down in browning reactions, or caramelize, at 335°F (168°C). For each 500 feet above sea level, subtract 1°F from every boiling point listed.

INSTRUCTIONS FOR TESTING SUGAR SYRUPS

The blown sugar: Boil your sugar longer than the former, and try it thus, *viz.* dip your scummer and take it out, shaking off what sugar you can into the pan, and then blow with your mouth strongly through the holes; and if certain bubbles or bladders blow through, it is boiled to the degree called blown.

The feathered sugar: It is a higher degree of boiling sugar; which is to be proved by dipping the scummer, when it hath boiled somewhat longer; shake it first over the pan, then give it a sudden flurt [shake] behind you; if it be enough, the sugar will fly off like feathers.

—from Hannah Glasse's *Compleat Confectioner* (ca. 1760)

between the fingers; as the concentration increases, the ball becomes harder. The most concentrated syrups turn into hard, brittle threads that make a cracking sound when shaken into water. Each of these stages has been correlated with a range of temperatures and a particular kind of candy.

The Effects of Weather

Obtaining the desired ratio of sugar to water in the syrup is only the first step in candy making, one that is relatively straightforward—at least on clear, dry days. On humid days, once a syrup has cooled enough to stop expelling water vapor it will actually re-absorb moisture from the air. When this happens, the syrup can dilute itself enough that the final candy will be softer than intended. Some cookbooks even recommend that certain kinds of candy not be attempted in humid weather. Less extreme advice calls for boiling the syrup to the upper end of the desired range. This allows for some dilution to occur without throwing off the final moisture content. Cool weather is always preferable to hot for making candy; the less time there is for a cooling syrup to be disturbed, the better.

Controlling Crystallization

The trickiest stage of candy making comes when the syrup cools from 250° or 300°F (121°–149°C) down to 80° or 70°F (27°–21°C). The rate of cooling, the movement of the syrup, and the presence of the smallest dust or sugar particles can have drastic effects. Here is why the solution is so sensitive. The amount of a given foreign solid that can be dissolved in a liquid is limited by the chemical properties of the two substances. When that limit has been reached, the solution is called *saturated*. At this point, the attractive forces between water and foreign molecules are in balance with the attractive forces between one foreign molecule and another. If the dissolved molecules are crowded the slightest bit more, they are more likely to be attracted to each other than to water and will begin to clump together and form into crystals. When the dissolved solid falls out of solution in this way, it is said to *precipitate*.

The saturation limit depends on temperature. A hot liquid whose molecules are moving rapidly can keep more foreign molecules from settling down with each other than a cold, sluggish liquid can. Now, the moment that a hot, saturated solution begins to cool, it becomes temporarily *supersaturated*. That is, it momentarily contains more dissolved material than it normally could at that temperature. And once the solution has become supersaturated, the smallest disturbance will induce precipitation, which removes dissolved molecules and continues until the new saturation point is reached.

Given a suitably concentrated syrup, the texture and appearance of the final candy is determined by the way in which the sugar precipitates. The secret to candy making, then, is to control this process. If a smooth, amorphous, glasslike drop is the aim, then the sugar molecules must be induced to form a solid mass without falling into individual, precisely ordered arrays. Such arrays are crystals, whose sharp edges and light-reflecting facets produce an entirely different sort of candy. Large crystals make for a coarse, grainy texture. A creamy fudge, on the other hand, can only be obtained by generating many thousands of microscopic crystals that are lubricated by just the right amount of sugar syrup. Fortunately, these strict conditions can be realized by fairly simple techniques.

The Influence of Temperature and Stirring

The first step in crystallization is the formation of a "seed," or nucleus, an initial surface to which the sugar molecules can attach themselves. The seed can be a few molecules that happen to come together during random motion, an event that occurs at a significant rate only in supersaturated solutions. But tiny crystals that form on the side of the pan from syrup spatterings or on a spoon that dries off between stirrings, as well as dust particles and even tiny air bubbles, can all serve as seeds in a cooling syrup and initiate crystallization. The formation of seeds by molecules in the solution is encouraged by agitation, which tends to push the molecules together, so that if a pan of supersaturated syrup is rocked or the spoon or thermometer moved around in it, crystallization is likely to begin. The use of a metal spoon can have the same effect by conducting heat away, cooling local areas of the syrup and so leaving them super-supersaturated. Because this interferes with controlled crystallization, the preferred utensil for candy making is the wooden spoon, which is a poor conductor of heat. At the other extreme, crystallization can be avoided altogether by cooling the syrup rapidly enough, in cold water or on a cold surface, that seed formation and crystal growth simply don't have the time to make much progress. This is what happens in the cold-water test and in the making of hard candy drops.

Because the size of sugar crystals determines the texture of candy, the cook must carefully control their growth from seeds. Generally, large crystals develop in hot syrups that are agitated only intermittently, while small crystals develop in cool syrups that are stirred continuously. Here's the logic. The rate at which a crystal grows depends on two basic factors: the ease with which the molecules can find their way to the crystal surface, and the opportunity they have to become properly oriented and then attached. Now the thickness, or viscosity, of a sugar syrup is a manifestation of how easily its component particles move. The hotter the syrup, the more energy its molecules have, the faster they move, and so the less viscous the liquid. Because more molecules will arrive at the crystal surface during a given time in a hot syrup than in a cold, lethargic one, crystals will grow more rapidly in hot syrups. At the same time, because stable crystal *seeds* are *less* likely to form at higher temperatures—an aggregate of a few molecules is more easily knocked apart in fast-moving surroundings—the total number of crystals formed in a hot syrup will be lower. Put these two trends together, and we see that if a hot syrup begins to crystallize, it will produce fewer and larger crystals than a cool one. This is why recipes for fudge or fondant, candies of fine, creamy texture, call for the syrup to be cooled drastically—from 235°F (113°C) down to around 110°F (43°C)—before the cook initiates crystallization by stirring.

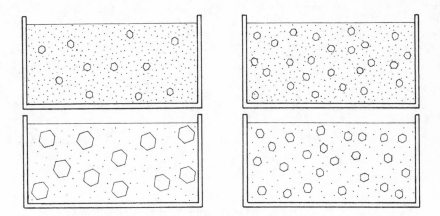

The more seed crystals a solution begins with *(top)*, **the more there are to share the remaining dissolved sugar, and the smaller each crystal will end up** *(bottom)*.

A similar kind of double trend is involved when a syrup is stirred. We have seen that agitation favors the formation of crystal seeds by pushing sugar molecules into each other. A syrup that is stirred infrequently will develop only a few crystals, while one that is kept in motion continuously will produce great numbers. Now the more crystals there are in a syrup, all competing for the remaining free molecules, the less material there is to go around, and so the smaller the average size of each crystal. The more a syrup is stirred, then, the finer the consistency of the final candy. This is the justification for wearing your arm out when making fudge: the moment you let up, the formation of seeds slows down, and the crystals you have made up to that point begin to grow in size. The candy is getting coarse and grainy before your very eyes. Try to imagine this setback when you are tempted to rest.

The Influence of Other Ingredients

So far we have been describing the behavior of pure sucrose solutions. But candies almost always contain other ingredients, and these will superimpose variations on the basic pattern. If, for example, you tried to make

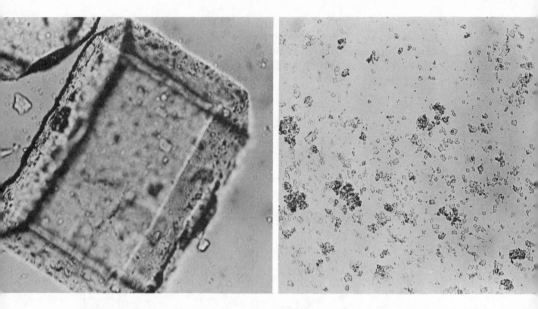

The influence of temperatures and beating on the texture of creamy candies. *Left:* the original granulated sugar from which the candy syrup is made. *Right:* the result when the syrup is cooked to 234°F (112°C), cooled to 104°F (40°C), and then beaten until stiff. This combination produces very fine crystals of sugar, and smooth, creamy candy. (From P. C. Paul and H. H. Palmer, *Food Theory and Applications.* Copyright © 1972 John Wiley and Sons, Inc. Reprinted by permission.)

candy with honey instead of table sugar, glucose, not sucrose, would be the first sugar to crystallize, and that process would be affected by the large quantities of fructose present. Because fructose is more hygroscopic than sucrose, it readily absorbs moisture from the air both during cooling and during storage, so that honey candies generally end up being on the sticky side. But the same basic rules about concentration and crystallization apply. Few cooks make candy out of honey, but no one uses pure sucrose either. Chocolate or cocoa, milk, butter, corn syrup, and various other ingredients all affect the crystallization of sucrose to some degree. Here is a summary of their particular influences.

Corn Syrup

Corn syrup is used in candy making specifically because it inhibits crystallization so well. The assorted chains of glucose units form a tangle that impedes the motion of sugar and water molecules and makes it more difficult for the sucrose to find a crystal onto which to fit. This property is useful if

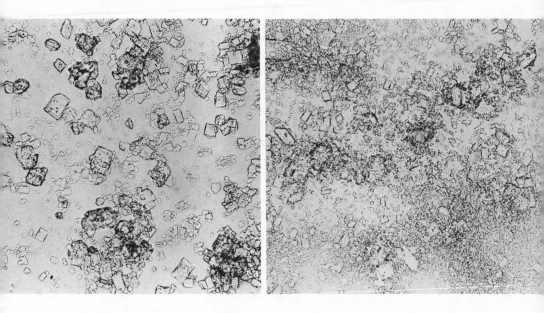

Left: this candy was made with the same syrup, but it was not allowed to cool before it was beaten. As a result, fewer seed crystals formed, and those that did grew more rapidly and became larger. *Right:* again the same syrup, this time cooled to 104°F, stirred a few times, and allowed to stand until it solidified. The short period of agitation produces fewer seed crystals and very uneven rates of crystal growth. Once more, the candy is grainy. (From P. C. Paul and H. H. Palmer, *Food Theory and Applications.* Copyright © 1972 John Wiley and Sons, Inc. Reprinted by permission.)

the cook wants to prevent crystallization altogether or to slow it down while he works to form seeds. Both hard candies and finely crystalline ones like fudge are often made with some corn syrup.

Cream of Tartar

Crystallization can also be impeded by "doctoring" sucrose: that is, by adding a small amount of acid to the syrup, usually in the form of cream of tartar, which has the advantage of being a solid—the potassium salt of tartaric acid—and so does not dilute the solution. In the presence of acid and heat, sucrose is "inverted," or broken down into its two components, glucose and fructose ("inversion" refers to a difference in optical properties between sucrose and a mixture of its component parts). Like the long glucose chains in corn syrup, invert sugar complicates matters for sucrose molecules by getting in the way during crystallization, thereby slowing the process down and giving the cook more time either to cool the syrup into a clear candy or to beat more seeds into existence. Honey is a natural source of invert sugar, and therefore can be used to impede crystallization as well as to contribute sweetness and its characteristic flavor.

Chocolate, Milk, Butter, Etc.

Corn syrup and cream of tartar are the only ingredients added specifically to control the behavior of the syrup; others make their principal contributions to taste and texture. This is obvious in the case of chocolate and cocoa. Milk solids thicken the texture and, because they brown easily, add a rich flavor to caramels and taffies. And butterfat lends smoothness and moistness to butterscotch, caramel, taffy, and fudge. These materials probably have the incidental effect of slowing down sugar crystallization, again simply by getting in the way, although fine particles of cocoa or powdered milk may act as seeds for crystal formation. Starch, pectin, gums, egg white, gelatin, and other ingredients that form jellylike solids are used to make various noncrystalline candies.

Types of Candy

It is convenient to divide candies into three groups: crystalline, noncrystalline, and miscellaneous gums, gels, and pastes. Here are brief descriptions of the principal candies made today.

Crystalline: Rock, Fondant, Fudge

The only candy in which large, coarse crystals are valued is rock candy—hardly a great favorite, but a vivid demonstration of crystal growth. Simply cook a syrup to the hard ball stage, then pour into a small glass, with a toothpick or swizzle stick to serve as a removable foundation, and let it sit for a few days. The resulting crystals can be preserved by washing the encrusted stick briefly under cold water, shaking off the excess, and letting it dry.

Fondant and fudge are the two most common crystalline candies, fondant being the basic cream candy and fudge being fondant with added milk, fat, and chocolate solids. In both cases, the aim is to produce very fine crystals, on the order of 10 microns, or 0.0005 of an inch across; crystals larger than about 35 microns feel gritty to the tongue. Since a pile of nothing but sucrose crystals would simply be fine table sugar, fondant and fudge are made to retain about 10% water by weight, which gives them their creamy consistency. Pure sugar fondant is actually made up of about equal proportions of solid—sucrose crystals—and liquid—a thick syrup of water and still-dissolved sucrose. Fudge is more complex, its syrup carrying milk solids and fat crystals as well as sugar crystals. The consistency of these candies will depend on how much water is left after concentration, cooling, and beating.

COMPOSITION OF COMMON CANDIES
(Percent)

Candy	Sucrose	Invert Sugar (% of Total Sugars)	Corn Syrup Solids	Moisture	Other Ingredients (% of Candy Weight)	
Fondant	95	0–5	0–5	10		
Fudge	30–70	0–15	10–40	10	Milk solids	5–15
					Fat	1–5
Hard	40–100	0–10	0–60	1		
Brittle	50–80		20–50	1		
Taffy	35–50	0–15	35–50	5	Milk solids	15–25
					Fat	0–10
Caramel	35–50	0–15	35–50	10	Milk solids	15–25
					Fat	0–10
Marshmallow	25–50	0–10	40–60	15	Gelatin	2–5
Starch jelly	25–50	0–10	30–70	15	Starch	5–10

If the syrup has been boiled too high and become too concentrated, the texture will be dry and crumbly; if it is not boiled high enough or absorbs moisture during cooling and beating, it will be runny. Small variations in water content—even 1 or 2%—make a noticeable difference.

Fondant and fudge are made with the help of corn syrup or an acid doctor, which favor the production of small crystals. After the syrup has been boiled and then cooled, it is beaten continuously for about 15 minutes, until crystallization is complete. This is hard work, since the syrup is very viscous to begin with and gets more so as the proportion of liquid decreases. For the first few minutes, the syrup becomes cloudy with trapped air bubbles and little else seems to be happening, but soon it becomes milky with tiny crystals, which scatter light and make the solution progressively more opaque. As the solid phase begins to match the liquid in volume, the candy loses its gloss and takes on a dull finish. By this point, the crystallization process is almost complete.

Fondant and fudge undergo a significant change in texture in the first day or so after they are made. If stored in an airtight container, they become softer and smoother. The nature of this "ripening," as it is called, remains unclear. It may be that very small crystals are dissolving back into the syrup matrix, thereby leaving more room for the average-sized crystals to move past each other. After a certain point, however, the surviving crystals tend to pick up more molecules from the solution and get larger, so that fondant and fudge get grainier after about a day of storage. And when the space between sugar crystals increases in fudge, the dispersed fats can more easily migrate and collect in large crystals that form a white layer—"fat bloom"— on the surface and change the candy's initially homogenous texture. The moral: eat your candy before it gets overripe.

Noncrystalline: Hard, Caramel and Taffy, Brittle, Foam

The noncrystalline candies are more various than the crystalline. The simplest is *hard candy,* the material of hard drops, clear mints, butterscotch, bonbons, and so on. It is made by boiling the syrup high enough that the final solid will contain only 1 or 2% moisture. It is then cooled rapidly without agitation before crystals have a chance to develop, usually by dividing it into small, easily chilled drops. Though the very high sugar concentration makes this syrup liable to crystallization at the slightest excuse, the extreme viscosity, together with substantial amounts of corn syrup or doctored sugar, is enough to produce an amorphous, glassy solid. Hard candy can be made opaque by "pulling," or stretching it just before it solidifies. This incorporates some air bubbles and forms tiny crystals at the last minute, long after they would have the chance to attain significant size.

Caramels and *taffies* end up with about the same moisture content as fudge—10%—and are also made with butter and milk solids, but they are not allowed to crystallize, and the result is a very different kind of candy. In the case of caramel, whose syrup is boiled only a few degrees higher than fudge's, the avoidance of agitation produces a moist, chewy solid, a random mixture of sugar and water molecules containing dispersed fat globules and milk particles. The characteristic caramel flavor comes from browning reactions between the milk proteins and milk sugar, lactose, as well as from the caramelization of lactose alone. "Caramelization" is the name given to the extensive chemical reactions that occur when any sugar is heated to the point that its molecules begin to break apart. More than a hundred different compounds, many of which have distinct flavors, can be produced from a mass of identical sugar molecules. Prolonging the period during which the syrup is cooked gives these reactions more time to proceed and develops a stronger flavor. Taffies are essentially caramels made from a more concentrated syrup and are therefore firmer. They are often pulled to incorporate air bubbles, which give them a somewhat lighter, chewier texture and a lighter color.

Brittles contain the same ingredients as caramels and taffies but are cooked to a very low moisture content, around 2%. This accounts both for their dry, brittle texture and their dark brown color; browning and caramelization reactions are extensive. Because some of the reaction products are themselves very reactive, it is advisable to cook brittle syrups only in stainless steel pans; some corrosion may occur on unprotected metal surfaces. Baking soda is often added to brittle syrup for several reasons: alkaline conditions favor browning reactions, help neutralize some of the acids produced thereby, and the bubbles of carbon dioxide that result from this neutralization become trapped in the candy, giving it a lighter texture.

The first *marshmallow*like confection, called *pâte de Guimauve*, was made in France from the gummy juice of the marsh mallow, a relative of the hollyhock. The juice was mixed with eggs and sugar and then beaten to a foam. Today, a viscous protein solution, usually gelatin, is combined with a sugar syrup concentrated to about the caramel stage, and the mixture is whipped to incorporate air bubbles. Whipping stretches the protein molecules out along the interface between liquid and air, causing partial coagulation; this gives the structure stability. *Divinity* is a close relative to the marshmallow that is made with egg whites. *Nougat* can be thought of as a cross between divinity and caramel; an aerated mixture of syrup and egg white is combined with a milkless caramel syrup, and chopped nuts are then added. The result is a candy that is chewier and denser than divinity or marshmallow, but lighter and tenderer than caramel.

Gels and Pastes

A host of candies is made by incorporating a sugar syrup into a solution of starch, gelatin, pectin, or plant gums and then allowing the mixture to solidify into a dense, chewy jelly. Prominent examples of such confections are *Turkish delight,* one of the most venerable and traditionally flavored with essence of rose, *jelly beans, jujubes,* and *licorice.* These days starch is the usual gelling agent, though pectin, a plant cell wall cement used to make jams and jellies, is most prized for its fine texture. These candies are relatively moist, being about 15% water. The solid phase in nut pastes like marzipan is provided by finely granulated sugar and the particles of nut proteins and carbohydrates, although gelatin or egg white is sometimes added to improve the binding.

Storage and Spoilage

Because of their generally low water content and because high external sugar concentrations draw water out of microbes by osmosis and dehydrate them, candies are seldom spoiled by the growth of bacteria or molds. Their flavor can be degraded, however, by the oxidation and consequent rancidity of added fats, whether in milk solids or butter. This process can be slowed down by refrigeration or freezing, but cold storage encourages another problem, "sugar bloom," in candies that are not tightly wrapped. Changes in temperature can cause moisture to condense on the candy surface, and some sugar will dissolve into the liquid. When it evaporates, the sugar crystallizes, leaving a rough, white coating.

Chewing Gum

This quintessentially American confection has quite ancient roots. It is thought that humans have chewed on the gums, resins, and latexes secreted by various plants for thousands of years. The idea of mixing gum with sugar goes back to the first sugar traders, the Arabs, who used the exudation of certains kinds of acacia, a substance now known as gum arabic. It and gum tragacanth are slightly soluble and eventually dissolve when chewed; they were used in early medicine as carriers that would release drugs slowly. Today, these complex carbohydrates are used principally as thickening agents, emulsifiers, and inhibitors of crystallization in such prepared foods as salad dressings, candies, and ice creams.

The Nature of Gummy Excretions

Gums, resins, and latexes are among the more common, more visible, and more mysterious of plant products. They are not involved in the primary processes of metabolism and growth, but appear to have the double function of removing waste materials and protecting the plant from mechanical injury by plugging up holes and discouraging the growth of harmful microbes. Gums are composed of long carbohydrate molecules with a limited capacity for holding water and are produced in the walls of certain plant cells; when tissue is broken, the gum flows to fill the injury, drying to a hard mass when it reaches the air. Resins, which are carried in specialized tree ducts, are mixtures of many different compounds, but nearly all of them are water-insoluble hydrocarbons. Latexes are water-based fluids produced by certain plants and carried in special tubes; dandelion stems, some kinds of lettuce, and banana peels all exude latexes when cut. Most of these fluids carry tiny globules of hydrocarbon polymers, very long chains of carbon atoms with hydrogen atoms projecting from them. And certain of these polymers have a coiled configuration that gives them the property of being elastic: they uncoil and stretch out when pulled but recoil and snap back when released. The best known of these special polymers is *rubber*, and a similar one was the basis of the first modern chewing gum. Once chemists were able to establish the molecular structures involved and once the price of latexes, which are found primarily in tropical plants, became high enough, synthetic versions were developed. Most chewing gums today contain little or no natural latex.

Gum in America

But we are getting ahead of the story. The commercial history of chewing gums begins around 1850, when the Curtises of Bangor, Maine, put out State of Maine Spruce Gum, the purified resin of black and red spruce trees. In those days, spruce resin was about the only gum chewed in either America or Europe, but its harsh taste and tough texture left it vulnerable to the challenge of chicle. It survives today only in the northern reaches of New England.

Various alternatives to spruce resin were tried, including paraffin, that waxy petroleum by-product, which was sweetened and flavored. The great breakthrough came in 1869, when a New York inventor by the name of Thomas Adams was introduced to chicle, the dried latex of the sapodilla tree, which is a native of Central and South America and valued by the natives for its fruit and hard wood. The story goes that after he was deposed in 1845, the Mexican general and president Santa Anna went into exile in New York, where he tried to sell the idea of using chicle as a rubber substitute. Thomas Adams tried unsuccessfully to vulcanize it—to improve its mechanical properties by treating it with sulfur and heat—and then got the idea of using it as a gum base. He patented chicle gum in 1871, and with sugar and sassafras or licorice flavorings, it quickly caught on.

In 1885 William J. White of Cleveland refined chicle gum by using corn syrup, which blends easily with the latex, as part of the sweetening, and flavored it with peppermint. He was the first to make millions from chewing gum. By 1900, Frank Fleer of Philadelphia had coated gum with a hard sugar shell to make Chiclets and gumballs. William Wrigley invented the Juicy Fruit and Spearmint flavors in 1893, and from then on showed a real genius for advertising, as C. W. Post did in the breakfast cereal business at about the same time. And in 1928 a Fleer employee named Walter Diemer perfected bubble gum by developing a very elastic latex mixture. A previous effort called Blibber-Blubber had not been terribly popular, but Diemer's gum produced larger bubbles and was more easily removed from nose and cheeks. The pink food coloring that had been arbitrarily used in the first mixture became canonical and remains the special mark of bubble gum. Picture cards of cowboys, Indians, and war heroes were an early merchandising strategy; Topps baseball cards came along in 1951.

Today, chewing gum is made mostly of synthetic polymers, especially styrene-butadiene rubber and polyvinyl acetate, though 10 or 20% of some brands is still accounted for by chicle or jelutong, a latex from the Far East. The crude gum base is first filtered, dried, and then cooked in water until syrupy. Powdered sugar and corn syrup are mixed in, then flavorings and softeners—vegetable oil derivatives that make the gum easier to chew—and the material is cooled, kneaded to an even, smooth texture, cut, rolled thin and cut again into strips, and packaged. The final product is about 60% sugar, 20% corn syrup, and 20% gum materials. Despite the advances of chemical science, there is still no foolproof way of removing gum from clothes or carpets.

SUGAR AND TOOTH DECAY

It is common knowledge today that sugary foods encourage tooth decay. Although it is only recently that this relationship has been understood, the relationship itself was recognized very early. In the Greek book of *Problems* attributed to Aristotle, the question is asked, "Why do figs, which are soft and sweet, destroy the teeth?" Nearly 2000 years later, there was a startling coincidence between the sudden availability of nearly pure sucrose in Europe and dental troubles on a previously unknown scale. As the West Indian sugar plantations geared up in the 16th century and the upper classes began to enjoy sweets in greater and greater quantities, noble teeth suffered. Queen Elizabeth herself, who had a legendary sweet tooth, was a victim. In 1598, a German visitor named Paul Hentzner saw her on her way to prayers one Sunday, and he later published his impressions:

> Next came the Queen, in the Sixty-fifth Year of her Age, as we were told, very majestic; her Face oblong, fair, but wrinkled; her Eyes small, yet black and pleasant; her Nose a little hooked; her Lips narrow, and her Teeth black; (a defect the English seem subject to, from their too great use of Sugar). . . .

A few decades later, this passing observation seems to have become accepted medical lore. In his *Klinike* of 1633, one of the first medical treatises since ancient times to relate diet and health, James Hart said that the "immoderate use" of sugar, as also of sweet confections, and sugar-plums, heateth the blood, engendreth obstructions, cachexias, consumptions, rotteth the teeth, making them look black; and withal, causeth many times a loathsome stinking breath. . . . And if ever this proverbe (sweet meats hath often sour sauce) was verified, it holdeth in this particular.

By the 18th century, the connection between sugar and tooth decay had become the material of cliché. But it was not until 200 years later that the connection was finally explained. Certain kinds of *Streptococcus* bacteria that thrive on sugar colonize the mouth and excrete acids that eat away at tooth enamel and so cause decay. (Part of this conclusion had been anticipated in 1773 by William Cullen, who asserted in his *Lectures on the Materia Medica* that sugar is composed of acids, and speculated that "it may, perhaps, by what adheres of it about [the teeth], turning acid, corrode them.") Clearly, the more food there is for the bacteria, the more active they will be, and the advent of pure sugar candies was no doubt a great boon to them. More important than the absolute amount of sugar eaten, however, is the length of time that sugar is in contact with the bacterial colonies harbored in dental plaque. Small, frequent snacks are more harmful than one large binge, and sticky candies that cling to the teeth and hard drops that are sucked on for many minutes may cause more trouble than say gum, whose sugar is quickly dissolved and washed away. Pure sugar is not the only culprit in tooth decay; starchy foods like potato chips, because they are broken down into sugars by enzymes in the saliva, can be just as harmful. A few other foods, notably some cheeses and tea, actually seem to inhibit the decay-causing bacteria. It may be that tannins in the tea and traces of antibiotics produced by cheese-ripening bacteria are the factors involved. While we wait for these mysteries to be resolved and for scientists to come up with an antidecay vaccine, there is no substitute for brushing the teeth.

CHAPTER 9

Wine, Beer, and Distilled Liquors

The subject of alcoholic beverages encompasses a long, colorful history, dozens of different liquids and techniques for making them, and familiar yet only partly understood effects on the human body. The enigma of fermentation—the transformation of fruit and grain into intoxicating fluids—attracted some of the best and some of the most headstrong scientists of the 19th century, and gave rise to the science of microbiology. The first microorganisms to be isolated in pure cultures were beer and wine yeasts, and the word *enzyme*, denoting the remarkable molecules that living cells use to transform other molecules, was coined from the Greek words for "in yeast," the place of their discovery. Of necessity this chapter will be limited to the two most important fermented beverages in the West, grape wines and grain beers, and to the distilled liquors made from them.

THE HISTORY OF ALCOHOL

Wines from Sweet Fruits

We will never know what the first alcoholic beverage was or how it was made, though it is possible to make a good guess. Yeasts produce alcohol as a by-product of obtaining chemical energy from sugars. But they generally cannot break down many-unit sugars or starch into usable food. The simple sugars glucose and fructose, and two-unit sugars like sucrose and maltose, are their preferred energy sources. In order to be fermented to any extent, then, a liquid must be fairly sweet. The juice of most wild fruits, including grapes, is high in acid and low in sugar, and probably could not have been made into wine until some selection and domestication had taken place. For this reason it is thought that the first alcoholic beverage was a wine produced perhaps 10,000 years ago when some honey was forgotten or stored for a week or two. The uniquely sweet date, as well as the sap of palm trees, may also have been exploited very early. Once agriculture had gotten underway and sweeter fruits had been propagated simply because they tasted better, many other kinds of wine were possible. The grape vine *Vitis vinifera*, a relative of Boston ivy and the Virginia creeper, had spread throughout the Middle East from its origins in central Asia, and grape wine was soon being made in Mesopotamia and Egypt. Because the vine is adaptable to various climates, is quick to mature and a heavy bearer compared to fruit trees, and because its fruit ferments smoothly, the grape became the fruit of choice in the Mediterranean area for wine making.

Wine would probably have had two attractions for its prehistoric discoverers. Like cheese, it is a partly spoiled yet edible food that resists further deterioration. The preservative influence is exercised by yeast-produced alcohol, a molecule related to the carbohydrates that can be tolerated by other organisms—humans as well as microbes—in only small amounts. This limited tolerance of ours would have made wine appealing for another reason: alcohol interferes with the normal operation of our nervous system. Surely it was alcohol's intoxicating power rather than its antiseptic properties that made wines and beers so popular so early.

By Homer's time, about 700 B.C., wine had become a staple beverage in Greece, one that was made strong, watered down before drinking, and graded in quality for freeman and slave. The Greek for "to breakfast," *akratidzomai*, meant literally "to drink undiluted wine" (it was used as a dip for pieces of bread). Dionysos (the Roman Bacchus), originally a composite god of ecstasy and of vegetation, became closely identified with the fruit that embodied both. The culture of the vine was not established in Italy until about 200 B.C., but it took hold so well that the Greeks took to calling southern Italy *Oenotria*, "land of the grape." And in the next couple of centuries, Rome advanced the art of wine making considerably. Pliny devoted a full book of his *Natural History* to the grape. He began by saying that there was an infinite number of varieties, and he knew that the same grape could produce very different wines under different conditions, that soil and climate have a large influence on quality. He described sweet, raisin, spiced, medicinal, vegetable, herb, honey, date, and fig wines, and named Italy, Greece, Egypt, and Gaul (later, France) as admired sources. The Romans knew enough about avoiding the tendency of wine to spoil or to sour into vinegar that aging was now possible. Varro, in his *Rerum rusticarum* ("Rustic Matters," around 50 B.C.) said "there are kinds of wine, for instance the Falernian, which when served are more valuable the more years they have been stored."

Pliny was not altogether pleased with the human accomplishment that his fourteenth Book covered. It seemed to him to be misplaced effort.

> There is no department of man's life on which more labor is
> spent—as if nature had not given us the most healthy of bever-
> ages to drink, water ... and so much toil and labor and expense
> is paid as the price of a thing that perverts men's minds and pro-
> duces madness, having caused the commission of thousands of
> crimes, and being so attractive that a large part of mankind
> knows of nothing else worth living for.

If the art and science of wine making had advanced considerably in Italy,
so had an awareness of the social and psychological dangers of excessive
drinking. Other ancient writers warned against inappropriate intoxication,
but also stressed the great value of wine as an anodyne, a means of soothing
the troubled mind. The Old Testament Book of Proverbs (around 500 B.C.)
advises kings not to befuddle their judgment, but then says: "Give strong
drink unto him that is ready to perish, and wine unto them that be of heavy
hearts. Let him drink, and forget his poverty, and remember his misery no
more." From a nearly contemporary but very different culture, Plato argued
in remarkably similar terms. His Athenian in Book II of the *Laws* proposes
that wine be prohibited to those under 18 and allowed only in moderate
quantities until age 30, since it made no sense for naturally ardent spirits to
be further inflamed.

> But when a man is verging on the forties, we shall tell him, after
> he has finished banqueting, to invoke the gods, and more partic-
> ularly, to ask the presence of Dionysus in that sacrament and
> pastime of advancing years—I mean the wine cup—which he
> bestowed on us as a comfortable medicine against the dryness of
> old age, that we might renew our youth, and our harsh mood be
> melted by forgetting our heaviness, as iron is melted in the fur-
> nace, and so made more malleable.

Whether or not alcohol is a medicine, it certainly is a drug, as we shall see.

Beers from Starchy Grains

Wines were not the only alcoholic beverages known to the ancient
world. Perhaps even before the grape was domesticated, people had discovered that starchy grains—barley, wheat, millet, corn—could be treated so as
to be fermentable. (Today, we know that it is necessary to break the starch
down into its component glucose units in order to support the growth of
yeasts.) Three such treatments were discovered. One is very rare today, but
may have been the sole source of alcohol for pre-Conquest Peru. *Chicha* was
and is made by chewing on ground corn and letting a human salivary
enzyme break down the starch into glucose and maltose (a double-glucose
sugar). Girolamo Benzoni saw *chicha* being made in the 16th century and
described the process in his *History of the New World*:

The women, taking a quantity of grain that seems to them sufficient for the wine (or *chichia*) intended to be made, and having ground it, they put it into water in some large jars, and the women who are charged with this operation, taking a little of the grain, and having rendered it somewhat tender in a pot, hand it over to some other women, whose office it is to put it into their mouths and gradually chew it; then with an effort they almost cough it out upon a leaf or platter and throw it into the jar with the other mixture, for otherwise this wine would have no strength. It is then boiled for three or four hours, after which it is taken off the fire and left to cool, when it is poured through a cloth, and is esteemed good in proportion as it intoxicates. . . .

Clearly a labor-intensive, small-batch process with little potential for development, though it is quite ingenious. A second technique, invented in the Far East, uses the *koji*, or a mold *(Aspergillus oryzae)* that secretes a starch-digesting enzyme as it grows on rice. Because each enzyme molecule can do its starch-splitting operation perhaps a million times before degenerating, a small quantity of moldy rice or chewed corn can reduce a portion of untreated cereal starch into sugars. Among other things, the *koji* is used to prepare rice for fermentation into *sake*, which is technically a beer despite its flatness and an alcoholic content comparable to that of strong wines. The use of molds to prepare grains for yeast has not caught on in the West.

The third technique of turning cereal starch into fermentable sugar, the one that is arguably the oldest and that predominates today, is *malting*. The grain is allowed to germinate for several days, during which time it generates enzymes that will break down the stored starch into glucose, the seedling's fundamental source of energy. This practice sounds and is identical to the currently popular sprouting of bean seeds, and it is speculated that malting first caught on because sprouted seeds are more palatable—sweeter, softer, and moister—than dry ones. Then perhaps the sprouts were left in water, or dried, ground, and made into a gruel that sat around for a few days and fermented. However it happened, it happened before the third millennium B.C., when barley and wheat beers were being brewed in Egypt and Babylon. The Babylonian Code of Hammurabi, from about 1750 B.C., lists punishments for sellers of beer who defraud their customers, and we know from Egyptian records of about the same time that the malted grain was preserved by baking it into a flat bread, which was then soaked in water when beer was to be made.

Wherever both have been available, beer has been the drink of the common people and wine the drink of the rich. The raw material for beer, grain, is cheaper than grapes and its fermentation is less tricky and drawn out. The Greek writer Athenaeus, in his *Deipnosophists* ("The Sophists at Dinner," about A.D. 230) set down just this reason for the invention of beer. Wine, he said, was first known and loved by the Egyptians, "and so a way was found among them to help those who could not afford wine, namely, to drink that wine made from barley" (Book 1). It is quite clear that to the Greeks and Romans, beer remained an imitation wine made by barbarians who did not cultivate the grape. Pliny, for example, describes it as a cunning if unnatural invention:

> The nations of the West also have their own intoxicant, made from grain soaked in water. There are a number of ways of making it in the various provinces of Gaul and Spain. . . . Egypt also has devised for itself similar drinks made from grain. . . . Alas, what wonderful ingenuity vice possesses! A method has actually been discovered for making even water intoxicated.

The knowledge of beer making seems to have passed from the Middle East through western Europe to the north, where with agriculture less advanced and the climate too cold for the vine, beer became the usual beverage. Wines had to be imported from the south and were simply too rare and expensive for any but the wealthy. (Among the nomadic tribes of northern Europe and central Asia who did not even cultivate grain, milk was fermented into the wines called *kefir* and *koumiss*.) To this day, beer remains the national beverage of Germany, the Low Countries, and Britain, and wine the beverage of Greece and Italy. The only Roman provinces to adopt the vine wholeheartedly were France and Spain, which border the Mediterranean and have suitable climates.

Wine and the Church

After the fall of Rome, the arts of viticulture and alcoholic fermentation were kept alive principally by the Christian Church in its many isolated European outposts, and especially in the monasteries. Local rulers endowed them with tracts of land, which they then cleared of forest and reclaimed from swamps, bringing systematic, organized agriculture to the sparsely settled north and west. Jesus had likened himself to the grape vine, people to his branches, and God to the farmer, and at the Last Supper had renewed the ancient connection between wine and blood by saying of the cup, "This is my blood of the new testament, which is shed for many" (John 15, Mark 14). Wine was therefore necessary to observe the sacrament of Communion, and the monasteries almost always tended vineyards, carrying the grape farther north than it had ever been. Wine and beer were also made for daily consumption, to serve guests, and to sell; it was a Benedictine, Dom Pierre Pérignon, who developed champagne as we know it. The Trappist beers of Belgium and the liqueurs known as Benedictine and Chartreuse ("Carthusian") are living remnants of this tradition. And the first European grape successfully planted in California was brought by a Franciscan, Fra Junipero Serra.

In addition to the ascetic orders, popes and cardinals also helped see viticulture through to modern times, not so much by engaging in husbandry themselves as by appreciating and encouraging the talents of others. Early in the 14th century, for mostly political reasons, the papacy was temporarily transferred from Rome to Avignon, now a city in the south of France. During his residence there, Pope John XXII established a vineyard in the district known today as *Châteauneuf du Pape:* the Pope's New Castle. When the time came to leave Avignon for Rome, many members of the papal bureaucracy were apparently reluctant to leave behind the wines of the Burgundy region, to the north of Avignon. In late 1367, the great humanist Petrarch wrote a letter from Venice to Urban V soon after the Pope had reached Rome, encouraging him to stay and not be convinced to return to Avignon. Petrarch had heard that certain of Urban's cardinals felt deprived in Italy, and considered the Rhône a river of Paradise

> because it brings them the wine of Burgundy, which they regard as the fifth of the natural elements [after earth, air, water, fire]. . . . It is unknown to all writers ancient and modern, nor ever numbered among the most precious wines, but has come to be exalted by these people, and placed almost on a level with the nectar of the gods. . . . (*Seniles,* Book 9, Letter 1)

It sounds as though snobbery and hyperbole have a venerable history when it comes to wine, though much of the exaggeration here may be Petrarch's. In any case, two centuries later the Church's expertise in wine, if fading, was still remembered. William Harrison wrote in his *Description of England* (1587) that "the stronger the wine is, the more it is desired; by means whereof in old times the best was called *theologicum,* because it was had from the clergy. . . ." In Harrison's day, the making of wine and beer had largely passed into the increasingly competent hands of lay society. He reports that the brewing of beer is "once in a month practiced by my wife and her maidservants," and brewers' guilds had existed in England since around 1400.

Distilled Liquors

The origins of distilled liquors, though much more recent than those of wine and beer, are just as obscure. Because alcohol is toxic to all organisms, even the yeasts that excrete it can not tolerate an environment that is more than about 15% alcohol, and this was as strong as wine or beer could get until a way of concentrating alcohol was found. The process of distillation is made possible by the fact that the boiling point of alcohol is about 173°F (78°C), or almost 40° lower than that of water. This means that if a mixture of water and alcohol is heated to the boil, more of the alcohol than the water will end up in the initial vapor. The vapor can then be cooled and condensed into a liquid—the word *distill* comes from the Latin *destillare,* "to drip," from the condensation of vapor on a cool surface—which will have a higher alcoholic content than the original beer or wine. The discovery of this phenomenon, and an understanding of how to exploit it, would clearly require a good deal more sophistication than did the taming of fermentation itself. The ancient Mediterranean knew something about distillation, though apparently not about its special effect on wine or beer. Here is Aristotle's somewhat confused account of distilling sea water and wine in his *Meteorology* (Book 2):

> Sea water, when it is converted into vapor, becomes drinkable, nor does it form sea water when it condenses again: this we say from experience. Other substances are influenced in this way. Wine and many liquids, when evaporated and condensed back into a liquid, become water. For these are only modifications of water which exist as a certain mixture, and it is the thing that has been mixed with the water that determines the flavor.

Surely Aristotle would have mentioned the remarkable properties of the "water" distilled from wine if they had been noticed. Instead, the Greeks seem to have treated both salt and alcohol as impurities. According to a book on the history of Indian food and drink, distilled rice or barley liquor was appreciated there around 800 B.C., several centuries before Aristotle. If so, the knowledge remained isolated for almost 1000 years.

Expertise in distillation made its way to Europe through the Persians and Arabs, who had trafficked in spices between India and the Middle East quite early in history. In the hands of these peoples, distillation became a technique not for concentrating the intoxicating power of wine, but for engaging in alchemical transformations of various substances. The alchemists had two great goals: the transformation of base metals into gold, and the prolongation of human life. In both cases, there was said to be a unique substance, the *elixir*, which was the key, and distillation was thought to be a powerful technique of separating mixed substances in the search for the elixir. A blend of indigenous philosophy and technology and of Greek scientific theory, one which received great impetus around A.D. 700 when the Moslems accorded it significant patronage, Arab alchemy had a strong influence on Western science, and could even be considered the spark that ignited it. Many of its terms—chemistry (from alchemy), algebra, alkali, camphor, elixir, talc, tartar—live on in modern languages, and even the kitchen knows its influence: the French word for double boiler, *bain-marie* (Mary's bath), is named after a legendary alchemist, Mary the Jewess. Another one of these linguistic borrowings is *alcohol*.

To the Arabs, *al kohl* was not the intoxicating ingredient of wine. At first, it named the dark powder of the metal antimony, which was used to color the eyelids. By a process of generalization, it came to stand for any fine powder, and then for the essence of any material. The "alcohol of wine" was probably discovered by distillation around 1100 at the medical school at Salerno, Italy, and the word *alcohol* was first used to mean the essence of wine by the 16th-century German alchemist and physician Paracelsus. This usage didn't become general until about 200 years ago. The circumstances of alcohol's discovery had an important effect on the early history of distilled liquors. To the School of Salerno wine was a useful medicine, and the success that distillation brought in concentrating the essence of its power and producing a stronger drug strengthened the belief that this was the way to find the elixir. Around 1300, the Catalan scholar Arnald of Villanova dubbed the spirit of wine *aqua vitae,* or the "water of life," whose Gaelic translation, *uisge beatha,* is the original of "whisky." And the first printed book devoted to distillation,. Hieronymus Brunschwygk's *Liber de arte distillandi* (1500), explained why this process is the necessary basis of the new medicine. Distillation

> is nothing else but the separation of the gross from the subtle and the subtle from the gross, the breakable and destructible from the indestructible, the material from the immaterial, so as to make the body more spiritual, the unlovely lovely, to make the spiritual lighter by its subtlety, to penetrate with its concealed virtues and force into the human body to do its healing duty.

If only this were what alcohol does! It is this connection between distillation and the pure and ethereal that gives us our synonym for liquor, *spirits.*

For several centuries after its discovery, alcohol or aqua vitae was thought of only as a potent medicine, and its manufacture was largely limited to apothecaries and monasteries. A relic of this time is the term *cordial,* applied to after-dinner drinks; it comes from the Latin for "heart" and in the Middle Ages meant a medicine that stimulates the circulation. Aqua vitae soon became the preeminent cordial, and eventually the only one. The 15th century appears to be the time when distilled liquor was liberated from the pharmacy and became a beverage to be enjoyed for its unique flavor and effects. The terms *Bernewyn* and *brannten Wein,* ancestors of our word *brandy* that meant "burnt," or distilled, wine, appear in German laws about public drunkenness beginning around 1450. It is guessed that the knowledge of distillation was carried to the British Isles, and particularly Ireland and Scotland, by monastic pilgrims between 1100 and 1300. Wine not being readily available there, they distilled barley beer instead and came up with whisky rather than brandy. Exactly when this momentous development took place is unclear, but the first documentary evidence comes from Scotland in 1494, when Friar John Cor received 1,100 pounds of malt "wherewith to make aqua vitae"—clearly a more than medicinal amount. Gin, distilled from rye, and rum, from molasses, were invented in the 16th and 17th centuries, and monastic specialties like Benedictine and Chartreuse date from about 1650 on.

THE NATURE OF FERMENTATION

The Yeast Controversy

People have been making wine and beer for roughly 5000 years, but we have *understood* what we were doing for little more than a century. Until well after 1800, brewers simply knew that whatever it was that caused fermentation could be found concentrated in the foam of newly made beer or wine. This "yeast" or "barm" or "leaven" was also useful in leavening bread, and brewer's yeast was the original source of the household starter. Like the fermentation of cheese, the fermentation of beverages was a matter of local conditions. Each region had its characteristic yeasts, which infected beer or wine spontaneously and were then passed on from batch to batch. Needless to say, poor quality was not uncommon in these centuries when nature, through human ignorance, was allowed to take its course. Microbes that cause acetous (vinegary) and putrefactive fermentations are no less natural than the yeasts that produce alcohol.

In 1836 came the first small steps toward understanding fermentation. Two scientists independently observing yeasts through the microscope saw small globules which, because they multiplied by splitting in two, seemed to be alive. Charles Cagniard de la Tour, a mechanical engineer who invented the siren, reported his findings in 1836. Theodor A. H. Schwann, a chemist working in Berlin and one of the proponents of the new cell theory, waited until 1837 to publish. In 1839, he quit his advanced research for more mundane work, because the scientific establishment of the time heaped ridicule on the idea that cells caused fermentation. Justus von Liebig and Friedrich Wöhler, the preeminent organic chemists in Europe, were convinced that fermentation was a purely chemical process that involved complex substances but no living beings of any kind. And in their influential review, the *Annalen der Pharmacie*, there appeared in 1839 an attack on the partisans of the cell theory. Immediately following a neutral report on the theory, an anonymous correspondent, since identified as Wöhler, reports that he has seen wine animals through the microscope:

> these animals have the form of a Beindorf still (without the cooling apparatus). . . . From the moment that they escape from the egg, one sees that these animals swallow sugar from the surrounding solution; one can see it arrive in the stomach quite clearly. It is instantly digested, and this digestion is instantly and most definitely recognized by the subsequent expulsion of excrement. In a word, these infusoria eat sugar, empty wine alcohol from the intestinal canal, and carbonic acid from the bladder. In the full condition, the bladder possesses the form of a champagne bottle. . . .

This parody of his observations was apparently too much for Schwann, and the prestige of Liebig and Wöhler was sufficient to bring this line of investigation to a halt, at least for a while.

Pasteur and Hansen

One of the reasons we remember Louis Pasteur is that he almost single-handedly defeated the chemical theories of fermentation and disease and invented the science of microbiology. In 1858, he published a study not of alcoholic fermentation, but of lactic acid fermentation: the type that sours milk. Rather than relying on purely visual observations, he developed a procedure that was, if not conclusive, at least strongly suggestive of the involvement of living organisms. He made a nutritious base of cooked yeast extract, sugar, and chalk, warmed it to about 90°F (32°C), and introduced a very small amount of material from an active, spontaneous fermentation. The result: an explosion in the globule population. Pasteur likened the situation to the fate of uncultivated land.

> What takes place in fermentation may be compared to what occurs in a plot of land that is not seeded. It soon becomes crowded with various plants and insects that are eventually harmful.

Clearly the way to control fermentation and so the quality of the resulting wine or beer is to "seed" the liquid: to culture grape or grain juice as we do the fields. This requires the ability to isolate the desirable seeds, or microorganisms, from the undesirable, and Pasteur did not live to see the techniques that would make pure cultures practical (the best he could do was to manipulate the growth medium to favor particular yeasts). But he did perfect a way of controlling undesired fermentation in *finished* wine or beer: the heat treatment now known as pasteurization.

The technical innovation that made pure cultures practical, and with them real progress in microbiology, came in a Danish brewery. Emil Christian Hansen studied the anatomy of various mushroomlike fungi, and then fermentation, before receiving the Ph.D. at Copenhagen in 1878. That same year, he took a job at the Carlsberg brewery, soon became director of the laboratory, and continued at that post until his death in 1909. He learned that several yeasts and bacteria could infect beer and give it bad flavors, and began trying to purify cultures of the desirable yeasts. He eventually succeeded in growing cultures from single cells and maintaining them indefinitely in special flasks, and found that there were actually several alcohol-producing yeasts, each with different characteristics. Hansen's expertise was first put to use in 1883. The original "Carlsberg Bottom Yeast I," which had been obtained in Munich in 1845, had spoiled. J. C. Jacobsen, the owner of Carlsberg, allowed Hansen to run a large-scale fermentation with one of the new pure cultures, and on November 12 the beer turned out to be excellent.

Jacobsen generously refused to patent the culture process, and Hansen published his method. Soon most breweries in Europe and America were using cultured yeasts. Our food and drink could now be "seeded," with more predictable results than were possible by trusting to the air or to a starter from the previous batch.

Yeasts and Alcohol

Yeasts are now defined as a group of about 160 species of single-celled, microscopic fungi. Not all are useful: some cause the spoilage of fruit and vegetables, some cause disease (for example, the fairly common "yeast infection" of *Candida albicans*). Most of the yeasts used in making bread and alcoholic drinks are members of the genus *Saccharomyces*, or "sugar fungi." We prefer them over other microbes for the same reason that we use only particular bacteria to sour milk: they metabolize sugar, produce carbon dioxide and alcohol (or in the case of milk bacteria, lactic acid), and do very little else. Strong odors or noxious substances are released, if at all, in only very small amounts. In this sense, they are *clean* microbes. Essential to the yeasts' production of alcohol is their ability to survive on very little oxygen, which most living cells use to burn fuel molecules for energy. The overall equation for the production of alcohol from glucose goes like this:

$$C_6H_{12}O_6 \rightarrow 2C_2H_5OH + 2CO_2 \qquad + \text{energy}$$

glucose   alcohol   carbon dioxide

No oxygen is needed to get energy out of the sugar. In fact, if plenty of oxygen is around, the yeasts do not deposit alcohol in their surroundings. Instead, the cells burn the sugar down all the way to water and carbon dioxide, thereby extracting more energy from the fuel.

$$C_6H_{12}H_6 + 6O_2 \rightarrow 6CO_2 + 6H_2O + \text{energy}$$

So in order to maximize the production of alcohol, brewers purposely limit the amount of contact the yeast has with air. Of course, these equations are great simplifications of many-stage biochemical reactions. A series of enzymes breaks glucose or maltose (a double glucose) down into successively smaller molecules, the energy released along the way being used by the cell to grow and divide. But the brewer's yeasts do not have the enzymes necessary to reduce larger glucose chains to their individual units, and this is why starchy grains must be pretreated with their own starch-breaking enzymes before they can be fermented.

Yeasts do introduce a variety of other compounds into the grape juice or cereal "wort" that contributes characteristic flavors; for example, they transform amino acids in the liquid into "higher," or longer-chain alcohols. And when, after producing its limit of daughter cells by budding, a yeast

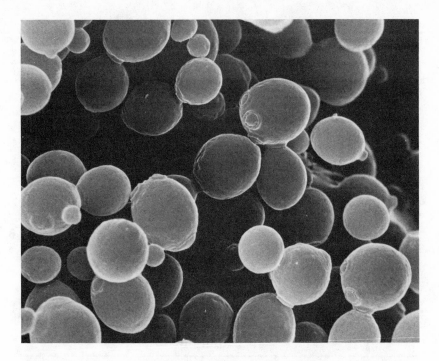

Cells of the yeast *Saccharomyces cerevisiae* as seen through the scanning electron microscope. Each is about 5 microns (0.0002 inch) in diameter. The cell at upper right center is in the process of reproducing, and bears the scars of previous buddings. (Courtesy Alastair T. Pringle, University of California, Los Angeles.)

cell dies, its enzymatic machinery digests the cell and releases its contents into the liquid. Because the yeast cells synthesize proteins and vitamins, especially B vitamins, as they grow, they can actually make a fruit juice or cereal mash much more nutritious than it was when uninfected. Unfortunately, we like our beer and wine sparkling clear and so remove this advantage when we skim or filter out the cloud of yeast cells. In some less affluent regions of the world, these beverages are consumed "whole" and are a truly valuable part of the diet.

Particular jobs are best done by particular kinds of yeast. Baker's yeast, for example, should ideally produce little alcohol and a lot of carbon dioxide to do the work of raising bread dough. Two different species are commonly used in making beer. One, the mainstay of Jacobsen's "Bottom Yeast I" now known as *Saccharomyces carlsbergensis* or *S. uvarum*, tolerates cold temperatures and falls to the bottom of the tank after fermentation; it is used in lager beers, which include most American brands, and in the bottle fermentation of champagne. The other, *Saccharomyces cerevisiae* (*cerevisia* is Latin for "beer"), does best at about 70°F (21°C), rises to the top of the tank, and is used for traditional English ales and beers. In wine making, the situation is more complicated. The important yeasts are usually strains of a variety of *Saccharomyces cerevisiae* called *ellipsoideus* for their shape. But better than 100 different strains of several yeast species have been found growing on the skin or in the released juice of European grapes, and most starters, when they are used, contain at least several strains that may work in sequence during fermentation. In some districts, the grapes are allowed to ferment without a starter. It is interesting that California vineyards apparently harbored very few useful yeasts until wine making had been well established there; starters had to be imported from Europe. Today, popular yeast cultures are produced by the University of California, and some American wine makers can now let their naturally infected grapes ferment unaided by starters. There is some controversy about the place of pure, single-strain cultures in wine making. Some people insist that a mixed population of microbes produces a more complex, more interesting flavor, while others doubt that the difference is significant.

WINE

Its Evolution

Our knowledge of wines in the ancient world is skimpy, but there is good evidence that we would find much of it undrinkable. What Egyptian wine tasted like is not known. The usual practice was to tread the grapes, squeeze them in a cloth bag, and strain the juice into a large clay jar, where it would ferment. The contents of the jars were eventually sampled and graded, and the jars marked and sealed until the wine was to be used. In Greece and Rome, tree resins or the pitch refined from them were used both to flavor and to preserve wines, either as a lining for the jar and stopper or as a powder added directly to the wine. The turpentiny taste of a wine so treated is probably best approximated by today's Greek *retsina*. Salt and other spices were also added to lengthen storage life, as the Latin writer Columella reports in his long discussion of wine preservation (Book 13 of *De re rustica*). From classical literature it appears that most wine was watered down before being served, and the late Latin cookbook of Apicius gives a recipe that can be interpreted as a way of changing red wine to white by the addition of egg white, bean meal, or potash (see page 179). While the ancients may not have had the most sensitive palates, they certainly were fairly sophisticated about cultivating the vine. Here is Pliny, nearly 2000 years ago.

> A vineyard with the single cross-bar is arranged in a straight row which is called a *canterius*. This is better for wine, as the vine so grown does not overshadow itself and is ripened by constant sunshine, and is more exposed to currents of air and so gets rid of dew more quickly, also is easier for trimming and for harrowing the soil and all operations. . . .

It was also about Pliny's time that wooden casks—an innovation of northern Europe—arrived along the Mediterranean as an alternative to earthenware pots. What use the Romans made of them is not clear, but they are crucial to the later development of fine wines since they make possible a controlled, limited contact with air, and thereby the process known variously as aging, developing, or maturing.

Very little of note seems to have occurred between the fall of Rome and the discovery of the New World, except that the vineyards of France slowly developed into the preeminent source of wine in Europe. Meanwhile Italy, the source of the vine for Rome's western provinces, fell far behind, a victim of political and economic circumstance. Until the middle of the 19th century it was little more than a collection of city-states, each with protective tariffs and little of the international trade that would bring competition and improvement. Most of the wine was consumed locally, and the grapes grown not in vineyards but in sharecroppers' plots, between rows of food plants or trained on trees. By the late 1600s the wines of France, and especially Bordeaux, which had the advantage of its own port, were so popular in England that they were paid the compliment of imitation by talented practical chemists. In the 1709 *Tatler* (No. 131), Joseph Addison described "subterraneous philosophers" who "rais[e] under the streets of London the choicest products of the hills and valleys of France."

The market for such imitations was expanded by the fact that the originals were becoming harder to get. In the 1670s, French trade protectionism provoked the English government to levy large duties on French imports, and political friction between the two countries led to the view that boycotting French wines was a patriotic act. Both economic and political conditions caused the English to rely increasingly on Portuguese wines, which soon became known as "port." In his poem "On the Irish-Club" (1730?), Jonathan Swift gave this sharp advice to Ireland's feckless senators and nobility:

> Be sometimes to your country true,
> Have once the public good in view:
> Bravely despise Champagne at Court,
> And chuse to dine at home with Port.

In Swift's time, port was simply a drinkable red wine. It was not until about 1840 that the quintessential English after-dinner port, a wine fortified with brandy, came into being.

Champagne

Strange to say, it may very well be that bubbly champagne was invented by the English at about the time that French wines were being discriminated against. Notice that Swift refers to champagne as a popular wine. The name, from an old province east of Paris, could simply refer to still wines from that region. But according to the *Oxford English Dictionary*, it is with the 1670s that the word *sparkling* comes to be applied frequently to wine. George Etherege's play *The Man of Mode*, written in 1676, includes a drinking song that celebrates the virtues of "sparkling Champaigne."

> It quickly recovers
> Poor languishing lovers,
> Makes us frolick and gay, and drowns all our sorrows.   (IV.i)

Champagne was the favorite drink of the English Restoration, but didn't catch on in the French Regency until nearly 40 years later. How is it that the English scooped the French with a French wine? The key appears to be cork. Cork stoppers, made from the inner bark of a certain oak native to Spain, had been used in English ale and wine bottles since the time of Shakespeare. For some reason, France continued to use plugs of hemp soaked in oil. Whereas any buildup of carbon dioxide in a French bottle would seep out through the fibrous hemp, it would be held back by the more elastic, tighter-fitting cork. Bubbly champagne was probably discovered when the English imported barrels of still champagne wine and bottled it themselves: residual yeasts produced enough gas in the bottle to make it sparkle. The French Benedictine Dom Pierre Pérignon, cellarer at a monastery in the Champagne region, is reputed to have been the first winemaker to use cork, perhaps in the 1690s. According to one story, he got the idea when he saw the cork-stoppered water bottles of visiting Spanish monks. Perhaps more important, he perfected the production of white wine from red grapes, and had a legendary palate when it came to blending grapes from different vineyards to obtain a well-balanced wine. The addition of yeast and sugar to induce a second fermentation in newly bottled still wine was probably a later refinement. Even if the English discovered sparkling champagne, it was the French who made it into more than a novelty.

Bottles

The great event of the 18th century was the gradual evolution of the wine bottle from a short, stout flask to the tall, thin bottle we know today. This was not merely a matter of fashion. For many centuries, wine bottles were simply used to convey the wine from cask to table or to store it for a day or two, and they were bulky enough that any number would have taken up a huge amount of space. All the aging of wine occurred in the cask, and since wood allows some oxygen to enter continuously, this period was limited to a few years. All wines had to be drunk within 4 or 5 years of their making. When the bottles had slimmed down enough that they could be stored on their sides, the contents in contact with the cork and preventing it from drying out, shrinking, and so admitting air, bottle aging had become possible, and with it the ability to keep wines from particularly good years—vintage wines—for many years after. The Greeks and Romans had been able to do this with the earthenware *amphora*, but then they did not have the advan-

tage of barrel aging. We may think of stoppers and containers as mere par-
aphernalia, but without wood, cork, and slim bottles, wine would not be what
it is.

Phylloxera

The 19th century brought both progress and disaster. The progress was
a growing understanding of the biochemistry of fermentation and of aging
and spoilage, largely contributed by Louis Pasteur. We will return to his
experiments in a few pages. The disaster was the introduction to European
vineyards of *Phylloxera vastatrix*, an insect pest that bores into the roots of
grape vines and thereby kills them. It was carried to Europe from North
America on some native American vines, and the European plantings, with

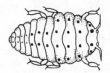

The root-boring stage of *Phylloxera vastatrix*,
the pest that devastated European and
Californian grape plantings in the 19th century

no previous contact and so no resistance to it, were gradually devastated,
from the 1860s through the 1880s. The infestation was eventually overcome
by planting resistant American or French-American hybrid rootstocks and
then grafting onto them the European varieties that had grown there before.
The phylloxera disaster had one positive effect: it gave the growers a chance
to bring some order to the older vineyards, reducing mixed plantings and
straightening out the rows for more efficient cultivation and harvesting.

Wines in America

Europeans didn't simply plant American grapes, because these varieties
belong to entirely different species and have a very different flavor, the so-
called foxy or Concord-grape flavor, which is so strong as to swamp any others. When either Leif Erikson or Bjarni Herjulfsson visited North America
around A.D. 1000 and named it "Vinland," he must have been impressed by
the growth of wild berries, for native American grapes had not yet been
domesticated. The major American species is known as *Vitis labrusca*,
labrusca being the word the ancients used for "wild grape." The European
species is *Vitis vinifera*, *vinifera* meaning "bearer of wine." There are more
species of *Vitis* native to North America than to any other continent. *Vitis
rotundifolia*, for example, also known as the scuppernong or muscadine, is
made into wine in the Southeast. The first English colonists tried to make
wine from the wild grapes, but failed because there wasn't enough sugar in
the juice for the yeast to ferment into alcohol. And imported European vines
couldn't survive the harsh climate and new range of diseases and pests. *Vitis
labrusca* was finally domesticated around 1780 and then widely planted, and
by 1840 some 15 states were producing considerable amounts of wine, with
Ohio, Missouri, and New York among the leaders. In the 1820s the Catawba

variety was discovered and made into wine in Maryland, and in 1849 Ephraim Bull, a neighbor of Nathaniel Hawthorne in Concord, Massachusetts, discovered the variety of *Vitis labrusca* that now bears the name of his town, and that remains the standard grape for juice and jellies in this country. Around 1860 Cincinnati was the center of the American wine industry, but in only a few years, epidemics of fungus and mildew wiped out the Ohio plantings.

Meanwhile, there was California. It is said that around 1770 Fra Junipero Serra brought the Mission grape, a rather characterless European variety that had been planted in Mexico by the Spanish in the 16th century, up the coast of California. Other European varieties were imported in the 1860s. The West Coast was free of most pests, and the climate similar enough to the Mediterranean that European vines thrive there. The wine industry, like the population, grew slowly at first, but the opening of the transcontinental railroad in 1869 expanded its market, and at the Paris Exposition of 1900, California's wines won 36 medals. This remarkable progress was cut short by the two plagues of phylloxera, which spread slowly from 1860 to 1900 (like the French, California growers put in American or hybrid rootstocks), and Prohibition. When the 18th Amendment was ratified in 1920, most vineyards were replanted in relatively bland juice grapes, and by the time of Repeal in 1933, hardly any wine grapes were left. California had to start from scratch. But this made it much easier for American vineyards and wineries to take advantage of new advances in breeding and in technology. In the 1940s and '50s, the University of California at Davis became a leading center of research in viticulture and oenology. From the 1950s on, tourism and interest in fine foods have encouraged an ever increasing production of wine in California and in several other states as well, which are at last succeeding in growing the European grapes. Still, the average American consumes less than a tenth of the wine that his French or Italian counterpart does.

In the last decade, wine connoisseurs have generally lamented steep price increases and the tendency of some producers to sacrifice long-term quality for the sake of immediate drinkability and sales. Still, considering Roman resin, 18th-century bottles, 19th-century plagues, and 20th-century Prohibition, there have been few better times in history to be drinking wine.

The Grapes

Why is it that certain grapes make better wines than others do? To some extent, this question is unanswerable, or answerable only with semi-circular reasoning. The *cépages nobles,* or noble vines from which the finest wines are made, are those with a distinctive flavor that takes well to the processes of fermentation and oxidation. Among the prized wine grapes are the Pinot Noir, Cabernet Sauvignon, Gamay—the chief red grapes of Burgundy, Bordeaux, and Beaujolais respectively—and Chardonnay, Sauvignon Blanc, and Riesling, the respective white mainstays of Burgundy and Champagne, Bor-

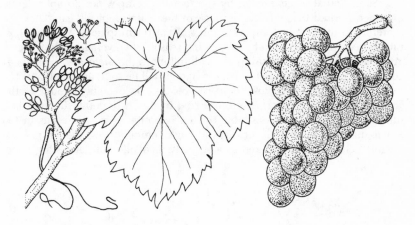

Vitis vinifera, the species of grape from whose many varieties most wines are made

deaux, and Germany. However, it is sometimes difficult to identify a particular variety. There are an estimated 10,000 different strains of *Vitis vinifera* vines, and there is plenty of opportunity for confusion about nomenclature. The picture is further complicated by the fact that French wines, which are classified by geographical origin, are usually a blend of two or more varieties, while American wines that are classified by variety need only take 75% of their volume from that variety. So about all that can be said is that certain grapes have traditionally been considered to be good wine grapes: and this judgment in turn has determined what other grapes will qualify.

There is less mystery about one characteristic of wine grapes: the balance between acid and sugar. One reason that the grape is well suited to controlled fermentation is that it contains enough sugar to reach a fairly high alcohol content, about 10%; and with a pH of about 3, it is acidic enough to discourage the growth of most other microbes. If either the sugar or acidity is low, the wine will not ferment cleanly or keep well, and it will taste flat. On the other hand, too much acid will give a sour wine, and too much sugar—more than the yeast can ferment to alcohol—leaves it undesirably sweet. So the first requirement of a good wine grape is that it reach the proper acid-sugar balance at maturity. Since the sugar content increases and the acidity decreases during ripening, the time of picking can also be varied in order to catch the grapes at their best. Generally, table wines are fermented until they are "dry," or fully depleted of sugar, so the question of balance comes down to whether, at a level of sugar that will produce sufficient alcohol, the acidity is appropriate for the wine's eventual flavor and stability. This is where the issue of climate becomes significant.

The Influence of Climate

Some initial limits are set by the fact that grapevines do not tolerate extended periods below freezing, nor can they take tropical heat. They are therefore limited to the temperate zones of both hemispheres. Within a hospitable region, the length of the growing season, the average daily temperature, and the number of sunny days all influence the sugar-acid balance. If cold weather comes before the fruit is ripe, it will be sour, as it will be if the season is cool or cloudy: low temperatures and little sunlight result in the accumulation of less sugar, even in a ripe fruit. On the other hand, a hot, sunny climate can be undesirable for the opposite reason: it results in too much sugar and too little acidity (at high temperatures, the grapes use organic acids for energy), and so a strongly alcoholic but flat-tasting wine.

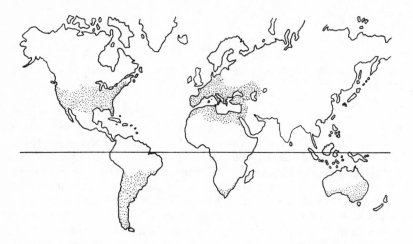

Regions of the world in which wine grapes are grown

And because such conditions hasten the maturation of the grapes, they generally turn out somewhat bland, without the full range of volatile substances that help contribute to the aroma of the wine. Finally, there is the fact that the best wine-making varieties, developed as they generally were in France, tend to prefer a relatively cool climate and mature more slowly than some other varieties. The upshot of all these considerations is that the best grapes—and so the best wines—can be produced only in a certain range of climatic conditions which may or may not be fulfilled in a particular place in a particular year.

A study of the average temperatures and sunlight in various wine-growing regions explains why it is that only certain regions of California have been associated with fine wines, despite the fact that nearly the whole state has a long, bright growing season. The relatively cool coastal valleys of Napa and Sonoma are climatically similar to Burgundy, Bordeaux, the Rhine, and Mosel, while the hot inland valleys that give us most of our jug wine resemble the regions of Spain that produce the sweetly bland sherry grapes. Even in California's cool valleys, the grapes generally ripen with more than enough sugar, and if anything may be low in acid, while Europeans have the opposite problem: plenty of acid, and not enough sugar. This is why California wines tend to be 1 to 3 percentage points more alcoholic than French wines. But this broad perspective can be misleading. Because rainfall, sunlight, and temperature depend on local topography—what direction the hillside faces, how high it is—and because the quality of the soil can vary over small distances, the microclimate of a particular planting and the grapes it produces may not be typical of the region. As Pliny observed long ago, "the same vine has a different value in different places."

The Anatomy of a Grape

We leave behind climatic zones and turn to the zones of the grape itself. The substances important to wine quality are not evenly distributed in the grape, nor is the juice from crushed grapes necessarily representative of the whole grape. Besides sugars and acids, there are other chemical compounds in the grape that affect the quality of wine. One such group is the phenolic compounds. These include the pigments, whether yellow or reddish, and the *tannins*, a group of molecules responsible for the astringency of red wines. The pigment molecules may also have an astringent effect. Grape berries consist of a skin surrounding the pulp and usually 4 seeds. The stems contain a large amount of tannins, acids, and bitter-tasting resins, and are usually separated from the grapes as they are crushed. The skin holds much of the fruit's phenolic compounds, both pigments and tannins—red skins are especially high in tannins—as well as most of the numerous compounds (aside from sugars and acids) that give the grape its characteristic flavor. As the grape matures, these compounds migrate to some extent into the pulp immediately below it. Like the stem, the seeds at the center are full of tannins, oils, and resins, and care is taken not to break them open during the pressing.

The pulp itself is divided into three zones: the first just below the skin, the third immediately surrounding the seeds, and the second in between. In a mature grape, zone I is lowest in acidity, zone III the highest, and zone II is the sweetest. And zones I and III will harbor some tannins, while zone II is free of them. (If grapes are picked before full ripeness, the distribution is somewhat different.) As a mass of grapes is crushed, the first juice to come out, the *free run*, is primarily from zone II, and it is the clearest, purest essence of the grape. With care, some 60% to 70% of the grape juice can be obtained as free run. As mechanical pressure is applied, juice from zones I and III, with their phenolic compounds, acids, and flavor components, augments the free run with a more complex character. The extent of pressing will have an important influence on the character of the final wine. Sometimes only the free run is used, sometimes the crush is well pressed.

The Stages of Wine Making

Wine making can be divided into three basic stages. In the first, the ripe grapes are crushed into juice. Exactly how much pressure is applied will have an important effect on the composition of the juice. Either sugar or acid may be added at this point to get the desired balance. In the second stage, the grape juice is fermented by sugar-consuming, alcohol-producing yeasts. The kinds of yeasts used, the temperature at which fermentation occurs, and the length of time that the juice is left in contact with the grape skins, all affect the eventual flavor of the wine. The third stage is variously called the aging, or development, or maturing, of the wine. This is essentially a period during which the chemical constituents of the grape and the products of

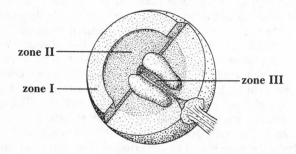

The zones of the grape. Each has a different chemical composition and makes a different contribution to the wine.

fermentation react with each other and with oxygen to form a stable system with an attractive flavor and body. Development may take place in large stainless steel tanks, large or small oak barrels, or in the bottle. Special wines—sherry, port, champagne—are treated in special ways, but they too go through the three basic stages.

As the science of wine making has advanced over the last few decades, many new techniques have been devised. Some wineries have adopted them, others have not. It is impossible to speak of *the* way in which wine is made today. With that qualification, we can look at a sketch of *some* of the ways in which wine is made.

Fermentation

The Initial Must

The transformation of the grapes begins with the harvest. Since the first bunches to be picked end up at the bottom of a heavy pile, they are soon crushed and may begin fermenting spontaneously in a matter of hours. For this reason it is important to get the grapes to the winery as quickly as possible. Once they arrive, they are crushed and the stems removed, usually in the same machine. The resulting crush is about 85% liquid, 10% skins, and 5% seeds. In chemical terms, the liquid portion, called the *must*, is 70 to 85% water, 12 to 27% sugars, mainly glucose and fructose, and about 1% acids, mainly tartaric, malic, and citric, 0.1% tannins, 0.5% minerals, and 0.01% nitrogen-containing compounds, mainly amino acids. In the case of white wines, the must is removed from the skins before fermentation, and so picks up little tannic material or pigmentation. Rosé musts are fermented in contact with the red skins for 1 to 3 days before separation, and red wine musts for 4 to 14 days, or until fermentation is nearly complete. The longer the must is in contact with skin and seeds, the deeper the color (whether yellow or red) and the more astringent the taste. If the must is pressed off the skins rather than being allowed simply to drain freely, then it will extract even more pigment, tannins, and minerals. The size of the fruit is also a factor. The smaller the grape, the greater the fraction represented by the skin, and so the more color and flavor compounds are concentrated in the must. The Cabernet Sauvignon grape is small, and this is one reason for the typically full color and flavor of its wines.

Before yeast is introduced and fermentation begins, two substances are added to the must in order to better control the outcome. One, the versatile sulfur dioxide, suppresses the growth of wild yeasts and bacteria that sour grape juice into vinegar (wine yeasts have adapted to this practice and tolerate the chemical), and prevents the oxidation of both flavor and pigment molecules by either the air's oxygen or newly released enzymes from the fruit cells (the same treatment is given to many dried fruits, and for the same reason). The wine maker must be careful with sulfur dioxide: too much and it will interfere with aging and give the wine an undesirable odor. Though this treatment may sound antiseptically modern, it is in fact centuries old. The Romans used sulfur only as an ingredient in cement for repairing cracks in storage jars, but Albert Henderson spoke in 1825 of adding some must impregnated with the fumes of burning sulfur to newly casked wine as a standard means of preventing further fermentation or spoilage (*The History of Ancient and Modern Wines*, London). The second additive is either sugar or acid, and it is used to correct the balance between these two substances, which is only rarely obtained naturally. In France, where the grapes tend to be acidic, sugar is the usual additive and again has been for centuries. The procedure is called *chaptalisation*, after Jean-Antoine Chaptal, one time

French minister of the interior, who gave it official sanction in his *Essai sur le vin* of 1801. The sugar content of the must can also be raised artificially by adding some must that has been boiled down and concentrated. In California and other warm climates the problem is usually not insufficient sugar, but insufficient acid. The remedy for this is the addition of pure tartaric acid, a natural component of the grape, or sometimes the sour juice of green grapes.

Air and Temperature

Fermentation can begin with or without the addition of a starter culture of yeast. In some areas, the fungi that naturally grow on the skin of the grapes are right for the job and go to work as soon as the grapes are crushed. The primary job of the yeast is, of course, the conversion of sugar to alcohol, but it also strongly influences the flavor of the wine by producing various volatile, aromatic molecules in the course of its growth, molecules that the grape itself cannot supply. Prominent among them are about 30 esters, a class of compounds that combine an acid with an alcohol or phenol, and the higher (longer-chain) alcohols, which the yeast produces when it processes amino acids in the must. There are a couple of conditions that must be carefully controlled in order to have a good fermentation. One is the degree to which the must is aerated. The more oxygen the yeast cells can get, the more they will multiply, but the less alcohol they will excrete. So the winemaker tries to begin the fermentation with a lot of oxygen—this ensures a good population of yeast—and to limit the contact with air thereafter.

The other critical variable during fermentation is temperature. The lower the temperature, the more alcohol *and* the more aromatic molecules are churned out by the yeast cells. The cells are able to tolerate higher levels of alcohol—which to them is a waste product—at low temperatures, and will continue to work past the level at which they would fail if warm. For some reason, they also produce more odorous molecules at low temperatures, and since a cool fermentation is also slower than a warm one, it provides a longer time for cells to die, disintegrate, and spill their contents into the must. Long, slow fermentation produces a more pronounced aroma in the finished wine. Two other advantages of cool fermentation are that unwelcome microbes are not as likely to flourish, and that less of the volatile material, including esters and alcohol, will evaporate away. A disadvantage in the case of red wines is that less pigment and tannin are extracted from the skins at low temperatures. In any case, winemakers now try to keep the fermentation going at 70°F (21°C) or below. Because fermentation itself produces heat as part of the energy liberated from the sugar, a cooling system is generally necessary; otherwise the yeast will pasteurize itself. Harvesting in the morning is also a good idea: the grapes are at their coolest then.

White or Red

The conditions of fermentation are varied according to the particular kind of wine being made. In the case of generally delicate white wines, the

juice is removed from the crush immediately, and the free run is often preferred to the pressed juice, whose tannic content and bitterness are not
desired. In warm regions, ordinary white wines are sometimes fermented on
the skins for a short period to improve their sugar-acid balance. Otherwise,
the must is fermented for 4 to 6 weeks at about 60°F (16°C), the quite low
temperature discouraging contamination by bacteria and the production of
easily noticed off-flavors. With the more robust red wines, the challenge is to
manipulate the skins in order to extract the greatest color and flavor. The
must is allowed to ferment at a temperature between 65° and 80°F (18–
27°C) in contact with the skins, which rise in the tank to form a cap. During
the early stages, the cap is punched back into the must at least twice a day.
When enough color has been extracted—after anywhere from 4 to 14 days—
then the must is drawn off the skins, cooled somewhat, and fermented for a
total of 2 to 3 weeks. The remaining mass of skins, or *pomace,* retains 10 to
15% of the must, which is pressed out and treated separately. This fraction
is rich in extracted materials, rough in flavor, and ages very slowly; it is
blended in small quantities with other wines.

Other, more rapid techniques for extracting color and tannins—including the use of heat and of carbon dioxide—are sometimes employed. Rosé
wines can be made in any of several ways: by blending red and white grapes
or wines, by using pale red grapes, or by fermenting normal red grape must
on the skins for a shorter time, perhaps one or two days. Europe has traditionally used the pale Aramon and Grignolino grapes, but these varieties do
not ripen well in California, where red grapes or a blend are generally
preferred.

The Composition of New Wine

Once fermentation is complete, the new wine is drained out of the fermentation tanks to begin the work of clarification and aging, in which the
usually cloudy, rough-tasting liquid develops into a clear, smooth wine. The
character of the grape juice has been transformed by fermentation. Not only
is nearly all of the sugar gone, converted into energy and alcohol, but many
other minor fruit components have disappeared, their place taken by new
yeast products. Here is a summary of the composition of new wine. Of the
sugar, perhaps 0.2% of the original 20% remains: an amount that is imper-
ceptible, and that will not support any significant fermentation during aging.
The alcohol content ranges from 10 to 14%. In addition, traces of methanol—
wood alcohol—derived from the breakdown of pectins in the grape cell
walls, and traces of various higher alcohols created in the yeast from amino
acids, are also present. These substances, known as *congeners,* are like ethyl
alcohol narcotic in large amounts, and in wine may contribute to hangover
as well as to bouquet. Then there is a large class of substances that is covered
by the term *extract content,* or the dissolved substances aside from sugar,
most of which are produced by the yeast. If the extract content is below 2%,
the wine is noticeably thin and watery; usually it is over 3% and contributes
to the wine's body. The acid content is largely due to the grapes themselves,

and it is also important to get right. If it is below 0.4%, the wine will taste flat and dull; if over 0.6%, it is fresh and tart. The ideal pH for dry wines (sweet wines need less acidity) is around 3: the wine tastes better, resists spoilage, and if red, has a better color (the anthocyanin pigment is sensitive to pH; see chapter 4 for a recipe that changes red wine to "white").

Then there is a whole host of minor constituents, present in small fractions of a percent. The most important among them are the tannins. In addition to giving the finished wine a slightly astringent feeling, the tannins contribute some body, stabilize the red pigments, and help clear the young wine of large suspended molecules that cloud it. The darker the wine, the more pigment *and* tannins have been extracted from the skins. White wines generally contain very little of these—perhaps 0.02%—while reds may have 0.1 or 0.2%. Some of the molecules that cloud wine and that tannins remove during aging are proteins, most of them released from dead yeast cells. Large-molecule carbohydrates from the grapes—cell wall constituents and such—also cloud the wine, as can trace minerals, especially iron and copper. Less troublesome, and in fact desirable for its heavy, smooth texture is glycerol, a product of fermentation; it constitutes about 1% of the liquid. Finally, there are the aromatic molecules—esters and so on—and some residual sulfur dioxide which, if present in concentrations over about 200 parts per million, will be detectable as an objectionable odor. To summarize: the new wine contains alcohol, acids, many different flavor- and body-contributing molecules, as well as large molecules and molecular clusters that make the wine cloudy.

Racking and Aging

Racking

Now begins the job of clearing, stabilizing the wine's appearance, and developing its flavor. Before anything can be done about chemical haze, the new wine must be cleared of yeast. This is done by racking the wine: letting the yeast cells and other large particles settle, carefully drawing the wine off, and repeating the process three or four times in the first year of aging, twice a year after that. The gradual, gentle action of gravity prevents any damage to the yeast cells that might release undesirably large amounts of cloudy protein or odorous molecules. Classically, racking is done in 50- to 1,000-gallon barrels at 50° to 60°F (10 to 16°C) for red wine, 32°F (0°C) for white. Today, in order to save labor and money—barrels are expensive—the initial rackings are done in large tanks. Because the young wine does contain a small amount of residual sugar, some fermentation continues and carbon dioxide gas is evolved. To prevent the casks from exploding, special valves, or *bungs*, allow the CO_2 to escape while keeping the outside air, with its oxygen and various microbes, from getting into the casks. Wooden casks allow some liquid to evaporate—the loss is known as *ullage*—and they must continually be topped off to prevent spoilage by oxygen-requiring, acid-producing bacteria (of the genus *Acetobacter*), which would turn the wine into vinegar.

WINE MAKING

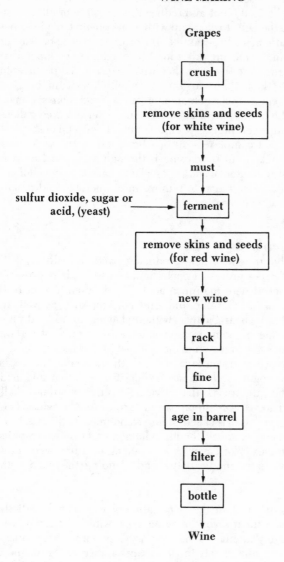

Grapes

crush

remove skins and seeds
(for white wine)

must

sulfur dioxide, sugar or → ferment
acid, (yeast)

remove skins and seeds
(for red wine)

new wine

rack

fine

age in barrel

filter

bottle

Wine

The cool, and in the case of white wines, cold storage temperatures not only suppress microbial activity, but also reduce the solubility of all the dissolved solids, and cause the wine to cloud up with a fine precipitate of various molecules. Excess tartaric acid, for example, forms crystals of potassium or calcium tartrate. Late in the racking procedure, the wine may be *fined:* that is, a substance will be added to the wine that attracts suspended particles to itself, and then settles to the bottom, carrying them with it. Gelatin, isinglass (purified fish collagen), egg white, casein, bentonite clay, and some synthetic materials are used in the proportion of about one ounce per hundred gallons to remove protein-tannin complexes, pectins, gums, hemicelluloses, yeast glucosans, and trace metal complexes. Any particles left in the wine after racking and fining are taken out by passing it through a filter before it is bottled. Today, asbestos, diatomaceous, porcelain, carbon, or cellulose-derived filters can be made so fine as to remove most microbes and practically sterilize the wine without heating it.

Malo-lactic Fermentation

In some wines, the growth of acid-producing bacteria in the cask is actually encouraged. Grapes grown in cool climates with a short growing season generally are picked with too much acid. This is often true in Bordeaux and Burgundy, and in German and Austrian vineyards as well. In order to improve the acid-sugar balance without having to add extreme amounts of sugar, winemakers sometimes encourage the growth of certain species of *Lactobacillus*, which take up malic acid—with tartaric, one of the major grape acids—and convert it into lactic acid (the same one that sours and coagulates milk in yogurt and cheese). While it is true that one acid simply replaces the other, the trick is that lactic acid is half as strong—half as sour—as malic, and so reduces the apparent tartness of the wine. This "malo-lactic fermentation," as it is called, is also sometimes used in California, where it is valued not so much for its reduction of tartness as for its contribution to the complexity of flavor. And inducing it in the barrel prevents it from occurring later in the bottle and clouding the wine with bacteria.

Important as they are, neither the clarification of wine nor the adjustment of its acidity is the primary work of aging. New wine has a raw flavor and a strong, simple, fruity aroma. As the wine sits, a manifold of chemical reactions slowly proceeds, and results in the disappearance of the rawness and the softening and complication of the bouquet. One way to think of this process is to see that the fermentation period has been a time of extreme and rapid chemical change in the grape juice. Once the yeast cells have stopped their work, the many new substances take a while to react with each other and reach some kind of equilibrium. There is one kind of reaction that stands out in importance: *oxidation,* or the combination of oxygen from the air with various molecules in the wine. Oxidation is the key to aging in barrels and in bottles, to the practice of letting some wines "breathe" before serving, and even to their deterioration once the bottle is opened. The first person to understand this was Louis Pasteur.

Pasteur and the Importance of Oxidation

In 1863 the Emperor Louis Napoleon asked Pasteur to study the "maladies" of wine that could cause such economic harm to the country, and three years later he published *Etudes sur le vin* (a second, enlarged edition appeared in 1873). In addition to describing the various microbes that were associated with bad wine, Pasteur discussed the proper development of wine, and came to this conclusion:

> In my view, it is oxygen which *makes* wine; it is by its influence that wine ages; it is oxygen which modifies the harsh principles of new wine and makes the bad taste disappear. . . .

Pasteur found that the making of wine was a slow, gradual process, that the oxygen does not do its work all at once.

> In the initial periods of the experiment, the wine is only flat, often quite disagreeable. Burgundy becomes bitter, loses its bouquet, its color blackens. But these effects are transitory, and in all cases, in order to appreciate the influence of the air, it is necessary afterwards to conserve the wine for a month or two in closed, full vessels, because there take place some considerable changes, generally favorable to the quality of the wine. It would be a great mistake to believe that the effect of oxygen is completed immediately.

But if oxygen slowly makes wine, it can also unmake it. Pasteur noted that wooden casks allow a slow aeration of the wine, but that this must not continue indefinitely.

> There is in the art of making wine a practice which is again directly in accord with the influence of oxygen on wine: that is the bottling. It is necessary to aerate the wine slowly to age it, but the oxidation must not be pushed too far. It weakens the wine too much, wears it out, and removes from red wine nearly all its color. There exists a period, variable for every sort of wine and for one such sort according to the year, during which the wine must pass from a permeable vessel to one nearly impermeable.

The elegantly simple method by which Pasteur illustrated the effects of very little and too much air on wine is easily duplicated today. He filled two small tubes all the way with wine, one red and one white, and filled two others

ACTION DE L'OXYGÈNE DE L'AIR SUR LES VINS ROUGES

Fig. 26

One of Pasteur's experiments demonstrating the influence of oxygen on wine. A full, airtight tube of red wine retains its color indefinitely, while in a tube that is half air, the pigment soon precipitates into a brownish sediment.

only half full, leaving a large pocket of air. He sealed the tubes and set them aside. In a short time, the wine in the half-filled tubes had thrown a sediment, and the red wine had turned brown, the white wine from yellow to brownish-amber. But the completely full tubes retained their color and clarity indefinitely.

Barrel Aging

Our understanding of oxidation has not progressed very far beyond Pasteur's. Oxygen is absorbed through the pores in wood casks, and is taken up first by metal ions in the wine, which then transfer it to various other molecules. The initially unpleasant effect of oxygen that Pasteur described is now called "bottle sickness." Eventually—sometimes long after the wine has left the barrel—a more pleasant population of aromatic molecules predominates, and the tannins and anthocyanin pigments begin to link up to each other, form particles, and precipitate. The loss of tannins reduces the astringency of the wine, and the loss of pigment lightens its color. If enough pigment is removed, the wine looks decidedly brown or tawny, because the brownish tannins are no longer fully masked.

The control of oxidation is a tricky business. The amount of time the wine spends in wood, the surface area of the casks, the frequency of racking and topping off, even the degree to which the wine splashes around when poured: all of these factors influence the amount of oxygen incorporated into the wine. In general, white wines are aged in the cask for 6 months to a year, red wine for a year or two. White or light red wines are generally bottled young with a fairly fresh, fruity bouquet, while more complex white wines and astringent dark reds require a long time and more oxidation to develop and smooth out.

The other factor to be considered in barrel aging is the development of a barrel aroma. The European species of oak traditionally used to make wine casks—*Quercus robur* and *Q. sessilis*—leak into the wine small amounts of vanillin, the major flavor constituent of vanilla which, together with other qualities of the wood, blends with the aromas of grape and yeast. The longer the wine is in contact with the wood, the more pronounced this barrel aroma, which can be overdone. The role of vanillin was discovered after California winemakers found that barrels made from typical American oaks—*Quercus alba* and *Q. lyrata*—didn't impart the same aroma. While we do import some European barrels, most American wine is aged in American oak, which some say interferes less with the qualities of the grape itself, and others say has too strong a character.

Bottle Aging

Wine continues to be affected by oxidation long after it leaves the cask. It picks up some air when it is poured into the bottle, and the bottle is sealed with a small space between wine and cork. So while oxidation is greatly reduced in the bottle, it does continue, and the winemaker must be careful not to splash the wine too much when pouring it or leave too large an air space. In the case of a red wine that should spend years in the cellar, less than ¼ inch of air in the thin neck may be appropriate. White wines and rosés benefit from about a year of bottle aging, during which time the aroma develops and the amount of free, odorous sulfur dioxide decreases. Ordinary red wines improve greatly after a year or two in the bottle, and those made largely of Cabernet Sauvignon grapes may get better for 5 or 10 years. It is important that bottled wine be stored on its side so that the cork stays moist and full and doesn't allow in any more air.

The Cork

Most wines are made by blending two or more different varieties, and this important test of the winemaker's art, about which little can be said, occurs just before bottling. The final wine is then filtered to remove any remaining microbes and haze, and run into the bottles. The corks are steril-

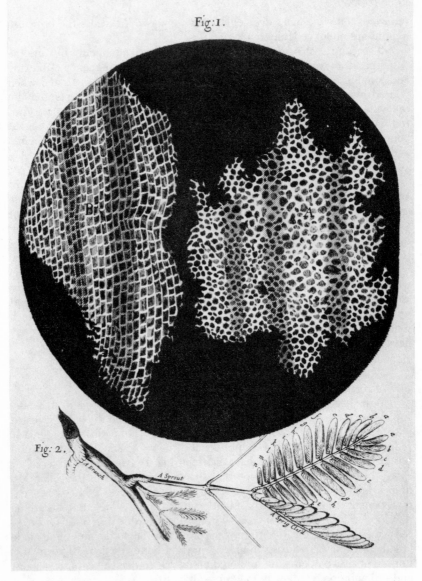

The first microscopical representation of cork, published by Robert Hooke in his *Micrographia* of 1665. This drawing is celebrated because it is associated with the first use of the word *cell* to describe the structure of plant tissue. Hooke likened what he saw to a honeycomb, and realized that the cells, or pores, explained cork's unique properties. "I no sooner discern'd these . . . but me thought I had with the discovery of them, presently hinted to me the true and intelligible reason of all the *Phaenomena* of Cork."

ized and softened in a mixture of sulfur dioxide and glycerol, and, if very porous, they may be waxed. Cork is the outer protective layer of an evergreen oak, *Quercus suber,* that is native to the western Mediterranean; Portugal supplies 80% of the world demand. Where most tree bark is fibrous, cork is composed of tiny air cells, about 0.001 inch across. Over half the volume of cork is empty space, and this is what makes it buoyant in water and so easily compressible. In addition, nearly 60% of the cell walls in cork is made up of *suberin,* a complex waxy substance similar to the cutin that coats many fruits, and this makes cork water-resistant and long-lived. We know from Horace's Third Ode that the Romans used cork to seal wine jars, and from Shakespeare (*As You Like It,* III.ii) that the Elizabethan English did the same, but it wasn't until the 18th century that it became the standard stopper. In the old days, beeswax was used to seal the cork in the bottle to keep it airtight and prevent insects and molds from infesting it. Today, metal foil and, increasingly, plastic are used instead.

The Bottle

And now a word about the bottle. Since the 18th century, certain shapes have become associated with regions and thereby with varieties of grape. The round-shouldered bottle is standard in Bordeaux, and in California holds Cabernet Sauvignon, Zinfandel, and Sauvignon Blanc. The slope-shouldered style is characteristic of Burgundy and in this country of Pinot Noir, Chardonnay, and "Chablis." German bottles are slope-shouldered but taller than the French, ordinary Chianti bottles short and squat (the Chianti classico comes in a Bordeaux-style bottle). And most wine bottles are tinted to screen out wavelengths of ultraviolet light that excite certain molecules in the wine

Traditional shapes of wine bottles. *From left:* **Bordeaux, Burgundy, German, Chianti, champagne**

and can spoil its flavor. Traces of metal oxides in the glass absorb these and other wavelengths and produce the characteristic green or brown tint.

Special Wines

The last few pages have described the making of dry table wines. Sweet, fortified, and sparkling wines are all treated in somewhat different ways. Here is a brief account of their production.

Sweet Wines

Table wines are generally fermented until they are *dry:* that is, until the yeast consumes nearly all the grape's sugar and converts it into alcohol. Some American rosés may be left with a 1 to 5% sugar content, but the European rosés are generally dry. Sweet or dessert wines are made from grapes that accumulate plenty of sugar, and are fermented only up to the point that something more than a desirable sweetness has been reached; the excess sugar is then slowly fermented in the barrel. White grapes—the choicest are Sémillon, Sauvignon Blanc, and Riesling—are used rather than red, since sweetness and astringency don't mix well together (port would seem to be the exception). Because relatively few grapes can ripen to the required sweetness in European districts, those that do often receive special distinction on the label. The German *Spätlese,* or "late-picked," is such an indication, and *Eiswein,* or "ice-wine," made from grapes left on the vine late enough that they freeze, is also a sweet wine. In California, fully ripe grapes are more easy to come by, and sometimes a dry wine is blended with some sweet must to raise the sugar content. Sweet wines may be aged for 2 to 4 years, sometimes more.

The most renowned sweet wines are made with the help of another useful fungus: the mold *Botrytis cinerea* (the Latin means "ashy cluster"), or what the French call *pourriture noble* and the Germans *Edelfäule,* "the noble rot." Under the proper climatic conditions, which are found in the Sauternes district of Bordeaux and the Rheingau in Germany, the mold first infests the grapes during periods of high humidity, thereby injuring the skin. Then, during later dry periods, water evaporates through the holes, producing sugar concentrations in the grape of 30% and more. The longer the grapes can stay on the vine—sometimes into October and November—the drier and sweeter they will be. The Germans call the choicest of these grapes *Trockenbeerenauslese,* or "select dried berries." Because botrytised grapes are rare, yield less than half the normal volume of liquid, and are difficult to ferment, the wines made from them are expensive. In addition to removing water, the mold leaves the grape with more glycerol which, with the sugar, gives the wine a rich, smooth consistency, less protein and acid, and a concentrated, distinctive flavor. California winemakers have had trouble matching the Sauternes and their Rhine equivalents because the climate is too dry for the initial infection, but plantings along rivers in the northern valleys are sometimes adequately humid. The artificial inoculation of picked ripe grapes has been tried, and there are some signs that more concentrated plantings on heavier soils are hospitable enough to support the noble rot in the manner to which it has become accustomed.

Sherries

A sherry can be a strong dry wine, a fortified dry wine, or a fortified sweet wine. The word comes from the name of the Spanish port city Jerez de la Frontera, which was Anglicized to "sherry" around 1600. The climate in southern Spain, like that of California's inland valleys, is very hot and bright—nearly 300 sunny days a year—and the crops differ very little from year to year, so there is no such thing as a vintage sherry. The principal sherry grape is the Palomino, which generally doesn't ripen with enough acidity to make a well-balanced table wine. Sherry must is fermented very fast at first, then very slowly, from September to December, until the wine is dry and reaches an alcohol content of about 14%. At this point, it is evaluated. If the wine has come out fairly light and delicate, it is made into a *fino* sherry. Fino is allowed to develop a special floating yeast composed mainly of *Saccharomyces beticus* and called the *flor*, which caps it in partly filled barrels, prevents premature oxidation by the air, but itself partly oxidizes some of the alcohol to the nutty-flavored compound acetaldehyde. Fino is then bottled and drunk very young. Because of its low acidity, it doesn't travel well, and is seldom found outside of Spain. Exported finos are usually fortified with grape alcohol to about 18%. Aged fino, which is more stable but stronger in the *flor* flavor and darker in color, is called *amontillado*, but so, confusingly, are blends of moderate quality.

If the new dry wine is heavy in character, it is destined to become an *oloroso* (Spanish for "fragrant," "perfumed"). The development of a *flor* is prevented by fortifying the wine immediately to 18%. The sherry is then aged in wood in a special system called the *solera*. This is a series of casks, each initially filled with the new wine of a particular year. As the water and alcohol evaporate, each cask is topped off by wine from the cask of the next year: the 1970 cask from the 1971; 1971 from 1972, and so on. What this system does is to blend the wines as they age, and to produce a final wine of consistent quality. In this dry climate, the wine loses more water than alcohol through the somewhat porous barrel as it ages, eventually reaching a strength of about 20% alcohol; and because it ages at a relatively warm temperature— the barrels are not cooled—it develops a characteristic concentrated, oxidized flavor that somehow suits it. The oloroso can be bottled as it comes from the *solera*—fortified but dry—or it can be sweetened with special wines often made from sun-dried Pedro Ximenez grapes. Cream sherries are especially sweet olorosos. In California, sherry wines may be "baked" to develop the flavor, or cultured with a "submerged flor": the wine and flor yeast are kept in large tanks and agitated, with air bubbled in to replace the normal contact of yeast with air at the wine surface. This technique produces more sherry in a much shorter time than the solera system.

Madeira

Like sherry, madeira was known to Elizabethan England, and like oloroso it is fortified, usually with brandy. It is made on the island of Madeira, where wines were first fortified as a practical matter to preserve them on long sea voyages, and to prevent the yeast from consuming all the sugar. Gradually, it was realized that the combination of brandy and long barrel aging in extreme kinds of weather produced a distinctive wine indeed. By 1700 ships were being sent to the Indies and back just to age the barrels of Madeira stored on board. In these less extravagant days, the new wine is sweetened, fortified, and held at 120°F (49°C) for four or five months, until it develops a distinctive note of caramel. It is then aged in a solera system. Madeira differs from sherry in being made from more acidic grapes, and has a tang that most other fortified wines do not.

Port

Today port is a sweet, fortified wine. Originally, the word stood simply for Portuguese wines, and it was not until about 1850 that the addition of brandy became regular. The red grapes are grown in a mountainous region called the Alto Douro, while the wine making goes on near Oporto, on the coast. Port is made by stopping the fermentation while about half the grape sugar is left, and pouring the wine into a barrel already a quarter full with brandy. The result: a strong, sweet wine. But because the fermentation is so brief, the grapes must be specially treated in order to extract the right amount of pigment, flavor, and tannins from the skins. Traditionally, this is accomplished by 12 hours of treading with the human foot. A recent innovation is to pump the fermenting wine continually over the cap of skins for several days. In order to avoid excessive oxidation, however, this must be done not with air pressure but with carbon dioxide. Once the wine is run off into the barrel, it is aged for anything from 2 to 50 years. Vintage port is made from especially good years—usually one in every three—and is bottled after two years, then aged in the bottle for perhaps 20 before the wine and brandy properly "marry." Tawny port, so named for its brown color (the result of precipitation of the red pigments), is aged longer in the barrel and oxidizes faster in that permeable container. Because so much of the aging of vintage port goes on in the bottle, the wine usually throws a good deal of sediment and requires very careful decanting. In order to ease the consumer's labor, the vintners have developed "late-bottled" port, which is aged in wood for 8 years before bottling and leaves little precipitate in the bottle. Because they are more alcoholic and in most cases sweeter than table wines, sherry, Madeira, and port keep well even after the bottle is opened. A good thing, since a little of each goes a longer way than table wine does.

Champagne

Sparkling wines can be the result of accident or design. In either case, they are bubbly because there has been some fermentation of residual sugar (or malic acid) in the bottle, and the carbon dioxide has been trapped by the tight seal of the cork. Accidental or incidental carbonation is always fairly weak, noticeable as a slight fizz in the glass and a slight tingling on the tongue. In the case of intentional secondary fermentation—and champagne is the prototype—the winemaker gives the yeast enough sugar to build up a gas pressure of 40 to 60 pounds per square inch at 50°F (10°C). It was in 1836 that a French pharmacist named François published a method for determining the sugar content of a wine, and correlated the amount of added sugar with the pressure of carbon dioxide built up from it by the yeast. Strictly speaking, champagne is the sparkling wine made in the area east of Paris that once bore that name. In the United States, the term is allowed for white wines that undergo secondary fermentation in a container of less than one gallon capacity—usually the bottle itself. In France, the major champagne grape is the Pinot Noir, which is preferred despite its red skin because its growing season is short enough that it almost always ripens to the right sweetness in this northernmost of France's major regions. The white Chardonnay grape is usually part of the blend. In California, the white Folle Blanche and French Colombard are also used. Ever since the time of Dom Pérignon, champagnes have been blended from several grapes, and several wines of different vineyards and ages.

The first stage in making champagne is to produce the base wine, which should have no sugar left and an alcohol content of about 12%. Much below this level, the wine will hold less carbon dioxide; much above it, the secondary fermentation will be inhibited. The base wine is then quickly clarified by cooling, settling, and racking. Because champagne is often served colder than other wines, it must resist the formation of haze even at 32°F (0°C), and so is stabilized at that temperature. Now comes the secondary fermentation, which must be carried out in a closed container in order to retain the gas (ordinary wines are fermented in an open vat, and the carbon dioxide simply bubbles to the surface and escapes). Sugar is added to the dry base wine as food for the yeast; between 3 and 4 tablespoons per gallon of wine will produce the desired gas pressure. Several different yeasts can be involved in the secondary fermentation, but all specialized champagne yeasts have in common the tendency to flocculate readily and so form a cohesive sediment that is easily removed from the bottle. One of these yeasts is *Saccharomyces carlsbergensis* (also called *S. uvarum*), a "bottom yeast" used in making lager beers. In the traditional method, the wine, sugar and yeast are put into individual bottles, corked, clamped, and kept at about 55°F (13°C) for 6 months. The cool temperature reduces the activity of other microbes, gives a cleaner flavor, and permits more carbon dioxide to be dissolved in the wine (this gas is most soluble in water at low temperatures, a fact that figures in the maple sap run; see chapter 8). The bottles must be quite thick to withstand the pressure raised during fermentation. Even then, breakage is not uncommon, and workers must wear masks and gloves when checking the progress of the wine.

Once the fermentation is over, the wine is left to age in contact with the yeast sediment for anywhere from 9 months to 5 years. During this time, most of the yeast cells die and their contents give the wine a distinctive flavor. The cold temperature and gradualness of the process prevent it from spoiling the wine rather than enhancing it. After aging on the yeast comes a clever process called "disgorging," whose purpose is to remove the sediment without losing much wine or gas pressure. First, the bottles are laid at a slant, the cork end down, and the temperature lowered to 25°F (−4°C). Then, over a period of a month or so, the sediment is concentrated in the neck of the bottles by "riddling" them, or periodically twirling them back and forth, replacing them slightly to one side of their original position. The sediment slowly walks down the sides of the bottles without mixing too much with the liquid. When the lees have gotten down to the cork, the "disgorger" freezes the neck of each bottle into a solid plug of sediment and a small amount of wine. He removes the cork, the gas pressure in the rest of the bottle forces the plug out, and the wine is thus clarified. After disgorging, the bottle is topped off if necessary, and the wine is usually sweetened with a small amount of aged wine mixed with sugar and brandy. The bottle is then recorked.

There are no strict definitions, but *brut* (or *nature*) champagne is generally less than 1.5% sugar, *sec* from 1.5 to 3%, *demi-sec* 3 to 6%. Brut champagne is especially esteemed because sugar can be used to obscure flavor defects in the original wine. Champagne needs little additional aging in the bottle before it is drinkable, though it may continue to improve somewhat over several years. The storage life of champagne is on the order of ten years, the limiting factor being the slow loss of carbon dioxide around the cork.

Wines that are labeled "champagne-style" or "champagne-type" are fermented for the second time not in individual bottles, but in large tanks. This innovation clearly saves a great deal of labor, but it changes the contact between wine and sediment and results in a wine of lesser quality. Because the sediment gets so deep, a series of anaerobic reactions that produces the odorous gas hydrogen sulfide takes place in a few weeks, and so the period during which the wine can be left in contact with the yeast is reduced drastically. A technique developed in Germany in the 1950s eliminates the riddling process by transferring bottle-fermented wine to a tank of 500 to 1,500 gallons, from which it is filtered back into bottles. This has the disadvantage of losing some carbon dioxide and allowing enough aeration that sulfur dioxide or ascorbic acid must be added to discourage oxidation. "Pop" sparkling wines are carbonated rather than twice-fermented.

APPRECIATING WINE

Apart from its physiological effects, our fascination with wine is probably attributable to the intricacy of its making, the diversity of results, and the fugitive pleasure that it brings. The varieties of grape, the soil they were grown in, the weather that year, the yeasts that ferment them, the skills of the winemaker in handling them, the years they spend in oak or in glass: all these elements and more enter into the quality of a wine and bear consideration. And because to analyze and enjoy a wine, to have it, is necessarily to lose it, the experience must be an intense, concentrated one. Wine connoisseurship, with its special lingo and inside information and quota of snobbery—experience takes money—can be as off-putting as it is appealing. Many of us would probably be content with the five F's proposed 800 years ago in the Regimen of Health from the School of Salerno:

Si bona vina cupis, quinque haec laudantur in illis:
Fortia, formosa, et fragrantia, frigida, frisca.

If you desire good wines, these five things are praised in them: Strength, beauty, and fragrance, coolness, freshness.

On the other hand, a more exact vocabulary of the basic characteristics of wine is useful for identifying and communicating what it is that pleases or displeases us in a particular bottle.

The Grapes

First there is the matter of the grape varieties used, and the basic question of whether the wine is red or white. French labels rarely indicate the kinds of grapes used, though particular regions of origin, which are shown— Chablis or Beaune or Sauternes—generally signify that varieties traditional to those regions have been used. On the other hand, many California wines are sold as "varietals": a Chardonnay, for example, or a Zinfandel. This usually indicates that certain highly regarded grapes constitute the greater part of the wine, and implies that the unique characteristics of such grapes will be noticeable. In fact, a varietal label guarantees only that a minimum of 75% of the grapes used are of that variety: and its qualities can be lost in the blending or other processing. Nor is an unblended wine necessarily better than a blend; nearly all wines, including the greatest, are blends of several varieties. Complexity and balance in flavor almost necessitate blending.

Of course, there are fundamental differences between white and red wines, regardless of whether fine or mediocre grapes went into them. White wines, with less phenolic and flavoring materials extracted from the skins, are more delicate and less complex in flavor than the reds, and so need less time in oak or bottle to develop. They also tend to be somewhat more acidic, and it may be this tartness that makes them seem the best partner for fish. White wines are served cooler than reds at least in part because warmth and tartness seem less suited to each other, and because a higher temperature will not liberate many more aromatic molecules from a white wine. In addition, if an excess of sulfur dioxide has been left in the wine—it is used more heavily with whites in order to prevent oxidation of the light yellow pigment— it is less noticeable at lower temperatures. The aroma of red wine, on the other hand, is noticeably more full at 60° to 68°F (16° to 20°C) than it is at 50° to 55°F (10° to 13°C) (the "room" and "cellar"temperatures of old that are recommended for red and white wines respectively; contemporary rooms and refrigerators usually deviate from these ranges). And given the complexity and astringency that come with long contact with the skins, red wines profit from the long, slow oxidation and marrying of flavors that develop in barrel and bottle aging. There are exceptions—sweet white wines profit from longer aging than most dry ones, and light, fruity reds like Beaujolais are best when young. But the general rule does hold good.

Clarity and Color

The appearance of a wine can give some important clues about when and how it was made and how it will taste. Regardless of color, if the wine is cloudy and will not settle, it has probably been contaminated by wild yeast or bacteria or by trace metals. In any case, the flavor is likely to be off. Tiny crystals (which *do* settle) are usually salts of excess tartaric or oxalic acid, and are not signs of spoilage. White wines actually range in color from straw yellow to deep amber. The darker the color, the older the wine—the yellow pigments quercetin and quercitrin turn brownish when oxidized—and underripe grapes may give the wine a slight greenish tint, though normally chlorophyll is destroyed during fermentation. Dessert wines like sherry and Madeira begin their life with a brownish tinge, the result of oxidation at high temperatures during aging, and may actually lighten with age as the pigments slowly precipitate. Most red wines retain a deep, rubylike color for about 5 years. As they age further, the anthocyanin pigments complex with some of the tannins and precipitate, leaving more of the brownish tannins visible, and the wine develops an amber or tawny tint.

Feeling and Taste

When we savor a wine, both touch and taste come into play. The *feel* of a wine is largely a matter of its astringency and viscosity. Astringency—the word comes from the Latin for "to bind together"—is the sensation we have when the tannins in wine "tan" our tongue and cheeks the way they do leather: they cross-link with the surface proteins and cause the tissues to contract slightly. This dry, constricting feeling, together with the smoothness and viscosity caused by the presence of alcohol, glycerol, other extracted components, and in sweet wines sugar, create the impression of the wine's body. In strong young red wines, they can be palpable enough that "chewy" seems a good description. Astringency is considered a defect in white wines.

The *taste* of a wine is mostly a matter of its balance between sweetness and sourness, though phenolic compounds in red wines can contribute a bitter note. Fructose and glucose are the predominant sugars in unsweetened wines, and one can begin to detect them at levels around 1%. In addition, both glycerol and alcohol itself are slightly sweet. Sweetness is sometimes used to mask the undesired tastes that result from a poor batch of grapes or problems with fermentation. The acid content of a wine is important in preventing it from tasting bland or flat, though too much acid will make it unpleasantly tart.

Aroma

The strongest impression wine makes on the senses is an olfactory one: the range of volatile molecules that reaches the nose is what fills out the flavor of a wine. It is in order to get more of these molecules into the air and up to the nose that we swirl the wine briefly in the glass before sniffing it and, less genteel, swish it around in the mouth before sniffing *out*. It has been estimated that wine contains several hundred different organic molecules, of which 200 have an odor: so it goes without saying that the classification of wine flavors is not simple. The volatile component of flavor has many names—aroma, bouquet, nose—and these terms mean slightly different things to different people. It is helpful, however, to divide the aromas of wine into those contributed by the grape itself, and those that are the result of fermentation, aging, or other treatment. It takes an experienced nose to detect the presence of Cabernet Sauvignon or Chardonnay, but most of us are able to make at least one such discrimination. All native American *Vitis labrusca* grapes, the best-known being the Concord, have an unmistakable aroma, usually called "foxy," which is due to the presence of methyl or ethyl anthranilate. Two other fairly obvious grapes are the muscat, the base for Asti spumanti, and the Gewürztraminer, both of which carry linalool, a distinctive, floral-smelling molecule also found in cinnamon and in orange flowers.

Then there are the odors that arise from the process of wine making, the host of volatile molecules produced by the growth or death of yeast cells, contact with wood, and slow oxidation in wood or bottles. These are the substances that make wine taste different from the fresh grapes. The outright yeasty smell—the odor of rising bread—usually disappears during clarification and aging, but the yeast live on in the complex of chemicals, especially esters and higher alcohols, that react with each other and with oxygen and blend in cask and bottle. With experience, the role of wood and the effects of bottle aging can be detected, as can a whole range of undesirable odors— among them, sulfur dioxide, and those called vinegary, moldy, corked (the result of *Penicillium* molds infecting the cork)—that are the result of faults in the grapes or in processing.

It is generally thought that the aroma of most wines, and especially young red wines, improves during a period of "breathing" after the bottle is opened. Simply opening the bottle and letting it sit exposes only a tiny fraction of the wine to the air and accomplishes next to nothing; the best way to aerate a wine is to pour it into another container. Breathing does help clear the bouquet by allowing some off odors and excess sulfur dioxide to escape, but it probably allows some desirable aromas to escape as well. The exposure to relatively large concentrations of oxygen can also bring about noticeable chemical changes. Young red wines may very well profit from some accelerated oxidation to make up for a short aging period. The best rule is to taste the wine after it is opened; if it is harsh, let it breathe for a while. In general, the younger the wine, the longer the desirable breathing period. Stories are told of young, intense wines improving for hours before they are served, and of century-old wines that turn stale, brown, and watery minutes after being decanted. The same factors are involved in how well a half-full bottle of wine takes to being kept in the refrigerator for a day or so. Dark, heavy wines are generally less hurt by oxidation in the opened bottle than light, delicate ones. The sweet fortified wines, which are often intentionally and heavily oxidized to begin with, are well protected against long-term spoilage by their high content of sugar and alcohol.

BEER

Wine and beer are distinguished from each other by their raw materials: wines are made from fruits, primarily the grape, while beers are made from grains, primarily barley. There are exceptions; sake is usually called a rice wine. The most important difference between the two raw materials is the form in which they store energy. Grapes use sugars and organic acids and are therefore sweet and sour, while mature grains stockpile starch, which is largely tasteless. Starch is also a form of sugar that yeasts cannot exploit, and this means that grain requires an extra step or two in the brewery. In order to break most of the starch down to single- and double-unit sugars which the yeast can feed on, the grain is allowed to germinate to a limited extent, and so produce starch-breaking enzymes that are designed to supply sugars to the growing seedling. The germination is arrested by drying out the grain, and the resulting *malt* is roasted to generate the color and flavor that it otherwise lacks. When it comes time for brewing, the malt is soaked in warm water to convert the starch to glucose, maltose, and some longer chains of glucose molecules. While it is true that grapes are much more easily fermented—as soon as they break open, the yeasts will begin to flourish in the sweet juice—grain has several definite advantages as a material for brewing. It is easier to grow, much more productive in a given acreage than the vine, and can be stored, either raw or malted, for months before being fermented.

The Evolution of Beer

We have seen Pliny's testimony that the barbarian provinces drank beer while the Romans drank wine, and in the centuries following the fall of Rome, beer continued to be an important beverage in much of Europe. Monasteries brewed it for themselves and for near settlements, and it may very well have been the most healthful liquid around. Many waterways were contaminated by human wastes, but the water in beer was boiled after soaking the malt and before fermentation, and the final beer was sweet and alcoholic enough to discourage many harmful microbes. By the 9th century, alehouses had become common in England, with individual keepers brewing their own. Until 1200, the English government considered ale to be a proper food, and did not tax it.

Hops and Lager

It was later in the Middle Ages that two great innovations made beer largely what it is today: hops were used to preserve and flavor it, and lager, or "bottom-fermented" beer was developed. For centuries, a mixture of spices had generally been added to beer both to give it flavor and to help delay spoilage. In Germany this mixture, called *gruit*, included bog-myrtle, rosemary, yarrow, and other herbs. Coriander was also sometimes used, and juniper in Norway. It was around 1100 that hops, the resinous cones of the vine *Humulus lupulus* (a relative of *Cannabis*) came into use, and by the end of the 14th century it had replaced *gruit*-like medleys in most of northern Europe. Hops made possible weaker beers by doing some of the preservative work of sugar and alcohol, and gave beer its familiarly bitter taste. Bottom fermentation, a process with some technical advantages over the traditional top fermentation—"bottom" and "top" indicate where the yeast collects in the tank—seems to have been developed around 1400 in a Bavarian monastery, and eventually gave the world a lighter-flavored brew. Unlike hops, this technique remained distinctly Bavarian until the 1840s, when the special yeast and techniques were taken to Pilsen, Czechoslovakia, and to Copenhagen. Pilsner lager, with its lightly roasted malt, is the prototype of most modern beers. England and Belgium are the only major producers of beer to persist with top fermentation.

England: Ale vs. Beer

England was slow to accept hops. Because northern Europeans made their hopped beer lighter, the English suspected hops to be an adulterant that disguised or encouraged poor quality. Although hops had been introduced to the island sometime in the 15th century, it was in 1524 that a group of Flemish immigrants settled conspicuously in Kent and began growing hops. A popular couplet of the time summarized the English attitude toward northern European innovations: "Hops, Reformation, Bays, and Beer/Came into England in one bad year." ("Bays" is baize, a kind of cloth.) Beer, a word of Germanic origin (but ultimately from the late Latin *biber*, "drink"), denoted a foreign beverage that by definition was hopped. The traditional, stronger English brew was *ale*. So John Taylor, the eccentric "Water Poet," wrote a silly pamphlet in 1651 called "Ale Ale-vated into the Ale-titude," and defined beer as "a Dutch Boorish liquor, a thing not knowne in *England,* till of late dayes an Alien to our Nation, till such time as Hops and Heresies came amongst us, it is a sawcy intruder into this Land. . . ." Laws were passed by some cities, including London in 1484, forbidding the addition of hops to ale. But by 1493 there were separate English guilds for ale brewers and beer brewers. And Reynold (or Reginald) Scot noted in 1574, in *A Perfite Platforme for a Hoppe Garden*, that the advantages of hops were overwhelming: "If your ale may endure a fortnight, your beer through the benefit of the hops, shall continue a month, and what grace it yieldeth to the taste, all men may judge that have sense in their mouths." Still, it was not until about 1700 that ale was hopped as a matter of course and ale and beer became pretty much identical.

The English tardiness in accepting hops was balanced by their pioneering in the making of bottled beer. Ordinary beer or ale was fermented in an open tank, and like wine, lost all its carbon dioxide to the air: the bubbles simply rose to the surface and burst. Some residual yeast might grow while the liquid was stored in a barrel, but it would lose its light gassiness as soon as the barrel was tapped. Sometime around 1600, it was discovered that ale kept in a corked bottle would become sparkling. Quite early on the discovery was attributed to Alexander Nowell, Dean of St. Paul's Cathedral from 1560 to 1602. Thomas Fuller, in his 1662 *History of the Worthies of England*, wrote:

> Without offense it may be remembered, that leaving a bottle of ale, when fishing, in the grass, he found it some days after, no bottle, but a gun, such the sound at the opening thereof: and this is believed (casualty is mother of more invention than industry) the original of bottled ale in England.

In any case, by 1700 glass-bottled beer sealed with cork and thread had become popular, just as champagne had during the same period, but both were largely novelties. Most beer was drunk flat, or close to it, from barrels. With the development of airtight kegs, of carbonation (around 1900), and the increasing tendency to drink beer at home instead of at the tavern, bubbly beer has become the rule.

Adulteration and Innovation

In 1516 the Duke of Bavaria published a *Reinheitsgebot*, or Edict of Purity, according to which the only ingredients allowed in beer were water, barley malt, yeast, and hops. That edict remains in effect today for the whole of West Germany, though only Bavaria holds to it for exported beers. England, however, never enforced such strict standards, with the result that the 18th century saw wholesale adulteration of beer and ale. Coriander, red pepper, tobacco, licorice, opium, ginger, linseed, molasses, the narcotic seeds of *Cocculus indicus*, various metal salts that improved the foam: all these are mentioned by contemporary writers as common additives. In 1821 the journalist and politician William Cobbett wrote in his *Cottage Economy* that the English workingman was being poisoned by both the beer of public brewers and the tea that many were now drinking instead. Forty years earlier, according to Cobbett, practically every household brewed its own beer, but taxes on malt had made this impossible, and had encouraged adulteration as a way of holding costs down. The malt tax, said Cobbett, "has done more harm to the people of England than was ever done to any people by plague, pestilence, famine, and civil war."

At the same time, the 18th century was a period of genuine innovation, and it was then that many of today's familiar British names—Whitbread, Worthington, Younger, Ind, Courage, Charrington, Guinness—got their start. By 1750, the greater control that coke and coal gave the maltster over roasting the malt made pale and amber brews possible. And because these drinks were light enough to see through, clarification became an important issue for the first time: suspended yeast and other particles couldn't be detected in a dark brown ale. Stout existed as a very strong beer, and porter, a dark beer made with extra hops and aged about six months, was the premium brew. Late in the century, the thermometer and hydrometer made it possible to control the different stages of the brewing process more closely; the introduction of steam and mechanization did the same. And in 1817, "patent malt" was developed. This was a relatively small amount of malt roasted just to give the desired color and body to beer. The darker malt is roasted, the less enzymes and starch remain in a condition to produce sugar for the yeast, so that if nothing but very brown malt is used, large amounts are necessary and the resulting beer is quite heavy. Patent malt was the beginning of porter and stout as we know them today: certainly darker and heavier than ordinary beer, but much lighter than they were 200 years ago. Since 1900, a tax system in England that gives an advantage to weaker beers, together with compulsory weakening during two world wars, and the pasteurization and refrigeration that make high alcohol and sugar levels unnecessary for good keeping, have brought about a general trend toward less alcoholic and less caloric beer.

Beer in America

If any country is known for its light—some would say characterless—brews, it would be the United States. This national preference would seem to be the result of our climate and history. Heavy beer is less refreshing when the summers get as hot as ours do. And the colonists weren't able to transplant refined English techniques directly to this country. As Robert Beverley described them in his *History and Present State of Virginia* (1705), Americans relied for their raw material either on England or on barely acceptable substitutes:

> Their richer sort generally brew their Small-Beer with Malt, which they have from *England*, though they have as good Barley of their own, as any in the World; but for want of the convenience of Malt-Houses, the Inhabitants take no care to sow it. The poorer sort brew their Beer with Molasses and Bran; with *Indian*

> Corn Malted by drying in a stove; with Persimmons dried in
> Cakes, and baked; with Potatoes; with the green stalks of *Indian*
> Corn cut small, and bruised; with Pompions [pumpkins]; and
> with the *Batates Canadensis*, or *Jerusalem Artichoke*. . . .

There was, then, no strong national tradition in the matter of beer, and in fact the majority of Scots and Irish who arrived late in the 18th and early in the 19th centuries probably preferred whiskey to beer in any case. So the way was clear for German immigrants to set the taste. In 1844, Frederick Lauer, a second-generation brewer in Reading, Pennsylvania, introduced the newly revealed lagering technique to the United States, and the distinctive brew caught on. Both Milwaukee and St. Louis were quickly established as centers of lager brewing: in the former, Pabst, Miller, Schlitz, and Blatz; in the latter, Anheuser and Busch; and Stroh in Detroit, Hamm in St. Paul, and Heileman in LaCrosse, all got their starts in the 1850s and 1860s. These names, their light, Pilsner-style beers, and now their "low-calorie" beers, remain dominant today, and the darker, richer brews of the English colonists appeal only to a small number of beer connoisseurs. The only indigenous, American style of beer we have is "steam beer," a rare relic of the California Gold Rush. Without the large supply of ice necessary to make lager beer, San Francisco brewers used the yeast and techniques appropriate to cool bottom fermentation, but brewed at top-fermentation temperatures. The result: a very gassy beer that gave off a lot of foam, or "steam," when the keg was tapped.

Today, most American beer is blander and more uniform than ever. The number of breweries in this country has fallen from over 2000 to less than 50 in the last century, and formulas include less malt and hops than they did 20 years ago. The logic seems to be that the more beer resembles water, the more of it we can and will drink.

How Beer Is Made

The brewing of beer can be divided into several stages. First, dry barley kernels are soaked and allowed to germinate in order to accumulate the starch-digesting enzymes, or amylases. Second, the partly germinated kernels, or malt, are dried to halt the enzyme activity, and kilned until they reach the proper color and flavor. Third, the malt is mashed in warm water, which revives the enzymes and results in a sweet, brown liquid called the *wort*. Fourth, hops are thrown into the wort, and the two are boiled together. This treatment extracts the hop resins that flavor the beer, inactivates the enzymes, kills any microbes present, and deepens the color of the wort. Fifth, the wort is fermented with yeast until the desired levels of sugar and alcohol are reached. Sixth, the new beer is filtered to remove most of the yeast, and then aged for some time. Finally, the finished beer is clarified, filtered again, sometimes pasteurized in bulk or bottles, and packaged for sale.

Germination

Beer begins with barley. Oats, wheat, corn, millet, sorghum, and other grains have also been used, but barley has become the grain of choice because it is best suited for malting. Each part of the barley grain except for the outer bran (which contributes tannins to the liquid during mashing) is involved in malting. As the embryo absorbs water, its cells begin to start up their biochemical machinery and grow. They produce various enzymes, including those that break down starch, proteins, and cell walls. These enzymes diffuse into the endosperm, where the embryo's food supplies are stored, break into its cells, and attack the starch granules and protein bodies. The embryo also secretes the hormone gibberellin, which slowly reaches the cells of the aleurone layer, just below the bran. Gibberellin stimulates the aleurone cells to produce digestive enzymes as well, and after about 5 days, aleurone enzymes are eating away at nearby endosperm.

Malting

The maltster's aim is to maximize the production of starch-splitting enzymes and the weakening, or "modification" of the endosperm, while minimizing the barley embryo's growth, which uses up sugar and so the yeast's future food supply. Modification has generally been completed by the time that the growing tip of the embryo reaches the end of the kernel, some 5 to 8 days after the grain is first soaked. If the shoot actually appears and grows out of the end, the embryo is consuming too much sugar. Even when the malt is kilned at just the right time, about 20% of the original starch has been lost to the embryo's growth. Malting is traditionally carried out in a 6-inch layer of barley on the floor of a large room, where it can easily be raked to keep it aerated and moist, but not wet. Today, continuously revolving drums are commonly used instead. It used to be that malting was restricted to the cool seasons, since heat increases the embryos' use of sugar and the possibility of spoilage by molds, but maltsters can now control the temperature and keep it at about 65°F (18°C). At slightly lower temperatures even more enzyme is conserved, but the time lost in slower germination is too costly.

Kilning

Once the barley has been malted, it is heated and dried. Dehydration kills the embryo, which stops consuming sugar, and inactivates the enzymes, whose natural medium is water. As long as the temperature is not raised too high, most of the enzymes survive intact and begin to work again when the malt is soaked in water. However, it is the kilning of the malt that contributes a large part of the beer's final flavor and color, and the higher the temperature, the more flavor and color compounds are produced via the browning reactions (see chapter 14). A 175°F (79°C) kilning produces a light malt that is appropriate for pale ales and lagers. In the United States, where very pale

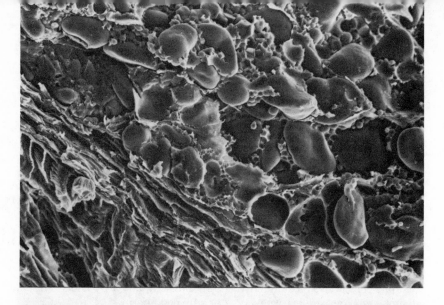

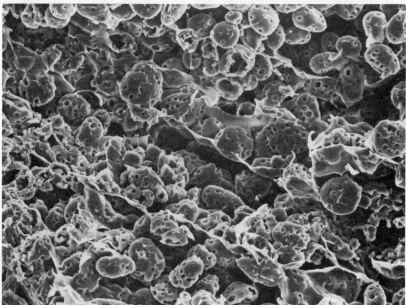

The modification of barley endosperm near the embryo during germination, seen through the scanning electron microscope. Mature barley kernel *(upper left)* has intact starch granules cemented into a protein matrix; sheetlike cell walls are also visible. After 48 hours of germination *(lower left)*, the protein matrix and cell walls have been degraded, and the starch granules pitted by the action of various enzymes. After 72 hours *(upper right)*, little or no protein matrix or cell wall material remains, and the starch granules have been digested extensively, revealing the rings in which the starch molecules had been deposited. Malting causes the barley to accumulate starch-digesting enzymes and to make its endosperm more vulnerable to attack by those enzymes during mashing. (Micrographs courtesy of A. W. MacGregor, Grain Research Laboratory, Canadian Grain Commission, Winnipeg, Manitoba.)

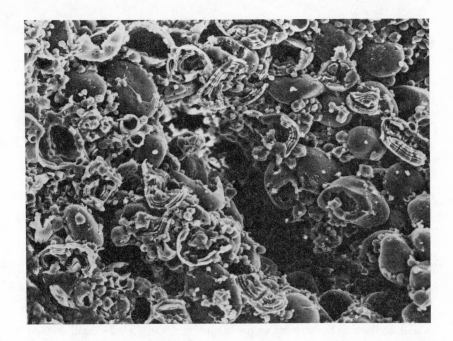

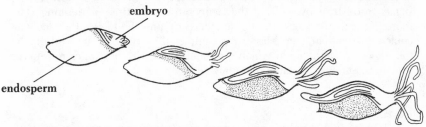

embryo

endosperm

Four stages in the malting process, the partial germination of barley kernels that develops starch-digesting enzymes. The shading indicates the progress of endosperm modification. Malting is stopped when the growing shoot just reaches the tip of the kernel. After this point, the embryo begins to use its starch up more rapidly, and so deprives the brewer and his yeast of the raw material for producing alcohol.

malt is desired, it is sometimes bleached with sulfur dioxide during kilning. At the other extreme, temperatures of 200° to 220°F (93°–104°C) will produce darker brown colors and more intense flavors, but the resulting malt will have correspondingly less enzymatic power to convert starch into sugar, and will contain less usable starch. The "chocolate" or black malts that give stout its sweet, acrid, heavy character cannot supply enough sugar to feed the yeast adequately, and must be supplemented with large amounts of blander malt. Once the malt has been kilned to the desired point, the now brittle rootlets are broken off and separated (they may be used for animal feed). The dried kernels can now be stored for several months until they are needed, when they are ground into a coarse powder.

Mashing

In the stage known as mashing, the ground malt is soaked in water at between 130° and 150°F (54°–66°C) for a couple of hours. The water, constituting as it does about 90% of the finished beer, has a definite influence on its quality. Burton-on-Trent, Munich, Pilsen, Dortmund, and other brewing centers were famous for their waters, and today most brewers carefully treat theirs. More of the malt can be extracted in soft water than in hard; a high carbonate content is helpful for dark beers; calcium ions are better for the taste than sodium or magnesium; traces of iron or copper cause a haze to develop: these are some of the considerations involved. In the process of mashing, much of the malt substance is leached into the water, and its composition changes drastically as the enzymes act. Fully 85% of the carbohydrate in malt was starch, while in the liquid wort, 70% is now in the form of various sugars: small amounts of glucose, fructose, sucrose, some 3-sugar remnants of starch, and about 40% maltose, or two-glucose molecules. Most of the remaining carbohydrates are the so-called *dextrins*, or chains of from 4 glucose units to hundreds. The yeast cannot feed on the dextrins, but these long molecules, together with some protein, give the final beer its full body and contribute to the stability of the foamy head. Only a very small amount of starch is left in the wort. The brewer can adjust the ratio of fermentable sugars to dextrins by varying the temperature and time of mashing. The more the enzymes work, the lighter the body and the higher the alcoholic content of the final beer. Mashing is completed by running the wort off the solid remains of the malt, which are then rinsed with hot water to remove some remaining extractable materials.

Cereal Adjuncts

Making the wort with nothing but malt and warm water is the "traditional" method, and is still standard in Germany, where the *Reinheitsgebot* still holds sway. The U.S. definition of beer, on the other hand, is a beverage "brewed or produced from malt, wholly or in part, or from any substitutes thereof." In this and most other countries, "adjuncts" are commonly added to the liquid to boost its carbohydrate content. Plain sugar is one obvious choice since it can be used directly by the yeast, and sugar has been added to beer wort for hundreds of years. More recently, as improved malting methods made possible high yields of enzymes in pale malts, cereal adjuncts—ground or flaked rice, corn, wheat, or barley—have been used as a cheap supply of starch, which the large amounts of malt enzymes break down to sugars and dextrins as if it were the barley's own. Both sugar and adjuncts lower the amount of malt needed, and so the brewer's production costs. They have the disadvantage, however, of contributing no flavor or color to the wort and are therefore mostly limited to pale, mild beers. Generally, British and Australian brewers are much more likely to use sugar than are Americans, who use equal amounts of malt and cereal adjuncts. Typical compositions of the *grist*, the solids added to the water during mashing, for three different English beers are listed below.

| ENGLISH BEER INGREDIENTS (PERCENT) | | | | | | |
	Black Malt	Mild Malt	Lager Malt	Sugar	Caramel	Wheat Flour	Flaked Maize
Mild Ale	2	73		15		10	
Malt Stout	4	93			3		
Lager			78				20

Hops

Once made, the wort is filtered through the malt husks and then boiled with the hops or, in this modern age, sometimes with hop extract, which is much more concentrated and so takes up less storage space. The hops are the female flowers, or cones, of a Eurasian-American vine, *Humulus lupulus*, which bears male and female flowers on different plants. In Britain it is traditional to grow a few male plants and produce seeded hops, while Europe and the United States prefer seedless ones. Native American varieties have a distinctive aroma which European brewers dislike, though more neutral American hybrids are now widely used. Hops are picked off the plant when mature, and then are dried, sometimes powdered, and stored until needed. They are valued for two things: their resins, which impart a bitterness to the beer, and their essential oil, which contributes to the aroma. In the past, highly hopped beers may indeed have been especially resistant to spoilage, but at today's ratios—between ½ and 3 pounds hops per 36 gallons beer, the higher figure typical of European brews—these materials seem to contribute little biological stability. Both the resins and oil are contained in small glands

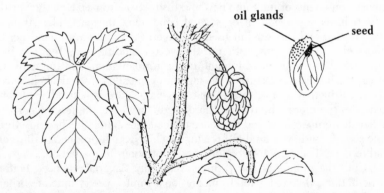

The hop vine, *Humulus lupulus*, and its cone, with a close-up of one of the cone's clustered leaves, or bracts

near the base of the hop's cluster of floral leaves, or bracts. The oil is a complex mixture of various hydrocarbons, esters, and alcohols, of which the major component is myrcene, a substance found also in bay and verbena. There are several forms of the major resins humulone and lupulone, some of which change structure during boiling and account for the bitter flavor. As the wort boils, these materials are extracted from the hops, while much of the volatile oil, together with the aromatic molecules formed by the malt, escape with the steam. This is fine with the brewer, since otherwise the aroma of the hops might be too powerful for the degree of bitterness attained. A recent innovation is the addition of pretreated bitter hop extract to the finished beer, a technique which avoids the losses in hoppiness during fermentation and conditioning.

During the hour or two of boiling in a large copper (traditional) or stainless steel tank (half an hour if it is done under pressure), several important things are accomplished in addition to the development of a hoppy flavor. The wort is sterilized so that the brewing yeasts won't have any competition during fermentation, and it is somewhat concentrated as part of the water boils off. The barley enzymes are inactivated, and so the carbohydrate mix of the wort—a certain amount of sugar for the yeasts, a certain amount of dextrins for body—is fixed. The color is deepened by browning reactions, mainly between the sugar maltose and the amino acid proline. And proteins and related long molecules that might cause the beer to become cloudy, complex with tannins from the barley hulls, coagulate into larger masses, and fall out of the solution. The precipitate that forms during boiling is called the hot break, and a cold break forms as the wort cools. The bed of hops can be used to strain out the breaks, though today the wort is usually centrifuged, then cooled and aerated. It is now ready for fermentation.

Fermentation

The brewer has finished transforming the bland barley grain into a rich sweet liquid, and now the yeast cells transform sugar into alcohol, and the minor components of malt into new and distinctive aromas. Exactly how the yeast behaves is dependent on several conditions: the particular strain of yeast, its age and the amount used (the average is around a pound per 36 gallons of wort), the composition of the wort, including the extent to which it is aerated, the amount of motion during fermentation, and the temperature of the liquid. For example, the higher, or so-called fusel alcohols, which give a distinctive aroma to alcoholic beverages, are produced in greater quantities the more amino acids there are in the wort, the less oxygen, the higher the temperature, and the more the cells are agitated. There are three basic sets of conditions used in brewing, each of which produces a different kind of beer. These are top, bottom, and continuous fermentation.

Until the middle of the 19th century, only Bavaria practiced bottom fermentation, while today top fermentation is limited pretty much to Britain and Belgium. The two processes employ different strains of yeast—*Saccharomyces cerevisiae* for top, *S. carlsbergensis* for bottom—which appear to

BEER MAKING

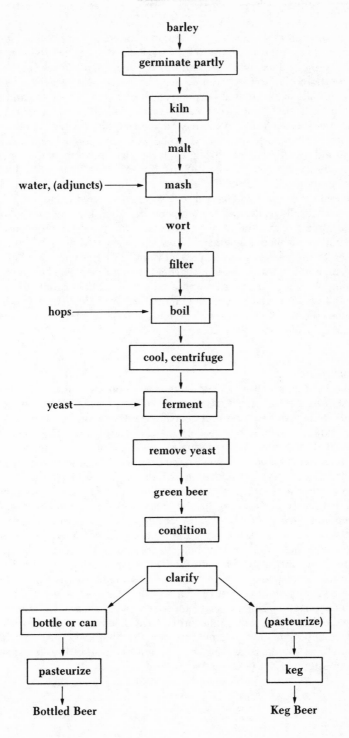

differ from each other mainly in the composition of their cell walls. Top yeasts tend to clump together, trap the carbon dioxide gas that they produce, and rise to the surface, while bottom yeasts remain separate and suspended in the wort, and fall to the bottom when fermentation is over. Top fermentation is usually carried out at between 58° and 68°F (14°–20°C) and takes 2 to 5 days, during which the yeasty foam is skimmed off several times. Because the yeast cells at the top have a good supply of oxygen and can be contaminated by other airborne microbes, top-fermented beers are often slightly acidic and relatively strong in flavor. Bottom fermentation goes on at distinctly lower temperatures, 39° to 54°F (4°–12°C), takes 5 to 14 days, and generally produces a milder flavor. Continuous fermentation is a highly accelerated version of brewing and makes possible a production of up to 360 gallons an hour. Its relatively high temperatures, short brewing period (about 4 hours), agitation, and the large quantities of yeast involved, produce a beer with a rather distinctive character. The wort passes from one fermenting chamber to the next, always in contact with an extraordinary number of yeast cells, and is released when it reaches the requisite alcohol content. The plumbing of such a system is complicated and must be scrupulously clean. As a result of this trickiness and the unusual flavor of the beer so produced, continuous fermentation is not widely practiced. Bottom fermentation is still the standard technique in the U.S.

Conditioning

The treatment of the beer after fermentation also varies according to the type of fermentation that has taken place. Top-fermented beer is cleared of yeast—today, this is done in a centrifuge rather than by skimming and settling—and then run into a cask or tank for conditioning. The green beer, as it is called fresh from fermentation, contains a little carbon dioxide, has a harsh, yeasty flavor, and is hazy with the detritus of dead yeast cells. In conditioning, a secondary fermentation is induced by adding either a small amount of yeast and some sugar or fresh wort, or some actively fermenting wort (this is called *Kräusening*) to the green beer. Inside the closed cask or tank, the liquid traps and absorbs the carbon dioxide produced. Undesirable odors can be forced out by opening the container briefly and allowing some gas to escape. These traditional techniques are sometimes replaced by simply pumping pure carbon dioxide into the beer—carbonating it—and thereby purging it of unpleasant odors. A few hops or some hop extract may also be thrown in at this point to adjust the flavor. A few days of cooling and the use of a "fining" agent—isinglass, clay, and vegetable gums are common—precipitate those materials that might form a haze. The beer is then centrifuged to remove any remaining yeast and precipitate, filtered, bottled or canned, and sometimes pasteurized. Keg beer, because it is generally kept cold in transit and storage, is usually not pasteurized.

Lagering

The conditioning process for bottom-fermented beer is somewhat different. Such beer is called lager, from the German *lagern*, "to store," "to lay down," because it was traditionally packed in ice for several months after fermentation in cool caves among the foothills of the Bavarian Alps. During this storage period, the yeast would settle to the bottom of the barrel and give the beer a character different from that of the quickly brewed, quickly conditioned ale (the similarity to champagne making is striking). Lagering was developed in the late Middle Ages, apparently as a way of maintaining a supply of beer throughout the summer, when the heat would interfere with a clean fermentation. Brewing took place in the winter and spring, and the beer was kept in the caves until needed. Its clarity, crisp flavor, and sparkle were happy side effects of the long, cool fermentation and cold storage. Today, because storage has the economic disadvantage of tying up money and materials, the tendency in the United States is to lager the green beer at temperatures just above freezing for a much shorter time, on the order of 2 or 3 weeks. As is true for top-fermented beers in Britain, carbon dioxide may be used to purge undesirable aromas; and centrifuges, filters, and additives, to help clarify it. As a replacement for wooden casks, some beech or hazelwood chips may be thrown into the tank for flavor. Most canned or bottled lager is pasteurized; keg beer is usually not. Better than 50 additives are permitted in American beer, including preservatives, foaming agents (usually vegetable gums), and enzymes—similar to meat tenderizers—that break down proteins into smaller molecules. Some companies avoid the use of preservatives, and usually advertise this virtue on the label.

The Final Composition

At the end of brewing, the dry, tasteless barley grain has been turned into a bubbly, bitter, somewhat acidic liquid (its pH is about 4) which is over 90% water, about 4% alcohol, a few percent carbohydrates, mainly maltose and the dextrins, and which contains traces of protein or amino acids and fats. Some B vitamins produced by the yeast also make it into the bottle. The exact fraction of alcohol depends on the brewing method used. Generally, American beers are more alcoholic, though more anemic in flavor, than European brews. One 12-ounce can or bottle of American beer contains about 140 calories, about a third of which are accounted for by the carbohydrates, and the rest by the alcohol. Low-calorie beers, a product of the 1970s, may have fewer than 100 calories, a savings produced by using less malt or cereal adjuncts to begin with. The alcohol content of light beer is generally only about 15% lower than that of ordinary beer, but there are many fewer carbohydrates left; much of the caloric reduction is accounted for by the near elimination of dextrins, which are broken down by added enzymes into fermentable sugars. It is for this reason that low-calorie beers have very little body to speak of: or, in the manufacturers' terms, they are "less filling." Here is a summary of the alcohol content (by volume) of various beers.

ALCOHOL CONTENT OF BEERS (PERCENT)			
British		American	
Brown ale	3.0	Low-calorie	3.75
Light ale	3.5	Lager	4.5
Lager	3.5	Malt liquor	5.6
Stout	4.8		
Strong ale	7.0		

The "Head"; Gushing

A few words about bubbles in beer. Brewing science has paid some attention to the matter of "head retention," or the length of time that the foam on a poured glass of beer lasts. Head retention is measured by the "half-life," or the number of seconds it takes for the foam to lose half its volume. A half-life of 110 seconds is considered to be very good. The nature of foams is discussed in detail in chapter 2; here, it suffices to say that a liquid foams most readily when it contains large molecules that can collect at the air-liquid interface—the bubble wall—and stabilize it by slowing the flow of liquid caused by gravity. In beer, these molecules are mostly dextrins, protein-dextrin complexes, and hop resins, whose concentration in the foam can give it a noticeably bitter flavor. It turns out that the use of cereal adjuncts, whose proteins are apparently more intact than malt's, improves the head retention of beer significantly. On the other hand, both alcohol and the usual haze-reducing treatments work against stable foams. The beer drinker, if he or she so wishes, can maximize the head retention by pouring a cold beer "vigorously" into a tall, narrow, scrupulously clean glass. Any traces of soap on the glass will cut short the life of the foam.

Many of us have had the experience of being inadvertently or purposely sprayed by our own or someone else's beer can: a sign of vigor and exuberance in commercials, but a mess in reality. For reasons that may have to do with the growth of an otherwise harmless mold on the grain, weathered barley makes beer particularly susceptible to gushing. The usual cause, of course, is agitation of the container, a common enough occurrence during normal transit and handling that antigushing agents are sometimes added to beer. Shaking the can sloshes some of the carbon dioxide from the empty head space into the liquid as bubbles, the smallest of which may remain in the liquid for quite a while and serve as collectors of more gas when pressure on the beer is suddenly released. (For similar reasons—minute air pockets providing molecules of carbon dioxide a space into which to diffuse— scratches on the bottom of a beer or champagne glass will produce a nice stream of bubbles.) These bubbles attract the same molecules to the gas-liquid interface—dextrins, hop resins, proteins—that stabilize the beer head in the glass, and when the bubbles eventually break in the still unopened can, microscopic associations of these molecules remain to act as bubble "seeds" when the top is opened. Beer gushes when the sudden release of pressure on opening the container causes some carbon dioxide to leave the solution (the lower the pressure, the less soluble the gas), collect in existing bubbles or form new bubbles on the seeds, and expand throughout the liquid, forcing a foam of gas and beer out the top. An undisturbed can of beer contains only a few bubbles or seed areas, and only the gas in the separate head space expands when pressure is released.

Storing and Serving Beer

A couple of points about serving beer. Lager beers are usually poured on the cold side, from 32° to 50°F (0°–10°C), as would befit the conditions of their making. Top-fermented beers, as visitors to Britain soon discover, are served at a cool room temperature, from 50° to 65°F (10°–18°C). This is mostly a matter of taste. One fact is worth remembering: no matter how it is made, the colder a beer is, the less full its flavor. Ice-cold beer may as well be guzzled, while warmer beers can be savored. Then there is the less subjective matter of the "sunstruck" beer sometimes encountered at a picnic where cups of the liquid have been left sitting in full sunlight. Certain wavelengths in this intense illumination cause the hop resin humulone to react with sulfur-containing molecules in the beverage and form isopentenyl mercaptan, an odiferous chemical which has relatives in the skunk's defensive arsenal. So keep beer in the shade, in cans or bottles that block out the light.

DISTILLED LIQUORS

Medicine, Pleasure, and Poison

Any alcoholic beverage can be distilled into a new liquor that is much stronger in both intoxicating power and flavor. (Like alcohol, the molecules that produce an aroma are more volatile than water.) Begin with new wine and you get brandy; green beer becomes whiskey. Tequila from the agave cactus, kirsch from cherries, calvados from cider, rum from sugar cane, skhou from milk, various spirits from potatoes and grains: these are just a few of the liquors that are made from the range of sugary and starchy fruits of nature. Then add sugar and flavorings to these liquors, and you have dozens of liqueurs. We will limit our attention to the more common spirits. *Brandy*, or "burnt wine," was probably the first distilled beverage in the West, and was used mostly as a medicine or common narcotic until the 16th century. It was then that distillers around Cognac, France, discovered that a second distillation, together with long aging in wood, smoothed out the harshness of the drink and made it pleasant as well as stupefying. The value of double distillation was also recognized in Ireland and Scotland, whose Gaelic equivalent of *aqua vitae, uisge beatha* or *usquebaugh*, was shortened and anglicized to *whisky* in the 18th century. *Gin*, an English abbreviation of *jenever*, Dutch for "juniper," was a medicinal concoction from rye, with juniper added for its flavor and diuretic effect, first formulated in 16th-century Holland. And *rum* was first made from molasses in the English West Indies around 1630. The American colonies were important distillers of rum, which they then used in the fur and slave trades.

Gin in England

The growing availability and drinkability of distilled liquors in the 17th century and after had consequences that the alchemists could hardly have foreseen. If alcohol is not the elixir, it is certainly a drug, and addiction soon became a problem. In England the principal scourge was gin, which like port was considered a patriotic alternative to French beverages in the late 17th century. Then a change in the taxation of beer made gin cheaper than that relatively harmless drink, and it quickly became a favorite of the city poor. In 1750, William Hogarth showed death lurking in every corner in his famous engraving of *Gin Lane*, and the next year, in his *Inquiry into the Causes of the Late Increase of Robbers*, the novelist-magistrate Henry Fielding blamed the rise in crime on "a new kind of drunkenness, unknown to our ancestors." Gin, he said, was the "principal sustenance" of "more than a hundred thousand people" in London alone.

Excess brought reaction: the government tried to control the price and supply of gin, and temperance movements were born. But official policy only encouraged smuggling and illicit distilling. When Charles Dickens came to write *Sketches By Boz* (1836–37), gin was now the water of forgetfulness for many castoffs of an industrial society.

> Gin-drinking is a great vice in England, but wretchedness and dirt are a greater; and until you improve the homes of the poor, or persuade a half-famished wretch not to seek relief in the temporary oblivion of his own misery, with the pittance which, divided among his family, would furnish a morsel of bread for each, gin-shops will increase in number and splendour.

Since Dickens's time, social conditions have improved, and beer and tea are once more the Englishman's beverages. Gin has been rehabilitated, and since the 1890s has been the prime constituent of the cocktail, its neutral, dry, astringent flavor mixing well with many other ingredients.

Whiskey in America

The story was different in the New World. For one thing, molasses was more plentiful than barley, and rum more common than beer. Then there was the background of the immigrants, the Irish and Scots who had a predilection for whiskey and were experienced at evading governmental attempts to regulate their cottage stills. How much a part of their life whiskey was, and how advanced their technique, is suggested in Dr. Johnson's account of his *Journey to the Western Islands of Scotland* (1775).

> A man of the Hebrides, for of the woman's diet I can give no account, as soon as he appears in the morning, swallows a glass of whisky. . . . I never tried it, except once for experiment at the inn in Inverary, where I thought it preferable to any English malt brandy. It was strong, but not pungent, and was free from the empyreumatick taste or smell. What was the process I had no opportunity of inquiring, nor do I wish to improve the art of making poison pleasant.

Rye and barley spirits were being distilled in the northern colonies by 1700, and Kentucky corn whiskey was established by 1780. The new American government tried to pay its war debts by taxing distillation in 1791, and in 1794 the largely Scots-Irish region of western Pennsylvania rose in the short-lived Whiskey Rebellion. President Washington had to call out federal troops to put it down. From then on, the rebellion went underground and moon-shining became entrenched, especially in the poor hills of the South where the small amount of corn that could be grown would fetch a better price if distilled. This evasion led to the formation in 1862 of the Office of Internal Revenue. Certainly the national taste for hard liquor was an important stimulus to the temperance movement that culminated in Prohibition. It has even been argued that Prohibition would never have come about if the grape had been easily grown in the East, and wine or beer had been the national drink. Thomas Jefferson thought as early as 1818 that wine was the best weapon against whiskey. In a letter to one M. de Neauville, he said

> I rejoice, as a moralist, at the prospect of a reduction of the duties on wine, by our national legislature. . . . No nation is drunken where wine is cheap; and none sober where the dearness of wine substitutes ardent spirits as the common beverage. It is, in truth, the only antidote to the bane of whiskey.

Given the last century or more of history, it is Dickens's prescription that seems closer to being adequate.

The Distilling Process

All distilled beverages are made in basically the same way. A weakly alcoholic liquid is heated to boiling. Because the alcohol and other aromatic constituents have lower boiling points than the water they are diluted in, the vapor rising from the hot mixture will be richer in these molecules than the original liquid. So, for example, all the alcohol in beer can be removed by

evaporating only one-third of the beer. Even more water can then be separated from the vapor by passing it over a surface that is kept *above* the boiling point of alcohol, but *below* the boiling point of water (that is, between
173° and 212°F, or 78° and 100°C). Much of the water vapor will condense
on the surface, while the alcohol and other aromatics remain in the gaseous
state. Finally, the concentrated vapor is itself condensed in a cooled stretch
of plumbing and collected as a liquid. It will range from 50 to 70% alcohol,
up from the original 5 or 10%. The process can be repeated one or more
times to obtain an even purer distillate. There is a practical limit to the concentrating effect of distillation: a 96% solution of alcohol will condense back
in the same proportions, owing in part to the mutual attraction (via hydrogen
bonds) that water and alcohol have for each other. A purer solution can be
obtained only by deliberately adding a third substance—often benzene—
that alters the balance of forces involved.

The Pot Still

Today, distillation is carried on in two kinds of apparatus. The first, the
pot still, was used exclusively until the 19th century. The beer or wine,
known as the *wash*, is heated to boiling in a large metal container. The
vapors rise into the neck, which is kept above 173° but below 212°F, where
water vapor condenses and falls back into the wash. The remaining vapor
passes into the condenser, a coiled tube that is cooled from the outside with
water, where it returns to the liquid state, and is collected in another container. In the making of whiskey, this liquid is called the "low wines," and
except for the purpose of moonshine, it is undrinkable. In addition to a large
proportion of alcohol, aromatic esters and aldehydes (the products of reactions between alcohol and acids, and alcohol and oxygen respectively), it contains a good deal of *fusel oil*, or the higher (amyl, butyl, propyl, and so on)
alcohols. These relatives of drinkable ethyl alcohol, or ethanol, which are
produced by the yeast from amino acids in barley or grape, are more potent
narcotics than it is, and even in the relatively low concentration of less than
a part fusel oil per 100 total alcohols, they give the low wines a pronounced
harsh flavor, oily body, and unpleasant aftereffects. It is the excessive presence of fusel oil, or "congeners" (congeneric substances) of alcohol, in moonshine and other primitive waters of life that make them potentially deadly.
The term *fusel*, in fact, comes from the German for "rotgut." It is in order
to reduce the level of fusel oil in the distillate that the low wines are distilled
a second time. This is the step that calls for real skill.

In the second distillation, the distiller pays careful attention to the
sequence of liquids that appear in the condenser. All of the different substances in the low wines have slightly different boiling points, with the result
that some will boil off before others. In the whiskey-maker's parlance, the
main run of relatively pure alcohol is preceded by foreshots and followed by
feints, both of which are too rich in congeners and other impurities. The
trick, then, is to divert the flow of the other alcohols to the side (to be redis-

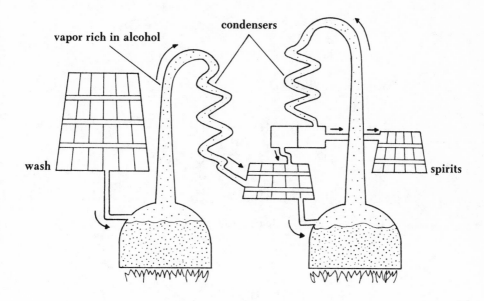

vapor rich in alcohol

condensers

wash

spirits

Pot distillation. The container at center receives both the distillate from the first pot, which must be further purified, and the undesirable portions—the foreshots and feints—of the second distillation, more of whose alcohol can thereby be extracted.

tilled), and collect only that mixture of ethyl alcohol, aromatics, and congeners that will result in a well-balanced whiskey. Some fusel oil is desirable for its contribution to flavor and body, but too much will make for a rough drink. The character of a particular whiskey depends, then, on the particular moments at which the distiller declares the main run begun and finished.

The Column Still

Double distillation in pots is the classic technique for making whiskeys, brandy, and rum. But in the 19th century a new apparatus was devised that eliminated an important disadvantage of pot distilling: namely, that the pots have a limited capacity, and must be drained of spent wash and cleaned before they can be used again. Between 1825 and 1835, Robert Stein in Scotland and Aeneas Coffey in Ireland perfected the patent, or continuous, or column still, a complex piece of plumbing that produces a continuous stream of distillate from a continuous input of wash. Here is a simplified description

methanol (wood alcohol) 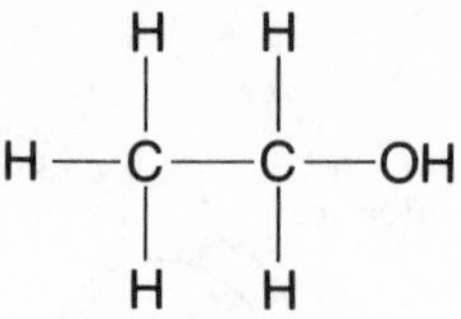   ethanol

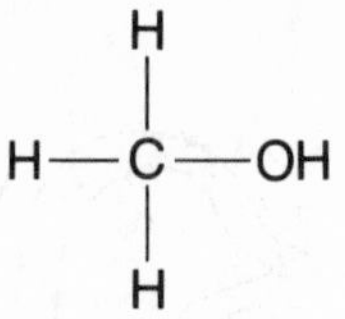

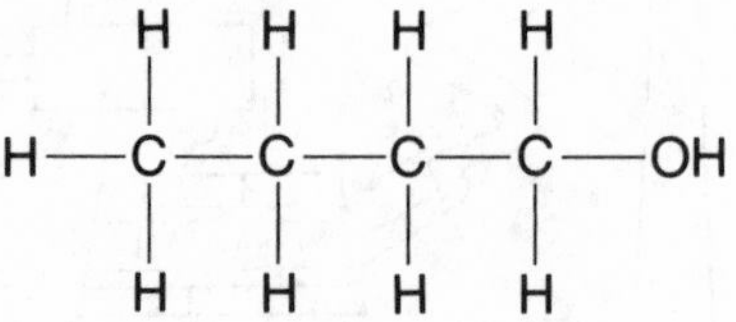

butyl alcohol

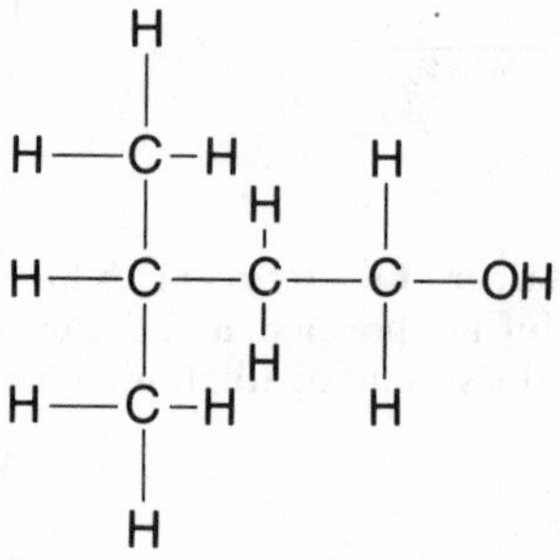

amyl alcohol

The structures of various alcohols. Methanol is a poison because our bodies convert it into formic acid, which accumulates and damages tissues in the eyes and brain. Butyl and amyl alcohol are two of the "higher," or longer-chain alcohols. These minor by-products of fermentation contribute a harsh flavor and, because of their fatty-acid-like hydrocarbon tails, an oily consistency to distilled liquors in which they have been concentrated.

of its operation. The liquid wash is dripped down one long column over a series of progressively hotter metal plates. This results in the concentration of substances with relatively low boiling points in a vapor at the cool top of the column. From here, the vapor moves to the bottom of a second column, again outfitted with hotter plates at the bottom, cooler ones at the top. The higher in this second column that the vapor rises without condensing on the plates, the lower its boiling point. The temperature of one plate is just right for condensing the desired ratio of alcohol to water, and this condensate is

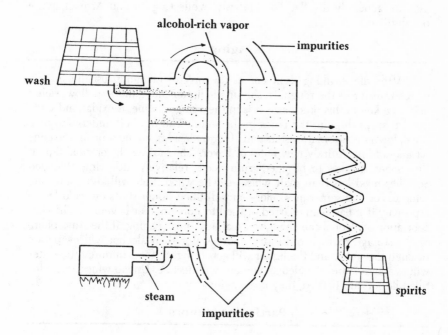

The column still. The plates in each column are hottest at the bottom and coolest at the top. Liquids with low boiling points are concentrated in the vapor that leaves the first column and rises in the second. As is the case with the pot still, the impurities are recycled to extract their fraction of desirable ingredients.

drawn off for aging. The substances that condense below this plate are the equivalent of the pot still's feints, and those that continue to rise above it would be the foreshots.

The disadvantage of the patent still is that the distiller has less control over the final mix of congeners and other aromatics. In order to get the single final mixture of sufficient strength, less of the flavor-contributing fusel oil is gathered and the distillate may come out relatively characterless. But precisely because few congeners make it through, the distiller need not care as much about the composition of the wash. For example, instead of brewing an all-malt beer, he can use just enough malt to provide the necessary starch-breaking enzymes, and replace the rest with adjunct materials—cheap, raw grain—that contribute plenty of starch but little flavor. Neutral grain whiskey is the name of this characterless distillate. Today, most Scotch is a blend of grain and malt whiskies, and is noticeably milder than the few straight malt whiskies that make it to market. American bourbon whiskeys and brandies are generally distilled continuously, while Cognac and Armagnac are pot-distilled.

Aging

After the distilling itself comes the process of aging, usually in oak, which transforms the rough, freshly made liquor into a more mellow, richer one. The key to this process, as it is in the aging of wine, is oxidation by the air that seeps through the pores in the wood. The fusel oil undergoes some breakdown into aromatic esters and other molecules, taking some of the edge off and adding entirely new notes of flavor. The harsher the original liquor, the longer it takes to be aged: rum needs relatively little time, blended Scotches need less than malt whiskies. The wood itself will also contribute some flavor and sometimes color—any distilled liquor starts out colorless—especially if it has been charred or used to store another beverage. The concentration of alcohol can change a great deal during aging. If the atmosphere of the storage room is quite dry, more water than alcohol will evaporate through the wood, and the liquor will get stronger. If it is humid, little water will evaporate, and the alcohol content will decline. Unlike wines, once distilled liquors are bottled, they cease aging.

Particular Liquors

Brandy

The two classic brandies are Cognac and Armagnac, the first named for a town and the second for a region in southwestern France, both not far from Bordeaux. Both are made from nondescript wines. The Cognac wine has a low 7½% alcohol content and a high acidity. It is double-distilled right after fermentation to an alcohol content of about 70%, then diluted with water to about 50%. During its aging in oak—a minimum of 2 years, and 5 years for Very Special Old Pale (improvement in flavor may continue for 15 to 20)—a great deal of evaporation occurs, the equivalent of about 2% of the barrel every year. When ready for bottling, ordinary Cognac is diluted if necessary to 40% alcohol, caramel is added for color, and a small amount of sugar for taste. Armagnac differs from Cognac principally in its dryness—no sugar is added—and a greater pungency that results from a less complete distillation (just once, to about 53% alcohol) and aging in a sappy, strong black oak. In California, column stills are used, and a small amount of caramel syrup generally added for color and flavor. Cheap brandies are made all over the world in both pot and patent stills, with color, sugar, and flavoring materials added.

Scotch

There are many kinds of whiskey made today, and they are made in various ways. Begin with Scotch malt whiskies, the singular of which is "whisky." Barley malt is the only grain used. Because the distiller's aim is to maximize his production of alcohol, he uses only a pale malt. The darker the color of the barley, the less enzymes and starch remain and so the less sugar there will be for the yeast to grow on. Rather than roasting it, he dries it slowly at about 120°F (49°C) for 2 or 3 days. He adds no hops to the wort because the resins will not end up in the distilled liquor. And he controls the mashing so as to reduce as much of the starch as possible to small, fermentable sugars (the brewer leaves some long chains, or dextrins, in order to give the beer body). Distiller's beer may get up to 10% alcohol, rather than the ale-brewer's normal 5%. After 3 days or so of fermentation, the beer is pot-distilled twice. The new whisky is then run into casks that have been used to store sherry, where over a period of years the alcohol concentration falls, the whisky flavor mellows, and color and flavor are extracted from the wood. Before bottling, the whisky is diluted to the desired strength (usually 40% alcohol) and, if necessary, colored with caramel. Though the vivid description of malting in his ballad "John Barleycorn" applies equally well to brewing beer, there is little doubt that Robert Burns celebrates the making of whisky in these stanzas:

> They filled up a darksome pit
> With water to the brim,
> They heaved in John Barleycorn
> There let him sink or swim.

> They laid him out upon the floor
> To work him farther woe,
> And still as signs of life appeared,
> They toss'd him to and fro.

> They wasted o'er a scorching flame,
> The marrow of his bones;
> But a miller used him worst of all,
> For he crush'd him between two stones.

> And they hae ta'en his very heart's blood
> And drank it round and round;
> And still the more and more they drank,
> Their joy did more abound.

It has been legal for Scotch whiskies to be blended since the 1860s, and today less than 2% of the Scotch consumed in Britain is "single malt," or the product of one pure malt distiller. There are about 115 such distilleries still operating, mostly in the Highlands, but they are generally licensed to the large blending and bottling companies—Dewar Bros., John Walker, Teacher, and so on—which mix several different malt whiskies with the cheaper and more neutral grain alcohol produced from corn by continuous stills in the lowlands. The better blends may be as much as 60% malt whiskies, the cheaper more like 30%. The original motive for blending was economic, but the public taste turned out to be waiting for just such a development. Strongly flavored Scotch whiskies were not at all popular outside of Scotland, but the milder, blended versions were there to fill England's need when the phylloxera disaster devastated the brandy vineyards in the 1870s and 1880s. It is from this period that the great esteem of Scotch dates. Malt whiskies owe a great deal of their character to the use of peat smoke for drying the malt, and peaty water for mashing. Peat, the mat of decaying and decayed vegetation which once was the cheapest fuel available in swampy areas of Britain, contributes volatile organic molecules to the brew that find their way into the distillate. The taste for this peaty character, even more than the taste for 40% alcohol itself, is an acquired one. The smokiest whiskies come from Islay and Campbeltown, on Scotland's west coast.

Irish and American Whiskeys

Now a hop and a jump westward. Irish whiskey (with an *e*) is made from a mixture of about 40% malted barley and 60% unmalted, and the malt is dried without peat smoke. For both these reasons, and because it is *triple*-distilled, Irish whiskey is milder than single-malt Scotch and even some blends. When we get to North America, the chief indigenous liquor is bourbon, which is made today from barley malt and ground corn. Once the grains have been mashed, they are fermented in one of two ways: as a sweet mash, in which fresh yeast works away for a couple of days, or as a sour mash, in which some yeast from the previous batch is reused, the fermentation lasting for 3 or 4 days. The beer is then run through a column still and aged in oak barrels that have been charred about ¼ inch deep inside; these give the liquor its color and some flavor. Where brandy and Scotch get less alcoholic with aging, the relatively dry atmosphere surrounding bourbon casks causes more water to be lost than alcohol. There are, of course, other American whiskeys, including corn, rye, and various blends. A liquor can legally be given the name of a particular grain if that grain constitutes 51% or more of the grist.

Gin

There are two principal styles of gin made today, Dutch and English. In its country of origin, a mixture of malt, corn, and rye is fermented and distilled 2 or 3 times in pot stills at low proof: that is, the distillate contains a fair amount of congeners. Then the distillate is distilled one last time, to about 48% alcohol, with juniper berries, whose aromatic molecules end up in the final gin. English-style, or "dry" gin, begins with mostly corn and malt; the beer is distilled in a continuous apparatus to 90 or 94% alcohol, and practically no flavor survives. This almost entirely neutral alcohol is then redistilled in a pot with juniper plus any of a number of other flavorings—among them lemon and orange peel, cardamom, coriander, angelica, anise, caraway, licorice—the exact combinations and ratios of which are trade secrets. This gin is diluted before bottling to a strength similar to the Dutch version's. American gins are distilled at even higher proofs than English. Because gins are chemically so uncomplicated compared to brandy and whiskey, they are generally not aged at all before bottling.

Rum

Rum too comes in different strengths. Most rums are distilled from fermented sugar cane juice or molasses to about 85% alcohol in a continuous still, and later diluted. Depending on the concentration of congeners allowed, a range of products is possible, from light- to heavy-bodied. Lighter rums are aged only for a year or so, heavier ones for up to 6. They are filtered, blended, and if a "silver" rum is desired, leached through charcoal. Caramel can be added to deepen the color and flavor, a procedure that seems appropriate for this particular liquor, which is made from sugar in the first place. Jamaica rum is generally stronger than the others because it is fermented with previously used, "sour" yeast for a longer time—5 to 20 days, as against 1 to 4 for others—and is pot-distilled to a much lower purity, around 70 to 80% alcohol.

Vodka

Vodka was scarcely known in the United States until the 1950s, but it now accounts for some 20% of the distilled spirits consumed here. The word is the Russian diminutive for "water." Vodka is distilled to a very high proof and so a minimum of flavor. Because it is powerful but anonymous, it has become very popular in mixed drinks. Vodka has traditionally been made from the cheapest source of starch available, usually grain, but sometimes potatoes. The source is immaterial, since none of its character survives distillation and a filtering through powdered charcoal that removes any remaining flavor.

Concentrated Alcohol: Proof and Tears

A couple of miscellaneous points. One is the use of the term *proof* to designate the alcoholic content of distilled liquors. In the United States, the proof number is just about double the percentage by volume of alcohol, so that 100 proof, for example, designates 50% alcohol. Actually, the proof number is very slightly more than double the percentage, because alcohol has the effect of causing a volume of water to contract on mixing. Fifty parts of alcohol added to 53.73 parts water gives not 103.73 parts of mixture, but 100.00. Such a measure of strength has only been possible since the 19th century, when adequately accurate means of measuring the specific gravity of liquids were developed. But the term *proof* dates from the 17th century, when less refined but nonetheless effective techniques were worked out to check on the quality of spirits. One was to burn the alcohol off the spirits; if the residual liquid was less than half the volume of the original, the liquor was up to strength. The proof test itself involved moistening gunpowder with the spirits, and then lighting it. If it burned slowly, it was at proof; if it spluttered or burst into flame, it was under or over proof respectively.

Regular drinkers of strong wines and spirits—at least those who drink from a glass—have probably noticed and mused upon the odd phenomenon known as "tears" or "legs" or "arches" of liquid on the inside of the glass, that seem to be in slow but continuous movement. First, a thin film coats the glass above the surface of the beverage. After a while, the upper edge of the film thickens and forms droplets which slowly fall back toward the drink. But this motion can be stopped and even reversed by gently tilting the glass so that the liquid surface gradually approaches the droplets. Just before the droplet meets drink, the droplet will often stop, poised a tiny fraction of an inch away from being swallowed up, and even begin to retreat back up the side of the glass, defying gravity. The first person to explain the tears of spirits was not himself a famous scientist, but related to one: he was James Thomson, a British engineer whose brother was the physicist William Thomson, Lord Kelvin. In the *Philosophical Magazine* of 1855, he published a short paper "On certain Curious Motions observable at the Surface of Wine and other Alcoholic Liquors," and attributed these motions to variations in the surface tension of the liquids.

Surface tension is a manifestation of the attractive forces among the molecules of a pure liquid, the forces that keep them from flying apart and becoming a gas. Water, because of its great capacity for forming hydrogen bonds with itself, has a particularly high surface tension. Now, when one liquid is diluted by a foreign substance that mixes evenly with it, its surface tension is reduced: the foreign molecules interrupt the network of its attractive forces. So the alcohol in wine or brandy lowers the surface tension of the water. But in an alcoholic beverage, alcohol molecules evaporate from the surface more rapidly than water does. So the attractive forces between water molecules are most effective right at the surface of the beverage, where the liquid is less alcoholic than it is in the center of the glass. The water is *most* pure, *most* concentrated, at the junction of liquid, air, and glass, which is slightly above the rest of the liquid surface because water clings strongly to, or "wets" glass.

And here a "surface-tension pump" goes to work. The purer water at the edge of the glass exerts a strong attractive force on the water molecules just below it, and draws them up the side of the glass. As the liquid rises, it continues to lose alcohol by evaporation, and because it is receding from the reservoir of alcohol in the body of the beverage, it becomes more and more purely water the higher it gets, and so pulls more and more strongly on both other water molecules and on the glass. The continuous gradient in surface tension pulls a thin film of the liquid up the interior of the glass, until it is supporting as tall a film as it can. The most watery liquid collects in a ring at the top of the film that gets larger and heavier, until the force of gravity takes over and pulls it down again in droplets that look like tears, and that trace out a series of arches as they fall. When a droplet gets close to the surface of the drink, it picks up some alcohol by diffusion across the tiny space of film separating droplet and drink, and therefore loses some surface tension. As a result, it feels a stronger pull from the now purer water above than from the more alcoholic film below, and so begins to climb the glass once more.

Because Thomson's paper was soon forgotten, these and related surface-tension phenomena have been named the Marangoni effects, after a later investigator. The higher the alcoholic content of the liquid, and the easier it is for alcohol to evaporate (warm temperatures and wide-mouthed shallow glasses are most favorable), the more pronounced these effects are.

DRUNKENNESS AND HANGOVERS

The *Problemata* is a work traditionally ascribed to Aristotle, though modern scholars suspect that it was patched together sometime in the first few centuries A.D. Whatever its exact provenance, it is quite old. The third Book of the *Problemata* is concerned entirely with drinking and drunkenness, and demonstrates that people have been puzzled for thousands of years by the less pleasant effects of alcohol. Among the questions it poses are these:

> Why is it that to those who are very drunk everything seems to revolve in a circle . . . ?
> Why is it that to those who are drunk one thing at which they are looking sometimes appears to be many?
> Why is it that those who are drunk are incapable of having sexual intercourse?
> Why is it that wine which is mixed but tends toward the unmixed causes a worse headache the next morning than entirely unmixed wine?
> Why has wine the effect both of stupefying and of driving to a frenzy those who drink it?
> Why is it that cabbage stops the ill effects of drinking?

Why are the drunken more easily moved to tears?
Why is it that the tongue of those who are drunk stumbles?
Why is it that oil is beneficial against drunkenness and sipping it
enables one to continue drinking?

Good questions! A more familiar relic of the ancients' interest in remedies
for excess is the word *amethyst*, which comes from the Greek for "not
drunken": a condition which the gem was supposed to guarantee. Today, of
course, much more is understood about the influence of alcohol on the
human body. And yet, because the effects are manifold and complex, a great
deal remains obscure. Worst of all there seems to be no miraculous amethyst—or cabbage.

Alcohol as Drug

The effects of alcohol can be grouped into two categories: direct and
indirect. The direct effects are what we usually call drunkenness, the indirect
both hangover and the physical problems associated with alcoholism: cirrho-
sis of the liver and so on. Alcohol is absorbed not just in the small intestine,
where most food is absorbed, but to some extent in the stomach and colon as
well. It may take anywhere from 2 to 6 hours for all the alcohol in a given
drink to be taken up by the body. The main factor seems to be the circumstances under which the alcohol is ingested. If the stomach is empty, the
alcohol is rapidly absorbed, while if there is other food present, it has to wait
its turn to come in contact with the walls of the digestive tract, and there is
some delay. Alcohol is absorbed more slowly from beer than it is from wine,
and more slowly from wine than from distilled liquors, because beer, wine,
and liquor contain progressively fewer other substances to compete with
alcohol for position along the stomach or intestinal wall. There is good reason
for oil to be singled out by the author(s) of the *Problemata*, because fats slow
the emptying of the stomach into the small intestine (see chapter 12), where
the most complete absorption takes place. Drinking whole milk or eating a
few rich cocktail hors d'oeuvres will get the alcohol into you more gently
and over a longer period of time. It is said that the carbon dioxide in champagne and carbonated mixed drinks has the reverse effect: it supposedly
encourages the valve between stomach and intestine to open. If true, this
would explain the extra kick that these drinks seem to have. In any case,
diluting alcohol and delaying its absorption have the desirable effect of evening out and to some extent moderating its effects on the body. But a heavy
drinker will end up with a lot of alcohol in his blood no matter what else he
eats or sips.

Once absorbed from the digestive tract, alcohol is rapidly distributed to
all body fluids and tissues, apparently crossing cell membranes without any
trouble. Because the degree to which we are intoxicated depends on the concentration of alcohol in the cells, large people can drink more than small
people without being drunker: they have a greater volume of body fluids in
which to dilute the alcohol. (Also, women generally have a lower tolerance

than men, apparently because fat accounts for a greater fraction of their body weight; see chapter 11.) This complete diffusion is what makes possible simple breath tests for determining the alcohol content of the blood. In the lungs, alcohol passes through the capillary walls into the air in a fixed proportion to its concentration in the blood, so that a measure of the alcohol content of expired air can be converted to blood content, and thereby to the degree of intoxication.

The direct effects of alcohol are manifestations of its nature as a *drug:* it alters the operation of the various tissues into which it diffuses. It is valued most for its influence on the central nervous system, on which it acts as a narcotic or depressant. The fact that it seems to stimulate more animated, excited behavior than usual is actually a symptom of its depressant effect on the higher functions of the brain, those that normally control our social behavior with various kinds of inhibition. Any "frenzy" to which we are driven is a result of this general "stupefying" influence. As more alcohol reaches the brain, memory, concentration, and thought processes in general are interfered with, muscular coordination suffers, the tongue stumbles, and the eyes see double. Sleep, and in extreme cases unconsciousness, may be the final stage. With regard to the idea that alcohol is an aphrodisiac, modern investigators continue to cite the authority of the Porter in Shakespeare's *Macbeth* (II.iii), who says that drink "is a great provoker of three things": "nose-painting, sleep, and urine. Lechery, sir, it provokes, and unprovokes: it provokes the desire, but it takes away the performance."

The Porter is also right about urine. Alcohol interferes with several pituitary and adrenal hormones, some of them antidiuretic. The result, while blood alcohol levels are rising, is increased water secretion, which necessitates more frequent trips to the bathroom, and which may leave a general feeling of dehydration the morning after. The presence of alcohol in the stomach also promotes the secretion of large amounts of gastric juice, whose acids can make trouble for people with ulcers (again, other foods will dilute stomach acid and so reduce irritation). And alcohol causes blood vessels all over the body to widen, or dilate. The increased flow of blood in capillaries just below the skin causes the familiar flushed, hot feeling, but this feeling should not be mistaken for evidence that alcohol "warms you up." The impression of heat comes from the transfer of deep body heat from the blood to the skin, but the skin transfers it to the air, so that the body as a whole is actually *cooled* by alcohol. There is some logic to drinking strong liquor in a hot climate, but the same practice can actually increase the risks of exposure in very cold weather.

How Alcohol Is Metabolized

Narcosis and losses of water and heat are the main effects of alcohol as a drug. But alcohol also qualifies as a food, because it is a source of energy. The body can break it down like a carbohydrate, and liberate about the same number of calories as carbohydrates provide. It is the metabolism of alcohol that is the key to recovering from drunkenness, since burning alcohol for energy removes it from circulation as a drug. In fact, this is really the only way to get rid of alcohol: a scant 5 or 10% is excreted in the breath or urine. The problem is that the body's biochemical machinery is not equipped to handle large amounts of alcohol. A single enzyme, alcohol dehydrogenase, or ADH, is required in the initial step of breaking alcohol down, but we have only small quantities of it, and this limits the rate at which we can digest alcohol. And if we overload the machinery, we can throw it out of balance.

Why is it that humans have an alcohol-digesting enzyme to begin with? After all, we have only been drinking beer and wine for a few thousand years, which is much too short a time to have evolved this complicated piece of biochemistry. Actually, ADH is present in the body not to deal with ingested alcohol, but to deal with alcohol produced *inside* the body. Like the yeasts who produce it for us, our bodies sometimes make small amounts of alcohol in the process of metabolizing carbohydrates. And the bacteria that inhabit our intestines excrete larger amounts, which we immediately absorb and must deal with. The ADH that is localized in the liver is responsible for cleansing the blood of this substance which, if allowed to accumulate, could poison us. In addition to the liver, some ADH is found elsewhere in the body, notably in the testes and the retina, where it is involved in intermediate stages of processing alcohol-related substances that are essential to the formation of sperm cells and to the biochemistry of sight. In each of these locations, the amounts of ADH present are relatively small.

This is the problem when it comes to dealing with ingested alcohol: there is only a limited amount of the enzyme necessary to break it down and remove it from the blood. When it is totally engaged, ADH works at a constant rate. In the average adult, this rate is about 7 grams of alcohol per hour; larger people generally have larger livers and so more ADH. In everyday measures, this means that the alcohol in 4 ounces of whiskey or a quart of beer will take 5 or 6 hours to disappear from the blood. To put it another way, someone who drinks more than about a gallon of beer or 16 ounces of Scotch a day will never clear his or her blood of alcohol.

Our absolute dependence on ADH to detoxify ourselves has certain unfortunate consequences. While exercise helps us burn food calories more rapidly than we would while sitting, alcohol is not available as an energy source to our muscles until ADH has worked on it: so exercise will not sober us up any faster than sleeping will. Nor does any other food, whether coffee or some other folk remedy. And because an excess of alcohol "saturates," or completely occupies a part of the liver's enzymatic system, some by-products tend to accumulate faster than they can be dealt with. One, lactate, enters the bloodstream as lactic acid, a substance that also accumulates in muscle tissue during vigorous exercise and is associated with the sensation of fatigue. Finally, there are genetic variations in the exact structure and operation of ADH and related enzymes, and these seem to be involved in the sensitivity, manifested by such symptoms as facial flushing, that some populations have to even moderate amounts of alcohol.

There are some exceptions to the rule that the metabolism of alcohol cannot be improved. Heavy drinkers do appear to increase their tolerance by bringing into action another detoxifying system in the liver. Certain drugs can either delay or speed the breakdown of alcohol, just as alcohol can alter the effects of many drugs. And for some reason large amounts of fructose introduced directly into the blood speeds the elimination of alcohol, though eating fructose doesn't work. But none of these exceptions is a pleasant one, a modern equivalent of cabbage or amethyst. And as we age, our livers actually become less efficient at dealing with alcohol, so that our tolerance slowly declines. Once you've imbibed a large amount of liquor, there's nothing to do but let nature take its slow course.

The Hangover

Then there is the matter of the hangover. The folk remedies for this affliction are many and ancient. In medieval times, the medical School of Salerno was already recommending the hair of the dog:

> *Si nocturna tibi noceat potatio vini,*
> *Hoc tu mane bibas iterum, et fuerit medicina.*

> If an evening of wine does you in,
> More the next morning will be medicine.

The insidious logic of this remedy is simple. The hangover is in part a mild withdrawal syndrome: the night before, the body adjusted to a high concentration of alcohol throughout its tissues, but by morning the drug is going or gone. Hypersensitivity to sound and light, for example, may be a left-over compensation for the general depression of the nervous system. The higher the peak blood alcohol level reached, and the longer that the peak is maintained, the more drastic the effects on the body, the stronger its compensation, and the more pronounced the aftereffects. Having another drink, then, restores the conditions to which the body had become accustomed, as well as lightly anesthetizing it. But this only postpones the hours of reckoning, and if repeated may eventually lead to alcoholism and the necessity of suffering a true drug withdrawal.

Only a few of the different symptoms that constitute a hangover can be directly treated. The dry mouth and to some extent the headache we wake up with can be due to the water loss that alcohol causes, so that drinking liquids may ameliorate them. Alcohol can also cause a headache by enlarging the cranial blood vessels; the caffeine in coffee and tea has the opposite effect, and may bring some relief. The general feeling of fatigue and malaise is attributable to derangement of the acid-base balance in our tissues (excessive acidity during alcohol intake is followed by excessive alkalinity, conditions which alter the most fundamental functions of our cells), the general stress that the entire body has borne as it narcotized itself and then removed the drug, and depressed blood sugar levels caused by inhibited glucose metabolism in the liver. The last of these three conditions can be treated simply by eating foods that contain some carbohydrates.

But the first two are less tractable; medical science has yet to come up with anything practical that will speed our recovery from them or from the other common ailments. Any residual fuzzy-headedness is due to alcohol that hasn't yet been cleared from the blood, and perhaps to the congeners, or higher alcohols, which are more potent narcotics than ethyl alcohol and may be significant despite their relatively low concentrations. An upset stomach is the result of alcohol's effects either on the brain or directly on the digestive tract itself: it stimulates gastric secretions and irritates any tissue it contacts. To a large extent, then, the price of alcohol's pleasures appears to be fixed, and the best medicine is the passage of time.

The consequences of regularly excessive drinking and alcoholism—general malnutrition, clearly harmful if as yet poorly understood effects on the central nervous system, cardiovascular system, digestive tract, and liver, and the risk of birth abnormalities in the children of pregnant women—are frequently publicized. It is good to remember that one common word for drunkenness, "intoxication," comes from the Greek *toxikon*, meaning "arrow poison." People who play carelessly with alcohol risk shooting themselves in several vital organs at once.

CHAPTER 10

Food Additives

This chapter will be brief, and will not include an annotated list of the many chemicals currently put in our food during processing. (Nor will it address the serious problem of *unintentional* contaminants, such things as pesticides, herbicides, and growth hormones.) The last decade has brought a great deal of concern about the possible ill effects of food additives on our health, and several reliable guides are already available. The problem is that new studies are constantly being done, much of our current information is only sketchy, and the most celebrated and well-studied cases—saccharin and nitrite—remain controversial. Rather than reviewing mainly inconclusive details, this chapter tries to outline what a reasonable attitude toward food additives might be.

Ever since it became a popular issue in the early 1970s, there has been a tendency for books about food additives to reduce the matter to a battle between innocent consumers and a profiteering combine of business, technology, and government. One early book, still in print, presents a chapter on the "Unholy Alliance" between science and the food industry, while another professes horror at the title of a journal called *Food Technology*, as if bread, beer, chocolate, "wild" rice, or a fresh trout grilled over the campfire were not the products of various technologies.

This is not the kind of tone that encourages rational discussion or action. Public debate about the proliferation of additives in our food, a debate which is important and necessary, is more likely to be constructive if we all recognize two basic facts: that our food has never been perfectly safe or free from additives, and that consumers share the responsibility for whatever risk there may be in eating today. We and industry are collaborators when it comes to determining the kind of world we live in. Distasteful as this idea may seem, it offers this element of hope: where malevolent enemies must be defeated, wayward allies are susceptible to persuasion.

ADDITIVES IN "THE GOOD OLD DAYS"

In many attacks on the business-science-government axis, there is the unspoken but unmistakable suggestion that before these predators were around to poison us slowly for profit, our ancestors enjoyed purer, healthier food. This is an illusion. Additives of one kind or another have been put into food for thousands of years. Initially, it was a matter of survival, a way of keeping some supplies for the lean times when fresh sources would not be available. Preservatives in the form of salt and smoke—the former containing various impurities, the latter full of toxic chemicals—were employed by the earliest civilizations. Heavy doses of spices like mustard and pepper served the same purpose—their "heat" is caused by chemicals that irritate living cells—and also masked the flavor of spoilage in a time when food was too valuable to discard when it went bad. It is only in the last 150 years or so that canning and refrigeration have offered alternatives to chemical preservation. But necessity was not the only motive in the ancient world for adding extraneous materials to food. We know that the Romans added potash to their wine and natural soda to vegetables in order to influence their colors. So the use of chemicals that contribute nothing directly to nutrition is a venerable tradition.

Nor is adulteration, or the secret use of additives for fraudulent purposes, an invention of the modern world. In the late 14th century, the English poet William Langland described the job of various city officials as in part

> To punish on pillories or on pining stools
> Brewers, bakers, butchers, and cooks;
> For these be the men on earth who do most harm
> To the poor people who buy piece-meal.
> They poison the people secretly and often. . . .
>   (*Piers Plowman,* Passus III)

The opportunity to adulterate food was greatest in the cities, which were supplied from outlying farm areas through middlemen. The practice flourished at least from the Middle Ages on through the time of the Industrial Revolution, when the cities grew rapidly. Just to get a whiff of the atmosphere at mid-18th century, consider this greatly abridged account of London foods by Tobias Smollett's country squire, Matthew Bramble, in *The Expedition of Humphry Clinker* (1771):

> As to the intoxicating potion, sold for wine, it is a vile, unpalatable, and pernicious sophistication, balderdashed with cyder, corn-spirit, and the juice of sloes. . . . The bread I eat in London, is a deleterious paste, mixed up with chalk, alum, and bone-ashes;

insipid to the taste, and destructive to the constitution. The good
people are not ignorant of this adulteration; but they prefer it to
wholesome bread, because it is whiter than the meal of corn
[wheat]: thus they sacrifice their taste and their health, and the
lives of their tender infants, to a most absurd gratification of a
mis-judging eye; and the miller, or the baker, is obliged to poison
them and their families, in order to live by his profession. . . .
they insist on having the complexion of their pot-herbs mended,
even at the hazard of their lives. Perhaps, you will hardly believe
they can be so mad as to boil their greens with a brass halfpence,
in order to improve their color; and yet nothing is more true. . . .
[Milk is] the produce of faded cabbage leaves and sour draff, lowered with hot water, frothed with bruised snails, carried through
the streets in open pails. . . . the tallowy rancid mass, called butter, is manufactured with candle-grease and kitchen-stuff. . . .
Now, all these enormities might be remedied with a very little
attention to the article of police, or civil regulation; but the wise
patriots of London have taken it into their heads, that all regulation is inconsistent with liberty. . . .

Smollett's book is fiction, but the veracity of this famous passage is attested
to by the chemist Fredrick Accum's 1820 exposé whose title alone casts suspicion on better than a dozen foods. "Indeed," said Accum in his introduction, "it would be difficult to mention a single article of food which is not to
be met with in an adulterated state; and there are some substances which are
scarcely ever to be procured genuine." After a group of physicians confirmed
Accum's charges, the first English Pure Food Laws were established in the
1860s. The United States followed suit 40 years later, after similar
investigations.

So there never really were any good old days when food was fresh and
pure, and adulteration and additives were only gleams in the chemist's eye.
If anything, conditions are much better today than they have been since the
cities arose, thanks to the modern technologies of canning and refrigeration,
medical science, and governmental oversight. Our system is by no means
ideal—adulteration, contamination, the use of possibly dangerous additives
do continue—but it is generally an improvement over the past. The fact that
we expect better is a sign that our standards are very high.

Even if it were possible to erase all of human culinary tradition and
reach back into prehistory to a time when foods were eaten as is, it must be
recognized that "pure" and "natural" are not necessarily synonyms for
"healthful." The various microbes that cause food poisoning and other diseases are as much a part of nature as we are, and would give us far more
trouble than they now do if they were not controlled by heat, cold, or preservatives. And as chapter 4 documents, many plants have developed chemical defenses against any animals, including humans, that might be tempted

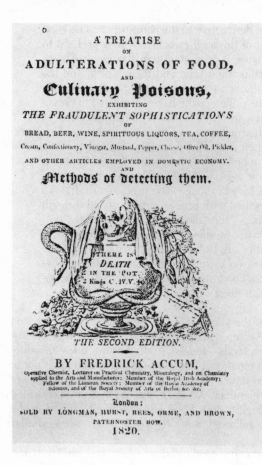

The title page of Fredrick Accum's 1820 exposé of food adulteration in England

to consume them. One important effect of the domestication of crop plants has been to reduce the toxic constituents of potatoes, lima beans, cabbage, and other foods to tolerable levels. Without at least some human intervention, then, much of our food supply would be of questionable safety. It may be that cooking itself arose as a primitive technique in part because heat generally makes food safer and easier to eat. Still, it is estimated that we consume several grams of natural, plant-generated "pesticides" each day. And the browning reactions that make fried and roasted foods so appealing also generate compounds that have mutagenic and carcinogenic activity. As a leading researcher in this field, Bruce N. Ames, has succinctly put it: "Nature is not benign."

THE RISE OF CHEMICAL ADDITIVES

Granted that we have always had additives, adulteration, and "food technologies," that some human manipulation of the food supply is inevitable and desirable, then how have we gotten to the point that several thousand additives are used routinely, and that some of our foods are little more than a mixture of additives: soft drinks, for example, or instant soups? Of course, advances in chemistry have made it possible to analyze, copy, and invent chemical structures, but chemistry has mainly served as a tool for dealing with the consequences of great economic and social shifts in the Western world: the movement of most of the population away from the country and into the cities, and the passing of a way of life that left women at home to cook all day. As farms were deserted for the factories, fewer and fewer people produced the food for more and more. Today, some of our eastern states import 90% of their food from the rest of the country. With this concentrated, centralized system of production came the need to protect foods from deterioration during the often long haul to the consumer. Hence the importance of preservatives and the development of bruise-resistant vegetables, blunt-ended carrots that don't puncture their plastic bags, and other qualities that have more to do with food as an item of commerce than as something tasty and nutritious. And because some fresh foods simply can't make it to market in good shape, freezing and canning—the latter usually involving some preservatives and texture improvers—have become more and more important.

Convenience and Consistency

Then there is the average consumer, who is now less willing or able to spend time on gathering and preparing food. Rather than making daily rounds to baker, butcher, and greengrocer, we go once or twice a week to the supermarket and expect our purchases to keep until we get around to using them. And because meals must be prepared quickly, we buy some dishes or whole meals premade to save time. It can't take more than 10 or 15 minutes every few days to keep well stocked with homemade salad dressing, and yet packaged dressings are a $650-million-a-year industry. These prepared foods too must survive on both market and home shelves for quite some time, and contain chemicals for that purpose. Convenience aside, we have also become accustomed to a wide variety of seasonal foods year-round—so that they must be grown in hothouses or shipped thousands of miles or stored for several months—and to a certain, consistent level of quality. Not necessarily excellent, but consistent. A friend of mine once tried to return a fresh bell pepper to the grocer because he had found a worm inside it! (A more organically minded consumer would have taken this as a good sign.) We have come to expect the kind of quality control that came into being with factory manufacturing, as if earth or farmer issued money-back guarantees. And yet we instinctively oppose the use of the pesticides that make the mass production of worm-free peppers possible.

As far as current knowledge in food chemistry will allow, the food industry has achieved consistency—and held down their own costs—by using chemicals tailor-made for particular tasks. Colors and flavors, sometimes copied from naturally occurring molecules, sometimes entirely synthetic, supplement or replace spices, fruits, and other ingredients that might vary in quality from crop to crop, or be temporarily unavailable, or just expensive, or that might deteriorate during processing and storage. Other chemicals slow the development of off-flavors in fats, retain moisture, firm the texture of canned produce, prevent oil-water mixtures like sauces and dressings from separating, either prevent or encourage foaming, and so on. Of course there are the preservatives that keep bacteria and molds from spoiling food before we are ready to eat it (it is estimated that fully 20% of the world's food production is lost to spoilage every year, though much of this figure is accounted for by rotting grain). And vitamins and minerals have been added to some foods since the twenties and thirties, when salt was fortified with iodine and bread with niacin to combat goiter and pellagra. Once real weapons in the fight against deficiency diseases, these additives are now treated as a selling point by the food industry. (Some otherwise nutritionally undistinguished breakfast cereals offer a whole day's worth of vitamins and minerals in one skimpy meal.) All told, there are several thousand chemicals now added to our food for one purpose or another: purposes which, with the possible exception of artificially coloring and flavoring food, sound reasonable enough, and which also tend to improve the profit of the food industry.

Other Advantages

Aside from the particular reasons that particular additives are used, a couple of points can be made in their favor. For one thing, food products made with additives are usually cheaper than the "all-natural" formulations—which themselves may contain chemicals extracted and purified from plants or animals—and so may be more available to more people. Process cheese is not exactly cheese, but not everyone can afford aged cheddar or gruyère. Additives, then, may enlarge the choice of foods to which people have access. And there are a few cases in which the ersatz is in some ways preferable to the real thing. Margarine is made from chemically altered vegetable oils, colored with synthetic carotene, and flavored with extracted milk solids, all in order to mimic butter. Yet its lack of cholesterol and lower proportion of saturated fats make it the better choice for one's cardiovascular health. Finally, there is evidence that some additives may actually prove to be beneficial to our health. The antioxidant preservative BHT (butylated hydroxytoluene) appears to have antiviral properties and is now actually being sold as a health food; and certain vegetable gums seem to lower blood cholesterol levels.

THE TROUBLE WITH ADDITIVES

Deception

Described in this way as historically inevitable and genuinely useful, food additives sound benign enough, and yet they have the reputation of being anything but benign. There are two basic reasons for this distrust. One is that additives can be used not simply to preserve and standardize and stabilize, but to replace traditional ingredients and conceal poor quality in the sole interests of profit. For example, thickeners and emulsifiers may be used in ice cream to create the smooth texture that would otherwise be the sign of a high proportion of cream, and flavorings and colors can replace real spices and fruits. To the extent that this is done without the buyer's knowledge—and many federal "standards of identity" which excuse a product from labeling allow certain levels of additives to go unlisted on the product labels—it is indeed an objectionable practice, and may even approach fraud.

On the other hand, it must be recognized that one of the most popular "foods" in the country today is almost entirely synthetic, and this doesn't seem to be held against it. Our annual consumption of soft drinks, which are nothing but carbonated water, sugar or saccharin, caffeine, a few organic acids, and some secret flavorings, is about 40 gallons per person. The original, Coca-Cola, was invented by an Atlanta pharmacist in 1886 as a hangover remedy (it contained extracts of coca leaves and cola nuts), and Pepsi was supposed to aid digestion (pepsin is a digestive enzyme, which Pepsi-Cola never contained). Soft drinks have no nutritional value whatever aside from the calories in sugar. But does their unnaturalness distinguish them in a meaningful way from coffee or tea, which are "natural" mixtures of water, caffeine, and flavors?

The Threat to Health

By far the most serious objection to the spread of additives in our food is that some of these new molecules may be slow and subtle poisons, that we may be purchasing economy, convenience, and pleasure with an increased incidence of cancer. Were the evidence clear-cut, the benefits of a particular additive could be weighed against its cost to health. Brighter color, smoother texture, cheaper flavor, or three more days on the shelf would clearly not count for much if these advantages also caused serious illness. Several widely used additives have already been banned because they appeared to pose a real threat: among them some coal-tar derived dyes, the artificial sweetener called cyclamate, and safrole, a component of sassafras oil and the original, natural flavoring in root beer.

Unfortunately, the evidence seldom is clear-cut, simply because it is very difficult to isolate the effects on people of a single, very minor part of their full diet. For one thing, researchers can't just feed a suspected carcinogen to human beings and see what happens, though back at the turn of the century, Harvey W. Wiley, who later became the first head of the Food and Drug Administration (FDA), had a young group of volunteers called "The Poison Squad" who actually were guinea pigs for such short-term studies. And the best that "epidemiological" studies, or surveys of selected groups in the population, can do is discover a *correlation* between the heavy use of certain foods and an increased incidence of disease. Even strong correlations are not the same as absolute proof of cause and effect, and they can be misleading. Recently, the finding of a correlation between high caffeine intake during pregnancy and birth defects was called into question because it turns out that heavy coffee drinkers also tend to be smokers.

In order to supplement inconclusive human studies, laboratory animals—usually rats—are used to determine more clearly the effects of particular substances. Because we want to know about the possible effects of long-term exposure—on the order of decades—to food additives, the animals are subjected to huge doses, many times the normal human consumption. In this way it is hoped that subtle, slow effects will be magnified enough to be detected in a few hundred animals (larger test populations would be very expensive), and in a matter of months rather than decades. But because rats are not humans, and the dosages of suspected chemicals far exceed the equivalent human intake, the results of these studies, positive or negative, again do not constitute absolute proof regarding the consequences for people. Arguments run both ways. The food industry usually emphasizes the artificiality of the test conditions, while public interest groups argue that any evidence is valuable: better safe than sorry. The general lack of certainty is reflected in the FDA's title for the list of approved additives, which are not classified as "Safe," but only "Generally Recognized As Safe."

Recognizing Risks

The issue of safety is also complicated by our curious, half-conscious attitudes toward both food and innovation. We have the idea that our food could easily be perfectly safe if only the producers were less mercenary and government more effective. The problems of spoilage and natural toxins rarely figure in this assumption. For some reason we are reluctant to see that a supply of plentiful, cheap, and convenient food entails certain risks, while we understand and generally accept the risks of driving or flying in exchange for the convenience they offer. (Of course, accepting risk is one thing; taking unnecessary risks is something very different.) Then there is the split mind we have about the benefits and risks of technology in general. An innovation is exploited in the first place because it promises to bring certain advantages, but it is only after the product or process has been in use for a while that all the disadvantages, the unforeseen side effects, become obvious. In the meantime, we become accustomed to the benefits, and tend to focus blame for the disadvantages on the most visible target: the industry that makes a profit while causing problems. We resist recognizing that solutions to old problems generate new problems, and that all of us, who also profit from the improvements, ultimately share responsibility for these new problems.

How we as individuals can exercise our responsibility for the condition of the world is a good question, and in the end a political one. But our more limited subject is food additives. Two brief case histories will help illustrate how much the forces of history and the vagaries of less-than-rational public demand can genuinely complicate the issue of food safety.

Saccharin and Nitrite

Saccharin: Free Sweetness

Saccharin is a synthetic chemical that did not exist before 1879 (for the circumstances of its discovery, see chapter 12), and has been used as an artificial sweetener since the turn of the century. Its great advantage is that the body cannot absorb it and use it, like sugar, for energy. It simply passes out of the body via the kidneys. So saccharin offers the satisfaction of a sweet taste without the calories that sugar involves. To begin with, artificial sweeteners were used by the food industry because they were cheaper than sugar; but in the last few decades they have become popular with a public obsessed with weight control, and the consumption of diet soft drinks and tabletop sugar replacements has skyrocketed. In 1969, sodium cyclamate was banned in the United States after animal studies linked it with cancer of the bladder. Then in 1977 a group of Canadian researchers found that saccharin had the same effect, and the FDA proposed a ban on it as well, except for medically prescribed use. But this time there were no other artificial sweeteners ready to fill the gap, and meanwhile many Americans had come to regard diet soft drinks as a necessity, a "food" that satisfies the sweet tooth without the inconvenience of providing any food energy at the same time.

There resulted the spectacle of various consumers' groups calling for a ban on saccharin while consumers themselves bought and hoarded diet soft drinks by the case, and while the American Diabetes Association warned of the "grave effects" such a ban might have on its constituency, which must limit its intake of sugar. A combination of general public pressure and a $1 million campaign by the Calorie Control Council, which represents the soft-drink and pharmaceutical industries, proved to be irresistible, and in October 1977 the House of Representatives voted 374 to 23 to place a moratorium on the FDA ban for 18 months. The decision of that regulatory agency, once accused of coziness with the food industry, was in effect nullified, and later withdrawn in favor of a warning label on products that contain saccharin.

This reversal may have been inevitable. Despite the fact that the rat experiments had shown saccharin to be a rather weak carcinogen—the estimate was that it increases the risk of human bladder cancer by 4%—the FDA was all but forced by law to call for the most drastic action. The "Delaney Clause" in the 1958 Food Additives Amendment to the 1938 Food, Drug, and Cosmetic Act requires that an additive be banned if it causes cancer in "appropriate" human or animal tests. The FDA was therefore an easy target for charges of overreacting and overregulating. Had a more moderate approach to saccharin been possible, one that could have taken into account the deep-seated, strong emotions involved in eating and dieting, then the controversy might have been less intense and more informative.

Still, was the eventual outcome a bad one? The fact that saccharin is consumed primarily by those who particularly want its unique properties puts it in the same class as cigarettes: a calculated risk that people take at their discretion. On the other hand, children do not smoke, but they may very well end up drinking diet sodas. And children run a higher risk from saccharin than adults because they have smaller bodies in which to dilute it. Should the FDA do something to protect children from saccharin, or is this the responsibility of parents? The answer is not altogether obvious.

Sodium Nitrite: Preservation, Color, Flavor

Then there is the matter of sodium nitrite, a simple salt ($NaNO_2$) that is used in the manufacture of such popular meat products as bacon, ham, hot dogs, and cold cuts. Nitrite has been the object of some concern since 1956, when it was shown that it can combine with nitrogen-containing groups on amino acids and related compounds to form nitrosamines, which are known to be powerful carcinogens in animals. But only traces of nitrosamines were found in cured meats, and while they can be formed by cooking and in the human digestive tract from the two separate ingredients (nitrite and secondary amines), cured meats account for only about 20% of the nitrite available in the body for this reaction. The other 80% are formed in the body from nitrates (KNO_3 or $NaNO_3$), compounds that are commonly found in many plants, and especially in vegetables raised on high-nitrate fertilizer. So the nitrosamine problem did not seem particularly urgent.

Then in 1978 an MIT study indicated that nitrite itself could be a carcinogen: it appeared to produce cancer of the lymph system in rats. This time, a year after the saccharin debacle, the FDA announced only that it was assessing the situation, not calling for an immediate ban. In the following months, the meat industry argued that nitrite is an important preservative, especially against the bacteria that cause botulism: the fact that cold cuts and some sausages are not cooked before eating makes the use of such an agent necessary. On the other side, consumers' groups said that nitrite is primarily a coloring agent, not a preservative, and that bacon and hot dogs are in any case cooked before eating and so need not be treated. The FDA ordered further study of the issue, and meanwhile the Department of Agriculture directed meat processors to lower the levels of residual nitrite in their products from an average 200 parts per million to 40, over a period of several years. In 1980, a review of the MIT study found fault with its method of diagnosing lymphatic cancer, and the FDA decided to take no further action. Some people continue to believe, however, that nitrite should be banned because botulism in meat is not a significant problem, while a 20% rise in nitrite available for the formation of nitrosamines in the body may very well constitute a serious and unnecessary hazard.

A Long History

A glance at the history of curing meats turns out to be instructive. Salt curing has been practiced for several thousand years as a means of preserving whatever meat or fish could not be consumed in a day or two. The primary effect of salt is to dehydrate the meat and inactivate any microbes on it by dehydrating them: the high concentration of dissolved material outside the cells draws water out of them, and they can no longer function. It also happened that heavy salting changes the color of the meat. Instead of turning grayish-brown, it develops the stable bright pink color that we associate with bacon, ham, and some sausages. The salt itself does not figure in the color change. In the earliest days, it was caused by an impurity in the salt: namely naturally occurring nitrate, which is reduced to nitrite by bacteria on the meat. The nitrite in turn undergoes an intricate series of reactions with the muscle pigment myoglobin, and the result is the pink compound nitroso-myoglobin. The chemistry of the change has been understood only in the last few decades, but people have actively manipulated it for centuries.

At some point in the Middle Ages, saltpeter—potassium nitrate (KNO_3)—was discovered as a useful and versatile substance. So called because it was found as a crystalline outgrowth on rocks, it was valued as an aid in extracting other minerals, as a fertilizer, and as an essential component in the new invention called gunpowder. Sometime in the 16th or 17th century, it was realized that saltpeter is also useful in the curing process for its effects on color and flavor. From that point on until the 1940s, saltpeter was a standard ingredient in the curing mixture: and a major one at that, as Hannah Glasse's proportions indicate. It is guessed that the levels of nitrate and nitrite in traditionally cured meats were from 10 to 50 times the average of recent years.

Two developments brought the nitrate and nitrite levels down so drastically. One was refrigeration. Cold temperatures slow or stop the activity of microbes, and so do much of the work that the curing mixture once had to do alone. The other was the gradual understanding, worked out by German chemists at the turn of the century, that nitrite rather than nitrate is the active ingredient. Once this was known, saltpeter could be eliminated from the mixture and be replaced by much smaller doses of pure nitrite. The first U.S. patent for the use of nitrite in curing was awarded in 1917, and the method was first permitted in commercial production in 1923. There was little or no consumer resistance to the reduction of the various salts used in curing; in fact, the trend was probably much appreciated. The traditional concentrations necessary for the job of preservation resulted in a meat so salty

SALTPETER IN CURING

Bacon, to dry: Cut the Leg with a piece of the Loin (of a young Hog) then with Saltpeter, in fine Pouder and brown Sugar mix'd together, rub it well daily for 2 or 3 days, after which salt it well; so will it look red: let it lye for 6 or 8 Weeks, then hang it up (in a drying-place) to dry.

—William Salmon, *The Family Dictionary:*
Or, Household Companion. London, 1710
(4th edition).

To Pickle Pork: Put a Pound of Salt-petre; and two Pounds of Bay-salt to a Hog.

—Hannah Glasse, *The Art of Cookery.* London, 1748.

To Corn Beef To Be Used The Next Day. Sprinkle the beef with saltpetre. A few minutes afterwards, rub it well with salt; repeat the rubbing four or five times in the course of the day, turning it every time it is rubbed. It must be well rubbed the next morning.

—*The Carolina Housewife.* Charleston, 1847.

Either saltpeter or nitrate of soda, closely-related mineral salts, are used to perform the following functions: First—Preserve the red of the meat. Second—There is a current opinion that they act to preserve the proper condition of the meat. The most important function of these salts, however, is to preserve the red color of the meat.

—*The Packers' Encyclopedia: Blue Book of the*
American Meat Packing and Allied Industries.
Chicago, 1922.

An attractive color is an important sales asset to a cured meat or meat product, and to secure this fresh bright pink or red color it is necessary to add saltpeter (KNO_3), sodium nitrate ($NaNO_3$), or sodium nitrite ($NaNO_2$) to the salt and sugar.

—P. Thomas Ziegler, *The Meat We Eat.*
Danville, Ill., 1944.

that, in order to be palatable, it was usually soaked in water overnight and then boiled for an hour or more.

This look backward reminds us of a couple of significant facts. First, salt-cured meats are a familiar, traditional part of our diet. Bacon and ham, unlike diet sodas and cheese spreads, existed long before big business. And second is that thanks to advances in food technology, we are probably now consuming minuscule amounts of nitrate and nitrite compared to our ancestors. We eat more fresh and less cured meats, and the meat packers put much less nitrite into cured meats than housewives used to in centuries past. Just in the last few years, the combination of required lower nitrite levels, and publicity about the possible risks involved, which appears to have affected per capita consumption of cured meats, has further reduced our exposure to nitrite. Given the outcry about that newcomer saccharin, the fact that these foods have been eaten for centuries without having aroused suspicion, and the fact that the now-suspect agent has been much diluted, a total ban on nitrite would seem to be excessively drastic. Consumers who are devoted to bacon or bologna would resist, the meat industry, with plants that produce several billion dollars' worth of such products a year, would be hurt, meat prices would rise: a series of consequences that would be difficult to deal with even if the evidence of risk were clearer.

Nitrite and Botulism
There remains some disagreement about nitrite's exact role in the making of cured meats. After all, if it were only incidental, then removing it would cause no great change in the products, and everyone would be happy. The adversaries of nitrite tend to the view that it does nothing but preserve the red color of meat. It is clear that saltpeter was valued largely for this reason, and food technologists held much the same opinion about nitrite well into the 1960s. Norman N. Potter said in his 1968 *Food Science* that "the prime purpose of curing is to preserve the red color of meat," and defined sodium nitrate and nitrite as "red color fixatives." It is also generally agreed that nitrite contributes a sharp, piquant flavor to cured meat, and retards the development of rancidity in the fat. But industry and government now make a claim for nitrite as an essential preservative that is especially effective against the bacterium that causes botulism. What truth is there to this seemingly rear-guard argument?

Apparently a good deal. In the last decade, independent researchers have shown that nitrite inhibits the growth of a wide range of bacteria. How this happens is not well understood, though it is a matter of a complicated interaction between nitrite and salt levels, acidity, and temperature. In fact, today's comparatively low levels of nitrite may be even more important than the high levels of the past, since the salt in cured meats is now less a preservative than a flavoring: we no longer have to boil our hams before baking them. It is thought that nitrite interrupts processes both within the microbe and on its cell membrane, inhibiting certain enzymes and interfering with the general production of chemical energy. And it does seem to be particularly useful against *Clostridium botulinum,* the bacterium that causes the sometimes fatal food poisoning called botulism. The very name of the illness is evidence of the risk from cured meats: it comes from the Latin *botulus,* meaning "sausage," and translates the German word *Wurstvergiftung,* or "sausage-poisoning," a malady that first attracted medical notice in late-18th-century Wildbad. To this day, the most common source of botulism in Europe is cured pork, and in Japan, salt fish. (Here, on account of the popularity of home canning, it is canned vegetables.)

The problem is that canned and cured foods are often not thoroughly cooked before eating, and may be stored at room temperature for some time. The spores of the botulism bacterium, which thrives only in oxygen-free surroundings, can easily tolerate boiling temperatures. If these spores manage to get into canned ham or vacuum-packed cold cuts, and these products are in turn allowed to reach a temperature of 50°F (10°C)—that is, sit at room temperature at any point during the long process of distribution, display, or transport to and storage in the home—then the spores will germinate into active bacteria, and these produce a deadly nerve toxin. The toxin is destroyed by temperatures above 160°F (71°C), but then cold cuts are seldom cooked, and the center of a ham may not get that hot. The virtue of nitrite is that it not only suppresses the active bacteria, but also weakens the heat-resistant spores so that they can be destroyed without resorting to pressure-cooking at the factory (which would change the meat's flavor and texture), and thus reduces the likelihood that germination will occur during careless handling of the food. Nitrite, then, is a triply useful additive: it flavors, colors, and preserves.

Continuing Progress

The nitrite saga is not over. It has recently been learned that ascorbic acid—vitamin C—and its close relative erythorbate, which have been permitted in curing solutions since the 1950s because they speed curing and stabilize both color and flavor, also improve the bacteria-inhibiting properties of nitrite, *and* reduce the formation of nitrosamines, during both cooking and digestion. So orange juice at breakfast and tomato in a BLT may be better for us than we had realized. And nisin, an antibiotic produced by the bacteria that sour milk and make cheese naturally resistant to spoilage, apparently also has a synergistic effect with nitrite, and may make lower doses of nitrite possible. These developments hold out the promise that the historical trend will continue: that thanks to a growing understanding of food chemistry we will be able to enjoy cured meats, once a necessity and now largely a pleasurable choice, with an even smaller exposure to possibly hazardous nitrite.

INFORMED CHOICE

As the stories of saccharin and nitrite illustrate, the issue of food additives is not a simple one. True, science has brought new hazards into the world, but it has also brought the knowledge necessary to recognize hazards and do something about them. The food industry has encouraged frivolous consumption, but it has also responded to shifting social arrangements and met the desires of a middle class preoccupied with convenience. We consumers want our food to be safe, but we also want fresh, worm-free produce in winter, and diet sodas. And many of us continue to smoke and maintain diets high in salt, sugar, and animal fat, despite the fact that such habits have been implicated in the most common and serious diseases of our time.

The value of some historical perspective on the subject is that it brings into view the conflicting values that complicate the matter of safety, and discourages unrealistic expectations. At the same time, it shows that public opinion is susceptible to education and can be powerful. By investing some of the time and effort that modern food technology has spared us in understanding that technology, in learning about the virtues and drawbacks of particular products and acting accordingly, we can exert some influence on producers. Companies are economic rather than moral entities, and respond more readily to changes in demand than to demands: witness the recent proliferation of prepared foods that loudly advertise their lack of preservatives or added sodium.

Of course, the idea of an intelligent choice presupposes the availability of the information required to make the choice. Current regulations leave a good deal to be desired on this count, since not all prepared foods are required to list their ingredients on the label. Only by means of more stringent labeling laws can hypertensives have full control over their sodium intake, can people with allergies to particular ingredients or additives avoid possibly dangerous surprises, can each of us buy what he or she wants, nothing more or less.

Common Sense

If in the meantime you wish on general principles to minimize your exposure to food additives, this is a straightforward matter. Avoid those prefabricated foods that usually contain additives: "instant" soups, gravies, salad dressings, dessert toppings, and the like; soft drinks, imitation fruit juices, snack or "junk" foods, cold cuts, precooked canned or frozen dishes, and so on. Replace them with fresh or plainly frozen fruit, vegetables, and meats, and cook these yourself. When choosing bread, cereals, cheese, condiments, ice cream, and other foods that we usually buy premade, check the label to make sure that the manufacturer has used ingredients that you approve of.

All this is nothing more than common sense, and the basic ideas—rediscover the advantages of preparing your own food, distinguish between worthwhile and frivolous innovations—were expressed quite clearly 200 years ago, when saccharin was not yet known, but adulteration and precursors of supermarket tomatoes were. In his book on education, *Emile*, Jean-Jacques Rousseau described how he would live if he were rich and could have anything he wanted. In the matter of foods, he said,

> I would always want those which are best prepared by nature and pass through the fewest hands before reaching our tables. I would prevent myself from becoming the victim of fraudulent adulterations by going out after pleasure myself.... I would lavish my own efforts on the satisfaction of my sensuality [that is, cook for himself], since then those efforts are themselves a pleasure and thus add to the pleasure one expects from them....
>
> It takes effort—and not taste—to disturb the order of nature, to wring from it involuntary produce which it gives reluctantly and with its curse.... Nothing is more insipid than early fruits and vegetables. It is only at great expense that the rich man of Paris succeeds, with his stoves and hothouses, in having bad vegetables and bad fruits on his table the whole year round. If I could have cherries when it is freezing and amber-colored melons in the heart of winter, what pleasure would I take in them when my palate needs neither moistening nor cooling?

Of course, agriculture has improved greatly since Rousseau's day, and "fraudulent adulterations" is too strong a phrase for inadequate labeling. But the choice between neglecting and accepting our responsibility for what we eat, between pale imitations of nature and improvements on it: these choices must still be made, and no longer just by the wealthy.

Part 2

Food and the Body

CHAPTER 11

Nutrition:
American Fads,
Intricate Facts

The well-intentioned cook has two principal aims in preparing a meal: to nourish and to give pleasure. Preceding chapters have dealt with particular foods, the ways in which we manipulate them, their nutritional and gustatory characteristics. This chapter and the next are devoted entirely to our nature as beings that eat: what materials we need, why we need them, how we extract and absorb them, what flavor is, and how we are affected by it. These questions fall into the three general and intimately related subjects of nutrition, digestion, and sensation.

Because these subjects are extensive ones, our treatment of them will necessarily be brief and incomplete, limited to information that might illuminate our everyday experience.

NUTRITIONAL FADS IN THE UNITED STATES

Why We Are So Gullible

Nutrition and diet are inescapable subjects today. Newspapers, supermarket tabloids, magazines, radio and television, and ever-expanding book racks bring us the latest word on macrobiotics, organic foods, virtuous vitamins, magic minerals, fiber, high-carbohydrate diets, high-protein diets, and on, and on. *Mental Health Through Nutrition, Nutrition for a Better Life, A Diet for One Hundred Healthy Happy Years*, and *Mega-Nutrients for Your Nerves* are only a few recent titles that promise to solve our problems with food. Americans looking for a path to healthful nutrition are offered a wilderness of choices.

It is worth asking why this should be: why, in this relatively sophisticated age, we should be spending billions of dollars a year on largely worthless dietary supplements and the latest revolutionary way to ward off cellulite or cancer. The answer, if there really is one, seems to lie deep in our national character, reflecting both its peculiar virtues and defects. No other country has been so fast and willing to experiment with its health on a large scale. And Americans are genuine pioneers in this regard. The nutritional fads of recent years actually mark the second such flowering in our history, the first having begun around 150 years ago. Sylvester Graham and his whole-grain flour, James Salisbury and his chopped steaks, and John Harvey Kellogg and his breakfast cereals are the best-known survivors of the first movement.

Our Knowledge Is Recent
     Perhaps we can claim general ignorance as an extenuating factor for 19th-century fads. Solid knowledge of human nutritional needs is barely a century old. From the time of Hippocrates, around 300 B.C., down to the 18th century, it was generally believed that all foods supply one basic nutriment. The science of nutrition begins late in that century with the French chemist Lavoisier, who was the first to assert that "Life is a chemical function," and who demonstrated that animal heat results from the process of oxidation (in fact, he gave oxygen its name). Around 1825 William Prout recognized three "staminal pinciples" in food, the "saccharina" (carbohydrates), "oleosa" (fats), and "albuminosa" (proteins), though how their roles differed remained a mystery. In the first comprehensive theory of nutrition, the great German chemist Justus von Liebig (1803–73) asserted that nitrogen-containing proteins form tissues and are consumed in mental, emotional, and muscular activity, while sugars and fats are oxidized to provide body heat. Liebig's student Carl Voit proved his mentor wrong about protein as a source of energy, and gradually the functions of the three chemical groups were clarified.
     By the turn of the century, around 16 different amino acids were known as the building blocks of proteins. By 1910 Frederick G. Hopkins in England, and Thomas Osborne and Lafayette Mendel in Connecticut, had demonstrated that different proteins have different nutritional value depending on their amino acid contents. And at the same time, Hopkins, the Dutch physician Christian Eijkman, and the American E. V. McCollum were isolating "accessory factors," or vitamins, for the first time, and showing that some severe diseases were in fact vitamin deficiencies. It was not until the mid-seventies that roughage, or fiber, attracted serious scientific interest.
     So the foundations for an informed view of human nutrition are very new. For this reason alone, it isn't surprising that our ancestors should have had peculiar ideas on the subject. Nor do we yet know the whole story, as the persistent debates about cholesterol, fiber, and vitamin C demonstrate. And ignorance, or half-knowledge, certainly does affect the gimmick of a fad. The pattern is predictable: scientists open up a new area of knowledge, and before it has been entirely explored and understood, the popular theo-

rists exploit it as the long-awaited Answer. When proteins, carbohydrates, and fats were the only known nutrients, fads centered on each of them. Vitamins are discovered, and out come Vitamin Bibles. Along comes fiber, and along comes the life-saving high-fiber diet. So ignorance may well determine the *content* of fads. But it really doesn't account for the impulse to develop them, or believe in them.

Optimism and Feverish Ardor

In the 1830s, just as Sylvester Graham's dietary ideas were catching on (see page 283), the French aristocrat Alexis de Tocqueville toured the country. His classic *Democracy in America* pays little attention to food habits themselves, but does give us a suggestive perspective on the subject. Tocqueville thought that the American insistence on the idea of equality had a massive effect on the general attitudes of citizens toward the conduct of life. In a society where all are equal, individuals will find the ultimate authority in matters of opinion and belief not in anyone else, or in the traditional values of class or nation, but in themselves. Corollary to this idea is a general optimism which this observer from the Old World found unwarranted. Americans, he said, have an exaggerated sense of the individual's control over his own destiny, and an equally mistaken sense of the perfectibility of human nature.

This begins to explain why Americans might ignore the advice of experts on one hand, and yet put their faith in the inexorable progress of "scientific" nutrition on the other. Tocqueville's analysis also suggests why it is that we should be especially interested in our diet. He described a distinctly American materialism, now long taken for granted, and argued that it too springs from the root idea of equality. In class societies, the rich take material well-being for granted, and the poor take its impossibility for granted. Only the middle class, because it is within reach of prosperity but not guaranteed it, must be obsessed with the struggle for wealth. The "classless" because "equal" Americans are blessed and cursed, by default, with the position of the middle class. Said Tocqueville, "It is strange to see with what feverish ardor the Americans pursue their own welfare, and to watch the vague dread that torments them lest they should not have chosen the shortest path which may lead to it." This is precisely what dietary fads offer: an easy shortcut to health and long life, which are the chief of material goods because the prerequisites for enjoying all others.

The desire for a more radical shortcut to happiness surfaces in the religious revivalism of the times. Various adventist groups in the 1830s and 1840s proclaimed the imminent return of Christ, leading Tocqueville to observe drily that "Religious insanity is very common in the United States." It is worth noting that Sylvester Graham was a Presbyterian preacher who was influenced by English Swedenborgians and became a cult figure himself, while John Harvey Kellogg ran a sanitarium for the Seventh Day Adventists (see pages 247–48). America's peculiar hospitality to dietary and religious revisionism might be explained by the fact that we have not had a single, strong, national tradition in either cuisine or belief. By contrast, the countries of the Old World have been much more resistant to change, much less willing to credit vitamin pills or bran flakes with the power of transforming their lives. American dietary gullibility, then, may in part be an unfortunate consequence of cultural pluralism.

The origins of American faddism are certainly more complicated and more noble than this. The toll in human misery taken by slavery, industrialization, and the rapid growth of the cities, gave rise to several important reform movements early in the 19th century, including abolition, suffrage, temperance, and the labor movement. The aim of dietary reform was to reduce the very high consumption of alcohol, meat (often salt pork), sugar, coffee, and tea, and to establish more balanced, moderate habits. This praiseworthy effort lives on today. But some prominent reformers were given at times to programmatic and rhetorical excesses—irrational and extreme prejudices against certain foods, promises of ideal health—which quickly became the mainstay of faddism. There is a line between prudent concern for the way one and one's fellows live, and the "feverish ardor" of self-interest that caught Tocqueville's eye. It is the latter that seems to animate the devotion to the strange prescriptions of the last 150 years.

Nutritional Fads, Past and Present

Enough of generalities. The point can be made more vividly by looking at a handful of examples, past and present, and realizing how little has changed in the last 100 years. Listen for the common notes struck by these writers, three from before the age of vitamins, and one from the enlightened seventies.

Salisbury Steak

Dr. James H. Salisbury, in *The Relation of Alimentation and Disease* (1888), proclaimed the discovery that bodily disorders of all sorts are largely caused by starchy foods. And why should bread and vegetables be bad for us? Because "By structure, man is about two-thirds carnivorous and one-third herbivorous." According to Salisbury, most of our teeth are "meat teeth," and our stomach is designed to digest lean meat; only the small intestine works on plant foods. It follows, then, that "healthy alimentation would consist in a diet of about one part of vegetables, fats, and fruits, to about two parts of lean meat." The trouble with even this amount of starch is that, since our digestive enzymes work only gradually on it, it "ferments" in the stomach and intestine to produce acid, vinegar, alcohol, and yeast: all substances which poison and paralyze the tissues and can cause heart disease, tumors, mental derangement, and especially tuberculosis. Salisbury's cure for all these ills, naturally, is a diet low in starch and high in lean meat, together with lots of hot water to rinse out the products of fermentation. Here is his prescription, or recipe.

> Eat the muscle pulp of lean beef made into cakes and broiled. This pulp should be as free as possible from connective or glue tissue, fat and cartilage. . . . The pulp should not be pressed too firmly together before broiling, or it will taste livery. Simply press it sufficiently to hold it together. Make the cakes from half an inch to an inch thick. Broil slowly and moderately well over a fire free from blaze and smoke. When cooked, put it on a hot plate and season to taste with butter, pepper, and salt; also use either Worcestershire or Halford sauce, mustard, horseradish, or lemon juice on the meat if desired.

So is born the Salisbury Steak, which in its contemporary manifestations often seems to be more glue tissue than muscle pulp.

Kellogg against Meat

Forty years later, one of the pioneers in breakfast cereals, Dr. John Harvey Kellogg, remarked that " 'Salisbury steaks' are now seldom seen or heard of": a fact for which Kellogg himself was partly responsible, he being a strong advocate of the vegetarian diet. He did share with Salisbury the belief, first made popular by Sylvester Graham, that diet was the determining factor in health, and said, "dyspepsia is unquestionably the foundation of the greater share of all chronic maladies," among which he included rheumatism, gout, tuberculosis, typhoid fever, "organic diseases of the spine and brain, and even insanity. . . ." This is how he put the argument for vegetarianism in *The New Dietetics: A Guide to Scientific Feeding in Health and Disease* (3rd edition, 1927).

> Flesh foods are not the best nourishment for human beings and were not the foods of our primitive ancestors. They are secondary or second-hand products, since all foods comes originally from the vegetable kingdom, being the product of the magic of the chlorophyll grain. There is nothing necessary or desirable for human nutrition to be found in meats or flesh foods which is not found in and derived from vegetable products.

Meats are especially unappealing because they necessarily contain the waste products of the animal's last muscular activity, so that "in the words of Professor Halliburton, the English Chemist, 'beef tea bouillon is simply an ox's urine in a tea cup.'" And muscle tissue itself, once inside the human intestine, encourages the growth of "putrefactive and other poison-forming bacteria" which can cause "auto-intoxication," or self-poisoning. The remedy, once more, is obvious: change your diet, this time by eliminating all meat.

Hay's Menus: Don't Mix

Striking a balance between Salisbury and Kellogg was Dr. William Howard Hay, who wrote an influential book called *Health Via Food* (1929). Hay thought, with Salisbury, that the fermentation of undigested starch caused poisoning from within, but agreed with Kellogg that meat is not a desirable food. As he put it, "ideal health cannot be attained with any other line of foods than those outlined by God to Adam and Eve in the Garden of Eden." Hay's gimmick was to assert that the digestion of starch "requires alkaline conditions throughout the digestive tract"—an unwarranted extrapolation from the fact that human saliva, which contains a starch-digesting enzyme, is alkaline—and that "acid at any stage [of starch digestion] will permanently arrest this. . . ." The arrest of digestion means the onset of fermentation, with disease not far behind. The solution to this dilemma is to avoid acid fruits, and acid-producing meats, in any meal that includes starchy foods: in other words, "scientific" menu planning.

> All of the foregoing is the result of 24 years of experience in the application to every sort of disease condition of the simple plan of treatment founded on the right selection and combination of foods, wholly without remedies of any kind whatever, the entire object being to arrest the formation of acids of adventitious character in the body. . . .

Don't eat starchy foods with anything else, and you'll have no need for medicine of any kind. The last two chapter titles in Hay's book indicate the drift of his appeal: "Everyone His Own Physician" and "A Medical Millennium." Hay's theory was quite popular for a while, and several menu books were put out by his disciples.

The Common Fallacies

Notice what the theories of Salisbury, Kellogg, and Hay have in common, aside from being wrong—even laughably so. First, each claims that most chronic diseases are caused by a diet that is faulty in one way or another. Second, each finds its justification in a religious or quasi-rational view of man's "natural" or proper diet. Hay refers to the biblical Eden, Kellogg to a "scientific" version of Eden, our primate ancestors, and Salisbury to the numerology of teeth and digestive organs. Finally, the key to health each claims to have found turns out to be a very simple one, a single aspect of diet. A contemporary observer, Alexander Bryce, described this tactic as "the elevation of some minor detail of eating and drinking into a cult." Avoid starch, says Salisbury; avoid meat, says Kellogg; avoid mixing starch and meat, says Hay. A simple rationale, a simple prescription, and the promise of a disease-free life: verily a Medical Millennium.

The fallacies of this kind of thinking are fairly obvious. First, while it is true that malnutrition can cause various diseases, it is by no means true that all, or even most diseases are caused by malnutrition. In fact, many necessary nutrients can be harmful when eaten in excessive amounts. Second, the appeal to the distant past for an indication of our "natural" diet presupposes that at some point the human body and human diet were in perfect coordination, and that over time we have deviated from this ideal state. As the three examples indicate, the choice of this point is more a matter of individual taste than of reasoned argument. And the very notion of such a point is an idealization. Human evolution, like all evolution, is a process of continual adaptation to a continually changing environment, more like compromise than a series of outright victories over circumstance. (More about this issue in a few pages.) Third, common sense should tell us that the existence of a panacea—a single substance which will prevent or cure all major disease— is highly unlikely. Both history, which records many claims but no confirmation of any, and our steadily improving knowledge of how the human body works—that is, in an extremely complicated way—are our best guides on this point. Proper nutrition is a matter of the inclusion and balance of many different substances, some of which, notably the vitamins, these early faddists were ignorant of. And no doubt we remain ignorant of some.

So we have come a long way since these Dark and Amusing Ages? Not really. Take a look at one of the more influential books of the seventies.

Linus Pauling and Vitamin C

In 1970 Linus Pauling, who had won Nobel Prizes for his efforts in both chemistry and the cause of peace, published a little book called *Vitamin C and the Common Cold*. This viral infection which "causes a tremendous amount of human suffering" can, Pauling believed, "be controlled almost entirely in the United States and some other countries within a few years, through improvement of the nutrition of the people by an adequate intake of ascorbic acid. I look forward to witnessing this step toward a better world." Pauling's evidence included personal experience and several reports in the medical literature, though there were and continue to be contradictory reports as well. And like Kellogg, he found an evolutionary rationale for his theory. Among animals, only primates, the guinea pig, a certain bat, and several birds require vitamin C in their diet; all others are able to synthesize the compound themselves from other food materials. Pauling postulated that these few animals suffered genetic mutations that prevented them from synthesizing the vitamin, but were able to survive because their plant diet was rich in ascorbic acid. The mutants then gradually replaced the normal stock because they lived more efficiently, dispensing with the enzymatic machinery necessary for production of vitamin C. Pauling then calculates the amount of vitamin C present in a mixed plant diet that would satisfy a human's caloric needs, and comes up with a figure of 2.3 grams, or nearly 40 times the recommended daily allowance. Because we no longer get most of our calories from fresh fruits and vegetables, we need to compensate by taking large daily supplements of vitamin C. Thanks largely to Pauling, taking extra vitamin C for a cold has become a standard home remedy in the United States, despite little clinical evidence in its favor.

Here, then, is a familiar strategy applied to a modern dietary detail: we can cure a great deal of human suffering by being careful about our intake of a nutrient singled out by an evolutionary fluke. And the resemblance to the fads of 50 or 100 years ago becomes stronger if we consider how Pauling's theory has developed. In 1976, on the eve of a predicted influenza epidemic, an updated edition of the book appeared, now called *Vitamin C, the Common Cold, and the Flu*. In its final chapter, called "Ascorbic Acid and Other Diseases," Pauling strongly suggests that vitamin C can be effective in combatting heart disease and cancer. Again, we feel a tug toward the promise of a panacea.

Similarly strong and broad claims have been made for a variety of substances, ranging from fiber (Dr. David Reuben's 1975 *Save Your Life Diet*) to DNA and RNA (Dr. Benjamin J. Frank's 1976 *No-Aging Diet*). And in 1981, a book called *The Beverly Hills Diet* reached the top of the best-seller lists. It was based on the proposition that carbohydrates and proteins must never be eaten together; proteins acidify, carbohydrates cannot be digested in acid conditions, and undigested carbohydrates are turned into fat. Sound familiar? This is Dr. Hay's 50-year-old combination theory, jazzed up for the eighties, and if anything more absurd now than it was when fewer people knew less about nutrition.

So to this day there continue to be doctors and scientists (and others) who discover various, previously neglected dietary keys to avoiding fearsome disease (or unfashionable figures). And to this day people buy books of this kind by the millions. Now to point out the similarities between modern fads on the one hand and the theories of Salisbury, Kellogg, and Hay on the other, is not to deny that vitamin C and fiber are important elements of the diet. It *is* to recommend a strong skepticism toward the suggestion that anything is a miracle nutrient. We should be ready to consider new ideas and information about diet and health, but we should be wary when they are accompanied by extravagant promises. Good nutrition is not a matter of finding the single key, but of a balance among all the materials we need to live: water, proteins, fats, carbohydrates, vitamins, minerals, fiber. And good health is not determined merely by diet. One's innate constitution, smoking and drinking habits, physical exercise, exposure to environmental pollution, and exposure to disease organisms are all important parts of the whole situation that determines health. So do eat wisely and well, but don't make nutrition into a fetish.

"Natural" Foods

The idea of so-called natural diets or foods deserves special attention. It is attractive in its suggestion of a fundamental harmony between the individual and the world at large, a condition of purity and simplicity. And it is hardly new. Our word *physician* has its root in the Greek word for "nature" or "origin," and there have been many people down through the centuries who have argued that a return to our original, proper way of life is the best guarantee of health. The modern lineage of this attitude goes back to Jean-Jacques Rousseau, who said in his *Emile* (1762), "The more we depart from the state of nature, the more we lose our natural tastes," and "All is good coming from the hands of the Author of things; all degenerates in the hands of man." The first notable application of this idea to nutrition is the English poet Percy Bysshe Shelley's pamphlet *A Vindication of Natural Diet* (itself based on a friend's tract, *The Return to Nature*), which appeared in 1813. Its first sentence is unequivocal. "I hold that the depravity of the physical and moral nature of man originated in his unnatural habits of life." Greek and biblical legend, and comparative anatomy of the digestive tract, both indicate that man was originally a herbivore, not a meat-eater. "There is no disease, bodily or mental, which adoption of a vegetable diet and pure water has not infallibly mitigated, wherever the experiment has been fairly tried." In the United States, Sylvester Graham popularized similar views, and in the countercultural currents of the sixties, they received new life and new forms. Perhaps most influential was the growing preference for foods that were relatively unprocessed, free of preservatives and other additives, and produced without artificial fertilizers and pesticides. Moral objections to eating meat received a newly sympathetic hearing. And interest in Eastern thought helped propel Georges Ohsawa's "Zen macrobiotic diet" into the attention of a wide public. Based on the classification of foods according to their balance of yin and yang, the macrobiotic diet offers several programs, the most advanced—because most elementary and "natural"—consisting entirely of whole grains.

"Nature" Is Usually an Idealization

In his prescient book of 1959, *Mirage of Health*, René Dubos pointed out that there is really no such thing as "Nature" in the sense of a single, stable, static entity, and that disease-causing microbes are no less a part of nature than we are. Beliefs to the contrary are our modern version of the Golden Age of the Greeks or the Hebraic Garden of Eden. All life is characterized by continual adaptation to particular, local conditions of climate, food availability, and predation, and these conditions change. Man has been a successful species largely because he is capable of adaptation to the most extreme variations in environment, and has covered the globe from the equator to the poles. At various stages, his ancestors have been fruit-eaters, more general herbivores, and hunters. It was only with the advent around 10,000 years ago of agriculture, which transformed wild plants into more productive crops with fewer toxic side effects, that we became largely dependent on grain. Is any one of these diets more natural than the others? The evidence indicates that human beings are omnivores, capable of thriving on a wide range of foods. It is certainly not true that all substances are equally good for us in all amounts, or that some foods may not cause more trouble than they are worth. On the other hand, such issues should be decided not by intuition or arcane cosmologies, but by patient and objective study. Otherwise, "Nature" becomes simply a convenient rationale for whatever code of life one happens to be attracted to.

Benjamin Franklin made this point wittily in his posthumously published *Autobiography*. In 1722, as a boy of 16, Franklin read Thomas Tryon's *The Way to Health, Long Life and Happiness, or A Discourse of Temperance* (1691), which advocated a vegetarian diet. He was convinced, and stopped eating meat of any kind. Then, a few years later, Franklin found himself on a ship becalmed off Block Island, and watched the crew haul in some cod and begin to fry them.

> on this Occasion, I consider'd with my Master Tryon, the taking every Fish as a kind of unprovok'd Murder, since none of them had or ever could do us any Injury that might justify the Slaughter. All this seem'd very reasonable. But I had formerly been a great Lover of Fish, and when this came hot out of the Frying Pan, it smelt admirably well. I balanc'd some time between Principle and Inclination: till I recollected, that when the Fish were opened, I saw smaller Fish taken out of their Stomachs: Then thought I, if you eat one another, I don't see why we mayn't eat you. So I din'd upon Cod very heartily and continu'd to eat with other People, returning only now and then occasionally to a vegetable Diet. So convenient a thing it is to be a *reasonable Creature*, since it enables one to find or make a Reason for every thing one has a mind to do.

The Original Macrobiotics

Such arbitrary reason making can be a very serious matter. The Zen macrobiotic diet, for example, has resulted in a rash of cases of severe malnutrition, ranging from scurvy to anemia. While a pure grain diet may do wonders for the yin-yang balance, it wreaks havoc with the body's biochemistry. Grains cannot supply complete proteins, and are deficient in several crucial vitamins and minerals.

In fact, it is worth recalling how the word *macrobiotic*—from the Greek for "long life"—originated. Christoph Wilhelm Hufeland, a German physician and educator, published a book in 1796 called *The Art of Prolonging Life*. It was very popular, immediately translated into most European languages, and from the third edition on was titled *Makrobiotik*, a word Hufeland coined in order to distinguish between prolonging life and curing disease, the latter being the primary role of medicine. Hufeland insisted that long life was to be obtained only by cultivation of the whole man. Good hygiene and moderate diet are important, but so are tranquillity of mind, genuineness of character, and goodness: "physical and moral health are as nearly related as the body and the soul." In this view, the improvement of life is not to be measured by regression to a suppositious past when man was in perfect harmony with a benign and static universe, but rather by advances in knowledge and in wisdom about the human condition. A healthy antidote to much modern nonsense.

THE ELEMENTS OF NUTRITION

Carbohydrates

Carbohydrates—sugars and long chains of sugar molecules like starch—are the most abundant organic compounds on earth. They are synthesized from water and carbon dioxide by plants in the process of photosynthesis, which transforms the sun's electromagnetic energy—sunlight—into chemical energy. They are the principal source of energy for all animals, including humans. Carbohydrates account for up to 80% of the caloric intake in many countries, and about half that figure in the United States. Even the animal fat that replaces carbohydrates in affluent countries derives ultimately from the plant starch in animal feeds. Among familiar foods, bread, breakfast cereals, potatoes, rice, pasta, fruits and vegetables, and desserts and candies are all largely composed of carbohydrates. Milk is the only food of animal origin to contain any significant amount of these compounds: lactose, or milk sugar, accounts for a few percent of its weight.

In the middle of the 19th century, the German chemist Liebig determined that carbohydrates are, like fats, oxidized in the body to produce heat. Subsequent work by his students showed that their primary role is to provide each cell with the chemical energy it needs in order to function, with heat a sometimes useful by-product. We now know that a relatively small amount of carbohydrate material forms complexes with other molecules and plays an important part in the structure of connective and nerve tissues, in the activity of hormones and the immune system, and in the general process by which one cell recognizes another.

There is a handful of important dietary carbohydrates. Our everyday table sugar is pure *sucrose*, a disaccharide extracted from sugar cane or beets. The sugars most commonly found in fruits and vegetables are the monosaccharides *glucose*, also known as grape sugar, corn sugar, or dextrose, and *fructose*, the major component of honey. Because it tastes sweeter than sucrose and yet has only half the number of energy-rich bonds, fructose offers fewer calories for the same sensation. *Starch* is a long chain, or polymer, of glucose molecules. Densely packed into granules in plant cells, these chains are the chief form of stored energy in plants.

Once we have eaten them, most carbohydrates are converted into glucose, which is the body's preferred fuel, and circulates through the blood to all cells in the body. While most tissues can also burn fats for energy, the brain and other nervous tissues can use only glucose. Because proper blood glucose levels are critical to the function of the body, disorders of the complex hormonal system that regulates these levels—for example, diabetes and hypoglycemia—are especially serious.

The human body does have a storage form of glucose called *glycogen*, which is similar to plant starch. But this energy reserve is relatively small. The average adult carries only about ¾ pound of glycogen, or half a day's energy, some of it in the liver to replenish the blood glucose levels, and the rest in muscle tissue. In a sense, however, much of the body is a potential source of glucose. Both proteins and the glycerol portion of fats can be converted to sugar in the liver; conversely, surplus sugar is transformed into fat and amino acids. The biochemical machinery in the liver, then, makes these three principal nutrients largely interchangeable, and thereby permits a greater flexibility in the diet.

But flexibility does not mean perfect freedom. Changes in the relative amount of carbohydrates in the diet have important consequences for the body. Carbohydrates are the preferred energy source, the first to be tapped, and when adequately supplied, they spare both fats and proteins from being consumed. When the carbohydrate intake is inadequate, we begin to metabolize fats very quickly. In extreme cases, the by-products of this process accumulate and can cause a disturbance in the body's acid-base balance, a condition known as acidosis. If fat reserves and intake are also low and dietary protein is marginal, then there will not be enough protein to replace body tissues, and these tissues begin to waste away. This situation, called protein-calorie malnutrition, is the most prevalent kind of malnutrition in the world today, and may affect up to one-quarter of our planet's population, or a billion people.

*Over*consumption of calories is the more common problem in industrialized countries. In this case, our glycogen reserves are topped off, and the remainder of excess energy is stored as fat. Though most of us try to avoid this outcome, the marathon runner makes selective use of it in the practice known as "carbohydrate loading." By eating lots of pasta and bread and drinking beer on the eve of a race, the runner maximizes his quickly mobilized glycogen stores, and postpones the need for fat metabolism.

Simple and Complex

Nutritionists distinguish between *simple* and *complex* dietary carbohydrates, or between sugars and starch, and advise us to get more of our calories from the latter, fewer from the former. Why so, if both end up as glucose or glycogen in the body? There are a couple of reasons. For one, table sugar and honey, because they are almost pure sugar, contain very little in the way of other nutrients, while the fruits and vegetables that contain both sugar and starch also give us vitamins, minerals, some protein, and fiber.

To the extent that refined sugars replace other foods in our diet—for example, a candy bar reducing our appetite at dinner—we trade a more balanced nutrition for pure, or "empty" calories. Valuable fuel, yes, but usually we have more than enough fuel, and not enough of the many other substances necessary for life. A second reason for preferring unrefined carbohydrates is that the body doesn't handle large doses of pure sugar very well. It is very quickly digested and passes immediately into the bloodstream, where it places great demands on the regulatory system. Many—but not all—starchy foods are more gradually broken down and assimilated. For some reason, legumes raise blood sugar levels less than either grains or potatoes.

ENERGY FROM FOOD

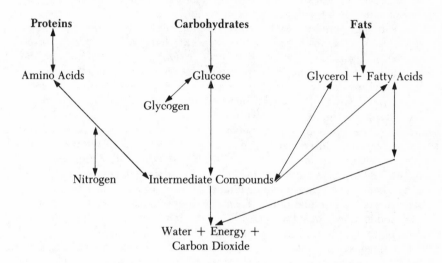

Proteins, carbohydrates, and fats can all be oxidized to obtain energy; all have caloric value. And with the exception of fatty acids, each kind of nutrient can be converted into the others.

Around the turn of the century, Americans got about twice as many calories from starch as from sugar. Today, sugar has pulled even, mostly through a decline in the consumption of bread and potatoes. In 1910 only 25% of the sugar we ate was hidden in commercial food products and beverages, while today the figure is more like 70%. Cutting down on our sugar intake, then, is more difficult than simply hiding the sugar bowl. We must also eat fewer prepared foods and drink fewer soft drinks.

Fats

Like carbohydrates, fats are used in the body primarily as an energy source, and have several other minor though important functions. Fats store about twice as much energy in a given weight as carbohydrates do, because many fewer of its chemical bonds have been oxidized (see chapter 13). One pound of fat will provide enough calories for 1½ to 2 days of activity. Cooking oils, meats, dairy products, eggs, sauces and dressings, and pastries are all significant dietary sources of fat, which, because it is a concentrated energy source, need not be present in large, obvious amounts in order to affect our caloric intake substantially. Fats are a universally popular element of food, probably on account of their various flavors, smooth texture, and the sensation of "fullness" they produce when eaten. In developing countries, they account for about 10% of the daily caloric intake, while in affluent societies like our own the figure is more like 40%.

Although some is dispersed invisibly among the body tissues, most of the fat we store as an energy reserve is concentrated in obvious *adipose* tissue made up of cells specialized for that function. Our glycogen reserves may be strictly limited, but there is no limit to the fat we can accommodate. People have survived starvation for up to a month thanks to their fat. Adipose tissue is also physically useful: several inner organs are cushioned by fat, and the layer just below the skin is a good insulator and reduces heat loss during cold weather. The average woman is about 25% fat by weight, the average man about 15%; the difference is generally regarded as an adaptation to the unique nutritional demands put on the mother by both pregnancy and breast-feeding. Like the rest of the body, adipose tissue is continually changing. When our weight remains constant, the breakdown and synthesis of fat are proceeding at equal rates. When our caloric intake exceeds our energy requirements, whether the excess be in the form of carbohydrates, fat, or protein, fat synthesis exceeds breakdown and we put on weight.

Unlike sugars, fats are not soluble in water, the body's chemical medium, and so must receive special treatment during digestion. They are emulsified, or broken down into small droplets, to increase the number of molecules exposed to the digestive enzymes in the small intestine. And in order to be distributed to the rest of the body, the resulting mixture of fatty acids, mono-, di-, and some triglycerides is coated with a phospholipid-protein envelope in the intestinal wall and released to the lymph fluid, where it complexes with blood proteins for the final journey to the liver or to adipose tissue. These lipid-protein complexes are called *lipoproteins*, and come in a couple of varieties, the high density and low density, or HDLs and LDLs. Recent research suggests that the relative levels of these substances may have something to do with the risk of heart disease; HDLs, which carry cholesterol to the liver, seem to be associated with lower risk.

The complexity of fat assimilation has one familiar consequence. Because it slows the digestive process, it delays the recurrence of hunger. This is one reason that fatty foods seem especially filling, or satisfying.

Essential Fatty Acids

Fats do much more for us than store energy. Certain fatty acids are converted by the body into phospholipids, those molecules with both fat-soluble and water-soluble ends that are the basic material of cell membranes. They also play a role in the transport of other molecules in and out of cells. The brain, nervous system, and liver are especially rich in phospholipids. A few of the fatty acids that go into these materials are, like some vitamins and amino acids, absolutely essential to the diet; the human body cannot synthesize them from related materials. We normally satisfy our requirement for essential fatty acids with the fairly common linoleic acid, an 18-carbon chain with two double bonds, which we can convert into the longer, more useful arachidonic acid. These polyunsaturated fatty acids were discovered as necessary growth factors around 1930, and researchers are only now beginning to understand exactly what roles they play. They are important components of cell membranes, and precursors of hormonelike substances called prostaglandins.

Cholesterol

Many of us probably associate the word *polyunsaturated* with advertisements for vegetable oils, and with the phrase "no cholesterol." Fat unsaturation—the presence of double bonds along the carbon backbones of its constituent fatty acids—became a commercial issue only when studies linked saturated fats with the development of heart disease. This is also the reason for the general notoriety of cholesterol, which is not a fat, but a different member of the chemical family called the lipids. It is a uniquely animal product; organ meats and egg yolk are especially rich sources. It is well known that excess cholesterol can be deposited on the walls of our blood vessels and so lead to circulatory problems. It is less well known that cholesterol is a very important molecule, and that the body synthesizes significant quantities of it. Cholesterol forms the nucleus of the vitamin D molecule, of various regulatory and sex hormones, and of the bile salts that emulsify fats in the intestine; it also contributes to the fluidity of our cell membranes. Most cholesterol synthesis goes on in the liver and intestine, but all cells can apparently supply their own immediate needs. The problem is that synthesis in the body proceeds independently of dietary intake (although it appears to be raised by the ingestion of saturated fats), and usually exceeds that intake. In addition, we have few mechanisms for getting rid of cholesterol. We cannot break the molecule down into its component atoms, and very little is excreted; in fact, bile salts are reabsorbed in the small intestine. There is also evidence that both physical inactivity and emotional stress can increase blood cholesterol levels. Because the situation is so complicated, there is still some debate as to whether reducing cholesterol in the diet by avoiding rich, fatty foods is an effective way to control its concentration in our bodies.

It is, however, indisputable that Americans today get many more of their daily calories—and with those calories more cholesterol—from fats than did our forebears: the figure was around 30% at the turn of the century, and is 40% today, or about what all carbohydrates contribute. It is estimated that this intake is from 6 to 8 times our actual nutritional needs; enough linoleic acid and fat to carry the fat-soluble vitamins are contained in one tablespoon of vegetable oil. And because we probably get much less physical exercise than people did 80 years ago, we need these extra calories even less than they did. But reducing the amount of fat in our diet is made difficult by the fact that two thirds of it is "invisible," or hidden from obvious view in lean meat, cheeses, fried foods, gravies and sauces, avocados, nuts and seeds, snack foods, and so on. And popular meat substitutes like quiche and granola often contain more fat than meat itself. An even more basic problem is that people *like* fats. But despite the controversies about saturation and cholesterol, despite the effort that reducing fat intake requires, such effort is probably worthwhile. It may not be a guarantee against heart disease or colon cancer, but the odds will certainly be shifted in the right direction.

The chemical structures of cholesterol and three related substances. For the sake of clarity, individual atoms are not shown; each junction of lines is occupied by a carbon atom. *Clockwise from upper left:* cholesterol, vitamin D, testosterone (male sex hormone), estrogen (female sex hormone).

Proteins

In 1838, about a decade after Prout had divided up food substances into the saccharina, oleosa, and albuminosa, the Dutch chemist Gerardus Mulder published his studies on the last group. Beginning with three very different materials—ox blood, egg white, and wheat gluten—he isolated what he thought was a single chemical compound with the formula $C_{40}H_{31}N_5O_{12}$. This compound, combining in different proportions with sulfur, phosphorus, and oxygen, appeared to constitute the fundamental material of most plant and animal tissue. Following the suggestion of the Swedish chemist Berzelius, Mulder named this material *protein*, from the Greek for "primary."

Within a few years, however, the picture had been complicated. Other chemists derived different formulas from different substances, and it was soon clear that protein was not a universal substance; rather, there were many different compounds of carbon, hydrogen, oxygen, and nitrogen. By about 1900 it was known that proteins are built up out of a few basic units, the amino acids, and that animals first digest dietary protein into its separate amino acids, and then synthesize their own proteins from them. Then between 1905 and 1915, Frederick G. Hopkins in England and the New Haven team of Thomas Osborne and Lafayette Mendel established that certain amino acids were required in the diet of laboratory rats, while many others could be omitted from the diet with no ill effects. Systematic studies in the forties and fifties led by William C. Rose, once a student of Mendel's, determined that eight amino acids are essential to the human diet. Deficiencies of one or more of these, which he found rare in this country outside of faddists, caused general fatigue and irritability. Our knowledge of the essential amino acids, then, is quite recent.

We need protein in our food because it constitutes the basic machinery of all life. Take away the water in our bodies, and most of what is left in our muscles, organs, blood cells, skin, nails, hair, even our teeth and bones, is protein. The enzymes that build up and break down other molecules, disease-fighting antibodies, oxygen-carrying hemoglobin, certain hormones like insulin: all these chemicals whose incessant activity keeps us going, are proteins. They are continually being used up or worn away, and protein from our diet is used to replace them, or, in growing children, to build them up. Like fats and carbohydrates, proteins can be burned for energy, but this happens only when supplies of these preferred fuels run low. Excess protein in an otherwise adequate diet will be converted to fat, but as one nutritionist has put it, this is analogous to buying fine furniture and then using it as firewood.

Complete and Incomplete Proteins

Twenty amino acids go to make up all human proteins. Of these, the adult needs a dietary supply of 8, the growing child 9 or 10. Our cells can synthesize the others, and from them the necessary proteins. Dietary protein is classified according to its provision of essential amino acids. Complete proteins include enough of them to allow normal bodily growth and function. All animal foods—meats, eggs, milk products—are complete protein sources, because all animals have the same basic biochemical machinery. Plants are organized in a very different way, however, and so plant proteins are generally *incomplete*. There are some exceptions to these rules, and human manipulation can make a difference as well. For example, wheat germ and soybean proteins are nearly complete, while gelatin, which is extracted from animal skin and bone, and zein, the major corn protein, are so incomplete that they cannot sustain life at all. Between these extremes, whole grains, beans, and nuts are barely adequate. Each typically lacks sufficient quantities of a couple of amino acids, and these deficiencies lower the total amount of usable protein. If a food contains more total protein than we need, but has only 50% of the necessary amount of one essential amino acid, then it can provide only 50% of our overall requirement.

If animal foods are the only fully adequate protein sources, then how do strict vegetarians survive? It turns out, fortunately, that plant proteins are often *complementary:* that is, the deficiencies of one food are counterbalanced by the strengths of another. Long before the world knew anything about amino acids, many cultures had adjusted their diets to take advantage of protein complementarity. Most kinds of beans are deficient in methionine, while grains usually lack lysine, and sometimes threonine and tryptophan. In the Americas corn and black or pinto beans, and in Asia rice and soybeans, have been staple combinations for a good thousand years. In Europe, peas and lentils were grown in association with wheat and barley in prehistoric times. Supplementation can also be achieved by including small amounts of animal protein in grain or legume dishes. Perhaps the most familiar instance of such a combination is macaroni and cheese.

How Much Protein Do We Need?

There has long been controversy about how much protein we actually need. Carl Voit estimated late in the 19th century that we need about 120 grams of protein per day, or the equivalent of nearly 2 pounds of meat (most meats range between 10 and 20% protein by weight), a figure that was enshrined as the Voit Standard. An American chemist named Wilbur Atwater came up with a similar figure, and Dr. Salisbury, whose chopped steak diet was in vogue at about this time, recommended 3 pounds of meat per day. It was not until 1902 that the dean of American physiological chemists, Russell Chittenden, halved this figure after seeing Horace Fletcher (whose theory of chewing we shall get to in the next chapter) maintain his fitness on 40 grams of protein, or ½ pound of meat per day. Chittenden confirmed the adequacy of 50 grams in studies on army volunteers, and this is only slightly lower than the (National Academy of Sciences) Food and Nutrition Board's current recommended daily allowance of 56 grams for the average-sized man. Women generally need about 10 grams fewer, but pregnancy and breast-feeding boost their requirements to around 70 grams. The World Health Organization, taking into account the fact that most of the world gets by on much less protein than we do, has established 40 grams as an adequate daily intake. Today, Americans average 90 grams of protein per day.

There remains a residue of bias toward protein in the American diet, particularly in certain weight-reducing regimens, and in the traditional meal of steak and eggs served to athletes in training and in competition. There are two misconceptions at work in this thinking: that foods high in protein are not as fattening as starchy or fatty foods, and that protein "builds muscle" and helps sustain muscular action. Not only is excess protein burned for energy or converted into fat—its caloric value is about the same as pure sugar's—but protein-rich foods like meat, even lean meat, usually contain substantial amounts of fat as well. And while a small added increment of protein may be needed by people who are actively developing muscle, most of the pregame steak is going to be used for energy, a role that is much more efficiently played by carbohydrates. Some football teams are taking a lesson from marathoners and have instituted high-carbohydrate meals instead, with pancakes or pasta the comparatively economical main course.

Vitamins

Vitamin deficiencies, like amino acid deficiencies, were being fought by trial and error long before their chemistry was understood. Scurvy, a debilitating disease characterized by infections, tenderness, general hemorrhaging, and anemia, plagued sailors on long voyages in the 16th and 17th centuries. Around 1750 a Scots physician, James Lind, discovered that oranges and lemons cured and prevented scurvy, and fifty years later the British navy ordered shipboard rations of limes: whence the nickname "limey." We now know that these fruits are effective against scurvy because they are excellent sources of vitamin C.

The two other major deficiency diseases, pellagra and beriberi, also become serious problems when normal dietary habits were suddenly changed. Pellagra spread through Europe when Indian corn was introduced from the Americas without the culinary tradition that improves its nutritional value, and beriberi accompanied the mechanized rice mill in Asia (see the section on grains in chapter 5). In fact, the study of beriberi gave us the term *vitamin*. In 1911 a Polish chemist, Casimir Funk, found that an extract of rice hulls prevented the disease, and guessing the chemical nature of the substance, called it "vitamine," a contraction of "vital amine." When, a few years later, it was discovered that the active material was not an amine—a particular kind of nitrogen-containing compound—the final *e* was dropped.

The single-letter designations of the vitamins also date back to the earliest days of investigation. In 1912 E. V. McCollum and Marguerite Davis at the University of Wisconsin found that rats whose only fat source was lard

failed to grow and developed eye trouble, while the addition of some butterfat or egg yolk corrected the problem. The antiberiberi factor, on the
other hand, was water-soluble. McCollum suggested calling the former, fat-
soluble factor "A" and the latter, water-soluble factor "B," since their actual
chemical compositions remained unknown. Even when they became known,
and when vitamin B turned out to have several different components, the
letters remained a convenient shorthand. The antiscurvy agent was assigned
the letter C, though it is also known as ascorbic—a contraction of "antiscor-
butic"—acid. Vitamins D and E were isolated and named in the thirties and
forties. Vitamin K skips further down the alphabet because its discoverer, the
Danish biochemist Henrik Dam, proposed the term *Koagulations Vitamin*
for its promotion of normal blood clotting in chicks.

Many Specific Roles

The vitamins differ from the three major kinds of nutrients in that they
are needed only in very small quantities, usually a few hundredths of a gram
per day. Their function is not to provide energy or to become part of our
physical structure, but rather to perform very specific roles in the metabolic
processes that regulate growth, tissue replacement, and general cellular activity. To understand this basic fact is to understand why it is that vitamins are
unlikely to be panaceas, and why excessive doses can be downright dangerous. To use a mechanical analogy, a machine's proper functioning will be
prevented if it is missing a cog or is low on oil, but one cog too many or an
overflowing crankcase will also foul the works. One particularly vivid example will serve. In the space of 10 days in February of 1974, an English health
food enthusiast name Basil Brown took about 10,000 times the recommended
requirement of vitamin A, and drank about 10 gallons of carrot juice, whose
pigment is a precursor of vitamin A. At the end of those ten days, he was
dead of severe liver damage. His skin was bright yellow.

Such drastic overdoses are made possible by the concentration of vitamins in pill form, and are encouraged by the idea, perhaps revived by the
few striking cures of deficiency diseases and the development of antibiotics,
that miracle chemicals do exist. Even the more moderate among us are
affected by this situation to some extent. It is estimated that some 50 million
Americans spend several billion dollars a year on vitamin and other nutritional supplements that are largely unnecessary. And the much-advertised
fortification of some breakfast cereals with a day's worth of several vitamins
and minerals, which goes far beyond making up for losses during processing,
is a profitable but meaningless achievement. Anyone who really needs to get
his vitamins in cereal would have such an otherwise impoverished diet that
he would face the more immediate problem of starvation. Further, a day's
dose of vitamins in a pill encourages the misconception that any diet supplemented in this way is healthful. A multivitamin washed down with a martini
is no substitute for a balanced dinner. Though there are people for whom
vitamin supplements are medically advised—those on very restricted diets

or suffering from intestinal malfunction, those taking certain drugs that interfere with vitamin metabolism, alcoholics, pregnant or breast-feeding women—their number is relatively small.

The Fat-Soluble Vitamins

Vitamins A, D, E, and K are fat soluble, and have several characteristics in common for this reason. Because we excrete very little fat, these vitamins will accumulate in adipose and other tissues if consumed in excess of our needs, and can reach toxic levels. They tend not to be lost from boiled vegetables because very little leaches into the surrounding water, but broiled or roasted meats that drip fat will lose some. These vitamins are also generally stable at normal cooking temperatures.

Vitamin A   is also known as "retinol" because it is essential to the workings of the retina. The rod and cone cells of the eye contain two different pigments, each composed of a retinol derivative, retinal, and a particular protein. When a pigment molecule is hit by a photon of light, it momentarily assumes a higher energy state, and then proceeds in several steps to separate into its component parts. This chemical change affects the concentration of ions in the cell, and the result is an electrical impulse which is transmitted to the brain. Although the retinal and protein components do rejoin to form active pigment, there is a continual degradation of the material, and a reliable supply of retinol is necessary to maintain optimal vision. One of the first symptoms of vitamin A deficiency is a loss of visual acuity at night due to the slowdown in the regeneration of pigment.

The chemical details are not known, but vitamin A is also necessary for the normal function of all our protective mucous membranes, including those in the eyes and the respiratory and digestive systems, of our skin ducts, and for the proper development of bones and teeth.

Principal dietary sources of vitamin A include milk, butter, cheese, egg yolk, and liver, the organ in which normal reserves accumulate. But the most significant source, and in fact the original source for all animals, is the dark green and yellow vegetables. Plant tissues that engage in photosynthesis— mostly the outer leaves—contain high concentrations of *carotenoids,* orange pigments of which two are the precursors, or provitamins, of A, and are converted into the vitamin in our intestinal wall, just after being absorbed. Carotenoid pigments play a subsidiary role in photosynthesis, and also protect the chlorophyll from damage by absorbing excess solar radiation. The darker green a vegetable is, the more chlorophyll *and* the more carotenoids that vegetable contains, and so the better source of vitamin A it is. Yellowish-orange vegetables like carrots, sweet potatoes, and winter squash, and fruits like peaches and cantaloupes, are also good sources.

At least in the United States, it appears that most cases of vitamin A deficiency are due not to an inadequate dietary supply, but to disorders of the digestive tract that interfere with its absorption. Overdoses are quite possible and potentially toxic; the provitamin can be stored in the body without being converted, and are less troublesome than the vitamin itself. Exactly what excessive amounts do to the body is not known, but typical symptoms of vitamin A overdose include headaches, dizziness, hair loss, and pain along the bones.

Vitamin D   designates a small group of chemical compounds related to, and in fact derived from, cholesterol. There is no standard dietary allowance for the average individual, because one form of the vitamin is synthesized in the human body when a cholesterol derivative in the skin is struck by ultraviolet wavelengths in sunlight and "activated." Deficiencies of vitamin D are rare in sunny countries. "Irradiated ergosterol" is the forbidding name given to the vitamin on the label of fortified milk.

Vitamin D was first isolated in the twenties by E. V. McCollum and coworkers at Johns Hopkins as an antirickets factor. Rickets is a kind of bodily deformation, most prominently bowlegs, caused by an insufficient deposit of calcium salts and other minerals in bone tissue. The bones therefore remain soft, and are easily deformed when forced to bear the weight of a child learning to walk. Vitamin D is converted in the kidneys into a hormone that then regulates the absorption of calcium in the small intestine and kidneys, and its incorporation in bone tissue. We now know that these processes are just as important for the adult as for the child. Even fully formed bone undergoes constant reshaping as the stress on it varies during life, and its minerals are constantly being replaced. For these reasons, an adequate dietary supply of minerals and of vitamin D remains important long after actual growth is over. Older people with a diet poor in vitamins and minerals and who stay indoors can develop osteomalacia, or "adult rickets," in which their bones soften, fracture easily, and cause rheumatic pains.

The largest contributor to our intake of vitamin D in the United States is fortified milk; most natural foods are only poor sources. Egg yolk, liver, fish, and fish-liver oils—the latter not really a food, but a supplement—all contain some vitamin D.

Deficiencies are generally limited to otherwise malnourished people in northerly, smoggy cities with little sun, to food faddists, people with digestive disorders, and to some pregnant or breast-feeding women and growing children. Consistent overconsumption of vitamin D leads to nausea and diarrhea, while extreme overdoses mobilize large amounts of minerals and can cause kidney damage and even the mineralization of blood vessels, the stomach, and other soft tissues. Ironically, too much activity of the vitamin D system can also cause net losses of bone minerals, just as the deficiency does.

Vitamin E,   or tocopherol, was discovered by the Berkeley endocrinologist Herbert M. Evans in 1922, when he found that a fat-soluble micronutrient was necessary for the successful reproduction of rats. Four different forms of the substance have now been isolated. While it was thought at first that vitamin E was required for the reproduction of all higher animals, we now know this to be untrue. Other evidence that it prevents anemia, liver and brain degeneration, and muscular dystrophy in various test animals, has proved to be inapplicable to humans. In fact, there is so far no clear evidence that there is such a thing as vitamin E deficiency in adults. And this very lack of an obvious role in human nutrition has made the vitamin a favorite with faddists, who simply extrapolate from animal experiments and recommend it for sexual and cardiovascular health.

This is what we do know about the tocopherols. They are most concentrated in the pituitary and adrenal glands and in the testes, though muscle and fat tissues store the greatest amounts overall. So far, their major role appears to be preventing the oxidation, and so degradation, of vitamin A and the polyunsaturated fatty acids. They may also be involved in the formation of red blood cells and in energy metabolism. In any case, the only documented cases of serious vitamin E deficiency come from newborns, who seem to receive little of the substance from the mother. Several blood cell disorders in premature infants respond to treatment with vitamin E. A few adverse effects, including general fatigue and malaise, have been associated with large daily doses. And there is recent evidence that large amounts may have a druglike activity and affect the body's hormonal system.

Animal foods are usually poor in vitamin E, while many grains and legumes are rich. Plants apparently synthesize tocopherols in order to prevent the spoilage of seed oils that nourish the embryo. The various vegetable oils together contribute about two thirds of our daily intake.

Vitamin K, Dr. Dam's *Koagulations Vitamin,* prevents hemorrhaging by transforming certain proteins into prothrombin and other blood-clotting factors. There are three groups of K compounds: one found in dark green, leafy vegetables, a second synthesized in animals, and a third synthesized in the laboratory which is more active than its naturally occurring models. The second group is our primary source of vitamin K, and is of especial interest because it is synthesized not *by* animals, but *in* them by various bacteria that inhabit our small intestine. This is a classic example of symbiosis: we supply the bacteria with a home and a supply of food, and they supply us with an essential micronutrient. The evidence suggests that they do more than their share, since we excrete large amounts of vitamin K. Partly because of this fact, there is no standard daily dietary allowance.

Deficiencies of vitamin K are rare but well documented. Newborns, because they arrive with a limited supply and a largely sterile digestive tract, are susceptible to clotting problems in their first few days of life. Patients taking antibiotics may devastate their helpful bacterial populations as well as those causing infections, and various intestinal diseases and disorders may interfere with absorption of the vitamin. Overdoses appear to be unknown.

The Water-Soluble Vitamins

Whereas the fat-soluble vitamins are little affected by heat and tend to remain in food during cooking, the water-soluble vitamins are easily lost to boiling water, and several are very sensitive to heat or light. Retaining these vitamins during cooking is therefore a matter of some care. On the other hand, excessive amounts can be excreted in the urine and so are less likely to cause trouble.

Vitamin C, or ascorbic acid, was first characterized chemically by Charles G. King and his students at the University of Pittsburgh in 1932 (for the story of its discovery, see pages 152–53). It is a six-carbon molecule closely related to glucose, from which it is synthesized by plants and by most animals, though not man. In plants, the role of ascorbic acid remains obscure; however, it is usually found concentrated in actively growing tissues such as sprouts and stem tips. In our bodies, it appears to be a cofactor for enzymes that form and maintain the collagen in our connective tissue, the "cement" material that surrounds individual cells, and the "ground substances" that form the structural matrix of bone, teeth, and of blood vessel walls. Serious deficiency of vitamin C, then, can damage the very fabric of the entire body. A few other roles—a general antioxidant effect, the metabolism of certain amino acids, the assimilation of iron, the activity of adrenal hormones—are less well established.

There continues to be some disagreement about the optimal intake of vitamin C. About 10 milligrams a day is enough to prevent scurvy. The recommended daily allowance in the United States is 60 milligrams, while Norway and Canada suggest 30, and West Germany 70. All of these figures are substantially below Linus Pauling's several grams per day. There are reports that smokers and those taking aspirin or birth control pills have somewhat higher requirements than normal, and that very high doses can in time contribute to the formation of kidney stones. Pregnant women are advised not to take vitamin C supplements because they appear to induce an artificially high requirement in infants, who then develop symptoms of scurvy on their normal diet.

We get most of our vitamin C from fruits and vegetables. Animal foods and grains contribute very little. Among fruits currants, strawberries, pineapple, and citrus fruits are especially rich in ascorbic acid; among vegetables sweet peppers, cabbage and its relatives brussels sprouts and broccoli, and spinach. When eaten in large quantities, potatoes can also be a significant source. However, cutting and cooking techniques affect the ultimate value of these foods. The vitamin is very soluble in water, and easily destroyed by oxidation, a process favored by exposure to the air, heat, light, alkaline pH in the cooking water, traces of metal, and by the presence of certain plant enzymes right in the tissue. In order to retain as much vitamin C as possible, foods should be eaten quickly after as little processing as possible. (See pages 179–80 for a more detailed discussion of cooking vegetables.)

The B Vitamins

Originally, Funk, McCollum, and others thought that the water-soluble B substance was simply the factor effective against beriberi. But subsequent research showed that it contained one chemical that was destroyed by heat, and another that was heat-resistant: and then that there were *two* heat-resistant components. So where once there was vitamin B, now there were B_1, B_2, and B_3. Today, 8 vitamins are collected under the heading of the "B complex." Most of them seem to have the same general function, which is to participate in the metabolism of carbohydrates, fats, and proteins throughout the body.

Thiamine, as B_1 has been named—the first three letters come from the Greek for "sulfur," which thiamine contains—is the vitamin that prevents beriberi. It is required for the production of acetylcholine, which is involved in the transmission of nerve impulses, and it plays a crucial role in the release of energy from all three major nutrients. Thiamine combines with two phosphate and then alters pyruvic acid, a molecule that is an intermediate between the original fuel and the eventual waste products CO_2 and H_2O, thereby pushing the oxidation process along. Without an adequate supply of thiamine, a supply that must increase as bodily activity increases, the biochemical machinery that liberates energy gets gummed up, with serious results for the muscles in particular. Beriberi means "weakness" in the native language of Sri Lanka, and its symptoms include muscular degeneration, heart irregularities, and emaciation that results in part from malfunction of the muscular walls of the stomach and intestine. Milder cases of thiamine deficiency cause fatigue, weight loss, and emotional disturbances.

Good dietary sources of thiamine include pork, organ meats, grains, dried beans and peas, and peanuts. Most of the B vitamins in grains are concentrated in a layer just under the hull, and this layer is frequently lost during processing. Beriberi did not become an acute problem in Asia until the improvement of rice milling techniques in the 19th century (see pages 238–39). The use of whole grains, or cooking fortified rice in a minimal amount of water so that none is discarded, will maximize thiamine retention. Deficiencies are still common in Southeast Asia, but in this country are found primarily among alcoholics, who consume many calories but very little of the substance so necessary to their metabolism.

Riboflavin, or B_2, had its chemical formula determined in the early thirties. Its name is a compound of ribose, a five-carbon sugar, and flavin, a class of chemicals that are fluorescent pigments (riboflavin fluoresces yellow-green). This vitamin is a component of two coenzymes that transfer hydrogen atoms during fuel oxidation, and of enzymes that oxidize amino acids. Riboflavin deficiency is thought to be one of the most common vitamin deficiencies, and is especially prevalent among the poor and among alcoholics. Early symptoms include cracked skin in the corners of the mouth, and eyes that are sensitive to bright light and easily tired. Dermatitis and more severe eye trouble may follow.

Organ meats are the most concentrated sources of riboflavin, and other meats, eggs, green vegetables, and enriched cereals and flours also contribute to our intake. In the American diet, the single most important source is milk. Like thiamine, riboflavin is stable to heat but sensitive to alkalinity, though it has a unique vulnerability: it is quickly degraded by both ultraviolet and visible light. This is one good justification for the replacement of clear glass milk bottles with opaque containers.

Niacin,   or B$_3$, is the pseudonym of *ni*cotinic *ac*id, which has been known as a plant constituent since the 19th century. It was identified as a vitamin in 1937, when Carl Elvehjem at the University of Wisconsin showed that it cured the disease in dogs that corresponds to pellagra in humans. Subsequent work demonstrated its effectiveness against the latter. Nicotinic acid became known as niacin in the nutritional field in the early forties so as to avoid confusion with nicotine. Antitobacco groups in the United States warned that eating B$_3$-enriched bread could induce an addiction to cigarettes.

Pellagra—Italian for "rough skin"—is the disease caused by niacin deficiency. It is characterized by inflammation of the skin, digestive problems, weakness, and mental disturbances, and can be fatal. It was first associated with the spread of American maize across Europe, became a serious problem in the American South after the Civil War, and is still common in corn-growing areas of the developing world, particularly Africa and Asia (see pages 242–43). Niacin deficiency is especially debilitating because like thiamine, niacin is crucial to the metabolic processes of every cell in the body. It is a component of two coenzymes, NAD and NADP, which play an essential role during the oxidation of fuel molecules. Another derivative is necessary in reactions that synthesize various fatty acids, cholesterol, and an amino acid. When these activities are interfered with, our tissues begin to degenerate.

Poultry, meats, fish, and to a lesser extent grains, bread, and vegetables are the principal dietary sources of niacin. The body can also synthesize niacin from the amino acid tryptophan, so that a diet adequate in protein is generally adequate in niacin as well. Corn has been a world villain because it is a poor source of both niacin and tryptophan, and those countries to which it was introduced knew nothing of the native American traditions of preparation and supplementation that guaranteed a proper amino acid balance. Niacin itself is a very stable chemical and little affected by cooking.

Pyridoxine   is the usual shorthand for B$_6$, though technically it names only one of several forms of the vitamin. When combined with phosphate, it is a coenzyme in the conversion of one amino acid to another, and the conversion of tryptophan to niacin. It also plays a role in the breakdown of glycogen to glucose, in the formation of antibodies, and in the synthesis of hemoglobin— all extremely important work. Fortunately, deficiencies are very rare, with meats and vegetables supplying most of the approximately 2 milligrams needed daily.

Cobalamin, or B_{12}, is a complex structure required for the generation of DNA precursors, and so is important for all cells, though bone marrow and the production of red blood cells are most seriously affected by dietary deficiency. Very tiny amounts, perhaps a few millionths of a gram daily, are needed, and deficiencies are usually found only in those with a defect in the body's mechanism for absorbing cobalamin; such people may suffer from anemia as a result. Strict vegetarians, including enthusiasts of Zen macrobiotics, can gradually develop B_{12} deficiencies and suffer damage to the digestive tract and to the spinal cord, though they do not develop anemia. The inclusion of eggs and dairy products in the diet will prevent this problem.

Folic Acid is involved in the synthesis of DNA and of the hemoglobin in red blood cells. Its name comes from the Latin for "leaf," and dark leaf vegetables are its richest sources. Most other foods are poor sources, and mild deficiencies are probably not uncommon. Severe deficiency can be manifested in anemia, digestive problems, and fetal abnormalities.

Pantothenic Acid and **Biotin** are vitamins that have never been observed lacking in humans. Biotin is probably produced by intestinal bacteria, and pantothenic acid is named for its universal presence in food (*pantothen* is Greek for "on all sides"). As a part of the molecule called Coenzyme A, it is required for the liberation of energy from carbohydrates; beyond this, it participates in the synthesis of fatty acids, cholesterol, hemoglobin, and acetylcholine. The fundamental nature of these functions may explain why it is ubiquitous in living things. Biotin is involved in the synthesis of fatty acids and the metabolism of carbohydrates.

Water and Minerals

We group these very different nutrients together because several minerals are required in order to regulate the balance of fluids in the body. Beyond this association, water and minerals have in common an inorganic origin. Unlike proteins, fats, carbohydrates, and vitamins, they would exist on earth even in the absence of life.

In fact, it has been proposed that our bodily fluids reproduce the inorganic environment that surrounded our single-celled ancestors in the primordial seas. This was the view of the Canadian biochemist A. B. Macallum, who suggested that our closed circulatory system and kidneys work to preserve in our blood plasma the concentrations of ions that prevailed in the ancient ocean. These internal concentrations are lower than those in today's ocean, and Macallum thought this good evidence for the kinds of conditions in which life had evolved. Such a theory is very difficult to verify, and it may seem unnecessarily tidy. But there is also considerable appeal to the idea that, having developed in the seas, we brought our old home with us when we colonized the land.

Our bodies are more water than anything else; 60% by weight on average, and we can spare very little. It may take us weeks to starve, but only a very few days to dehydrate. About three quarters of our water is contained within individual cells. Of the rest, most is "interstitial" fluid surrounding the cells, with a small amount in the blood plasma, and even less in the secretions of our various glands. Outside the cells, water is a means of transporting nutrients and wastes, a physical cushion, a lubricant. When exhaled or perspired, it removes excess heat energy from the body. Within the cell, it is the solvent in which all chemical interactions occur.

As one might expect, water has been a favorite instrument of the diet reformers. The water cure was big business in 19th-century America; Graham and Kellogg both relied on it to some extent. It had its beginnings not in pseudoscience, however, but in the veterinary experience of Victor Priessnitz, an illiterate native of a small mountain village near Nurnburg who, around 1820, began to cure animals of various problems with water compresses, applied the same treatment to himself after a bad fall, and then ministered to others. His reputation grew rapidly, and eventually he was sought after by the princes of Europe. One element of his regimen was the consumption of 20 or 25 glasses of cold water every day. Priessnitz himself was not interested in explanations, but his disciples argued that water is an excellent solvent and so cleanses the system of all impurities and disease-causing materials. Unlikely as it may seem, the same rationale has been used to condemn water as the scourge of mankind: water is so powerful a solvent that, given the chance, it will eat away anything solid, even metal.

Today, about 2 quarts of liquid a day are recommended for the adult who doesn't work up much of a sweat.

Minerals and Water Balance

Because there is a continual interchange of molecules across the membrane of every cell in the body, some kind of control mechanism is needed to maintain stable chemical conditions so that imbalances do not build up. *Sodium* and *potassium*, in the form of positively charged ions, *chlorine*, as a negative ion, and *phosphorus*, as a component of several negatively charged complexes, are the elements principally involved in regulating the concentrations of other chemicals inside and outside cells. Their duties are strictly circumscribed. Sodium and chloride ions are the principal minerals in the blood plasma, potassium and phosphate in the cellular cytoplasm. They regulate the general concentration of bodily chemicals by the principle of *osmosis*, or the diffusion of certain molecules across the semipermeable cell membranes in order to equalize the ion concentrations on either side. For example, an increase of sodium ions in the blood will draw water from the cells into the plasma, raising the potassium concentration within and lowering the sodium concentration without. By controlling the movement of water, the body controls the activity of substances dissolved in it. Normally, the adrenal glands and kidneys control these mineral balances, but when an excess of sodium, usually from large amounts of table salt, is habitually ingested and remains in the body, so much water is drawn into the plasma that high blood pressure can develop and put a dangerous strain on the heart and vascular system. It is thought that the system of sodium regulation and transport in many (but not all) people breaks down under high sodium intake, and that this failure is the principal cause of hypertension.

Salt Because it is a very concentrated, indeed nearly pure source of two essential minerals, sodium chloride—salt—has always been a highly prized commodity, though its current cheapness and availability mask this fact. In the Old Testament, salt is prescribed as an offering to God, and there is an intriguing reference in Ezekiel (16:4) to a custom of rubbing newborns with salt. In Roman times, soldiers were paid a special allowance to buy salt called the *salarium*, a word that gives us our "salary." And according to Pliny, wit was called *sales* because "all the humor of life, its supreme joyousness, and relaxation after toil, are expressed by this word more than by any other." Though this usage does not survive, such phrases as "salt of the earth" (coined by Jesus) and "worth his salt" derive from similar esteem. Historically, the salt trade has often been controlled by kings on the grounds that minerals are a product of the earth, not of men. A popular revolt against the salt tax, or *gabelle*, was an important force in the French Revolution. In culinary matters, salt is valued not only for its flavor—one sign of its importance in the body is the fact that we have a specialized taste sensation for detecting it—but also as a preservative. A highly concentrated salt solution will draw water out of bacterial and mold cells and so inhibit their growth.

It is estimated that the average American consumes 6 to 15 grams of salt every day, a dosage that probably supplies 5, 10, even 25 times as much sodium as we actually need. Baking soda and powder and other food additives also contribute to our sodium intake, as do meats, milk, eggs, and a few vegetables such as spinach and celery. It is widely recommended that we cut down on our sodium consumption to minimize the risk of developing hypertension. Salt also provides us with more than enough chlorine, which is an important constituent of our gastric juice as well as a regulator of fluid balance.

Potassium too is generally consumed far in excess of bodily needs, but only because it is a significant element in most foods, especially meats, whole grains, potatoes, tomatoes, carrots, celery, citrus fruits, and bananas. Potassium deficiencies are usually the result of general disturbances in eating or digestion. This mineral has several functions in addition to osmotic regulation. It is a participant in various enzymatic reactions. And a small but essential amount circulates in the extracellular fluids to aid in the transmission of nerve impulses, including those that cause the contraction of muscle cells. One serious result of potassium deficiency is irregularity in the action of the heart.

Other Minerals

We turn now to those elements which, with the exception of phosphorus, are not involved in fluid balance. Phosphorus and calcium together account for three quarters of all minerals in the body, and are present in five times the amounts that sodium, chlorine, and potassium are. Their principal role is a structural one, reinforcing the protein-carbohydrate matrix of our bones. Several other minerals are crucial components of various enzymes, hormones, and other organic compounds.

Calcium   Quantitatively speaking, the major function of calcium in the body is hardening the bones and teeth. Practically all of the nearly 3 pounds of calcium we each carry is deposited in these tissues in the form of various hydroxide and phosphate salts. Once deposited there, it is not fixed forever, but is constantly taken back into the blood plasma and reused. There is also some loss of calcium through excretion. If dietary intake is not sufficient to replace lost stores, bone will lose the mineral faster than it can be replaced, and serious weakening of the skeleton can result. Insufficient dietary calcium, together with a shortage of vitamin D, which helps the body absorb and then deposit calcium in bone, are often responsible for the susceptibility of older people to fractures and broken bones. The mineral is lost much more slowly from teeth than from bones, but once lost from teeth, it is irreplaceable.

About 1% of our calcium is put to less palpable uses. It is a catalyst in blood clotting, activates certain enzymes that digest fats and proteins, affects the passage of molecules across cell membranes, carries nerve impulses, and is an integral part of the machinery of muscle cells. Given this range of responsibilities, it makes good sense that bone would act as a huge calcium reservoir, that the body would trade structural strength for the continuation of even more essential services.

Far and away the best source of calcium is milk and milk products. Clams, oysters, canned salmon with its bones, and collard, turnip, and mustard greens are all good sources, but are not as frequently consumed as milk is. Other foods can actually hinder the absorption of calcium. The oxalic acid in certain plants, notably spinach, and phytic acid in the outer layers of whole grains, both form insoluble compounds with calcium ions, thereby binding the mineral in a form that cannot cross the membranes of intestinal cells. And certain fatty acids will combine with calcium to form insoluble soaps; this is also what happens when soap is used in hard water and forms a scum. In any case, severe calcium deficiency is rare, and usually involves a shortage of vitamin D as well. The recommended daily allowance for adults—and this goes for the elderly as well—is the equivalent of 2 or 3 cups of milk, or 3 to 4 ounces of cheese. Growing children and child-bearing women need even more.

Phosphorus   The body contains about half as much phosphorus as calcium by weight, again with the great preponderance in the mineral filling of bones and teeth. But the biochemical functions of phosphorus are even more extensive. It is a constituent of the genetic materials DNA and RNA, of the phospholipids that make up cell membranes, and of inorganic compounds that control the pH of various fluids. And it is at the heart of energy production in every single cell: adenosine triphosphate, or ATP, carries in its phosphate bonds the energy liberated by the oxidation of fats and carbohydrates. ATP is the universal currency of energy in living things. Generally, a diet adequate in protein and calcium is also adequate in phosphorus. Animal foods, including milk, and whole grains are good sources.

Sulfur The body contains about 6 ounces of sulfur, most of it in three particular amino acids which are major constituents of skin, nails, hair, and the connective tissues. It is also an element of such substances as insulin, the hormone that controls blood sugar levels, heparin, an anticoagulant in the blood, and polysaccharides that contribute to the matrix of cartilage, bones, and teeth. A balanced protein intake will include enough sulfur in its amino acids. Animal foods and (despite their deficiency in sulfur-containing amino acids) dried legumes are among the richest sources.

Magnesium We carry about an ounce of this mineral, half of it in bone salts, and most of the rest inside individual cells, where it activates various enzymes involved in producing chemical energy. A very small amount circulates in external fluids and participates in nervous and muscular activity. Magnesium deficiency is usually the result of a generally inadequate diet or of disease, and can show up in muscular tremors. Dairy and grain products, legumes, nuts, and green vegetables all contribute useful amounts of magnesium to the diet.

Trace Elements

The rest of the minerals to be mentioned are found in the body in very small amounts, perhaps a few grams or less. In many cases their precise functions, and the amounts we need in our diet, remain obscure. This is true of, among others, cobalt, chromium, nickel, selenium, tin, and silicon. There is even disagreement over some better understood elements, including fluorine, copper, and zinc.

Iron is unique among minerals in having specialized absorption and storage mechanisms that limit our intake and set aside reserves for future needs. At any given time, about a quarter of our iron is bound to ferritin, a protein complex, in the liver, spleen, and bone marrow. Another protein, transferrin, carries iron in the blood plasma. Its saturation with the mineral seems to determine how much the intestinal wall will absorb from our food.

About 85% of the active iron in the body is a component of hemoglobin, the red pigment in blood cells that carries oxygen from the lungs to other tissues. Some 5% is contained in myoglobin, a related protein that stores oxygen in muscle tissue; it is myoglobin and not hemoglobin that gives red meats their color. Iron is also required by various proteins, enzymes, and vitamins for proper operation. Iron shortage results in anemia, a condition in which the body cannot get enough oxygen and is therefore easily fatigued.

Iron deficiency is said to be the most common form of malnutrition in the United States. It is especially prevalent among women, who lose iron during menstruation and who generally eat smaller meals than men to begin with. One cause of this problem is that our intestines absorb iron very inefficiently. The average figure is around 10%, while we assimilate from 50% to 90% of most other nutrients. Lean and organ meats, green leafy vegetables, whole-grain foods, and legumes are among the best sources; dairy products are iron-poor. Not too long ago, the American diet was probably well supplemented by the practice of cooking foods in cast-iron pots and pans. One study showed that this treatment of such acidic foods as spaghetti sauce and apple butter multiplied their iron content by a factor of 30 to 100. Today, our preference for stainless steel, aluminum, and enamelware has greatly reduced the opportunity for such fortification.

Iodine We harbor very small quantities of this element—a few hundredths of a gram—but they are absolutely essential. In fact, iodine was the very first nutrient to be added to foods as a supplement. It is a component of the thyroid hormone, which controls the rate of energy production in all cells. It thereby influences the growth and general activity of every organ. A deficiency of iodine leads to goiter, an enlargement of the thyroid gland that becomes visible as a fleshy bulge around the front of the neck, and to a reduced capacity for activity. A severe deficiency during pregnancy will cause mental retardation and muscular and skeletal malfunction in the child.

Iodine deficiency is not at all uncommon in the world, and is especially troublesome in mountainous and inland areas, though it is nearly unknown along seacoasts. The reason for this pattern is that iodine is found principally in the very soluble salts of sodium and potassium. Any iodide salts in the soil tend to be washed by rainfall into rivers and eventually into the seas, and glacial activity in the Northern Hemisphere ten and twenty thousand years ago has covered ancient sea deposits with a thick layer of debris, or scraped layers of soil away. So seafood and vegetables or animals raised near the seacoast are good sources of iodine, while inadequate soil levels elsewhere support only iodine-poor stock.

The glacier-scoured American Midwest became known as the "Goiter Belt" early in this century for precisely these reasons. Then, in a series of studies conducted between 1907 and 1920, Drs. David Marine and O. P. Kimball at the Western Reserve University in Cleveland established the connection between goiter and iodine deficiency. The problem then became the means by which iodine intake could be increased. For a while, an iodine fetish raged: people wore bottles of the gas around their necks, and some ingested so much that they induced goiterlike enlargement of the thyroid gland. In 1924 Michigan began the experiment of supplementing table salt with sodium iodide, and Rochester, New York, added it to the municipal water supply. Today, many countries required the iodization of all salt, though it remains optional in the United States. Goiter has been all but eradicated in developed countries.

Copper   About ⅒ gram of this element is incorporated into the body, with highest concentrations in the liver and brain. It plays a role in the formation of hemoglobin and of phospholipids, and is also involved in bone development and energy production. Organ meats, shellfish, grains, and most other seeds are good sources. Dietary deficiency of copper is rare, and excessive intake can cause damage to the liver, kidney, and brain. For this reason, and because copper metal readily reacts with many foods, the use of unlined copper utensils is not recommended.

Fluorine   There is an ongoing debate as to whether fluorine should be considered an essential nutrient, whether it plays any role in our general metabolism. It is very well established, on the other hand, that small amounts of fluorine, which we ingest as fluoride ions, are incorporated into the calcium-phosphate minerals of tooth enamel and, by increasing the crystal size and so lowering the exposed surface area of the minerals, strengthen the enamel against the dissolving action of decay-causing bacteria in the mouth. There is also recent evidence that fluoride may slow the loss of bone calcium during old age. Among foods, tea leaves are notably rich in fluoride.

The correlation between fluoride intake and reduced tooth decay dates from the first decade of this century, when it was noticed that inhabitants of towns with naturally high fluoride levels in their water supplies—so high that they caused unsightly mottling of the teeth—had unusually healthy teeth nevertheless. In 1945, four cities in the United States and one in Canada participated in a controlled study of the effect of boosting fluoride levels in drinking water to one part per million. The results were clear-cut, and in 1950 the U.S. Public Health Service officially endorsed water fluoridation. The World Health Organization followed in 1958.

But there has been persistent opposition to fluoridation in this country, and on several grounds: that it represents an invasion of the individual's civil rights, a kind of compulsory medication; that fluoride has harmful side effects; even that it is part of a conspiracy to ruin the country. One pamphlet, for example, claimed that fluoride is a "nerve poison" meant "to paralyze, demoralize, and destroy our great republic from within." To date, no harmful effects of these very low levels of the mineral have been found. And several courts have upheld the constitutionality of fluoridation; the public good is as much the point here as it is in the compulsory inoculation of schoolchildren against polio. In any case, less than half of the country's population drinks fluoridated water, and tooth decay continues to cost us many millions of dollars a year.

Zinc   The body contains 2 or 3 grams of zinc, with high concentrations in the liver, pancreas, kidney, and brain. It is a component of several enzymes involved in energy production and the digestion of proteins, and is required for the normal function of several metal-protein complexes. Animal foods are the best sources of zinc; deficiencies are rare.

Fiber

Although dietary fiber is mainly a collection of complex carbohydrates, we separate it from sugars and starch in this discussion because it is not really a nutrient: it is by definition *not* digested and used by the body for energy or building materials. We do not have the particular enzymes necessary to break these carbohydrates down into absorbable units. But fiber does provide the valuable service of softening and giving bulk to the waste products of digestion, thereby easing their elimination from the body. Many other claims have been made for fiber, and some of them may be justified, but, as is usually the case, the situation turns out to be complicated.

The general association of "roughage" and "regularity" goes back at least to the days of Graham and Kellogg, but the issue was given new importance in 1974, when a group of British physicians led by Dr. Denis Burkitt published a report in the *Journal* of the American Medical Association. They noted that noninfectious diseases like cancer and heart disease have become predominant in the industrial West while remaining rare in developing countries. Burkitt and his colleagues presented evidence that the diet of African villagers includes much more fiber than that of English or urban African populations, and that fiber decreases the time that food is retained in the intestine, and increases the weight of material passed. They then suggested that through various mechanisms, both direct and indirect, dietary fiber could reduce the risk of diverticulosis, appendicitis, gallstones, varicose veins, hemorrhoids, colon and rectal cancer, and atherosclerosis, or hardening of the arteries. For example, possibly carcinogenic products of intestinal bacteria would have less time to accumulate and be in contact with the colon if fiber decreases "transit time." And fiber may complex with bile salts, which are synthesized in the liver from cholesterol and are reabsorbed in the small intestine. The more bile salts that are excreted, the more must be synthesized, and so the less cholesterol would be left to clog bloodstreams.

These arguments were quite plausible, received a great deal of attention from newspapers and magazines, and made whole grain foods more popular than ever before. Today, several years after all the excitement, there is considerable uncertainty in the medical community about what fiber actually does, and even what it actually is. At first, fiber designated the indigestible components of plant cell walls, especially cellulose. But subsequent research has shown that a fair amount of cellulose is broken down by stomach acids and intestinal bacteria, long before it gets to the colon. And the fates of the many other components—hemicelluloses, pectins, gums, and other "unavailable carbohydrates"—are unclear and probably vary widely. Further, high-fiber foods—for example, wheat kernels and carrots—differ so much in their composition that generalizations about the effects of "fiber" are very hard to make. As an illustration, it has recently been determined that only certain gums complex with bile acids to interfere with their reabsorption, and neither bran from grains nor most other common foods contain much of these materials.

Beyond these uncertainties about the nature of fiber itself, it now appears that some conditions thought to be altered by dietary fiber—the number and timing of bowel movements, the colon's bacterial population—are highly individual traits, and depend on many other living conditions as well. No clear causal relationship has been found between fiber intake and the incidence of diverticulosis, appendicitis, or atherosclerosis. And in the case of bowel cancer, it appears now that the important factor is the proportion of meat and animal fat in the diet. The seemingly helpful effect of dietary fiber may simply be an indirect reflection of the fact that the more fiber-containing foods one eats, the less meat one is likely to have room—or money—for.

Exactly how fiber is to be defined, what foods contain what kinds of fiber, and what such materials do in the body, all remain matters in need of clarification. Whole grains, quickly cooked vegetables and raw fruit, and bran cereals all probably do us some good. But many of us have an exaggerated sense of their benefits. Whatever it is, fiber is no miracle.

Final Observations

In the likely event that this bolus of information should prove difficult to swallow, to say nothing of digestion, here is a distillation of its essential lessons. First, nutrition is an intricate matter, and the more we come to know about it, the more intricate it gets. Dietary fads, with their master keys to health, rarely take account of this complexity, and so never become anything more than fads: they never work. Second, good nutrition is a matter of balance, of inclusiveness rather than exclusiveness. No single kind of food can supply the dozens of substances essential to the workings of our bodies. And several important nutrients can become poisons if consumed in excessive quantities.

It should also be pointed out that the recommended daily allowances for various elements of our diet are not engraved in stone. They represent the results of a limited number of studies on a limited number of people. René Dubos has pointed out that these numbers, and even the very attempt to define the adequacy of our diet in terms of particular quantities of nutrients, are called into question by the many obvious exceptions to the rules. People lead long and physically demanding lives in African, Asian, and South American cultures that provide diets severely inadequate by our standards. And there are very wide variations in the diets of healthy people even in our own country. It seems likely that we are "conditioned" nutritionally in our early years, that within limits we can adapt to the food supplies that are available to us.

The point, then, is to avoid making a fetish out of nutrition. It is worth giving some careful thought, worth trying to change habits that seem likely—given our current knowledge—to cause unnecessary trouble. But it is not worth keeping track of tens of calories, or milligrams of potassium, or International Units of vitamin E. Nutrition by the numbers is no guarantee of the good life, and probably precludes it by overriding the calls of appetite and pleasure.

CHAPTER 12

Digestion and Sensation

In this chapter we shift our attention from foods themselves to the ways in which the human body deals with food: how we extract nutrients from it, how we feel the urge to consume it, how we recognize and enjoy it.

THE NATURE OF DIGESTION

The gastrointestinal tract is a continuous tube that includes the mouth, esophagus, stomach, small intestine, and colon. It has four basic functions: to break foods down physically into small pieces, to reduce them chemically to their basic constituents—fatty acids, amino acids, sugars—to absorb these molecules into the bloodstream, and to get rid of the unusable residue. In a sense, the digestive tract is an inner extension of the body's exterior; it segregates food from our true insides until that food is fit for our use. This is clearly essential work, and all animals have some version of this tube. The systems of higher mammals, however, are much more elaborate than those of, say, fish and reptiles. Over evolutionary history, an initially straight, undifferentiated passage has been fitted with a series of antechambers and an increasingly complex lining. Most of these changes can be attributed to the great nutritional demands made by agile, warm-blooded bodies and large, versatile brains.

Here are three examples of such digestive adaptations. We humans and many mammals have teeth that make chewing possible, while reptiles, fish, and even some predatory mammals use theirs primarily as weapons, and swallow their prey in large chunks. But because digestive enzymes must make direct contact with individual food molecules to break them down into absorbable fragments, large chunks of food will take much longer to be

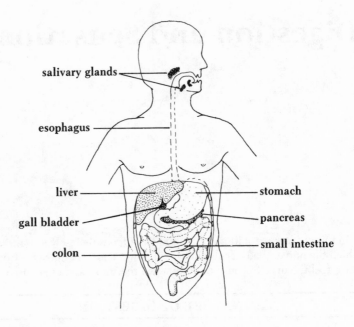

salivary glands

esophagus

liver

gall bladder

colon

stomach

pancreas

small intestine

The digestive system

digested than will finely divided particles. Chewing, then, gets food to the rest of the body more rapidly and efficiently, and so helps make it possible for animals to be continuously active. A similar adaptation is the stomach, a holding tank that first developed in the fishes, and came to assume the role of muscular "teeth." It crushes food by contracting its walls. Later on in evolution, some intestinal enzymes became associated with this antechamber, so that our stomachs act on food both physically and chemically.

A final example of the streamlining of digestion is the lining of our small intestine, where nutrients are actually absorbed into the body. The more sites there are along the intestine where food can be absorbed, the faster and more completely it will be available to the organs that need it. Along with our advanced and demanding bodies, there evolved an intricately folded—and therefore greatly expanded—intestinal surface. On top of this, the surface area of each of many million nutrient-absorbing cells is multiplied by means of about a thousand tiny buds. An average human has something approaching a football field in intestinal surface area.

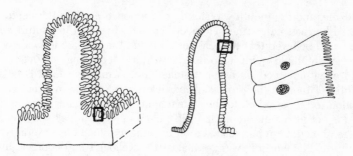

The multiplication of surface area in the intestine. Folds in the intestinal wall *(left)* are divided into many little buds, or villi *(center)*, each of which is covered with cells whose outer borders are in turn budded with microvilli *(right)*. The nutrients in our food are absorbed through the microvilli.

These are some of the anatomical developments that have given us the nutritional cushion to be warm-blooded, active, and cerebral animals. The cultivation and improvement of crops, cooking, which increases the digestibility of many foods, and even the scientific investigation of nutrition, can all be considered cultural adaptations to the same end.

The Human Digestive System

The Mouth

The first thing we do to our food is to chew it into small pieces and moisten it for its passage to the stomach. Teeth originate not in bone, despite their appearance, but in a hardening of skin tissue. The dentition of a particular species depends on its diet, so that some carnivores have no grinding molars, and strict herbivores may have no piercing canines. The human mouth includes several different kinds of teeth, a fact that testifies to our nature as omnivores.

In addition to lubricating, saliva dissolves small amounts of our solid foods and so makes it possible for the taste buds to sample them. It also contains ptyalin, an enzyme that digests starch into maltose, a glucose-glucose disaccharide. We can notice the activity of ptyalin when we chew a bland, starchy food like potatoes or bread long enough that it begins to taste sweet. The principal function of ptyalin appears not to be serious digestion (it is not very concentrated and is inactivated when it reaches the stomach), but the dissolving of any starch left in the mouth after we swallow. In any case, its presence in our saliva is a sign of the importance of plant foods to our evolutionary ancestors; strict meat-eaters have no need for such an enzyme.

American dietary fads have not limited themselves to particular foods or nutrients; portions of the anatomy have also been singled out for praise and blame, and the mouth was stoutly championed early in this century by a retired businessman named Horace Fletcher. He called it "Nature's Food Filter," and believed that the sense of taste and the impulse to swallow constitute an infallible guide to proper nutrition. According to Fletcher, we should chew our food until it is tasteless, for "while any taste is left in a mouthful of food in the process of masticating or sucking, it is not yet in a condition to be passed on to the stomach; and what remains after taste has ceased is not fit for the stomach." Fifty chews per mouthful was Fletcher's average, though he reported that a bit of pungent onion once exercised him 722 times. "Fletcherism" was popular at Dr. Kellogg's Battle Creek sanitarium, and English disciples held "munching parties" in his honor. A distinguished physiologist of the time actually found something of value in Fletcherism: by making eating such hard work, it forced people to cut their caloric intake drastically.

The Stomach

Chemical processing of our food begins in this muscular sac, where the enzyme pepsin is secreted and digests proteins into somewhat smaller chains of amino acids. Very limited amounts of other enzymes are also present in the stomach, but seem to have little effect; only proteins are affected in any quantity. Nutrients are not absorbed across the stomach wall, but two other common substances—alcohol and aspirin—are, and this is the reason for their swift effects on the rest of the body. In fact, it is often recommended that neither alcohol nor aspirin be taken on an empty stomach. The sense of this practice is confirmed by the disposition of food as it enters. The first material to arrive forms a thin layer on the wall, and subsequent arrivals coat the preceding layer. Accompanying alcohol and aspirin with food, then, delays their contact with the stomach wall, thereby slowing alcohol's absorption and diluting aspirin's irritating effects on the stomach lining.

The presence of food in the stomach stimulates involuntary movements in the walls of the gastrointestinal tract called peristaltic waves, which mix the stomach contents thoroughly with the highly acidic, pepsin-containing gastric juice and eventually propel the mixture into the beginning of the small intestine. It takes an average of four hours for the stomach to be emptied. Several factors influence the rate at which food passes to the intestine, and so the period during which one feels "full" after a meal. Solid foods, and particularly foods that reach the stomach in lumps rather than in finely divided particles, take longer to be readied for the small intestine than do liquids. And the presence of fat in the mixture that reaches the beginning of the small intestine causes a hormone to be released that *inhibits* the peristaltic actions of the stomach walls. Why? Because fats, unlike proteins and carbohydrates, are not water-soluble, and so must be divided into small droplets, or emulsified, before the water-based digestive enzymes can begin to act efficiently. This extra step means that fat digestion is a relatively slow process, and the rate at which food passes into the small intestine must be retarded.

So the stomach retains its contents longer when fats or oils are a significant part of the meal, and such a meal therefore seems especially satisfying.

The Small Intestine

Most digestion and nutrient absorption takes place in this long, narrow tube. The length of the intestine relative to body size varies from species to species, and seems to depend on diet. Plant foodstuffs, with their tough cell walls, take more work than the soft muscle tissues of animals, and the intestinal tracts of herbivores tend to be longer than those of carnivores. The human intestine is middling, which indicates that it evolved to handle a mixed diet. The muscular walls churn and mix the food with digestive juices, and specialized cells in the lining absorb the nutrients that are liberated.

When the highly acidic contents of the stomach enter the small intestine, the sudden change in pH stimulates the release of a group of hormones, which in turn stimulate the pancreas and liver to secrete various substances. The liver and its storage pouch, the gall bladder, deliver *bile*. Bile salts, which are derived from cholesterol, are like soaps in that they are emulsifiers; they break up droplets of fat and oil into much smaller particles, increasing the number of fat molecules exposed to the surrounding enzymes and so speeding the work of digestion. The pancreas delivers a set of enzymes that acts on all three major nutrients, together with bicarbonate ions that neutralize the pH of the stomach contents. But only fats are digested into components basic enough that the body can then absorb them directly. The final stages of carbohydrate and protein digestion occur right at the surface of the intestinal lining, where cells secrete their own enzymes and complete the job of getting down to single-unit sugars and individual amino acids. Something like 90% of the nutrients that pass through the small intestine are absorbed there. This figure includes not only the amino acids, simple sugars, fatty acids, and monoglycerides, but also water, minerals, and vitamins, which are complex organic molecules that are rapidly absorbed and not affected by digestive enzymes.

The Colon

This portion of the digestive tract does no digesting; its job is to absorb a little more water, a few more minerals, especially potassium, and to excrete excess sodium. The waste materials left over after absorption usually remain in the colon for about a day. A meal may take from one to a few days to pass completely through the digestive system; the "transit time" varies considerably according to the food eaten and individual habits. Certain substances such as capsaicin, the hot principle in red peppers, irritate the intestinal lining and increase its movement, thereby cutting down on the transit time of that meal. Fiber in large quantities appears to have the same effect; this is why it is used as a restorer of "regularity."

Intestinal Bacteria; The Yogurt Mystique

The human digestive tract harbors a "resident" bacterial population (as opposed to the occasional, transient troublemakers) in the millions, with most of them in the lower reaches of the intestine and colon. When we are born, our gut is nearly sterile, but as we feed, we foster colonies of lactose-fermenting *Lactobacilli*, and eventually a wide range of flora, including *Lactobacilli, Streptococci, Staphylococci*, coliform bacteria, and yeasts. The particular mix varies from person to person, and depends on the exact composition of the digestive secretions, the kinds of antibodies we have developed, and our diet. At least some of these microbes are more than mere parasites; they synthesize enough vitamins B_{12} and K that we are largely freed from the necessity of finding dietary sources for these nutrients.

As it happens, misconceptions about the nature of the colon and its bacteria are responsible for one of the more durable claims to the secret to health. Ilya Metchnikoff (or Mechnikov) was a distinguished Russian embryologist and pathologist who won the 1908 Nobel Prize for his discovery that white blood cells fight bacterial infection. Taking his lesson from Darwin, Metchnikoff pointed out in his book *The Nature of Man: Studies in Optimistic Philosophy* (1904) that evolution does not guarantee perfection, that most organisms survive in the world despite various imperfections and disabilities. And man is one species that is in especially serious disharmony with nature. The pain of childbirth is one of many "disharmonies of reproduction"; both suicide and a love of life in the face of necessary death are symptoms of the "disharmonies of self-preservation." Then there are the "disharmonies of digestion," the chief example being the colon, which is useful for animals that live on bulky plant foods, but which is "certainly useless in the case of man." Worse than useless: positively harmful. For it harbors great numbers of bacteria, some of which produce toxins and slowly poison the body. According to Metchnikoff, the colon brings on our premature death.

Metchnikoff put his faith not in nature, but in science and its ability "to amend the evolution of the human life, that is to say, to transform its disharmonies into harmonies. . . ." And science, he later wrote in *The Prolongation of Life* (1908), offers several options in the case of the colon. Surgical procedures to remove or bypass it, though popular among doctors, were potentially hazardous, as was the frequent use of antiseptics. Metchnikoff opted instead for a treatment based on traditional culinary practice. Bacteria that excrete lactic acid render pickled vegetables and soured dairy products inhospitable to many other microbes by lowering the foods' pH. Why not do the same job in the human intestine? Metchnikoff thought that the best method was to ingest lactic bacteria by eating uncooked pickles, sauerkraut, sour milk, and yogurt. And his workers found examples in France, Bulgaria, Russia, and America of remarkable longevity in people who lived on such foods. Thus began the yogurt mystique that persists to this day. One early sign of Metchnikoff's influence in the United States is the title of a book published in 1926 by a Dr. James Empringham: *Intestinal Gardening for the Prolongation of Youth.*

To be fair, Metchnikoff admitted that his theory was only a theory, and he looked to the future for a more exact understanding of human aging. And there is still uncertainty about the possible harmfulness of intestinal bacteria. A distinction is made between two general types of flora: the fermentative, which feed on carbohydrates almost exclusively, and the putrefactive, which prefer proteins. The waste products of the former are mostly carbon dioxide and water, with some alcohol and lactic acid, while the latter also excrete nitrogen- and sulfur-containing residues that offend the nose; hence their name. Because there is a correlation between bowel cancer and diets high in animal protein and fats, it has been suggested that the putrefactive bacteria may somehow produce carcinogenic compounds from cholesterol and bile salts. So there may be something to Metchnikoff's theory of "autointoxication." But eating lots of yogurt is not likely to be much of a remedy, because *Lactobacillus bulgaricus* is not a species that can colonize our digestive tract, and yogurt contains animal protein and fat. Probably the best way to reduce our putrefactive population—if in fact this is worth worrying about—is to cut down on meat, eggs, and dairy products, and eat more carbohydrate-rich fruits and vegetables, grains and legumes.

HUNGER AND THIRST

The sensations of hunger and thirst are often, but not always, the immediate cause of our eating and drinking. These sensations seem to be centered in our stomach and mouth respectively, and result in a drive to remove or satisfy them. Like other biological drives, these are innate. We know instinctively what we have to do. But physical, cultural, and psychological influences make human behavior in this area quite variable. We can want and eat food even when our bodies don't need it, or we can refuse it despite desperate need. Illness, vanity, grief, neurosis, misinformation, or ignorance can interfere with the body's innate good sense.

Despite our impression that hunger and thirst are localized in single organs, this is not the case. Consider our desire for liquid. The human body can tolerate only small fluctuations in the 60-odd % of our weight that is taken up with water. If the average body loses more than about a cup and a half by perspiration or excretion, thirst will develop. But the dry mouth we identify with thirst is only one symptom, and a secondary one at that: rinsing the mouth with water will not alleviate the sensation. It turns out that cells located in the hypothalamus, a region at the base of the brain, monitor the sodium and potassium concentrations in the body fluids and trigger the feeling of thirst when they become too high. Because increased concentrations of sodium can be caused by a high salt intake as well as loss of water, a very salty meal will make us thirsty even if we drink some fluids with it. Thirst can be satisfied, then, only by the actual ingestion of water.

The case of hunger is somewhat more complicated. Once more, the stomach movements we associate with hunger are only a secondary signal. It has recently become apparent that hunger is triggered by a remarkably intricate chemical mechanism. It involves a delicate balance among nutrient levels in the blood, digestive hormones, and a range of substances active in the nervous system. The hypothalamus probably helps coordinate the different parts of this mechanism and translate its workings into signals—sensations of hunger—for the conscious mind.

Knowing When to Stop

An interesting problem comes up when we try to explain how we know when to *stop* eating or drinking. Common sense would tell us that hunger and thirst would cease when blood nutrient levels and ionic concentrations returned to normal. The trouble is that the body takes time to assimilate its food—a much longer time than it takes for hunger and thirst to diminish. This phenomenon is known as *preabsorptive satiety*, and seems to be controlled by yet another special set of receptor cells. It is not yet clear exactly how the intake of water is sensed; it may be that distension of the stomach walls is one clue. Much more is known about the detection of food, which occurs at several points along the digestive tract. Mechanoreceptors in the stomach walls register pressure, and chemoreceptors in the small intestine sense the presence of sugars and amino acids. The most interesting signal, however, comes from the mouth itself. Somehow—perhaps via the taste buds—the brain senses the fact that we are eating, and then is able to raise the blood glucose levels temporarily. The blood glucose receptors detect this change and the body responds with an almost immediate lessening of hunger sensations. In one experiment, the mouth was rinsed with a solution of saccharin, which is sweet but has no food value whatsoever, and still the blood glucose levels rose immediately, as if in anticipation of the arrival of sugar.

All told, we understand little about the complex, exquisitely tuned biochemical systems that control our sense of hunger and satisfaction. But it is evident that in wealthy societies with abundant supplies of food, these fundamental controls are frequently ineffective. When they malfunction or are overridden by cultural or psychological factors, serious problems, from anorexia to obesity, can result.

TASTE AND SMELL

The Nature and Function of the Chemical Senses

Of our traditionally numbered five senses—most physiologists would add balance, temperature, and pain to the list—taste and smell are the *chem-*

ical senses, the means by which we sample the chemical composition of our surroundings and our food. Because nutrition is a matter of finding and ingesting particular chemical compounds, some such sense has been a necessity from the very beginnings of life. Even single-cell organisms must have a way of discriminating between useful and harmful molecules. Protozoa, for example, will move toward a source of sugar, but will avoid poisonous alkaloids. As the animals evolved, they developed a specialized network of cells, the nerves, to obtain information, code it into electrical signals, and transmit it to other parts of the organism. The first chemical sense was generated by bare nerve endings exposed to the oceanic environment. Humans may retain the vestiges of this primordial system in what has been called the "common" chemical sense, which detects the presence of certain harmful substances and causes reflex actions—choking, sneezing, weeping—to expel or wash them away. Chemicals that irritate or burn on contact or when inhaled, including particular components of red and black pepper, onions, and garlic, stimulate sensory cells in an indiscriminate way, rather than giving rise to ordinary sensations of taste or smell.

Specialized senses of taste and smell that respond in a selective way to various chemicals first appeared in the insect. The two senses are distinguished by the kind of connections they have with the brain. Smell is the simpler and perhaps the more primitive arrangement, with the receptor cell being the end of a single nerve fiber that extends directly to the brain. The taste impulse, on the other hand, is generated in a separate receptor and then must be transferred to a nerve fiber. Given these two basic structures, different animals have deployed them in very different ways. Fish can have taste buds all over their skin, while land animals, because their surfaces are dry, concentrate their chemical receptors in specially moistened organs. Many insects smell with their antennae and taste with their feet. Both taste and smell in aquatic animals detect water-borne substances. For the land-dweller, smell has become the detector of chemicals wafted in the air, and is in this way the long-distance chemical sense, though it also cooperates with taste by sampling food in the mouth.

The Importance of Smell

Smell is a versatile sense that has played a very important role in the development of the animals, and especially of the mammals, many of which are nocturnal and so not very dependent on sight. Smell is used to detect both prey and predator, to recognize trails, territories, social groups, and to initiate sexual contact. It is thought that the primates, and especially humans, have much less sensitive olfactory systems than other mammals because the development of the upright posture removed our collective nose from the odor-laden ground and made sight the dominant sense in daily life.

One indication of smell's role in our development is the fact that the frontal lobes of our brain, the area in which association and the higher mental processes take place, arose from the original junction between brain and olfactory nerve. Most of the information about the world crucial to the survival of the animal came to this point, and this is where the increasingly complex machinery necessary to make the best use of that information was set up. In a way, then, smell gave birth to the mind. And even now, long after its dominance in our daily life has passed, it can still be tremendously evocative. The poet laureate of this experience is of course Marcel Proust, whose narrator in *In Search of Lost Time* reflects on the power of the chemical senses:

> But when from the ancient past nothing stands, after the death of beings, after the destruction of things, odor and taste alone survive, more fragile but more lively, more insubstantial, more persistent, more faithful: like spirits above the ruins of all the rest, they remain to recall, to await, to hope, and to bear unbendingly, in an almost impalpable droplet, the immense edifice of recollection.

Because the nose shares an airway, the pharynx, with the mouth, we smell and taste our food simultaneously, and what we call the "flavor" or "taste" of food is really a combination of these two sensations. The 18th-century French gastronome Brillat-Savarin put it colorfully: "smell and taste form a single sense, of which the mouth is the laboratory and the nose is the chimney; or, to speak more exactly, of which one serves for the tasting of actual bodies and the other for the savoring of their gases." With taste and smell, then, we first decide whether a particular food is edible, and then go on to sample its chemistry simply to enjoy it.

Flavor Preferences

It appears that almost all of our preferences and aversions in this area are learned: that is, we must find out by experience what constitutes acceptable food. Humans do have an innate liking for sweetness, probably a relic of our ancestors' fruit-eating days millions of years ago, and an innate dislike for bitterness, which is a valuable response in a world where many plants contain poisonous and generally bitter alkaloids. But beyond these two reactions which human newborns and other mammals demonstrate, our attitude toward smells and tastes is molded by social custom, opportunity, and often private associations with pleasant or painful moments. Even our disgust at rotting food, which would have a protective advantage, appears to be learned at some point in childhood—how or why remains mysterious—and is not shared by our mammalian relatives. This general openness and flexibility is very likely a manifestation of our nature as omnivores, or animals who learn to take advantage of whatever resources are available. The value of a neutral attitude toward the smell of decomposition is demonstrated among various animals, including rats, which can exploit the bacterial synthesis of vitamins and amino acids in their colon by using their own feces as food.

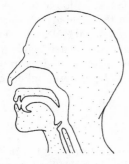

The mouth and nose share an air passage, so that both tongue and nose can sample the chemical characteristics of food as it is eaten.

To say that our flavor preferences are largely learned does not mean that there are not significant biological influences at work. In fact, the influences of the sensory, digestive, and general metabolic systems on each other are remarkable and still only partly understood. We have known since the time of Pavlov (around 1910) that pleasurable flavors in food result in a greater flow of saliva and gastric and pancreatic juices, compared to that elicited by bland or unpleasant flavors. Whether or not the difference in volume affects the efficiency of digestion is unclear, but it is possible that we may actually get more out of good-tasting food for this reason. In addition, the sense of taste can be affected by the nutritional status of the body, and hungers for particular nutrients can develop. This effect is best documented in the case of salt. A report in the 1940 *Journal* of the American Medical Association describes the case of a 3½-year-old boy who died a week after being admitted to the Johns Hopkins Hospital. An autopsy determined that he had suffered from damaged adrenal glands, a condition that resulted, among other things, in the uncontrolled excretion of sodium from the body. An interview with the parents revealed the fact that, from the age of twelve months on, the boy had eaten large amounts of salt, dipping other foods in a bowl of it. When put on the hospital regimen and so a normal salt intake, his supply of sodium was fatally inadequate.

A less extreme manifestation of the malleability of our taste preferences is the common experience that flavors can be very pleasing at the beginning of a large meal, but are much less so at the end. Nor is this just a matter of "having our fill" of these flavors. The satisfaction we get from eating is normally determined by our need to eat, our need for a fresh infusion of essential nutrients, and when that need has been met, our response to sensations of taste and smell becomes more neutral. Such variation in our response to fixed stimuli, together with other evidence—for example, the fact that we are aware of contractions in our stomach only when it is empty, even though it is also active when full of food—have led to the conception of basic physical pleasures as being determined by the body's "internal milieu." There is a certain set of optimal physical and chemical conditions—temperature, ion concentrations, glucose levels, and so on—for the body's successful operation, and we experience feelings of pleasure when we take action to restore these conditions or maintain their equilibrium. When the same action, if continued, will disturb the internal milieu, then we cease to take pleasure in it. Of course, this fundamental control mechanism can easily be detuned by other, interfering influences, such as fashion, gluttony, anxiety. And gastronomes would insist that their pleasure in food goes far beyond the influences of mere appetite.

Spices and Culinary Traditions

Finally, a few words about the role of taste and smell in the development of distinctive cuisines. In recent years a psychologist, Paul Rozin, has been investigating the nature of food preferences, the ways in which they are formed, and their function in our life. His tentative conclusions are fascinating and persuasive. According to Rozin, omnivorous animals are faced with a dilemma, or paradox. On one hand, they are equipped to try new food sources and have a vast range of choices, which enhances their chance of survival; on the other, any strange food is potentially harmful and may cause discomfort, even death. The omnivore's capacity for eating nearly everything, then, is tempered by an innate distrust of novelty. This fundamental aspect of our biological heritage, says Rozin, accounts for a couple of universal human characteristics: the tendency of cuisines, or styles of cooking—with "style" including both the choice of foods and the means of preparation—to be very conservative, and the great value placed on those special flavorings, usually of little nutritional importance, that we call spices.

Rozin points out that the agricultural revolution, the domestication of various plants and animals around 10,000 years ago, brought about a tremendous reduction in the variety of foods that humans eat. About 50 animal and 600 plant species were cultivated, out of edible varieties numbering many thousands that our hunter-gatherer ancestors, with less control over their food supply, had to choose from. And in any particular region of the world, a mere handful of plants and animals forms the core of the cuisine. Such cores are quite resistant to change over time. Corn, beans, and tomatoes have been the staple foods of Central and South America for several thousand years; the same is true of rice and soybeans in China, and of wheat, rye, oats, lentils, and peas in Europe. Only in the last 500 years, since the discovery of the Americas, does the picture become complicated. Europe adopted American natives, including corn, tomatoes, and the potato, while wheat and other grains have taken strong root over here. In any case, the advent of agriculture in human development has meant a triumph of distrust over our capacity to find and exploit new food resources. Culinary tradition is a guarantee of safety, and usually of nutritional wisdom: if it worked for your parents, it will work for you.

Cuisines are also given to characteristic flavorings, which are often their most distinctive aspect, and which have little, if any, nutritional significance. The chili pepper of Mexico, the fermented soybean sauce of the Far East, the turmeric-cardamom-etc. mixture of Indian curries, are a few of the most venerable and striking examples. Rozin notes the fact that these flavorings are usually carefully maintained among immigrant groups whose normal staple foods may be unavailable in their new home, and suggests an analogy to the well-known habit among mammals of "marking" their territories with their own scents. Strong, distinctive flavors mark our foods as familiar and so acceptable, and may help ease the introduction of new staple foods into the cuisine. And, on the other hand, the shared preference for certain flavorings can help reinforce a sense of social solidarity, a feeling of community. The popular interest in exploring novel, exotic cuisines, so evident in the current variety of restaurants and cookbooks, is a development only of the last three or four decades.

Taste

Taste Buds

Our sensations of taste arise from the activity of specialized *epithelial cells*—cells that line the body's cavities—on the tongue. From 40 to 60 of these cells cluster in groups to form *taste buds*, which are embedded in small projections, called *papillae* (Latin for "nipples"), on the tongue's upper surface. In everyday speech, we tend to call these visible bumps the taste buds, when in fact a particular papilla may support several taste buds or none at all. The average adult has a few thousand taste buds, about half of which are located on a handful of large *vallate* ("surrounded by a wall") papillae across the back of the tongue. The rest are found on *foliate* ("leaflike") papillae, which are concentrated at the back edges of the tongue, and on *fungiform* ("mushroom-shaped") papillae, which are scattered evenly across its surface. We also have many *filiform* ("thread-shaped") papillae which carry no taste buds; in other mammals, these bumps often become hardened and are used

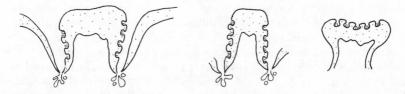

Left to right: vallate, foliate, and fungiform papillae, the small bumps on the tongue that support taste buds

for grooming fur. Interestingly, human fetuses and children have a larger population of taste buds than do adults; the back of the throat, the underside of the tongue, and the inner surface of the cheeks are all sensitive to taste early in life. Our supply of taste buds gradually decreases with age, especially after about 45 years. The individual sensory cells that make up the taste bud have a rather short life—around 10 days—and are continuously regenerated. We can be grateful for this turnover whenever we burn our tongue.

The sensory cells are arranged in little globules and are guarded by ordinary epithelial cells. The latter leave only a small pore through which dissolved food molecules must pass in order to be detected. The sensory cells extend small projections, or microvilli, into the space beneath the pore. Just like the equivalent structures in the small intestine, these folds in the cell membrane offer a large surface area to the environment, and so maximize the possibility of contact with food molecules and thereby the sensitivity of the cell. The salivary glands, together with serous glands on the tongue itself, supply the solution in which food is dissolved, and which washes the microvilli free of old material.

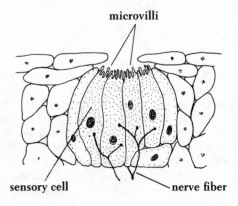

microvilli

sensory cell   nerve fiber

Close-up of a taste bud

Each taste bud is penetrated by about 50 nerve fibers, which transmit electrical impulses from the sensory cells to one of several nerves, and thence to several parts of the brain: the medulla oblongata, where salivation is controlled, the hypothalamus, which regulates feeding behavior, and the cerebral cortex, where our consciousness of sensation arises. The dynamics of the tasting process—how an individual food molecule generates an electrical signal in the sensory cell, how this signal codes the identity of the molecule, how the brain makes use of this information—all of these fundamental activities remain obscure to us. It is thought that certain proteins on the surface of the microvilli act as receptor sites for specific molecules, by virtue of their particular shape. When the matching molecule attaches itself momentarily to the membrane, the protein somehow changes configuration, thereby changing the permeability of the membrane to certain ions, and so gives rise to a small electrical current which is picked up by the nerve fibers. The nature of the current, and the overall pattern of such currents produced by the whole tongue, are thought to be the information by which the brain determines the particular taste being experienced. The ultimate question of how electrical impulses, whether from tongue or nose or eye, are transformed into subjective *sensations*, whether sweetness or floweriness or greenness, remains one of the great mysteries of human biology.

The Basic Tastes

By most accounts, there are four basic taste sensations: sweet, sour, bitter, and salty. Some would include two others, the alkaline or soapy taste, and the metallic taste, which we encounter when our mouth bleeds or comes in contact with such reactive metals as iron and copper. The metallic and alkaline sensations have been little studied. We do know, however, that the other four are somehow associated with the topography of the tongue. Any of the four primary tastes can be sensed anywhere on the tongue, but taste buds are most sensitive to sweetness along the front tip, to sourness along the sides, to saltiness along the front edge, and to bitterness across the back. It seems that individual receptor cells and the nerve fibers that connect with them show different sensitivities to different taste qualities, and the way in which the whole system is "wired" together results in the pattern of response by which we can map out the tongue.

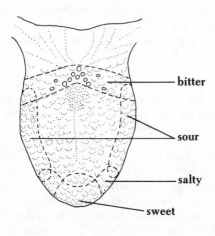

A map of the tongue, showing the areas that are most sensitive to particular sensations

Why do we taste these four particular qualities, and what do they indicate about the chemical nature of the food we eat? Good questions without good answers, at least right now. Only in the case of sourness is the stimulus consistent: it is always an acid of some kind. But sourness is not just a measure of the hydrogen ion concentration; the rest of the molecule from which the hydrogen breaks off also affects the intensity of the sensation. A salty taste can be caused not only by sodium chloride, but by other chlorides, and by sodium fluoride: in other words, by many ionic compounds. Here too, both the positive and negative ions affect the taste. Bitterness is most frequently caused by alkaloids, those plant products meant to discourage animals from feeding by upsetting their biochemical machinery. Quinine and caffeine are the two most familiar examples of bitter alkaloids we have come to enjoy in low concentrations. But certain amino acids and ionic compounds also evoke a bitter sensation. And sweetness is produced not only by various sugars, but by some amino acids, some ionic compounds, and such synthetic substances as saccharin and the cyclamates.

We can clear up some of this confusion by appealing to common sense. The human animal in which these sensations arise meets up with very few of the substances enumerated above in the state of nature. By and large, sweetness has been an indication of the presence of sugar, sourness of acid, saltiness of sodium chloride, and bitterness of poisonous alkaloids. Given this simplification, we can guess at the rationale for at least some of these sensations. Sugar is a useful fuel, and is usually found in vitamin-rich plant tissues. Our collective sweet tooth—of the four tastes, only sweetness is innately preferred—is perhaps a relic of our distant ancestors' career as fruit-eaters. Salt is an essential nutrient, and we have seen evidence that a specific hunger for salt can arise in extreme cases of sodium loss or deprivation. It would be useful to be able to detect alkaloids before they poison us, though evidence that a very primitive class of sea dwellers, the coelenterates, have an aversion to bitterness, suggests that the dangers of water contaminated by heavy metals may have been the original impetus for this sensation. As for acidity— perhaps this goes back to the time, billions of years ago, when an estimate of the pH of our watery home would have been a help.

Foods are chemical mixtures, so that we seldom encounter any of the basic taste sensations in isolation. And certain compounds can indirectly affect the taste of others by inhibiting the activity of taste buds, or enhancing other sensations by contrast. Monosodium glutamate, or MSG, seems to increase the intensity of salt and bitter flavors, while a handful of odd compounds, including the artichoke component cynarin, will make just about anything else taste sweet, apparently by blocking the other receptors.

The temperature of food also affects our sensitivity to its taste, though a definite pattern has not yet been found. Because taste is a matter of a brief interaction between food molecule and protein on the membrane of the sensory cell, and because the energy of these molecules will determine how likely and how long their interaction will be, we would expect that low temperatures would decrease the rate of detection, while high enough temperatures would make it difficult for the receptor molecule to capture the food molecule. The experimental results generally bear this prediction out. Figures for maximum taste sensitivity range from 72° to 105°F (22° to 41°C), and there is some evidence that sweet and sour sensations are enhanced at the upper end, salty and bitter at the lower. At any given temperature, however, we are much more sensitive to bitter substances than we are to sweet, sour, or salty ones, by a factor of about 10,000. For some reason, synthetic sweeteners are effective at concentrations nearer to bitter substances than to table sugar.

Saccharin and Aspartame

The history of artificial sweeteners began in 1879 in the Johns Hopkins laboratory of the chemist Ira Remsen, later the second president of that university. Constantin Fahlberg, a German student working under Remsen on a particular class of organic chemicals, ate a piece of bread one day and found it overpoweringly sweet. He traced the taste to his hands, and then to a compound, $C_6H_4CONHSO_2$, which he had just handled. He dubbed it *saccharin* and went on to patent it—without Remsen. (Cyclamates, no longer in use since their implication with cancer in the late sixties, were discovered in much the same way. In 1937 a chemist at the University of Illinois, Michael Sveda, lit a cigarette and noticed that the paper had absorbed some sweet substance.) The industrial production of saccharin began in 1900. It became an important sugar substitute during the shortages of World War I, and attained general popularity in the 1950s. Because saccharin is 200 to 700 times as sweet as table sugar, it ends up being much cheaper per unit of sweetness, and because it is not absorbed and used for energy, as is sugar, saccharin is a calorie-free flavoring.

The structures of sodium saccharin *(top)*, aspartame *(right, divided into its aspartate and phenylalanine portions)*, and monosodium glutamate *(left)*

Aspartame, the newcomer among artificial sweeteners, was discovered by chance in 1965 when another research chemist, James M. Schlatter, noticed that his fingers were sweet (these stories make one wonder about the standards of laboratory hygiene). It is a combination of one amino acid, aspartic acid, with a derivative of another, phenylalanine. Like all proteins and sugars, it delivers 4 calories per gram, but because it is some 160 times sweeter than table sugar, a teaspoon's worth of sweetness costs only a tenth of a calorie. It has the disadvantages of breaking down and losing its sweetness at cooking temperatures and in acid foods, and of being unsafe for people who are born with the metabolic defect called phenylketonuria and who must limit their intake of phenylalanine.

MSG

Monosodium glutamate, or the sodium salt of glutamic acid, a common amino acid, has been an ingredient in Japanese cooking for a very long time, but was not actually produced as a flavoring material until 1909. In 1908 Kikunae Ikeda, a Japanese scientist, was intent on isolating a basic taste substance that was neither sweet, salt, sour, nor bitter. While studying a particular kind of kelp, a seaweed traditionally used to make soup, he isolated the glutamate ion and found it to have its own taste, which he described as "savory," and to alter the tastes of other foods in an appetizing way. Initially, MSG was manufactured from wheat proteins, which are rich in glutamic acid. In the 1950s, a bacterium was isolated that synthesizes and excretes glutamic acid when fed an excess of ammonium ions (NH_4^+), which supply the necessary nitrogen. Today, nearly all MSG is produced by the fermentation technique that this discovery made possible. World production now stands at around 250,000 tons. The Japanese use MSG to coat table salt, thereby preventing it from caking and expanding its contribution of flavor.

MSG is a mysterious substance. For one thing, practically nothing is known about its ability (or that of several other compounds called nucleotides) to "improve" or "enhance" other flavors. Whether it acts at the receptor site, in the taste cell, or at the junction of taste cell and nerve, whether it increases receptor sensitivity or alters the pattern of electrical signals, are questions still unanswered. MSG was tracked down in the late 1960s as the culprit in the so-called "Chinese restaurant syndrome," in which a burning sensation on the torso, a feeling of pressure behind the forehead and eyes, and chest pain suddenly afflict susceptible people who begin a Chinese meal with MSG-laden soup. It seems that anyone given a large enough dose on an empty stomach will develop these symptoms, which disappear in an hour or so. Because glutamic acid is found in especially large concentrations in nervous tissue and plays a role in the transmission of nerve impulses, it has been suggested that a temporary dietary excess may disrupt parts of the nervous system. This theory has yet to be verified.

Smell

When we speak in everyday conversation of the taste or flavor of food, we are usually referring not just to the responses of our taste buds, but also, even mostly, to the food's odor. This fact is demonstrated by the common experience that food has no "taste" when we have a cold, or when we intentionally block the connection between mouth and nose. Tasting food, then, is a cooperative effort of taste and smell, with smell the dominant partner. Perhaps this is why we speak of a connoisseur as someone with a fine palate: that anatomical feature contributes no sensations of its own, but lies midway between tongue and nose, and so may stand for both senses.

While the tongue provides us with the means of distinguishing a handful of chemical qualities, the nose can discriminate among many hundreds of different substances, and does so with far greater sensitivity. It is estimated that the activity of only 40 olfactory cells, each stimulated by 8 molecules, is sufficient for us to detect the presence of some chemicals. This virtuosity is probably a relic of the many important roles smell has played in mammalian life, most of which have ceased to be as important for humans. Most other animals have a much keener sense of smell than we do. The average human has some 5 or 10 million olfactory cells, while the dog, for example, has nearly 50 times this number.

The Olfactory Cells

The sensory cells in the nose differ in a couple of important ways from those on the tongue. First, the portions of the cell to which the odor molecules become attached are not simply folds in the cell membrane, but rather *cilia*, or distinct, whiplike organs that are used in other cells—certain single-celled animals, for example—as a means of locomotion. The cilia extend from each cell into a thin layer of mucus, a mixture of water, proteins, and carbohydrates that is liquid but viscous, and that is secreted by special cells interspersed among the sensory cells. Like saliva, mucus is the body's portable version of the sea that originally dissolved molecules and brought them into direct contact with the delicate cell membrane.

The second unique feature of the olfactory cells is the fact that they are themselves nerve cells with a direct line to the brain, rather than epithelial cells serviced by nerve fibers. Branches of the nerve cell reach right into the cilia, and the other ends extend directly into the olfactory bulb of the brain, a few centimeters away. From here, information is transmitted to several other areas, including the hippocampus, where the emotions arise, and the hypothalamus, among whose many duties is the regulation of feeding behavior. There are also free endings from fibers of one of the major cranial nerves, which surface in the olfactory region and contribute to our sense of smell independently of the olfactory cells proper. Olfactory cells are the only nerve cells capable of regeneration; they are replaced every month or two.

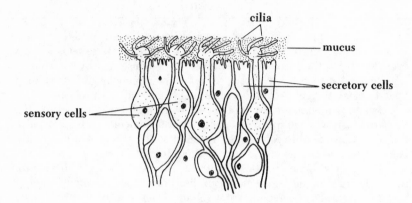

The sensory cells of the nose, which are nerve cells with specialized endings.
The actual receptor sites may be on the cilia, on the bulb that supports the
cilia, or both.

Both the anatomical location of the olfactory area and the fact that it
detects airborne rather than solid or liquid molecules have very familiar consequences for the operation of this sense. In humans, the olfactory cells
occupy about 2 square inches in the upper reaches of the nasal cavities, not
directly in the path of air flow during breathing, but somewhat above. It is
for this reason that we sniff when consciously trying to detect faint odors: the
sudden and irregular blasts of air eddy up into the olfactory region and permit a better sampling of its chemical content. By contrast, our sensitivity to
the flavor of food in our mouth is greatest when we breathe out with the
mouth closed; air from the lungs passes along the back of the mouth on its
way to the nose and brings some food vapors with it. And because molecules
must be in the vapor phase in order to be smelled, and the number of molecules in that phase depends on their temperature, warm or hot substances
are more odorous than cold ones. It is no mixed metaphor, then, for us to
speak of a scent or trail going cold. The mouth helps us out in this respect
by warming even cold foods to some extent. Ice cream, for example, doesn't
smell like very much, but once in the mouth it can have a very complex
flavor.

Theories of Olfaction

Our understanding of the molecular basis of smell is still rather primitive. It does seem that, as is the case with taste, electrical signals are produced
by the momentary interaction between odor molecule and receptor membrane, and that the coding of signals from many cells taken together—the
rapidity of interactions, the area in the olfactory region from which the

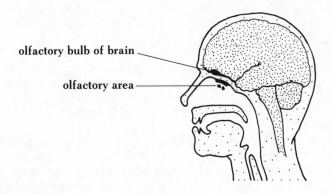

olfactory bulb of brain

olfactory area

The olfactory area is located high up in the nasal cavity where little air normally flows. Sniffing helps to pull more odorous molecules over the olfactory cells, and so increases our awareness of scents.

signals come, the pattern, both spatial and temporal, that they produce—somehow gives the brain enough information to determine the kinds of compounds involved, and their approximate concentrations. Several theories have been proposed regarding the initial encounter of molecule and membrane, each with a different idea of the means by which the odor is recognized. Though none so far is entirely convincing, one seems to be on the right track. The "stereochemical" model of J. E. Amoore suggests that the shape of the molecule determines the odor we assign it. "Primary" odor molecules fit into unique receptor sites on the membrane, while "secondary" molecules can fit into two or more kinds of receptors, and generate a composite signal analogous to the mixture of primary colors. This is also the general model for the sense of taste, and as a model it is not really new. The early Roman poet Lucretius subscribed to an atomist theory of the world, and in *De Rerum Natura*, "On the Nature of Things," gave a primitive but prescient account of the senses:

> those bodies which can touch our senses pleasantly are made of smooth and round atoms, but contrariwise all that seem to be bitter and rough are held in connection by atoms more hooked, and are therefore accustomed to tear open their way into our senses and to break the texture by their intrusion.

The problem with Amoore's application of the principle of shape to smell is that the number of different receptors proposed—around seven—seems to be too small to code the several hundreds of odors we can actually identify. By contrast, the severely limited number of tastes makes that sense a good candidate for the shape theory. But an indication of the fundamental validity of Amoore's model of smell is the fact that two molecules with precisely the same chemical makeup but with different structures—one is the mirror-image of the other—have very different odors. The difference in odor between lemons and oranges is due to such a pair of molecules, *R*- and *S*-limonene. A more striking example is the substance carvone, one of whose forms smells like spearmint, the other like caraway. It is likely, then, that molecular shape is an important factor, if not the only factor, in the discrimination of smells.

One challenge to students of olfaction has been to define the "primary odors" which, unlike tastes, are not terribly self-evident. So far, only various groups of general odor "classes" have been developed. Amoore has seven: floral, pepperminty, musky, pungent, putrid, ethereal, and camphoraceous. As the terminology indicates, the distinctions are still somewhat imprecise. Recently, Amoore has proposed that the phenomenon of *anosmia,* or the inability of some people to detect certain smells, might be used to define primary odors. If a person has a generally normal olfactory system with only one or two "blind spots," then these are probably indications that specific receptor systems have failed. This in turn implies that such specific systems exist in normal humans. In order to guide his choice of test odors, Amoore limited himself to those which are naturally associated with the human body, and which therefore may have played a significant role in our precultural past as reproductive or territorial signals. He found that there were four anosmias of this type, to what he called sweaty, spermous, fishy, and musky odors. He generated these odors by synthetic chemicals that deliver a single type of molecule to the olfactory cell. These results are suggestive, but confirmation is needed that the analogues of these chemicals are in fact produced by the human body.

Adaptation and Fatigue

A final word about an aspect of sensory physiology common to both taste and smell, and for that matter to the other senses as well. If the receptor cells are exposed continuously to a particular stimulus, the intensity of the cells' response to that stimulus gradually diminishes, with the result that an even larger stimulus is needed to trigger *any* response. The sensory system, then, adapts to an everpresent stimulus by reacting less and less to it: we "get used" to the smell or taste and eventually fail to notice it. Exposure to very strong smells results in an even more dramatic kind of adaptation called "fatigue," which can occur in a matter of minutes and take minutes to recover from; in the meantime, those receptors are completely incapacitated. We can generalize about these phenomena by saying that our senses are so designed as to be most sensitive to change, rather than to continuity or monotony. Here, then, is a physiological argument for varied menus. Our palate will be sharper, more attentive to nuance if it is entertained, kept off guard, given something new every few bites. Contrast and variety are biologically certifiable culinary principles.

Part 3

The Principles of Cooking: A Summary

CHAPTER 13

The Four Basic
Food Molecules

This chapter introduces the four chemical protagonists of the cooking process. Their names and nutritional significance are familiar enough, but a somewhat deeper acquaintance with them can be very useful. *Water* is the major component of nearly all foods, and one particular property of water solutions, their acidity or alkalinity, has an important influence on the behavior of the other food molecules. *Carbohydrates*, the specialty of plants, include sugars, starch, cellulose, and pectic substances. They are involved in, among other things, the thickening of sauces, the staling of bread, the softening of vegetables during cooking, and the notorious effect of beans on our digestive systems. *Proteins* are the sensitive food molecules. Salt, acid, heat, and even air can change their shapes and thereby their behavior drastically. Curdled milk, solidified egg white, baked bread, and broiled meats are all the result of alterations in proteins. Finally, there are the *fats* and *oils*. Why these substances don't mix with water, why fats are solid when oils are liquid, how they go rancid, and why they are useful in making soaps, can all be understood by knowing something about their basic structure.

WATER

Water is surely the most familiar chemical compound. But its most important influences are often the least visible. Water moderates our climate and carves our landscapes. And all life, including our own, exists in a water solution: a legacy of life's origin, billions of years ago, in the oceans. Our bodies are 60% water by weight; raw meat is about 75% water, and fruits and vegetables up to 95%. In part because water is ubiquitous, we take its behavior for granted, and fail to realize how peculiar a substance it really is.

The Importance of Hydrogen Bonds

The anomalous properties of ordinary water can be understood as various manifestations of a single fact: that water is the medium par excellence for the formation of hydrogen bonds, or weak but important bonds between a hydrogen atom on one molecule and an oxygen or nitrogen atom on another molecule (for more details, see the Appendix). Each water molecule is composed of one oxygen and two hydrogen atoms. The covalent bonds between oxygen and hydrogen are asymmetrical: oxygen has a greater affinity for electrons than hydrogen, and so the shared electrons are held closer to the oxygen atom. Now if all three atoms were arranged in a straight line, the molecule as a whole would be electrically symmetrical. The net positive charges on the hydrogen atoms would balance each other, and the center of positive charge would fall in the same place as the center of negative charge, right on the oxygen atom itself. But because of the disposition of other electrons in the oxygen atom, its covalent bonds are oriented not at an angle of 180°, but at about 105°, the molecule therefore assuming a V-like shape. As a result of the bonding angle and the electrical asymmetry of each bond, the water molecule as a whole is electrically asymmetrical: the center of positive charge does not coincide with the center of negative charge. Water molecules are polar.

The electrical imbalance between hydrogen and oxygen is unusually pronounced, so that the attraction between an oxygen atom on one molecule and a hydrogen on another is stronger than that between hydrogen and either carbon or nitrogen atoms, the other principal ingredients of organic molecules. Because water is nothing more than a combination of oxygen and hydrogen, every single atom in water can conceivably participate in relatively strong hydrogen bonding. And the V-shape of the molecule makes it easy for hydrogen atoms to approach the oxygens on other molecules, which are "on the point" and unprotected by the positive charge on their own hydrogens. So favorable are the conditions for hydrogen bonding that molecules in ice and liquid water are probably participating in from one to four hydrogen bonds at any given moment. Indeed, the liquid phase is viewed not as a mass of single, independently moving particles, but as a mixture of large, hydrogen-bonded assemblages, each involving from two to dozens of molecules.

Liquid and Gas

With this brief introduction to water's chemical nature, we are in a position to appreciate some of its physical peculiarities, properties that we make use of every day in the kitchen. To begin with, if water were to behave in accord with the hydrogen compounds of the other elements in oxygen's column of the periodic table—sulfur, selenium, tellurium—then we would expect it to freeze at −150° and boil at −110°F (−101° and −79°C). That

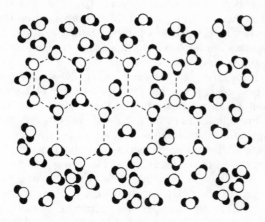

Liquid water probably includes many highly organized, hydrogen-bonded regions that are constantly breaking apart and re-forming.

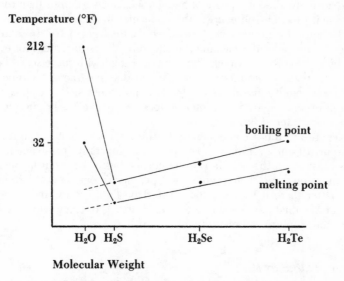

The melting and boiling points of the hydrogen compounds of oxygen, sulfur, selenium, and tellurium. Water deviates from the predicted pattern *(dotted lines)* because the oxygen-hydrogen covalent bond is much less balanced than the others, and so permits strong hydrogen bonding between molecules.

is, it would be a gas at normal earth temperatures, and life as we know it would not exist. But the covalent bonds in H_2S, H_2Se, and H_2Te are not as asymmetrical as they are in H_2O, and these compounds do not form strong hydrogen bonds. The extensive hydrogen bonding in water makes it more difficult to separate one molecule from another, and results in abnormally high melting and boiling points.

The stabilizing effect of hydrogen bonding also gives water an unusually high "latent heat of vaporization," or the amount of energy that water absorbs without a rise in temperature as it changes from a liquid to a gas. Plants and animals have taken advantage of this property in their mechanisms for temperature regulation; when overheated, they secrete water onto their surfaces and the water evaporates, in the process absorbing large amounts of energy from the organism and carrying it into the air. Ancient cultures have used the same principle to cool their drinking water and wine: they stored them in porous clay vessels that evaporate moisture continuously from their surfaces. In the absence of refrigeration—at market for example, or on a picnic—any food can be cooled to some extent by misting it with water or covering it with a damp cloth.

Ice

Normally, the solid phase of a given substance is denser than the liquid phase: with less kinetic energy, the molecules move around less and settle into a compact arrangement determined by their geometry. In solid water, however, the molecular packing is dictated by the requirement for even distribution of hydrogen bonds. The result is a solid with more space between molecules than the liquid phase has. Water does contract in volume and so increase in density just above the freezing point, but as ice crystals begin to form, it expands again by a factor of about one-eleventh. Because ice is less dense than water, it floats.

There are more consequences to this anomaly than are immediately apparent. In our everyday experience, it is generally inconvenient: water pipes burst when the heat fails in winter; bottles of beer put in the freezer for a quick chill and then forgotten will pop open; containers of leftover soup will shatter there if they are too full for the liquid to expand freely. And plant and animal tissues may be damaged during freezing when expanding ice crystals rupture cell membranes and walls.

Water Is Slow to Heat

Finally, water has an abnormally high *specific heat*, the amount of energy required to raise its temperature by a given amount. That is, water will absorb quite a bit of energy before its temperature rises. For example, it takes 10 times the energy to heat an ounce of water 1° as it does to heat an ounce of iron 1°. In the time it takes to get an iron pan too hot to handle on the stove, water will have gotten only tepid. One natural substance alone

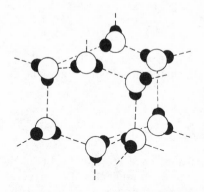

The crystalline structure of ice, which is determined by the disposition of hydrogen bonds between molecules. Ice is an anomalous solid because it occupies more space, and so is less dense than its liquid phase.

has a higher specific heat than water, and that is liquid ammonia. Again, the explanation lies in the hydrogen bond. Recall that liquid water is composed of large assemblages of molecules, all held together by hydrogen bonding. We can think of these assemblages as similar to the solid phase: they are more organized and less free in their movements than the unassociated molecules around them, though their structure and size are continually changing. This means that before these masses can express added energy as heat, which is molecular movement, the constituent molecules must be free to move. The first increment of energy, then, goes toward distorting and then breaking the hydrogen bonds that hold the aggregates of molecules together. In a sense, liquid water undergoes continual phase changes as it is heated, even long before the actual boiling point is reached.

The consequences of this peculiarity are considerable. It makes it possible for living organisms, which are from 50 to 95% water, to moderate the effect of sudden environmental temperature changes. Our bodies can absorb *or* lose substantial amounts of energy without becoming dangerously hot or cold. In this, we are like miniature oceans or lakes or humid atmospheres, which even out climatic changes by soaking up or releasing energy in extreme conditions without reaching extreme temperatures themselves. We take advantage of this same kind of buffering when we cook something delicate, for example a custard, in a water bath rather than exposing it to direct heat. It also means that a pan of water will take more than twice as long as a pan of oil to heat up to a given temperature; oil has a much lower specific heat.

Acids, Bases, and the pH Scale

Despite the fact that the molecular formula for water is H_2O, even absolutely pure water contains other molecular structures. As we have said, bonds are continually being formed and broken in matter, and water is no exception. It tends to "dissociate" to a slight extent, according to this general equation:

$$H_2O \rightleftharpoons H^+ + OH^-$$

But because a single hydrogen ion, which is simply a proton, is very reactive, it probably attaches itself immediately to another nearby molecule. The dissociation is therefore usually written as

$$2H_2O \rightleftharpoons H_3O^+ + OH^-$$

Under normal conditions, a very small number of molecules exists in the dissociated state, something on the order of 2 ten-millionths of a percent. A very small number, but a very important one: for the presence of relatively mobile protons, which are the basic units of positive charge, can have drastic effects on other molecules in solution. A structure that is stable with only a few protons around may be very unstable when many protons are in the vicinity. So significant is the proton concentration that we have a specialized taste sensation to estimate it: sourness. Our term for the class of chemical compounds that release protons into solution, *acids*, derives from the Latin *acere*, meaning to taste sour. The complementary chemical group that accepts protons and neutralizes them we call *bases* or *alkalis*.

The properties of acids and bases affect us continually in our daily life. With bases we make soaps and detergents. And practically every food we eat, from steak to coffee to oranges, is at least slightly acidic. Further, the acidity of the cooking medium can have great influence on such characteristics as the color of fruits and vegetables and the texture of meat and egg proteins. Some measure of proton activity, or the strength of acids and bases, would clearly be quite useful. A simple scale has been devised to provide just that.

The standard measure of proton activity in solutions is *pH*, a term suggested by the Danish chemist S. P. L. Sørenson in 1909. The idea was to get numbers more convenient to work with than the minuscule percentage just cited. So pH is defined as "the negative logarithm of the hydrogen ion concentration expressed in moles per liter." The logarithm of a number is the exponent, or power, to which 10 must be raised in order to obtain the number. For example, the hydrogen ion concentration in pure water is 10^{-7} moles per liter, so the pH of pure water is 7. (Because our atmosphere contains carbon dioxide, which dissolves in water to form an acid, the pH of ordinary distilled water is more like 5.5.) Pure water is both a mild acid and a mild base—the number of protons equals the number of alkaline OH^-, or hydroxyl groups—so pH 7 is defined as *neutral*. A pH lower than 7 indicates a greater concentration of protons and so a net acidic solution, while a pH above 7 indicates a greater prevalence of proton-*accepting* groups, and so a net basic solution. Each increment of 1 in pH signifies an increase or decrease in proton concentration by a factor of 10; so there are 1,000 times the number of hydrogen ions in a solution of pH 5 as there are in a solution of pH 8. Here is a list of common solutions and their usual pH, for the sake of orientation.

human gastric juice	1.3–3.0
lemon juice	2.1
orange juice	3.0
black coffee	5.0
milk	6.9
egg white	7.6–9.5
baking soda in water	8.4
household ammonia	11.9

Soft and Hard Water

Water is by far the most effective solvent of any common liquid, because its molecules are well-suited to surrounding both positive and negative centers of charge, thereby separating ions or molecules from each other and so dissolving a substance. For this reason, it is really never found in anything like pure form. Tap water is quite variable in composition, depending on its ultimate source (well, lake, river) and its municipal treatment (chlorination, fluoridation, and so on). Two common solutes in tap water are carbonate (CO_3^{-2}) and sulfate (SO_4^{-2}) salts of calcium and magnesium. Calcium and magnesium ions are troublesome because they interfere with the action of soaps by forming insoluble scums, and because they leave crusty precipitates on showerheads and teapots. Such so-called hard water can also affect the color and texture of vegetables, and the consistency of bread dough. Hard water can be softened either city-wide or in the home, usually by one of two methods: precipitating the calcium and magnesium by adding lime, or using an ion-exchange mechanism to replace the calcium and magnesium with sodium. Distilled water, which is produced by boiling ordinary water and collecting the condensed steam, is fairly free of impurities.

The Boiling Point

The boiling point of water is one of the few reliable reference points in the kitchen. The sign that it has been reached is unmistakable, and for the great majority of cooks, those who live fairly close to sea level, it occurs at very close to the same temperature. The boiling point decreases 1°F for every 500-foot gain in elevation above sea level, and will be slightly lower under a low-pressure weather system than it is under a fair-weather high (for more about water as a cooking medium, see chapter 14).

The sure sign of 212°F (100°C) water is bubbling. Why? When the water in a pan is heated near boiling, molecules at the very hot bottom vaporize and form regions that are less dense than the surrounding liquid. They therefore rise in the liquid. Large bubbles do not make it to the surface until vapor formation is rapid enough and molecular velocities high enough

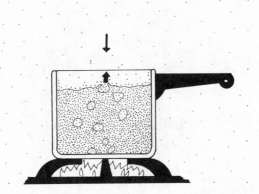

The bubbles in boiling water are pockets of water vapor formed on the heated pan bottom. They cannot reach the surface unless the vaporized molecules absorb enough energy to overcome the combined weight of the liquid water and the atmosphere. Otherwise, the vapor pockets simply dissipate in the liquid.

to resist the combined pressure of liquid and atmosphere (the small bubbles that form very early on are pockets of air that had been dissolved in the cold water but became less soluble as the temperature rose). The water temperature at a full, rolling boil is only slightly higher than in a gently bubbling pot, and will not get any higher until the phase change from liquid to gas has been completed (steam can be much hotter). Generally, a gentle boil does less physical damage to food, while a full boil is most useful in reducing the volume of liquid.

The Effect of Sugar and Salt

When salt, sugar, or any other nonvolatile compounds are dissolved in water, the freezing point of the resulting solution is lowered and its boiling point raised. We take advantage of this effect by using rock salt to melt ice on roads, and to freeze ice cream. As far back as the 18th century, solutions of calcium chloride were used to reach temperatures of $-27°F$ ($-33°C$). The helpfulness of solutes at the other end of the scale is, however, more limited. It takes one ounce of salt to raise the boiling point of a quart of water by a mere 1°. A Denverite who wanted to boil water at 212°F would have to add more than half a pound of salt to that quart of liquid.

Both the depression of the freezing point and the elevation of the boiling point are due to the fact that the water molecules have been diluted with other particles, which interfere with the process of phase change. Imagine that we add salt to some water, so that the number of sodium and chloride ions adds up to 1% of the total number of particles, including ions and water molecules. Now we heat the solution close to the boiling point, and many water molecules are receiving enough kinetic energy to escape into the vapor phase. But every hundredth particle in the solution will *not* escape, despite the fact that it too is absorbing energy: sodium and chloride ions cannot exist as free particles in space except at extremely high temperatures. So at a given temperature, there will be fewer water molecules escaping the liquid phase from a solution than there would be from pure water. In order to free the *same* number of molecules, the temperature of the solution must be raised to compensate for the retarding effects of the solute. The same logic—a low-ered rate of phase change—applies to freezing as well. Here, every hun-dredth molecule to reach the face of the accreting ice crystals will not be water, and therefore will not become assimilated. So the freezing process is impeded, and the temperature must be lowered further in order to compensate.

CARBOHYDRATES

The name for this large group of organic compounds comes from the early idea that they were made up of carbon and water, with the chemical formula $C_n(H_2O)_n$, n representing a whole number. This formula does hold for most carbohydrates, though the oxygen and hydrogen atoms are never found as intact water complexes within the molecule. Unlike water, carbo-hydrates are produced only by living organisms. They serve two basic func-tions: energy storage, and, in plants, structural support as well. Energy is stored in the various sugars, in plant starch, and in glycogen, or "animal starch." Plant cell walls are strengthened by cellulose, the hemicelluloses, and by pectic substances.

Sugars

Sugars are the simplest carbohydrates. There are many different kinds of sugar molecules, each distinguished by the number of carbon atoms it contains, and then by the particular structure it assumes. Five-carbon sugars are especially important to all life because two of them, ribose and deoxyri-bose, form the backbones of ribonucleic acid (RNA) and deoxyribonucleic acid (DNA), the carriers of the genetic code. The sugars most commonly encountered in the kitchen are 6-carbon sugars, both singly and in combi-nation. Glucose and fructose both have the same chemical formula, $C_6H_{12}O_6$, but fructose takes the form of a 5-atom ring, while glucose has 6 corners. Our table sugar is nearly pure sucrose, which is a *disaccharide* ("double-sugar") composed of one glucose and one fructose molecule joined together. Milk, both human and bovine, contains a disaccharide called lactose, which is built up from glucose and galactose, yet another 6-carbon sugar. With the

Top: the molecular structures of glucose and fructose, sugars with the same chemical formula ($C_6H_{12}O_6$). *Bottom:* sucrose is a combination of glucose and fructose. A molecule of water is left over when the separate molecules fuse to form sucrose. (For the sake of clarity, carbon groups off the main rings of carbohydrate molecules are given in shorthand.)

exception of lactose, all of our familiar sugars can be obtained from plants of various kinds. Fructose is a major component of honey, and sucrose is extracted in quantity from sugar cane and beets.

Sugars store energy in the form of relatively weak covalent bonds, which can be broken with the liberation of energy during oxidation:

$$C_6H_{12}O_6 + 6O_2 \rightarrow 6CO_2 + 6H_2O + \text{energy}$$

More on foods as fuel in the section on "Fats and Oils," and in the Appendix.

As can be deduced from the hydroxyl (OH) groups projecting from the rings of these molecules, sugars participate readily in hydrogen bonding, and so are quite soluble in water. In fact, sugars have a strong affinity for water, and will absorb it from the atmosphere if given the chance. It is for this reason that an apple put in a breadbox with a loaf of bread will dry the bread out, while an apple confined with a cake will help keep the cake moist. In each case, the object with the higher sugar content draws water molecules through the air from the other. Single-ring sugars are more hygroscopic, or moisture-retaining, than disaccharides; this is why candies made with honey tend to get quite sticky.

Oligosaccharides

The oligosaccharides ("several-unit sugars") raffinose, stachyose, and verbascose are 3-, 4-, and 5-ring sugars, respectively. They are commonly found in the seeds and other organs of plants, where they make up part of the energy supply. These sugars would hardly be worth mentioning if it weren't for the fact that they all affect our digestive tracts. Humans do not secrete any digestive enzymes capable of breaking these molecules down into their component single sugars, and only single sugars can be absorbed by the intestine. As a result, the oligosaccharides pass intact into the colon, where various bacteria that *can* digest them do so, producing large quantities of carbon dioxide gas in the process. (This subject is discussed in detail in chapter 5.)

Polysaccharides

Polysaccharides are sugar *polymers*, or molecules composed of many thousands of individual sugar units. Usually only one or two kinds of sugars are found in a given polysaccharide, and seldom more than four. With such huge molecules, it is impossible to separate and identify all the minor variations in structure and composition. Instead, polysaccharides are classified according to a general size range, an average composition, and a common set of properties. Several properties are fairly widespread. Because polysaccharides, like the sugars of which they are composed, have many exposed oxygen and hydrogen atoms, they can form hydrogen bonds and will absorb water. However, depending on the attractive forces among the polymers themselves, they may or may not be soluble in water. Then there are differences in structure. Polysaccharides that exist in long, linear chains tend to form a viscous meshwork in water that can solidify into a gel when cooled. Another common molecular configuration, the compact, branched form, resists gel formation. These considerations are important in the making of starch-thickened sauces and pie fillings.

Starch

By far the most important polysaccharide for the cook is starch, the compact, unreactive polymer in which plants store their supply of sugar. Starch is simply a chain of glucose molecules. It can exist in two different configurations: a completely linear chain called *amylose,* and a highly branched form called *amylopectin,* each of which may contain many thousands of glucose units. Most plant starches are a mixture of these two kinds, which are deposited in a series of concentric layers to form starch granules. The relative proportions of amylose and amylopectin in a given plant starch will determine its physical properties and so its culinary uses. The more amylose, the stronger the gel that will be formed, while the greater the amylopectin content, the more viscous the solution that can be made without its setting into a gel—a desirable characteristic in a pie filling, for example. "Waxy" corn or rice starch, which is almost entirely amylopectin (because of its irregular shape, amylopectin has a soft, mealy, "waxy" consistency, while amylose is harder), does not gel, while ordinary corn, rice, and wheat starches do.

The same characteristic of amylose that makes gelling possible—the ability to form hydrogen bonds with many other starch molecules, thereby creating an extensive mesh—has less fortunate consequences as well. Once they have gelled in water, starch molecules tend to settle into stronger and more numerous association with each other, forming more hydrogen bonds among themselves, and fewer with the incorporated water. As a result, the gel shrinks into a denser, more compact structure, and some water is squeezed out. This process is responsible for the eventual separation of starch-based sauces, and even more important, for the staling of bread. It takes place most rapidly at temperatures just above freezing, and so is encouraged by refrigeration and during the actual process of freezing (once frozen, bread stales very slowly).

Glycogen

Glycogen, or "animal starch," is a polymer similar to amylopectin, though much more highly branched. It is a fairly minor component of animal tissue and so of meats, although its concentration at the time of slaughter will affect the ultimate pH of the meat, and thereby its flavor and texture (see chapter 3).

Cellulose

Cellulose is, like amylose, a linear plant polysaccharide made up solely of glucose sugars. Yet the two compounds have very different properties, and are put to very different uses by plants. Starch is a food storage depot, but cellulose is a structural support which is laid down in cell walls in the form of tiny fibers analogous to steel reinforcing bars. Both the plant itself and animals regularly digest starch into its component sugars to obtain chemical energy, but the plant clearly does not want to break down its own body, nor can many animals digest cellulose. Those that do—hay-eating cattle, wood-eating termites—can do so only because their guts are populated by cellulose-digesting bacteria. And in the kitchen, hot water breaks apart and dissolves starch granules, while cellulose remains unaffected. Because cellulose is a material remarkably resistant to change and is therefore so useful as a building material, it is far and away the most plentiful organic compound on earth.

How is it that two compounds with identical chemical compositions have such different properties? The key is a minor difference in structure. Starch is made up of molecules of so-called α-D-glucose, in which, following a standard convention, the first carbon atom is bonded to a hydroxyl group below the ring, and to a hydrogen atom above. Cellulose is a polymer of β-D-glucose, in which the positions of hydroxyl and hydrogen are reversed. This simple difference in molecular geometry changes the geometry of the glucose-glucose bond, and of the deployment of hydrogen-bonding atoms around the molecules. The result: cellulose polymers tend to form much stronger bonds among themselves than do amylose polymers, and in the aggregate become an insoluble mass. In addition, plant and animal enzymes designed to break the α-glucose linkage and depolymerize starch have no effect on the β-glucose linkage of cellulose. This is why the cellulose in vegetables and grains passes through our digestive system as fiber.

Starch *(top)* and cellulose *(bottom)* are both long strings of glucose molecules. On account of small differences in the configurations of the glucose units, these two substances have very different properties. Starch is one of our most important sources of energy, while cellulose is largely indigestible.

Hemicelluloses and Pectic Substances

These rather vaguely defined sugar derivatives are found in conjunction with cellulose in the plant cell walls. If the cellulose fibrils are like reinforcing bars, the amorphous hemicelluloses and pectic substances are a sort of jelly-like cement in which the bars are embedded. The significance of these compounds for the cook is that, unlike cellulose, they are soluble to varying degrees. Hemicelluloses are markedly soluble in alkaline conditions, and this means that vegetables will tend to remain firm when boiled in neutral or acidic water, but will become mushy in alkaline water. The dissolving of soluble pectin is another factor in the softening of cooked vegetables, and fruit becomes more tender during ripening because insoluble protopectin is converted by enzymes into pectin. Pectin is also extracted from citrus fruits and apples and used to thicken fruit syrups into jams and jellies; in acid conditions, the molecules complex with each other to form gels.

PROTEINS

While the carbohydrates are dominated by starch and cellulose, substances remarkable for accomplishing two very different tasks with very similar materials, the proteins astonish with their sheer variety and versatility. Just for starters, proteins are the major constituents of the protective skin, the propulsive muscles, and the vital organs of animals. Hemoglobin and myoglobin, the molecules that carry and store oxygen in animal tissues, are mostly protein. And all plant and animal enzymes, those molecules that tear other molecules down and build them up according to the needs of the organism, are proteins. Proteins are at the heart of all organic movement, change, and growth: the most important characteristics of life itself.

As one might expect, given this range and complexity of function, protein chemistry is one of the more arcane disciplines. Still, most proteins follow a few rules of behavior which, once understood, can help make sense of many culinary experiences.

Protein Structure

Amino Acids

Like starch and cellulose, proteins are large polymers of much smaller molecular units. These units are called *amino acids,* and there are about 20 of them that occur in significant quantities in food. Unlike the carbohydrates, which rarely incorporate more than 3 or 4 different sugars, proteins can and often do contain most of these 20-odd amino acids. Fortunately we need not worry very much about particular molecular structures.

The amino acids are a diverse family of compounds, but most have one corner in common that gives the family its name. The presence of nitrogen

Top: the corner of the amino acid molecule that includes the amine group, NH₂, and the acidic carboxyl group, COOH. *Bottom:* in water, one oxygen atom loses a hydrogen ion, while the nitrogen atom picks one up. The molecule has both acidic and basic properties.

in the NH_2, or *amine* group, distinguishes the proteins chemically from carbohydrates and fats. In the large range of pH between acidic 4 and alkaline 9, the amine group picks up a hydrogen ion, thereby behaving like a base, and becomes NH_3^+. But in the same pH range, the COOH group loses its hydrogen, becoming COO^-, and thereby behaves like an acid. This double character of the molecule, both acid and base, partly accounts for the versatility of amino acids in chemical reactions.

The Peptide Bond

Proteins are formed from amino acids when the nitrogen of one molecule and the carbon of another join together in what is called a peptide bond. Because the outer electron clouds of the carbon atom are directed toward the corners of a tetrahedron, the carbon-nitrogen backbone forms a sort of zigzag pattern, with the "side groups"—the X, Y, and Z that represent the rest of each amino acid—sticking out to the sides.

Due to the relative electron affinities of oxygen, carbon, and nitrogen, the peptide bonds, those between N and C=O, are unusually strong, and resemble the double bond in the difficulty with which the atoms can be rotated along their axes. Rotation is, however, very easy around the other backbone bonds. This combination of properties affects the ultimate shape of the protein as a whole, and so its behavior in groups or in reactions with other compounds.

$$\left[\text{X}\right]-\underset{\underset{\text{H}}{\overset{|}{\underset{|}{\text{N}}}\,^{\text{H}}}}{\overset{\text{H}}{\underset{|}{\text{C}}}}-\overset{\overset{\text{O}}{\|}}{\text{C}}-\text{OH}\quad+\quad\left[\text{Y}\right]-\underset{\underset{\text{H}}{\overset{|}{\underset{|}{\text{N}}}\,^{\text{H}}}}{\overset{\text{H}}{\underset{|}{\text{C}}}}-\overset{\overset{\text{O}}{\|}}{\text{C}}-\text{OH}\quad\longrightarrow$$

$$\left[\text{X}\right]-\overset{\text{H}}{\underset{\underset{\text{H}}{\overset{|}{\underset{|}{\text{N}}}\,^{\text{H}}}}{\underset{|}{\text{C}}}}-\overset{\overset{\text{O}}{\|}}{\text{C}}-\overset{\text{H}}{\underset{|}{\text{N}}}-\overset{\text{H}}{\underset{\underset{\left[\text{Y}\right]}{|}}{\underset{|}{\text{C}}}}-\overset{\overset{\text{O}}{\|}}{\text{C}}-\text{OH}\quad+\quad\text{H}_2\text{O}$$

When the nitrogen atom on one amino acid reacts with a carbon atom on another, the result is a peptide bond *(center)*. Peptide bonds form the backbone of protein molecules, which are long chains of amino acids with their "side groups"—the remainders of the amino acids, here abbreviated to X, Y, and Z—projecting from the chain *(bottom)*.

The Protein Helix

One effect of the peptide bond is the regular alternation of C=O and N—H groups, whose projecting oxygen and hydrogen atoms are well suited to hydrogen bonding with each other. This regularity and suitability help to stabilize the shape of protein molecules. There is a natural tendency for the rotatable bonds along the backbone to rotate, and the molecule as a whole twists around to form a spiral, or helix, with the C=O and N—H groups on neighboring turns, four amino acid groups apart, bonding to each other. Very few proteins exist as regular helices, as we shall see, but those that do tend to join together in strong fibers. Animal protective tissues, including

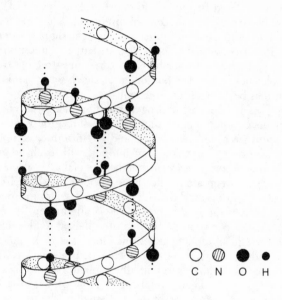

The regularly spaced hydrogen and oxygen atoms projecting from the protein backbone readily engage in hydrogen bonding *(dotted lines)*. As a result, the backbone (which consists of nitrogen and carbon atoms) twists into a spiral, or helical shape.

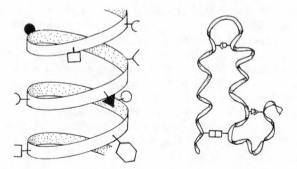

The side groups (here represented in a shorthand of geometrical shapes) that project from a protein's backbone can bond with each other and thereby throw kinks into the molecule's shape.

skin, hair and fur, nails and claws, are composed of helical proteins, as are connective tissue collagen, an important factor in meat tenderness, and one of the components of muscle protein. The slight elasticity of wool and human hair is a consequence of helical structure: tension on the fiber breaks some of the hydrogen bonds between turns and stretches out some sections of the molecules. When the tension is removed, some bonds re-form, and the fiber shrinks back to a limited extent.

As it happens, hydrogen bonding along the backbone is not the only influence on protein structure. The side groups on the amino acids—the groups we have labeled X, Y, and Z—introduce complications. These groups can be quite large, and can interact with each other in various ways: via hydrogen bonds, covalent bonds (especially between sulfur atoms), or ionic bonds between atoms that have lost or gained electrons in the surrounding water. And because the side groups are seldom as regular as the backbone, the helix is rarely the most stable configuration for them. Protein structure is, then, a compromise between the helical tendencies of the backbone and the irregular tendencies of the side groups. It can range from the long, extended, mostly helical molecule with a few kinks or loops, to compact, elaborately folded, substantially nonhelical molecules that are called "globular" proteins. There are also very large proteins that turn out to be regular clumps or aggregates of several smaller chains, each containing perhaps hundreds of amino acids, which are held together by *inter*molecular bonds.

Proteins and Water

In living systems and in most foods, proteins exist in a water solution, so it is important to know how these two substances tend to interact. Because all proteins are capable to some extent of hydrogen bonding, they will absorb and hold at least some water, although the amounts vary greatly according to the kinds of side groups present and the overall structure of the molecule. Water molecules can be held "inside" the protein, along the backbone, and "outside," on polar side groups. If nonpolar, water-repelling side groups predominate in a particular protein, then it will absorb little water.

Whether or not a protein is *soluble* in water depends on the strength of the bonds between molecules, and on whether water can separate the molecules from each other by hydrogen bonding. The wheat proteins that form gluten when flour is mixed with water, and those that make up muscle fiber in meat, are two kinds of protein that absorb considerable amounts of water without actually dissolving. Strong covalent disulfide bonds (S—S) between molecules hold them together. Many of the proteins in milk and eggs, on the other hand, are quite soluble.

Protein Denaturation

A very important characteristic of proteins is their susceptibility to *denaturation*, or the alteration of their structure by chemical or physical means. This change involves breaking any or all of the bonds that maintain the molecule's shape, with the exception of the strong backbone bonds: these can be broken only in extreme conditions or with the help of enzymes. Denaturation is not a change in composition, only a change in structure. But structure determines behavior, and denatured proteins are quite different from their originals.

Proteins can be denatured in many ways. The most common culinary techniques are the addition of salt (cured meats), of acid (pickling), and the application of heat. In the first two cases, the introduction of many positive and negative ions increases the likelihood that bonding among protein side groups will become disrupted. When this happens, the protein molecule unfolds, thereby exposing even more of its atoms to the reactive surroundings. Heat has much the same effect, though by different means. It increases the random motion of the protein atoms to the point that they are more energetic than the hydrogen and covalent bonds that hold them together. More bonds break than can be re-formed, and again the molecules unfold, with only the strong backbone bonds holding them together.

Coagulation

There are three general consequences of denaturation that follow for most food proteins. Each has to do with the unfolding of the molecule. First, because the molecules have been extended in length and so are more likely to come in contact with each other, and because the side groups are now more exposed and so more reactive, denatured proteins tend to bond with each other, often via disulfide bonds, and form solid clumps: that is, they *coagulate*. This is why egg whites become hard and opaque when cooked.

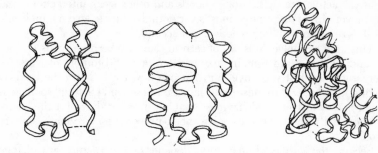

In the native protein, many side groups are involved in internal bonding *(left)*. When heat or chemical changes denature the molecule, then bonds break and the side groups are exposed to the surroundings *(center)*. Bonding between *different* molecules now occurs much more easily, and large masses of coagulated protein can form *(right)*. This is what happens when meat is cooked, eggs are scrambled, or hollandaise sauce curdles.

Second, because coagulated proteins have bonded to each other rather than to water molecules, and now have a much smaller surface area in contact with their environment, they tend to hold much less water on both outer and inner structures than do native, undenatured proteins. This is why eggs and meat lose water during cooking, even if they are boiled or steamed. Some heat-induced denaturation is reversible, however, and this has practical significance. A roast fresh out of the oven will lose a substantial amount of juice when its tissues are squeezed during carving, but if it is allowed to cool somewhat, the proteins will reabsorb some water that had been released to the tissue's interstices, and the roast will become firmer, easier to carve, and moister to the palate.

Finally, because they are extended, denatured proteins are more vulnerable to attack by protein-breaking enzymes, which depend on access to the backbone chain. For this reason, cooked meats are more digestible than raw.

Enzymes

Since we have just mentioned enzymes, this is a good place to make a few general points about this particular class of proteins, which can be so troublesome in the kitchen. Enzymes are organic catalysts. That is, they selectively increase the rate of chemical reactions that otherwise would occur only very slowly, if at all. Human digestive enzymes, which break up proteins into individual amino acids, and starch into individual glucose units, are obvious examples. Enzymes are proteins that have very complicated shapes, and these shapes are determined by the kind of reaction the molecule is to promote. There is always an "active site," a position on the surface where the molecules that are to be altered can fit. They are attracted to and held at the active site by hydrogen bonds and other weak interactions, and the turnover of reacting compounds is very rapid. A single enzyme molecule can catalyze up to a million reactions per second.

This kind of molecule is of concern to the cook because foods of both plant and animal origin contain enzymes that once did important work for the organism when it was alive, but that can now easily harm the food by changing its color, texture, taste, and nutritiousness. And bacterial spoilage is partly a matter of bacterial enzymes breaking the food down for the bacteria's own use. By and large, the cook wants to prevent *any* enzymatic activity in food.

Because the activity of an enzyme depends on its structure, any change in that structure will destroy its effectiveness. So cooking foods will inactivate any enzymes they may contain. One vivid example of this principle is the behavior of raw and cooked pineapple in gelatin. Pineapples, like figs and papayas, contain an enzyme that breaks proteins down into small fragments. If raw pineapple is put in gelatin for a dessert or fruit salad, this enzyme digests the gelatin molecules and liquefies the water-protein gel. Canned pineapple, because it has been heated, no longer contains the active enzyme, and cooperates quite well with gelatin.

There is a complication, though. The reactivity of most chemicals increases with increasing temperature: the molecules move faster and so collide more frequently, and they have more energy with which to change or be changed in some way. The rule of thumb is that reactivity doubles with each rise of about 20°F (10°C). The same tendency goes for enzymes, up to a range in which they become less effective and finally inactive. Many are most active around 100° and denature around 170°F (38° and 77°C), though some must be exposed to boiling temperatures. In general, the best rule is to heat foods as rapidly as possible, thereby minimizing the period during which the enzymes are at their optimum temperatures, and to get them all the way to the boiling point. Conversely, storing foods at very low temperatures slows active enzymes down and delays spoilage.

FATS AND OILS

Fats and oils are members of the same class of chemical compounds, the *triglycerides*. They differ from each other only in their melting points: oils are liquid at room temperature, fats solid. Rather than use the technical *triglyceride* to denote these compounds, we will use *fats* as the generic term, and will consider oils as liquid fats.

The fats, like the proteins, are an extremely heterogenous group of molecules—again, we will not worry about exact formulas—but unlike the proteins, they serve only one fundamental purpose: to store energy. As we have seen, the sugars and starches also have this function. So what is the point of this duplication? The answer lies in the single largest difference between carbohydrates and fats: fats store more than twice as much energy as carbohydrates in a given weight, and so are a much more efficient supply. While plants generally rely on carbohydrates because economy in weight is not very important, their seeds, which must include enough food to support the embryo until it is self-sufficient, often contain oil as well as starch. Soybeans, corn kernels, safflower seeds, and peanuts are all major sources of cooking oils. Animals, though they store small amounts of glycogen in the muscles and liver, and rely on glucose in the bloodstream for immediate energy needs, bank most of their energy reserves in fat. Economy in weight is important for an organism that must work not only to move, but simply to stand up against the force of gravity. The body of a typical woman is about 25% fat by weight (a man's body is closer to 15%). This means that if her fat supply were converted to its energy equivalent in carbohydrates, a 120-pound woman would weigh 150 pounds. In a world where swiftness and agility increase an animal's chances for survival, fat is clearly the preferable means of energy storage.

Once extracted from plants and animals, fats are remarkably versatile materials and serve many more culinary purposes than do carbohydrates or proteins. They are frequently used as a cooking medium, lubricating the food and transferring heat from the pan. Fats are an especially happy alternative to water in this regard, because they can be heated to very high temperatures, upward of 400°F (205°C), and so make possible crisp textures and intense flavors (see the section on browning reactions in chapter 14) that boiling simply cannot produce. Different cooking fats also contribute characteristic flavors, and can come to define a whole cuisine, as Waverly Root demonstrates when he divides France into regions that prefer butter and those that rely on olive oil. As actual ingredients, fats smooth and moisten the textures of many foods. In pastry dough, solid fat "shortens" or tenderizes the gluten structure; in bread dough, it increases loaf volume and lightness; in cake batter, it incorporates air bubbles and helps leaven the mixture. Because oil is more viscous than water and leaves a film on food surfaces, it is used in water-oil emulsions to coat foods evenly with flavorful ingredients. Finally, the soap with which we wash the dishes is derived from fats.

Unfortunately, fats can be as big a problem as they are a help. Changes in their structure and composition constitute one of the major causes of flavor deterioration during the storage of foods, especially meats.

The Structure of Fats

Glycerol and Fatty Acids

The major constituent of natural fats and oils is the triglyceride, a combination of three *fatty acids* with one molecule of *glycerol*. A given fat will be a consistent mixture of several different triglycerides.

Glycerol is an alcohol, a carbohydrate that behaves like a base by accepting protons. In fats, it acts simply as a common frame to which three fatty acids can attach themselves. Glycerol seems to have few functions in plants and animals aside from this binding, though in purified form—it is a natural by-product of soap making, in which the fatty acids are freed—we use it in candies and icings to prevent sugar crystallization, as an ingredient in cosmetics, and as a moistening agent in tobacco, among other applications.

The fatty acids are so named because they consist of a long hydrocarbon chain attached at one end to an acidic carboxyl group

$$\begin{matrix} & O \\ & \| \\ HO - & C \end{matrix}$$

For example, the most common fatty acid, oleic acid, is eighteen carbon atoms long. Many fatty acids are considerably larger than this, the largest occurring naturally in any significant quantities being about 35 carbon atoms

Top: glycerol, a molecule to which three fatty acids attach themselves to form a triglyceride. *Bottom:* the most common fatty acid, oleic acid. (The geometry of the double bond has been simplified; see figure on page 603.)

The formation of a triglyceride, the basic material of fats, from one molecule of glycerol and three fatty acids. Three molecules of water are left over from the reaction. Notice that the acidic heads of the fatty acids have been capped and neutralized by the glycerol, so that the triglyceride as a whole is much less polar.

long. Because of the indefinitely extendible carbon chain and variations in bonding (such as the double bond at the center of oleic acid), there are innumerable possible fatty acids, though in fact about 50 seem to account for most natural fats. Triglycerides, in turn, can be composed of any combination of three fatty acids. The properties of the triglyceride depend on the structure of its three fatty acids and their relative positions on the glycerol frame. And the properties of a fat depend on the particular mixture of triglycerides it contains.

We will not worry about these intricacies. The *characteristic* behavior of culinary fats and oils is determined by the characteristic behavior of the fatty acids, which is not at all difficult to understand. The only real significance of the glycerol frame is that it joins three initially semipolar, or electrically asymmetrical molecules—the acidic carboxyl group easily gives up a hydrogen ion—into an almost entirely nonpolar, electrically symmetrical molecule.

Fats, Water, and Soaps

Perhaps the most familiar property of fats is that they don't mix with water, even when they are liquids themselves. Oils minimize the surface at which they contact water by coalescing into large blobs, and resist being divided into smaller droplets. Fats are *hydrophobic*, or "water-fearing," because their nonpolar molecules cannot form hydrogen bonds with water. Instead, they associate only with each other or with other lipids (see page 606). Another consequence of their relatively weak mutual attraction is that all natural fats, solid or liquid, float on water. Water is denser on account of its extensive hydrogen bonding, which packs its molecules more tightly together.

Another kind of molecule with which fats will associate is the fatty acid, three of which are combined with glycerol to make a fat in the first place. But fatty acids have a split personality, and it is this characteristic that allows us to clean our greasy dishes in a sink of water. Fatty acids have both a nonpolar hydrocarbon tail *and* a polar head; the carboxyl end readily loses a hydrogen ion and becomes negatively charged. By combining with a positive metal ion, usually sodium, to form a salt, the fatty acid becomes a *soap*. When a soap is mixed with water, the fatty acid is released as a negative ion. As it comes in contact with fats or hydrocarbon chains like gasoline, it is automatically oriented with its hydrocarbon tail immersed in the other chains, and its polar head sticking out in the polar water. If this happens to two separate droplets, the droplets will now be unable to coalesce: their surfaces are covered with negative charges, and so they repel each other. This chemical action, together with the mechanical agitation that breaks up masses of fat into smaller particles, eventually disperses the fat in the water solution and allows us to rinse it away. (The same principle, the *emulsification* of oil and water, is exploited to make various sauces; see below, page 607, and chapter 7.)

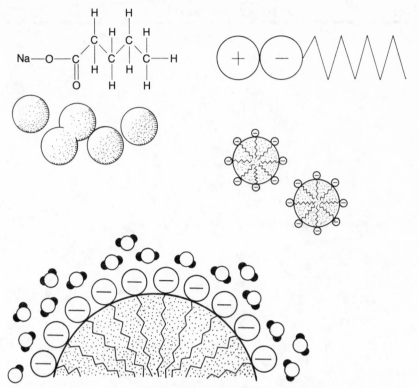

How soaps disperse grease. A soap is a combination of a fatty acid and a metal ion, usually sodium (Na) *(top)*. When dissolved in water, the fatty acid and metal ion separate. Left to themselves, fat droplets in water coalesce in large puddles *(center left)*. Fatty acids prevent this from happening by burying their non-polar tails in the fat, and leaving their negatively-charged, polar heads out in the water *(center right)*. A layer of water is held at the droplet surface by hydrogen bonding *(bottom)*.

The Romans appear to have been the first to make soaps from animal fats, which they broke down into fatty acids by using wood ashes, a concentrated source of the alkaline salt potassium hydroxide (lye). Today's detergents are simply hydrocarbon chains with synthetic polar heads that can be engineered according to the applications planned.

Why Fats Are Good Fuels

Another important thing about the structure of fatty acids and so of fats is their strong resemblance to other, simpler hydrocarbons: gasoline, for example, which is a mixture of 6- to 12-carbon chains, or paraffin waxes, with molecules more than 20 carbons long. The obvious similarity between octane and fatty acids suggests why it is that fats are such good, efficient fuels: they are, in fact, animal gasoline. As is explained in the Appendix, energy is liberated from organic compounds by the process of oxidation, in which $C-C$

Octane, one of the hydrocarbon components of gasoline

Glucose *(left)* and a 10-carbon fatty acid. The molecules have nearly the same weight, but the fatty acid contains many more oxidizable bonds, and so stores more chemical energy.

and C—H bonds are broken to form relatively stronger C—O and H—O bonds. Gasoline and fats are large molecules that contain many oxidizable bonds—octane, as you can see, contains nothing but such bonds—and therefore they are ideal fuels.

Now we can understand why it is that sugar and starch are only half as efficient as fat. Compare the structures of glucose and a 10-carbon fatty acid, which has nearly the same molecular weight as the sugar (both carbon and hydrogen weigh less than oxygen). In glucose, 7 of the 24 carbon bonds are already given over to oxygen, and so cannot be oxidized further. In the fatty acid, all but three carbon bonds are available for oxidation, making 37 compared to glucose's 17. For the same weight, then, the hydrocarbon chain offers twice the number of energy-rich bonds—twice the calories—that glucose does.

Saturation

Both vegetable oil advertisements and nutritionists have frequently asserted the benefits of using unsaturated rather than saturated fats, though the differences between the two are rarely explained. A *saturated* hydrocarbon is one whose chain, like octane's, is filled to capacity with hydrogen atoms: there are no double bonds between carbon atoms. An *unsaturated* hydrocarbon has one or more double bonds along the carbon backbone. The common fatty acid shown on page 599, oleic acid, which has a double bond in the middle of the molecule, is unsaturated. A fat molecule with more than one double bond is called polyunsaturated.

Saturation is important because double bonds significantly alter the geometry and the regularity of the chain, and so its chemical and physical properties. Recall that carbon bonds are directed toward the corners of a tetrahedron, so that the carbon chain, like the backbone of a protein, is actually a long zigzag. Because the double bond distorts the usual bonding angles, a carbon-carbon double bond has the effect of adding a kink to the chain.

A saturated portion of a hydrocarbon chain *(left)*, and an unsaturated portion *(right)*. The irregularity in geometry and bonding caused by the double bond makes it easier for other molecules—in particular, O_2, or oxygen gas—to get at the chain and disrupt it. In culinary fats, off-flavors may result from such reactions.

Notice that the double bond opens a space unprotected by hydrogen atoms on one side of the chain. In this configuration, some carbon atoms are exposed to the approach of stray reactive molecules. Atmospheric oxygen is just such a molecule, and is one of the major causes of flavor deterioration—*rancidity*—in foods containing fats. How this kind of fat oxidation occurs is not fully known. It is thought that the electron-hungry oxygen replaces the hydrogen on a carbon adjacent to the double bond, thereby forming a very unstable complex which decomposes and produces various other reactive compounds. These include very short-chain fatty acids, which are volatile and can be noxious. The more unsaturated the fat, the more prone it is to this sort of deterioration. Beef has a longer freezer life than chicken, pork, or lamb because its fat is less unsaturated, and so more stable.

It turns out that the tendency for highly unsaturated fats to be quickly oxidized can be very useful. The so-called drying oils—linseed oil is a common one—are used in paints and furniture polishes precisely for this quality. These materials oxidize so rapidly and completely on contact with air that they form a dry, hard, protective film of hydrocarbon polymers.

The kinks in unsaturated chains have physical as well as chemical effects. It makes sense that a group of identical and regular molecules would fit more neatly and closely together than different and irregular molecules. Fats composed of straight-chain saturated fatty acids fall into an ordered solid structure—the process has been described as "zippering"—more readily than do kinked unsaturated fats. Animal fats are about half saturated and half unsaturated, and solid at room temperature, while vegetable fats are about 85% unsaturated, and are liquid oils in the kitchen. Even among the animal fats, beef tallow is noticeably harder than pork, lamb, or chicken fat because more of its triglycerides are saturated.

Double bonds are not the only factor in determining the melting point of fats. Short-chain fatty acids are not as readily "zippered" together as the longer chains, and so tend to lower the melting point of fats. And the more variety in the structures of fatty acids incorporated, the more likely the mixture of triglycerides will be an oil.

Hydrogenation

If it is true that plants produce oils for the most part, then how is it that we have solid margarine made from corn or soybeans, or vegetable shortenings? Because seed oils can be artificially saturated—a process called *hydrogenation*, for obvious reasons—in order to obtain both the desired texture and improved keeping qualities. A small amount of nickel, on the order of 0.1%, is added to the oil as a catalyst, and the mixture is then exposed to hydrogen gas at high temperature and pressure; afterward, the nickel is filtered out. Vegetable oils are never completely saturated because such products turn out with a very hard, even brittle, texture. By one estimate, common shortenings contain 22 to 50% saturated, 45 to 75% unsaturated, and 5 to 15% polyunsaturated fats. About 20% of the fat in hard margarine is saturated, 13% in soft margarine.

Fats and Heat

Most fats do not have sharply defined melting points. Instead, they soften gradually over a range of 10° to 20°F (5° to 10°C). If it weren't for this fact, we would not be able to spread butter on bread or toast, but would either have to pour it on or chip it into little pieces for each bite. This gradual softening is a consequence of the presence of several different triglycerides in a given fat. As the temperature rises, different fractions melt at different points and slowly weaken the whole structure. An interesting exception to this rule is cocoa butter, which is remarkably uniform in composition and structure: only three fatty acids appear in the majority of its triglycerides, with the same one always occupying the middle position. Pure cocoa butter is quite brittle up to about 93°F (34°C), at which point it melts quite quickly. Because the phase change from solid to liquid occurs just below body temperature, cocoa butter will absorb heat from the mouth without itself warming up to body temperature. As it melts, then, it cools the mouth.

The mixture of triglycerides in a given fat will also affect the kind of structure the fat assumes when it cools down and solidifies. If only a few kinds of fat molecules account for a significant proportion of the total number, then they will be able to form orderly arrays, with molecules of the same structure clumping together. The result is a fat with noticeable crystals embedded in a more amorphous phase. Both bacon fat and lard, if allowed to harden slowly, develop this grainy texture. On the other hand, a more random mixture of triglycerides will solidify into much smaller, less stable crystals. Crystal size is an important consideration when fats are used to shorten pastry or to leaven cakes (see chapter 6).

As that cliché of ancient battle, boiling oil, would suggest, fats do eventually change from a liquid to a gas: but only at very high temperatures, from 500° to 750°F (260° to 399°C). This high boiling point, far above water's, is the indirect result of the fats' large molecular size. While they do not participate in hydrogen bonding, fats do form another, weaker kind of bond which is named after the Dutch physicist van der Waals. Van der Waals bonds are formed when *any* two atoms on different molecules approach close enough to induce a slight charge imbalance in each other; this imbalance in turn causes a very weak electrical attraction. Because fat molecules are capable of forming so many van der Waals bonds along their lengthy hydrocarbon chains, the individually small interactions have a large net effect: it is hard to knock the molecules apart from each other.

The Smoke Point

The high boiling point of fats makes it possible to cook food very quickly, but not as quickly as one might think. Most triglycerides begin to decompose at temperatures well below 500°F and may even spontaneously ignite. These factors limit the maximum useful temperature of cooking fats. The characteristic temperature at which a fat breaks down into visible gaseous products is called the *smoke point*. Not only are the smoky fumes obnoxious—glycerol, for example, is converted into an acrid irritant aptly named *acrolein*—but other materials that remain in the liquid, including chemically active free fatty acids, tend to ruin the flavor of the food being cooked.

The smoke point seems to depend on the initial free fatty acid content of the fat, which is generally lower in vegetable oils than in animal fats. The former begin to smoke around 450°F, the latter around 375°F (232°, 191°C). Fats that contain other materials, such as emulsifiers, preservatives, and in the case of butter, proteins and carbohydrates, will smoke at lower temperatures than pure fats. Decomposition during deep fat frying can be delayed by using a tall, narrow pan and so reducing the area of contact between fat and atmosphere. The smoke point of a fat is lowered every time it is used, since some breakdown is inevitable even at moderate temperatures, and trouble-making particles of food are always left behind.

Rancidity

We have already looked at the most important cause of flavor deterioration in fats: namely, oxidation by the oxygen in the air. It should be noted that trace metals (including the iron in animal myoglobin), salt, and light all accelerate fat oxidation, while phenolic substances, both natural—vitamin E, or tocopherol—and synthetic—BHT and BHA—and such spices as sage and rosemary, all inhibit oxidation. But there are several other causes of fat rancidity. Fats stored in the refrigerator tend to absorb whatever odors are around; this happens because odor molecules are usually nonpolar and so are especially soluble in fats. Plant and animal enzymes that break fats down into their component fatty acids will eventually spoil the flavor of vegetables and meats unless they are disabled by sufficient heating. And both bacteria and molds decompose fats. Prompt consumption, or at least air- and light-tight wrapping, are the best defenses against rancidity, whether the food be milk, nuts, vegetables, or candy. Rancidity in cooking fats is caused by moisture in the food being cooked; the combination of water and heat energy can hydrolyze a fat into its constituent fatty acids and glycerol, which then decompose further. Certain metals, especially iron, accelerate this process, so that iron pots should be avoided in deep fat frying (aluminum and stainless steel are fine). A combination of hydrolysis and oxidation is responsible for the increased viscosity of used cooking oils and the formation of gummy deposits on the pot; certain reaction products in the oil act like the drying oils, linking together into long polymers.

Related Compounds: Lecithin and Monoglycerides

Fatty acids, fats, and oils are all members of a large chemical class called the *lipids*. These are organic molecules that are insoluble in water, but soluble in nonpolar solvents like benzene. Waxes, carotenoid pigments, vitamin D, cholesterol, and steroid hormones like cortisone are all lipids. Two other kinds of lipids deserve brief mention. The *phospholipids* are di-, rather than tri-glycerides: the third place on the glycerol molecule reacts with a molecule of phosphoric acid (H_2PO_4) to form a complex with a polar head.

Lecithin, a diglyceride with a polar head. It acts as the culinary equivalent of a soap by stabilizing oil droplets in foods.

Because the two hydroxyl groups on the phosphorus atom can each lose a hydrogen ion in water, the phospholipid resembles a fatty acid in that it is water soluble at the head and fat soluble at the tail. One important phospholipid is *lecithin,* which makes up about 30% of egg yolk and contains a carbon-nitrogen complex at the polar end in place of one hydroxyl group.

Phospholipids like lecithin, and *monoglycerides,* which contain only one fatty acid, are commonly added to foods because they are excellent emulsifiers. By acting as a soap and preventing fat droplets from coalescing in water, emulsifiers stabilize such mixtures as sauce béarnaise and salad dressing and prevent them from separating into two phases. Such "surface-active" molecules have many other applications as well. For example, monoglycerides have been used for decades in the baking business because they help retard staling, apparently by complexing with amylose and blocking starch retrogradation.

CHAPTER 14

Cooking Methods and Utensil Materials

Each of the basic methods of cooking, from broiling over a fire to irradiating in a microwave oven, has its own unique influence on food. This chapter explains how these methods work, and describes the properties of the various metals and ceramics with which we apply them. However, we must begin with a very important kind of reaction that food molecules undergo when subjected to certain conditions. The browning reactions come up in nearly every chapter of Part 1. They have remarkable effects on both the flavor and appearance of food, and only particular cooking methods can generate them.

BROWNING REACTIONS AND FLAVOR

The browning reactions produce flavors that are characteristic of the cooking process, rather than enhancing or slightly modifying flavors that are intrinsic to the food. These reactions are named for the typical coloration they leave behind; in fact, colors ranging from yellow to black may result, depending on the circumstances. We will pass over one kind of browning that is altogether undesirable and a sign of deterioration: the discoloration of fresh fruits and vegetables. This is caused by the action of certain plant enzymes on phenolic compounds in the tissue, and results from cell damage; it is discussed in detail in chapter 4.

Caramelization and the Maillard Reaction
The simplest useful browning reaction is the process of caramelizing sugar, and it is actually anything but simple. When sucrose—table sugar—

is heated, it first melts into a thick syrup. It then slowly changes color, becoming light yellow and progressively deepening to a dark brown. At the same time, its initially sweet flavor takes on a kind of richness, the change we associate with roasting marshmallows or baking meringue. The chemical reactions involved in caramelization are very numerous and not very well understood. If glucose, an even simpler sugar than the disaccharide sucrose, is browned, this single species of molecule breaks down and recombines to form at least 100 different reaction products, among them sour organic acids, sweet and bitter derivatives, many fragrant volatile molecules, and brown-colored polymers. It is a remarkable transformation, and a fortunate one for the palate.

Even more fortunate is the browning reaction responsible for the color and flavor of bread crusts, chocolate, coffee beans, dark beers, and roasted meats and nuts: foods that are *not* primarily sugar. It begins with the reaction of a carbohydrate unit, whether it be a free sugar or one bound up in starch, with the nitrogen-containing amine group on an amino acid, which may also be free or part of a protein chain. An unstable intermediate structure is formed and then undergoes further changes, producing many different by-products. Again, a brown coloration and full, intense flavor result. This sequence is known as the Maillard reaction, after its French discoverer.

High Temperatures Required

Both caramelization and the Maillard browning proceed at a significant rate only at relatively high temperatures: caramelization, for example, becomes noticeable at around 310°F (154°C). This is because large amounts of energy are required to force the initial molecular interactions. The practical consequence of this limitation is that most foods brown only on the outside and during the application of dry heat. Recall that water cannot be heated above 212°F (100°C) until it is vaporized, unless it is under high pressure. So foods that are cooked in hot water, and the moist interiors of meats and vegetables, will never exceed 212°F. But the outer surfaces of foods cooked in oil or in an oven quickly dehydrate and reach the temperature of their surroundings, perhaps 300° to 500°F (149° to 260°C), and even more in the case of broiling. Maillard browning normally occurs rapidly enough to make a difference only at these elevated temperatures, though alkaline surroundings and artificially concentrated sugar-amino mixtures can lower the threshold.

So it is that boiled or steamed foods are plain-looking and bland compared to their broiled, baked, or fried counterparts: the moist techniques don't let them get hot enough to brown. This is a very useful rule to keep in mind. For example, the secret of a rich-tasting stew is to be sure to brown the meat, vegetables, *and* flour quite well before adding any liquid. On the other hand, if you want to emphasize the natural flavors of the food, avoid the high temperatures that create the intense but less individualized browning flavors.

Because purified sugars and amino acids are available and inexpensive, the food industry has used them to synthesize, or at least approximate, flavors that in their natural form are much more costly. Imitation maple, chocolate, coffee, tea, honey, mushroom, bread, and meat flavors have been produced in this way. Sometimes a mixture of sugar and different cereal flours, heated to about 400°F (205°C) for a few minutes, will accomplish similar effects. The coffee substitute Postum, which was marketed by the breakfast-cereal tycoon C. W. Post around the turn of the century, is simply a mixture of roasted wheat, bran, and molasses.

Drawbacks

Browning reactions are, like most things in life, a mixed blessing. There are three minor drawbacks to the Maillard reaction. First, many dehydrated foods and many fruit juices are prone to gradual browning even at room temperature, probably because of the concentration of carbohydrates and amine-containing molecules (enzymic browning is also a factor). It has been determined that small amounts of sulfur dioxide will block these unwanted color and taste changes, and such treatment is now common for these foods. Then there is the fate of the amino acids that participate in browning reactions. Generally, they are either broken down into various intermediate compounds, or are incorporated into the polymeric pigment molecules. In either case, they are no longer available to our bodies as amino acids (the pigment molecules are indigestible). And the amino acid most likely to react, lysine, is one of the "essential" amino acids whose concentrations determine the nutritive value of proteins. Color and flavor—visual and gustatory pleasure—are produced, then, at the expense of nutrition. Fortunately, this is a significant problem only when food is terribly overcooked and so is distasteful as well. Finally, there is evidence that some products of the browning reactions can damage DNA and may be carcinogens. The ubiquity of browned foods, both today and through thousands of years of history, would suggest that they do not constitute a major threat to public health. But it is probably prudent not to char too many meals on the barbecue.

COOKING METHODS

Cooking can be defined in a general way as the transfer of energy from a heat source to food. Our various cooking methods—boiling, broiling, baking, frying, and so on—achieve their various effects by employing very different materials—water, air, oil—and by drawing on different principles of heat transfer. There are three such principles, and an acquaintance with them will help us understand how particular culinary techniques affect foods they way they do.

Heat Transfer

Conduction: Direct Contact

When thermal energy is exchanged from one particle to a nearby one by means of a collision or a movement that induces movement (through electrical attraction or repulsion), the process is called *conduction*. Though it is the most straightforward means of heat transfer in matter, conduction takes different forms in different materials. For example, metals are by and large good conductors of heat because, while their atoms are fixed in a latticelike structure, the outer electrons are very loosely held and tend to form a free-moving "fluid" or "gas" in the solid. This same electron mobility makes metals good electrical conductors. But in nonmetallic solids like ceramics, conduction is more mysterious. It seems that heat is propagated not by the movement of energetic electrons—in solids of ionic- or covalent-bonded compounds, the electrons are not free—but by the vibration of individual molecules or of a portion of the lattice, which is transferred to neighboring areas. This is a much slower and less efficient process than electron movement, and nonmetals are usually referred to as thermal or electrical *insulators*, rather than conductors. Liquids and gases, because their molecules are relatively far apart, are very poor conductors. Experience might lead us to believe that water is an exception to this rule, but it isn't, as we shall see in a moment.

The conductivity of a material will determine such things as how quickly it will heat up and cool off, and how evenly heat will be distributed in the material: both important properties of pots and pans.

Convection: Heat Transfer in Fluids

The transfer of heat brought about when the molecules in a fluid, whether liquid or gas, move from a warm area to a cooler one, is called *convection*. Convection is a process that combines conduction and mixing: energetic molecules are displaced from one point in space to another, and then collide with slower particles. Convection is an influential phenomenon, contributing as it does to winds, storms, ocean currents, the heating of our homes, and the boiling of water on the stove. It occurs because air and water take up more space—become less dense—when their molecules absorb energy, and so rise when they heat up and sink again as they cool off. That convection rather than conduction is the primary factor in boiling water can be demonstrated by fixing a long tube of water at an angle and heating it with a flame near the top, thereby preventing warm water from rising (an experiment more easily done in the laboratory than in the kitchen). The water above the flame will boil long before the bottom of the tube gets hot.

Radiation: Pure Energy

We all know that the earth is warmed by the sun. How does solar energy traverse the 93 million miles of nearly empty space between it and the earth?

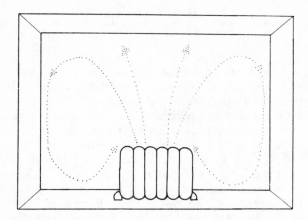

"Radiators" heat rooms primarily by convection. The steam-heated pipes warm nearby air molecules, which rise, carry their energy to the rest of the room, and then fall again as they cool.

Clearly neither conduction nor convection can explain the continuous transfer of heat that makes life possible on this planet. The answer is thermal *radiation*, a process that does not require physical contact between source and object. All matter emits thermal radiation all the time, though normally we can detect it only when something is very hot. The glow we feel when our hand is near a hotplate or a room radiator or in direct sunlight, is the result of thermal radiation. It is emitted by atoms and molecules which, having absorbed energy, are now releasing it again in the form of "waves" or "particles"—both useful models for a very abstract idea—of pure energy.

As unlikely as it may seem, radiated heat is close kin to radio waves, visible light, and X rays. Each of these phenomena is a part of the *electromagnetic spectrum*, a wide range of energy radiation caused by the motion of electrically charged particles. Such motion creates electrical and magnetic fields, and conversely, such fields alter the motion of particles in the matter on which they impinge. One of the first to recognize that heat radiation is related to light was the English oboist and astronomer William Herschel, who noticed in 1800 that if a thermometer was moved from one end of a prism-produced light spectrum to the other, the highest temperatures would register below the red band, where no light was visible. Because of its position in the spectrum, heat radiation is called *infrared* (*infra* is Latin for "below").

The electromagnetic spectrum is a manifestation of the range of energies carried by radiation, and its energy determines the kind of effect radiation will have on matter. At the bottom end of the scale, radio waves are so weak that they can only move free electrons: this is why metal antennas, with their "electron gas," are necessary to transmit and receive such radiation.

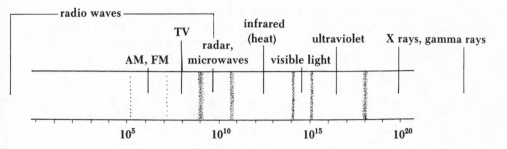

Frequency, cycles per second

The electromagnetic spectrum. We use both microwave and infrared radiation
to cook our foods. (The scale employs a standard abbreviation for large
numbers; 10^5 means a 1 followed by 5 zeros, or 100,000.)

Next come microwaves, which are energetic enough to set polar molecules
like water into vibration, and so have some usefulness in the kitchen. Then
there is heat radiation, which causes and is caused by the vibration of nonpolar molecules—including carbohydrates, proteins, and fats—as well as
polar ones. (How can nonpolar molecules set up an electromagnetic field? As
they vibrate, their bonds are slightly distorted, and oscillating charge imbalances are the result.) Visible light is capable of altering the orbits of bound
electrons, and so can initiate such chemical reactions as those that cause flavor deterioration in fatty foods. At the top of the spectrum, ultraviolet, X,
and gamma rays carry increasing amounts of energy and have correspondingly drastic effects on molecular structure. Ultraviolet rays from the sun can
ruin the flavor of milk and beer as well as burn our skin.

Because all matter has a finite temperature and so all molecules are
vibrating to some extent, all objects are emitting at least some infrared radiation all the time. The hotter they get, the more energy will be radiated in
higher regions of the spectrum. So it is that glowing metal must be at a higher
temperature than metal that does not radiate visible light, and that blue-hot
metal is hotter than red-hot. It turns out that the *rate* of infrared radiation
is relatively inefficient below about 1800°F (982°C), or the point at which
objects begin to glow visibly red. The practical consequence of this fact is
that heat radiation is a relatively slow process except at very high cooking
temperatures, those characteristic of broiling over glowing coals, electrical
elements, or gas flames. (The so-called "radiators" used to heat rooms with
steam or hot water do their work principally by air convection.) At typical
boiling, baking, and frying temperatures, conduction and convection tend to
be more significant than infrared radiation.

Methods of Heating Food

Pure examples of the three different means of heat transfer are seldom found in everyday life. All matter radiates to some degree, and solids and fluids are not as a rule segregated from each other. This is certainly true of cooking. As simple an operation as heating a pan of water on the stove involves radiation and conduction from an electrical element (radiation and convection from a gas flame), conduction through the pan, and convection in the water. Still, one mode usually predominates and, together with the cooking medium, will have a distinctive effect on the food. The most important variables are the typical temperature attained in the surrounding medium, and the rate at which the food is heated.

Broiling

Broiling is the modern, somewhat controlled version of the oldest culinary technique: roasting over an open fire or glowing coals. Though air convection no doubt has a certain role, especially as the distance between heat source and food is increased, broiling is largely a matter of infrared radiation; otherwise, gas flames or electrical heating elements mounted *above* the food would work very slowly. The heat sources used in broiling all emit visible light and so are intense radiators of infrared energy. Glowing coals or the nickel-chrome alloys used in electrical appliances reach about 2000°F (1093°C), and a gas flame is closer to 3000°F (1649°C). The walls of an oven, by contrast, rarely exceed 500°F (260°C). The total amount of energy radiated by a hot object is proportional to the fourth power of the temperature (measured in °Kelvin, or °C + 273), so that a coal or metal rod at 2000° is radiating about 3^4, or 81 times as much energy as the equivalent area of oven wall at 500°.

This tremendous amount of heat is at once the great advantage and the principal problem of broiling. On one hand, it makes possible a rapid and thorough browning of the surface, and so produces characteristically intense flavors. On the other, there is a huge disparity between the rate of radiation at the surface and the rate of *conduction*, via water, within the food. This is why it is so easy to end up with a steak that is charred on the outside and cold at the center. The aim, of course, is to position the food far enough from the heat source to match the browning rate with the inner conduction. (Infrared radiation weakens quickly with distance not because the rays lose energy, but because they spread out in all directions and so become less concentrated.) This is a very tricky matter, depending as it does on the nature and thickness of the food and the intensity of the heat. Either long experience or an innate knack are the best guides; the fifteenth of Brillat-Savarin's introductory aphorisms is, "We can learn to be cooks, but we must be born knowing how to roast."

Baking

In the technique called baking, we surround the food with a hot enclosure, the oven, and rely on a combination of radiation from the walls and, to a lesser extent, air convection to heat the food. Typical baking temperatures are well above the boiling point, from 300° to 500°F (149° to 260°C), and yet baking is nowhere near as efficient a means of heat transfer as is boiling. A potato can be boiled in less time than it takes to be baked at double the temperature. This is so because neither radiation at 500° (despite a surface area of radiating walls that is quite large compared to coals or broiler elements) nor air convection at such high temperatures—the hotter a gas is, the thinner it is—and in an enclosed space, is very efficient. The air currents are less violent than bubbling water, and because air at room temperature is about one-thousandth as dense as water, the collisions between medium and food are much less frequent in the oven (this is also why we can reach into a hot oven without immediately burning our hand). *Convection ovens* increase the rate of heat transfer by using fans to force more air movement, and baking times are accordingly much reduced. But such efficiency is not necessarily desirable. Meat can profit by the slow, gentle heating that regular ovens provide. Finally, baking easily dehydrates the surface of foods, and so will brown them nicely, as we all know from experience.

Because baking requires a fairly sophisticated container, it was probably a rather late addition to the culinary repertoire. The earliest ovens seem to accompany the refinement of bread making around 3000 B.C. in Egypt; they were hollow cones of clay that contained a layer of coals, with the bread stuck onto an inside wall. As a relatively compact metal box easily installed in individual homes, the modern oven dates from the late 19th century. Before then, most meat cookery was done over the fire. The term *roast* for a big piece of baked meat dates from this era.

Boiling and Steaming

In boiling and its variants, braising and stewing, food is heated by the convection currents in hot water. The maximum temperature possible is, of course, the boiling point, 212°F (100°C). So it is that the "moist" cooking methods cannot trigger browning reactions, and this is why meat, vegetables, and flour are normally browned in a frying pan before they are stewed. Yet despite this upper limit, which is substantially lower than the temperatures characteristic of broiling, baking, and frying, boiling is a very efficient process. The entire surface of the food is in contact with the cooking medium, which is dense and turbulent enough that the water molecules continuously and rapidly impart their energy to the food. *Steam*, though it is less dense than liquid water and so makes less frequent contact with the food, compensates for this loss in efficiency with a gain in energy. The vaporized molecules are slightly more energetic than those in the liquid phase, and the slight pressure developed in the closed container will raise the boiling temperature a little.

One of the great advantages of water as a cooking medium is that it takes no finesse at all to reach and maintain its boiling point, no matter how long or hard you heat it. Getting a steady temperature in oil, on the other hand, requires a thermometer and continuous monitoring. Actually, whenever a substance is undergoing a phase change, from solid to liquid or liquid to gas or the reverse, its temperature remains constant until the phase change is complete, even though a flame may be pouring energy into it. Temperature is a measure of molecular motion. Just below the boiling point of a liquid, the molecules are moving just as much as they can and still remain associated with each other. If any further increment of energy is absorbed, it will push some molecules into the gaseous phase, and they will leave the liquid along with that extra energy. At the boiling point, then, all added energy goes into the work of vaporizing the molecules, and none is left over to change the temperature of the liquid. (The same idea applies at the melting point, and to the reverse processes of condensation and freezing.) So the boiling point of water is a useful, easily recognized and achieved standard temperature. (Oils too boil, at much higher temperatures, but not before they begin to break down; see chapter 13.)

Of course, the boiling point of water is constant given constant conditions, but it can vary from place to place and even in the same place. This is because the boiling point of a liquid depends on the atmospheric pressure bearing down on its surface. The greater the force exerted against the liquid by molecules in the air, the greater the energy needed by the liquid's mol-

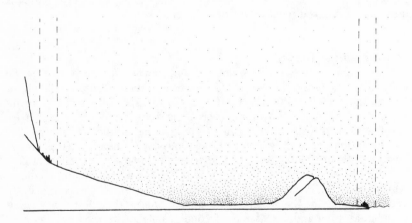

A column of air over Denver contains many fewer molecules, and so exerts less pressure on a pan of water, than does a similar column over Boston.

ecules to escape from the surface, and so the higher the temperature at which the liquid boils. Atmospheric pressure decreases with increasing elevation—there is less and less air to exert pressure as you rise—and this means that the boiling point of water is significantly lower in Denver than it is in Boston. Every 1000 feet in elevation above sea level lowers the boiling point about 2° below the standard 212°F (1° below 100°C). And food takes longer to cook at 200° than it does at 212°. Even a low-pressure front over Boston can lower the boiling point, or a high raise it, by as much as a degree or two. The same principle is put to use to *speed* cooking in the pressure cooker. This appliance reduces cooking times by trapping the steam that escapes from boiling water, thereby increasing the pressure on the liquid, and so raising its boiling point—and maximum temperature—well above 212°.

As a culinary technique, boiling probably followed roasting and preceded baking. It requires containers that are both water- and fire-proof, and so probably had to await the development of pottery, around 10,000 years ago.

Frying

Pan frying is a method that depends for the most part on conduction and convection. The layer of oil between pan and food is very thin, and so the path for convection currents is very short. The oil has several roles to play: it brings the uneven surface of the food into uniform contact with the heat source, it lubricates and prevents sticking, and it supplies some flavor. Because oils can be heated well above the boiling point of water, they also produce the intense flavors of the browning reactions. As is the case with broiling, the trick in frying is to prevent the outside from overcooking before the inside is done. The surface is quickly dehydrated by the high temperatures—odd as it sounds, frying in oil is a "dry" technique—while the interior remains largely water and never exceeds 212°F (100°C). In order to reduce the disparity between outer and inner cooking times, we generally fry only thin cuts of food. It is also common practice to fry meats at a very high initial temperature—to sear them—in order to accomplish the browning, and then to reduce the heat while the interior heats through.

Deep frying differs from pan frying by employing enough oil to immerse the food altogether. As a technique, it resembles boiling more than pan frying, with the essential difference that the oil is heated to as much as double the boiling point of water, and so will cook the food more rapidly and brown its exterior.

How far back frying goes is hard to tell. The rules for sacrifice in Leviticus 2, which date from about 600 B.C., distinguish between bread baked in an oven and cooked "on the griddle" or "in the pan." Pliny, in the first century A.D., records a prescription for spleen disease that calls for eggs steeped in vinegar and then fried in oil. And by Chaucer's time, the 14th century, frying was common enough to serve as a colorful metaphor. The Wife of Bath says of her fourth husband

> That in his owene grece I made hym frye
> For angre, and for verray jalousye.
> By God! in erthe I was his purgatorie,
> For which I hope his soule be in glorie.

The image of someone frying in purgatory or hell became a popular one, perhaps because of the vivid crackling and sputtering (caused by the sudden vaporization of water in the food) characteristic of the process.

Microwaves

Microwave ovens transfer heat via electromagnetic radiation, but in a band somewhat lower in energy than the infrared region. This simple shift makes for a unique effect. Whereas infrared waves are energetic enough to increase the vibratory movement of nearly all molecules, microwaves tend to affect only polar molecules, whose electrical imbalance gives the radiation a handle, so to speak, with which to move them. This is why the air, composed of nonpolar molecules like N_2, H_2, and O_2, and nonpolar materials like plastic (made of hydrocarbon chains) and glass, do not heat up in a microwave oven, while foods, which always contain some water, heat up quite rapidly.

Here is how a microwave oven works: a transmitter, very much like a radio transmitter, sets up an electromagnetic field in the oven which reverses its polarity some 2 or 5 billion times every second (it operates at a frequency of either 915 or 2450 million cycles per second, compared to wall socket currents at 60 cycles, and FM radio signals at some 100 million cycles per second). All polar molecules in the food, water being the predominant one, are forced to orient themselves with the surrounding field, but because the field is constantly changing, the molecules oscillate back and forth. This motion is transmitted to neighboring molecules as kinetic energy, and the temperature of the food quickly rises.

Microwave radiation has one great advantage over infrared: the fact that it cooks food much faster. Oven microwaves can penetrate food to a depth of a couple of inches, while infrared energy is almost entirely absorbed at the surface. Because heat radiation can affect the center of foods only by the slow process of conduction, it is easily beaten by waves with a substantially deeper reach. This reach, together with the microwaves' concentration on heating the food and not its surroundings, results in a very efficient use of energy. By comparison, baking is rather wasteful.

Several disadvantages of microwave cooking should be noted. One is that, in the case of meat cookery, speedy heating can cause greater fluid loss and so a drier texture, and makes it more difficult to control the doneness of a roast. Another problem is that microwaves cannot brown foods, since the surface gets no warmer than the interior. And our expectations for many foods, including breads, baked potatoes, and roasts, include the flavors produced during browning. This objection has carried such weight that manufacturers market hybrid appliances that include a microwave transmitter, an electrical element for browning, and in some cases a convection fan. Finally, there remain some questions about appropriate limits on microwave leakage from such ovens and the permissible exposure of humans to this kind of radiation, though only continuous and careless operation of kitchen ovens is likely to be harmful.

Microwave ovens are, needless to say, a very recent invention. In 1945 Dr. Percy Spencer, a scientist working for Raytheon in Waltham, Masachusetts, filed a patent for the use of microwaves in cooking after he had successfully popped corn with them. This kind of radiation had already been used in diathermy, or deep heat treatment for patients with bursitis and arthritis, as well as in communications and navigation. Microwave ovens became a popular appliance rather suddenly in the 1970s.

UTENSIL MATERIALS

We end this chapter with a brief discussion of the materials from which we make our pots and pans. We generally want two basic properties in a utensil. Its surface should be chemically unreactive so that it will not change the taste or edibility of food, and it should conduct heat evenly and efficiently so that local hot spots will not develop and burn the contents. Unfortunately, these requirements are not fulfilled ideally in any single material.

The Behavior of Metals and Ceramics

As we have seen, heat conduction in a solid proceeds either by the diffusion of energetic electrons, or by vibration in crystal structures. A material whose electrons are quite mobile is likely to donate those electrons to other atoms at its surface: in other words, good conductors like metals are usually chemically reactive. But inert compounds, by the same token, are poor conductors. Ceramics are mixtures of metal- and semimetal-oxides whose covalent bonds hold electrons tightly, and which therefore transmit heat by means of slow and inefficient vibrations. If subjected to the direct and intense heat of the stovetop, these materials cannot distribute the energy evenly. Hot areas expand while cooler areas do not, mechanical stresses build up, and the utensil cracks or shatters. This is why ceramics are used only as ovenware, which is subjected to very slow and diffuse heat, or as thin coatings on the surface of metals, which do the job of distributing the heat evenly. But because thin layers of nearly any material are vulnerable to erosion or mechanical damage, emaneled pans are far from ideal.

As it happens, most of the metals commonly used in kitchen utensils naturally cover themselves with a very thin layer of ceramic material. Metallic electrons are mobile, and oxygen, as we know from fuel and food oxidation (see the Appendix), is electron-hungry. When metal is exposed to the air, the surface atoms undergo a spontaneous reaction with atmospheric oxygen to form a very stable oxide. (The discoloration on silver and copper that we call tarnish is due to metal-sulfur compounds; the sulfur comes mainly from air pollution.) These oxide films are both unreactive and fairly tough; aluminum oxide, when it occurs in crystals rather than on pans, makes up the abrasive called corundum, and is also the principal material of rubies and sapphires (the colors come from chromium and titanium impurities respectively). The problem is that these natural coatings are only a few molecules thick—on the order of 50 angstroms, or about 2 millionths of an inch—and are relatively easily scratched through or worn away during cooking. The film over aluminum can be made up to a thousandth of an inch thick, and so fairly impervious, by the process of *anodizing*. This involves making a sheet of the metal the positive pole in a solution of sulfuric acid and so forcing the oxidation of the metal.

Stainless Steel

The important exception to the rule that metals form protective coatings is, of course, iron, which we all know rusts in the presence of air and moisture. The orange complex of ferric oxide and water ($Fe_2O_3 \cdot H_2O$) is a loose powder rather than a continuous film, and so does not protect the metal surface from further contact with the air. Unless it is protected by some other means, iron metal will corrode continuously (this is why pure iron is not found in nature). Efforts to make this cheap and abundant element more resistant to rusting resulted late in the 19th century in the development of *stainless steel*, an alloy that contains about 15% chromium (steel, like cast iron, is alloyed with carbon to improve on the mechanical properties of pure iron). Chrome and chrome plating are synonymous with bright and permanent shininess because chromium is extremely prone to oxidation and naturally forms a thick protective oxide coat. In the stainless steel mixture, oxygen reacts preferentially with the chromium atoms at the surface, and the iron never gets the opportunity to rust.

"Seasoning" Cast Iron

Stainless steel has the drawback of being substantially less efficient a thermal conductor than iron. This is true of most alloys. The addition of large numbers of foreign atoms to a metal apparently interferes with electron mobility by causing structural and electrical irregularities. Stainless steel is also a fairly expensive material. For these reasons, cast iron is still widely used in kitchen utensils. In order to forestall rusting, cooks try to build up an artificial protective layer on cast-iron pans by "seasoning" them, or coating them with cooking oil and heating them for several hours. The oil penetrates into the pores and fissures of the metal, sealing it from the attack of air and water. There may also be an effect analogous to the behavior of the drying oils, which are largely unsaturated, prone to oxidation, and which polymerize to form a dry, hard layer when exposed to air. Since both heat and metallic ions accelerate fat oxidation, it may be that a similar layer is produced during seasoning with cooking oils. To avoid removing the protective oil layer, users of cast iron clean their pans with mild soaps and an abrasive like salt, rather than with detergents and scouring pads.

To summarize, there is no single material that is both chemically unreactive and an excellent conductor of heat, that will leave the food unaltered and yet heat it evenly. Even among the metals, there is a wide range of properties. Here is a list of the principal materials from which kitchen utensils are made today, and a few words about their particular merits and deficiencies.

Aluminum

Aluminum has been in use for even less time than steel has, despite the fact that it is the most abundant metal in the earth's crust. It is never found in the pure state, and it was not until around 1890 that a method was developed for separating the metal from its ore in any quantity. In cookware, it is usually alloyed with small amounts of manganese and sometimes copper. Aluminum's prime advantages are its relatively low cost, lack of toxic effect on humans, a heat conductivity second only to copper's, and a low density that makes it lightweight and easily handled. Its ubiquitous presence in the form of foil wrappings and beer and soft drink cans testifies to its usefulness. But because unanodized aluminum develops only a thin oxide layer, reactive food molecules—acids, alkalis, the hydrogen sulfide evolved by cooked eggs—will easily penetrate to the metal surface, and a variety of aluminum oxide and hydroxide complexes, some of them gray or black, will be formed in minute quantities. This may become a problem when very light-colored foods are being cooked, since they may be noticeably discolored. Aluminum also seems to intensify the sulfurous odor of cabbage, broccoli, and related vegetables, especially if they have been cooked overlong.

Ceramics

Ceramics are varying mixtures of a number of different compounds, notably the oxides of silicon, aluminum, and magnesium. *Glass* is a particular variety of ceramic whose composition is more regular, and usually includes a preponderance of silica (SiO_2). Until fairly recently, these materials were made from naturally occurring mineral aggregates. The molding and drying of simple clay pottery, or *earthenware*, dates from about 9000 years ago, or about the time that plants and animals were first domesticated. Less porous and coarse than earthenware, and much stronger, is *stoneware*, which contains enough silica and is fired at a high enough temperature that it vitrifies, or becomes partly glass. The Chinese invented this refinement sometime before 1500 B.C. *Porcelain* is a white but translucent stoneware made by mixing kaolin, a very light clay, with a silicate mineral, and firing at high kiln temperatures; it dates from the T'ang Dynasty (A.D. 618–907). This fine ceramic was introduced to Europe with the tea trade in the 17th century, and in England was first called "China-ware," and then simply "China." The first glass containers were not molded or blown, but laboriously sculpted from blocks, and date from 4000 years ago in the Near East.

The outstanding characteristic of ceramic materials is chemical and mechanical stability: they resist corrosion very well. In their modern forms they are nontoxic, although lead-containing vessels probably plagued many early cultures, among them the Romans. But ceramics have poor thermal characteristics. They are poor conductors of heat, and so tend to be used only in slow, uniform processes like baking and our modern version of braising, crockpot cookery. On the other hand, poor conductivity can be turned to advantage if the aim is to keep food hot. Good conductors like copper and aluminum quickly give up heat to their surroundings, while ceramics retain it well. Another problem is that ceramics expand and contract significantly when warmed and cooled, and sudden local changes in temperature cause mechanical stresses that can fracture the material. Pyrex glass incorporates an oxide of boron that has the effect of reducing thermal expansion by a factor of about 3, and for this reason is much more resistant, though not immune, to thermal shock.

So-called *enamelware* is made by fusing powdered glass onto the surface of iron or steel utensils. This was first done to cast iron early in the 19th century, and today enameled metal is widely used in the dairy, chemical, and brewing industries, as well as in most of our bathrooms. These very thin coatings are subjected to extreme stress in the kitchen; it is a good idea, for example, not to quench a hot enameled pan too quickly in cold water. The other major protective layers are *synthetic coatings* developed by industrial chemists. Teflon and Silverstone are carbon-fluorine polymers that were developed in the 1960s. They are all but inert and have the special advantage of a surface to which nothing will stick; because of their very low friction characteristics, these materials are used extensively on bearings and other moving machine parts. Their great disadvantage, familiar to all who have used them, is that they are easily scratched, and food sticks to the scratches.

Copper

Copper is unique among the common metals in being found naturally in the metallic state, and for this reason it was the first to be used in tool-making, about 10,000 years ago. In the kitchen, it has one thing going for it: its unmatched conductivity, which makes fast and even heating a simple matter. But it is also relatively expensive, since its conductivity has made it the preferred material for millions of miles of electrical circuitry. It is troublesome to keep polished, because it has a high affinity for oxygen *and* sulfur, and forms a greenish coating when exposed to air, especially polluted air. Most important, copper cookware can be harmful. Its oxide coating is sometimes porous and powdery, and copper ions are easily leached into food solutions. They can have attractive effects: the green color of cooked vegetables is improved by their presence. But the human body can excrete copper in only limited amounts, and excessive intake may cause gastrointestinal problems and, in more extreme cases, liver damage. No one will be poisoned by the occasional zabaglione whipped in a copper bowl, but clearly copper is not a good candidate for everyday cooking. To overcome this major drawback, manufacturers line copper utensils with tin, but tin, as we shall see, has its own limitations.

Iron

Iron was a relatively late discovery because it exists primarily as oxides in the earth's crust, and had to be discovered in its pure form by accident, perhaps when a fire was built on an outcropping of ore. Iron artifacts have been found that date from 3000 B.C., though the Iron Age, when the metal came into regular use without replacing copper and bronze (a copper-tin alloy) in preeminence, is said to begin around 1200 B.C. *Cast iron* is alloyed with about 3% carbon to harden the metal. The chief attractions of cast iron in kitchen work are its cheapness and safety; excess iron is readily eliminated from the body, and most people can actually benefit from additional dietary iron. Its greatest disadvantage is its tendency to corrode, though this can be avoided by assiduous seasoning and gentle cleaning. Like aluminum, iron can discolor light foods. And iron turns out to be a poorer conductor of heat than copper or aluminum. But precisely for this reason, and because it is denser than aluminum, an iron pan will absorb more heat and hold it longer than an aluminum pan of the same thickness. We might characterize the thermal behavior of cast iron as steady.

Stainless Steel

Stainless steel, as we have said, was developed to improve the corrosion resistance of iron and steel. Stainless steel for kitchen use is typically about 15% chromium, and may also contain some nickel. But chemical stability is bought at a price. Stainless steel is more expensive than cast iron, and it is an even poorer heat conductor. For these reasons, stainless utensils are made very thin, and the result of this practice is their tendency to develop hot spots and to burn food. The transfer of heat can be evened out by coating the underside of the pan with copper, or by inserting a copper or aluminum plate in the pan bottom, but of course these refinements add further to the cost of the utensil. Still, these hybrids are probably as close to the chemically inert but thermally responsive ideal as we have.

Tin

Tin was probably first used in combination with copper to make the mechanically tougher alloy called bronze. Today, outside of the lined steel can, tin is generally found only as a nontoxic, unreactive lining in copper utensils. This limited role is the result of two troublesome properties: a low melting point, 450°F (232°C), that can be reached in extreme cooking procedures, and a softness that makes the metal very susceptible to wear. The tin alloy called *pewter*, which used to contain some lead and now is made with 7% antimony and 2% copper, is not much used today.

Appendix

Appendix

A Chemistry Primer: Atoms, Molecules, Energy

ATOMS AND MOLECULES

The Atom

When we try to understand those parts of experience that are not immediately visible or tangible, we rely on models, or analogies to everyday objects and experiences. The idea of natural "laws" is one such analogy. The atomic model had proponents in Greece 2500 years ago, philosophers who thought that all matter was made up of particles of earth, air, water, and fire. In its modern form, the atomic theory dates from the last century. In essence, it states that all matter is a mixture of a few pure substances called the *elements*. An atom is the smallest particle into which an element can be subdivided without losing its characteristic properties. The atom too is divisible into smaller particles, *electrons, protons,* and *neutrons,* but these are the building blocks of all atoms, no matter of what element. The different properties of the elements are due to the varying combinations of subatomic particles contained in their atoms. The Periodic Table arranges the elements in order of the number of protons contained in one atom of each element. That number is called the atomic number.

Protons and Electrons
The atom is divided into two regions: the nucleus, or center, in which the protons and neutrons are located, and a surrounding "orbit," or more accurately a "cloud" or "shell," in which the electrons move continuously. Both protons and neutrons weigh about 2000 times as much as electrons, and

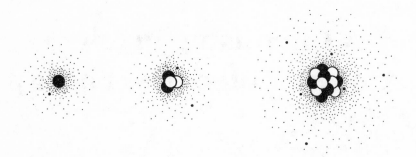

Left to right: atoms of hydrogen (one proton and one electron), helium (two protons, two neutrons, and two electrons), and carbon (six each of these three particles). The shading indicates the general region in which electrons are likely to be found. The electron cloud is huge compared to the nucleus; if this drawing were to scale, the shaded area would be close to a mile in diameter.

so practically all of an atom's mass is concentrated in the nucleus. Protons and electrons carry one unit of positive and negative electrical charge respectively, while the neutron carries no charge. Stable forms of the elements are electrically neutral, and their atoms contain equal numbers of protons and electrons.

If like charges repel each other and opposite charges attract, then why is it that the protons in the nucleus don't push each other away and the orbiting electrons fall straight into the nucleus? It turns out that there are forces besides electricity at work. The protons and neutrons are bound together by very strong nuclear forces and can exist at rest, while it is the nature of electrons to be in continual motion. So protons and electrons will always be attracted to each other, but will never consummate their attraction.

The Chemical Bond; Molecules

It is hard to overestimate the importance of the electron's mobility in the behavior of matter. Chemistry can be defined with some reason as the science of electron movement, and there is even a saying that in order to become a good chemist, you must learn to think like an electron. Electrons are arranged around the nucleus in an elaborate hierarchy which determines how strongly any particular electron is held in orbit. This attraction in turn determines the chemical behavior of the element. For example, the metals along the far left of the periodic table hold their outermost electron very weakly, and easily lose it to atoms in the third- and second-to-last columns, which tend to pick up available electrons. Elements in the middle of the table have intermediate properties, while those at the far right were originally named the "noble" gases because they resist both the loss and gain of

Periodic Table of the Elements

I   II   VI   VII

1 H 1.0080																	2 He 4.003
3 Li 6.940	4 Be 9.013											5 B 10.82	6 C 12.011	7 N 14.008	8 O 16.000	9 F 19.00	10 Ne 20.183
11 Na 22.991	12 Mg 24.32											13 Al 26.98	14 Si 28.09	15 P 30.975	16 S 32.066	17 Cl 35.457	18 Ar 39.944
19 K 39.100	20 Ca 40.08	21 Sc 44.96	22 Ti 47.90	23 V 50.95	24 Cr 52.01	25 Mn 54.94	26 Fe 55.85	27 Co 58.94	28 Ni 58.71	29 Cu 63.54	30 Zn 65.38	31 Ga 69.72	32 Ge 72.60	33 As 74.92	34 Se 78.96	35 Br 79.916	36 Kr 83.80
37 Rb 85.48	38 Sr 87.63	39 Y 88.91	40 Zr 91.22	41 Nb 92.91	42 Mo 95.95	43 Tc (99)	44 Ru 101.1	45 Rh 102.91	46 Pd 106.4	47 Ag 107.880	48 Cd 112.41	49 In 114.82	50 Sn 118.70	51 Sb 121.76	52 Te 127.61	53 I 126.91	54 Xe 131.30
55 Cs 132.91	56 Ba 137.36	57 La 138.92	72 Hf 178.50	73 Ta 180.95	74 W 183.86	75 Re 186.22	76 Os 190.2	77 Ir 192.2	78 Pt 195.09	79 Au 197.0	80 Hg 200.61	81 Tl 204.39	82 Pb 207.21	83 Bi 208.99	84 Po 210.	85 At (210)	86 Rn 222.
87 Fr (223)	88 Ra 226.05	89 Ac 227.0															

58 Ce 140.13	59 Pr 140.91	60 Nd 144.27	61 Pm (147)	62 Sm 150.35	63 Eu 152.0	64 Gd 157.26	65 Tb 158.93	66 Dy 162.51	67 Ho 164.94	68 Er 167.27	69 Tm 168.94	70 Yb 173.04	71 Lu 174.99
90 Th 232.05	91 Pa 231.	92 U 238.07	93 Np (237)	94 Pu (242)	95 Am (243)	96 Cm (247)	97 Bk (249)	98 Cf (251)	99 Es (254)	100 Fm (253)	101 Md (256)	102 No (253)	

The Periodic Table of the elements, so-called because similar chemical properties recur at regular intervals in the atomic number (the number of protons in the atom). The metals in Groups I and II *(at left)* tend to give up their outermost electrons, while the elements in Groups VI and VII *(at right)* are likely to capture loosely held electrons in other atoms. Hydrogen, carbon, nitrogen, oxygen, phosphorus, and sulfur constitute the handful of elements that play leading roles in the chemistry of cooking. Aluminum (13), silicon (14), tin (50), and a few of the transition metals—chromium (24), iron (26), and copper (29)—provide us with materials for our cooking utensils.

electrons, and so are chemically unreactive and inert. The imbalance in electrical properties among different elements is the basis of all chemical reactivity, and of the *chemical bond,* which is the *sharing* of one or more electrons between two different atoms. When two or more atoms are bonded together by electron sharing, they form a *molecule.* The molecule is to a chemical compound what the atom is to an element: the smallest unit that retains the properties of the whole. Most matter, and practically all food, is a mixture of different chemical compounds, and its behavior is determined by the kinds of bonds that hold these compounds together.

Ionic Bonds; Salt

There are three kinds of chemical bonds that are especially important in the kitchen, as they are throughout nature. One is the *ionic bond,* in which one atom completely captures the electron(s) of another, so great is the disparity between their affinities for electrons. This situation is typical of compounds formed between metals and the gases of the second-to-last column of the periodic table. Salt, our most common seasoning, is a compound of sodium and chlorine held together with ionic bonds. In the salt crystal, positively charged sodium *ions*° (*atoms* are by definition electrically neutral) alternate with negatively charged chloride ions, the sodium having lost its electrons to the chlorine. But because several positive sodium ions are always in a state of attraction to several negative chloride ions, we cannot really speak of individual molecules of salt, with this particular sodium atom bonded to that particular chlorine atom.

An ionic bond results when one atom completely captures one or more electrons of another atom. (The nucleus is now represented by a simple black disk.)

Covalent Bonds: Stable Molecules; Sugar

Molecules are the product of a second kind of bond, called *covalent* (from the Latin, "of equal power"). When two atoms have roughly similar affinities for electrons, they will *share* them rather than gain or lose them entirely as ions have done. In order that sharing occur, the electron clouds of two atoms must actually overlap, and this condition results in a fixed arrangement in space between two particular atoms. The bonding geometry determines the overall shape of the molecule, and molecular shape in turn defines the ways in which one molecule can react with others. The elements

In the covalent bond, atoms share electrons and thereby form stable groups called molecules. Shown here are three different ways of representing a molecule of water, which is formed from one oxygen and two hydrogen atoms (left).

° The word *ion* is Greek for "going," and was coined by the pioneering student of electricity, Michael Faraday, to designate those electrically charged particles that migrate to one or the other pole when an electrical field is set up in a solution.

most important to life on earth—hydrogen, oxygen, carbon, nitrogen, phosphorus, sulfur—all tend to form covalent bonds which make possible the complex assemblages that constitute our bodies and our foods. The most familiar pure chemical compound is probably sucrose, or table sugar, a covalent complex of carbon, oxygen, and hydrogen atoms. Covalent bonds are generally strong and stable: that is, they are not broken in large numbers unless subjected to heat, electrical force, or other reactive chemicals. One very general definition of cooking might be "the rearrangement of covalent bonds in food molecules."

Hydrogen Bonds; Water

Less strong and stable than covalent bonds, but no less important in the chemistry of life, is the *hydrogen* bond, one of several "weak" bonds that do not form molecules, but do make temporary links between different molecules or different parts of one large molecule. Weak bonds come about

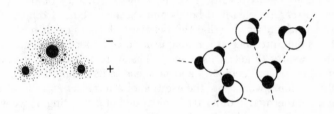

Water molecules are electrically asymmetrical. The separation of positive and negative centers of charge leads to the formation of weak bonds between oppositely charged centers on different molecules (shown here by the dashed lines). These are called hydrogen bonds.

because most covalent bonds leave at least a slight electrical imbalance among the participating atoms. When two atoms of the same element are involved, as is the case with hydrogen gas, H_2, or oxygen gas, O_2, the electrons are shared absolutely equally. But consider water, whose chemical formula is H_2O. The oxygen atom has a substantially greater affinity for electrons than the two hydrogen atoms, and so the shared electrons are held closer to the oxygen than to the hydrogens. As a result, there is a net negative charge in the near vicinity of the oxygen, and a net positive charge around the hydrogen atoms. This unequal distribution of charge, together with the geometry of the covalent bonds, results in a molecule with a positive end and a negative end. Such a molecule is called *polar* because it has two separate centers, or poles, of charge. The distinction between polar and nonpolar molecules is a very useful one, and helps explain why water and oil do not mix (see chapter 13).

But back to the hydrogen bond. It results from the attraction between a covalently bonded hydrogen atom with some net positive charge, and a covalently bonded atom with a net negative charge: usually oxygen or nitrogen. The hydrogen bond is important because it is very common, because it brings different molecules into close association, and because it is weak enough that these molecular associations can change rapidly. Many of the chemical interactions in plant and animal cells occur via hydrogen bonds. And liquid water is one great collection or network of constantly shifting hydrogen bonds, a fact that makes water one of the more remarkable substances on earth. More on this in chapter 13.

ENERGY

We have spoken in passing of "strong" and "weak" bonds, and the ease or difficulty with which bonds are formed and broken. These are important considerations for us because most cooking is a matter of the systematic breaking of certain chemical bonds and the formation of others. The key concept by which we describe the behavior of chemical bonds is *energy*. Despite its omnipresence as a public issue today, energy becomes very elusive when we try to get at its essence. The word is a Greek compound of "in" and "force" or "activity," and now has as its standard definition "the capacity for doing work," or "the exertion of a force across a distance." The intricacies of thermodynamics are fascinating but distracting, so we will sidestep them and make this generalization: energy is that property of physical systems that makes possible most *change*. Objects in a universe without energy—already a contradiction in terms, since Einstein showed that mass is itself a form of energy—would be largely unchanging. Things in motion would never slow down or speed up, things at rest would never move. Conversely, the more energy available to an object, the more likely that object is to be changed, or to change its surroundings. Our kitchens are ordered around this principle: stoves and ovens change the qualities of food by subjecting it to heat energy, while the refrigerator preserves these qualities by removing heat and so slowing down the chemical changes that constitute spoilage.

Different Forms of Energy: Temperature
Atoms and molecules can absorb or release energy in several different forms. One, the nuclear energy that is released when a subatomic particle loses part of its mass, plays no role in cooking. A second, the electrical energy involved when protons and electrons interact, is a measure of bond strength, as we shall see in a moment. Three others are all different kinds of motion,

or "kinetic" energy, and are manifestations of *heat*. "Translational" energy moves a molecule from one place to another; "rotational" energy spins it in one place, and "vibrational" energy sets it vibrating. The common kitchen measurement we call temperature is an indicator of molecular motion: the higher the temperature, the faster the food molecules are moving, and the more likely they are to be changed when they collide with each other.

Bond Strengths

When two atoms form a bond with each other—and thereby become a molecule—by sharing electrons, they are attracted and brought together by an electrical force. That is, in the process of forming the bond, some of their electrical energy is transformed into energy of motion. And the stronger the electrical force, the more rapidly they accelerate toward each other. Notice what this means. The stronger the electrical force that brings the atoms together, and so the stronger the bond, the more energy is released—lost— from the system in the form of motion, or heat. Strong bonds, then, "contain" less energy than weak bonds. This is another way of saying that they are more stable, less susceptible to change, than weak bonds.

Bond strength is defined as the amount of energy released from the participating atoms when the bond is formed. This is the same as the amount of energy that is required to break that bond. So if the atoms in a molecule are moving with the same kinetic energy that they had released when the bonds were formed, then those bonds will begin to break apart. Covalent bonds are very strong, on the order of 10 times as strong as hydrogen bonds. It takes about 100 times the average kinetic energy possessed by molecules at room temperature to break covalent bonds, which means that they are broken very rarely at room temperature. Hydrogen bonds, however, are constantly being formed, broken, and re-formed under everyday conditions. This turnover, like all chemical activity, increases as the temperature rises.

The Nature of Fuels: Oxidation

The relative bond energies among three common elements—oxygen, carbon, and hydrogen—have determined the way in which nearly all life stores and liberates energy. Take the human exploitation of such fossil fuels as natural gas, oil, or gasoline. These are all carbon-hydrogen molecules containing anywhere from a few to hundreds of atoms. Now oxygen has a substantially greater affinity for electrons than either carbon or hydrogen, and so will form stronger bonds with these two elements than they will between themselves. This means that if oxygen is allowed to react with hydrocarbons, bond energy will be transformed into heat energy. This reaction is called *oxidation*, and results in the formation of oxygen-carbon and oxygen-hydrogen products. The chemical equation looks like this:

$$2 \ C_8H_{18} + 25 \ O_2 \rightarrow 16 \ CO_2 + 18 \ H_2O + energy$$

> (*This means that 2 molecules of octane plus 25 molecules of oxygen react to form 16 molecules of carbon dioxide and 18 of water, and produce energy in the process.*)

Fifty weak carbon-hydrogen and carbon-carbon bonds are broken to form 50 strong carbon-oxygen and hydrogen-oxygen bonds. A flame or spark supplies the initial energy needed to break a bond in the fuel. The first reaction with oxygen in the air then liberates enough heat to break several other fuel bonds, and the chain reaction is underway: the fuel continues to burn on its own, in the process releasing large amounts of energy.

Plants and animals take advantage of the same principles to supply their own cells with energy. The most efficient foods, like the best fuels, are those molecules with the most extractable energy, which is to say the most weak covalent bonds. Organisms either build up or ingest those carbon-hydrogen molecules we call sugars, starch, and fats. Fats are structurally quite similar to the hydrocarbons in gasoline, and in fact all fossil fuels derive from initially living matter. When an organism needs energy, it uses these food molecules as fuel. Because it does not need heat so much as it does the bonding energy to build up new molecules, the living organism "burns" its fuel chemically. A series of enzyme-controlled transfers of electrons (any reaction that removes an electron from a molecule is called oxidation) gradually breaks the fuel down until an eventual reaction with oxygen forms the usual waste products. When we inhale oxygen and exhale carbon dioxide and water vapor, then, we are truly keeping the home fires burning.

A few words about the reverse process. Animals cannot build up their fuel reserves from scratch. They must begin at least with sugar molecules, which can then be stored as glycogen ("animal starch") or fat. Plants, on the other hand, can synthesize sugars from carbon dioxide and water. What accounts for this vegetable virtuosity? Well, the amount of energy it takes to form fuel is as great as the energy stored in and released by the fuel when it is burned down to its final waste products. Normally, living cells cannot supply enough energy to break up the strong bonds in a molecule of water. The plant can, however, harness solar energy. Chlorophyll converts sunlight into chemical energy and makes possible *photosynthesis:* the construction of fuel molecules via light. All animals depend ultimately on photosynthesis for their energy. The food chain begins there.

THE PHASES OF MATTER

Generally, matter can exist in three different phases, or states, depending on the kinetic energy of its atoms or molecules: as a solid, a liquid, and a gas. The temperatures at which a material melts—changes from solid to liquid—and boils—changes from liquid to gas—are determined by the bonding forces among the atoms or molecules. The stronger the bonds, the more energy needed to change them, and so the higher the temperature at which they shift from one phase to another.

Solids

At low temperatures, atomic motion is minimal, and atoms or molecules can settle into closely packed, well-defined structures. In a crystalline solid, the particles are arranged in a regular, repeating array, while in amorphous solids they are randomly oriented. Salt, sugar, and ice are examples of crys-

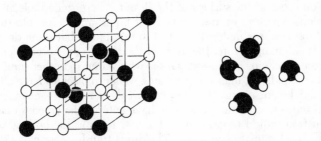

Crystalline solids, such as salt, are made up of atoms or molecules in highly ordered, regular arrays *(left)*. Amorphous solids, such as glass, are masses of atoms or molecules that have been fixed in a random arrangement *(right)*.

talline solids, and glass is amorphous. Large, irregular molecules like proteins and starch often form both crystalline and amorphous regions in the same chunk of material. Ionic bonds, hydrogen bonds, and other, "weak" forces may be involved in holding the particles of a solid together.

Liquids and Gases

At a temperature that is characteristic of each substance, the motion of individual molecules becomes great enough that the intermolecular forces are frequently overpowered. The fixed structure then breaks up, leaving the molecules free to change position to some extent. However, most of them are still close enough to others that they remain influenced, though not immo-

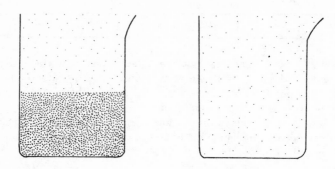

Left: the fluid and cohesive phase of matter that we call a liquid, together with a few evaporated molecules. *Right:* the fluid but dispersed phase that we call a gas.

bilized, by these forces and so remain in loose association with each other. This cohesive but fluid phase is, of course, a liquid. The exceptions to this description occur at the surface of the liquid, where some molecules reach speeds high enough to escape the body of liquid and move into the atmosphere. Most of these molecules return to the liquid, but some do not, and the higher the temperature, the more of them that are energetic enough to escape completely, or *evaporate*. This is why water and other liquids evaporate more rapidly at high temperatures. When molecules have attained a high enough kinetic energy that they can break away from each other's influence and move freely into space, they have entered the gaseous phase. The transition from liquid to gas is most familiar to us as the boiling of water, one of the basic culinary operations. The direct transformation of a solid into a gas is called "sublimation," and is the cause of that deterioration in foods known as "freezer burn." Freeze-drying is a more useful outcome of the same process.

Decomposition

Various substances, among them several complex organic compounds that figure in food preparation, do not change from one phase to another when heated, but instead decompose into entirely different kinds of molecules. For example, sugar will melt, but rather than vaporize, it breaks down into more than 100 other compounds in the process called caramelization. Starch, which is a long chain of sugar molecules joined together, will not even melt. Neither will proteins.

Mixtures

Finally, brief mention and definition of several simple kinds of mixtures that are important in the kitchen; more detail will be provided for particular instances. A *solution* is a system in which individual ions or molecules are dispersed in a liquid. Salt and sugar dissolved in water are the most common culinary examples. When the added material is dispersed in groups of several molecules, the mixture is called a *colloid*, and is often slightly cloudy and more viscous than either the pure liquid or a solution. Milk, cream, and mayonnaise are progressively thicker colloids, all made up of small oil droplets dispersed in water. Flour-thickened sauces are colloids in which long starch molecules are dispersed in water, while tapioca is a solid colloid, a *gel*, that consists of water molecules dispersed in a network of starch.

Bibliography

The historical and scientific literatures on food are vast, and the following list of sources is by no means comprehensive. It is intended only to document the more important facts and ideas presented in this book, to give credit to translators and researchers, and to give interested readers some leads for more detailed information. The bibliography for each chapter is divided into two sections: first, works of a general nature or meant for a general audience; and second, more specialized and technical studies.

A BASIC LIBRARY

The following encyclopedias, dictionaries, and armful of single volumes have been valuable sources of information for every chapter in this book.

Original Sources

These books contain firsthand accounts of past culinary practices.

Apicius, M. G. *De re coquinaria: L'art culinaire*. Edited by J. André. Paris: C. Klincksieck, 1965. In English: *The Roman Cookery Book*. Translated by B. Flower and E. Rosenbaum. London: Harrap, 1958. Apicius was a gourmand of the 1st century A.D.; the text attributed to him probably dates from a century or two later.

Cato, M. P. *On Agriculture*. Translated by W. D. Hooper. Cambridge, Mass.: Harvard Univ. Press, 1934. Written in the 2nd century B.C. by a Roman statesman.

Columella, L. J. M. *On Agriculture,* 3 vols. Translated by H. B. Ash. Cambridge, Mass.: Harvard Univ. Press, 1941–55. Dates from the 1st century A.D.

Pliny the Elder. *Natural History*. 10 vols. Translated by H. Rackham and others. Cambridge, Mass: Harvard Univ. Press, 1938–62. This encyclopedic treatise was compiled by a Roman naturalist who died in A.D. 79 while observing the eruption of Mt. Vesuvius.

Brereton, G. E., and J. M. Ferrier, eds. *Le ménagier de Paris*. Oxford: Clarendon, 1981. An unfinished handbook on managing a household, compiled around 1390 by its anonymous author for his young bride.

Sass, L. *To the King's Taste*. New York: Metropolitan Museum of Art, 1975. Medieval English recipes.
Tirel, G. [Taillevent]. *Le viandier*. Edited by J. Pichon and J. Vicaire. Paris: Techener, 1892. A book of recipes put together around 1390 and attributed to the cook of France's Charles V.
Warner, R., ed. *Forme of Cury*. In *Antiquitates Culinariae*. London: 1791. An anonymous collection of recipes from the English court of Richard II, roughly contemporaneous with *Le viandier*.

Brillat-Savarin, J. A. *The Physiology of Taste*. Translated by M. F. K. Fisher. New York: Harcourt Brace Jovanovich, 1978. Originally published in Paris in 1825.
Glasse, H. *The Art of Cooking Made Plain and Easy*. 4th ed. London: 1751. Perhaps the first best-selling cookbook.

Histories of Diet, Food, and Cooking

Braudel, F. *The Structures of Everyday Life: The Limits of the Possible*. Translated by S. Reynolds. New York: Harper and Row, 1981.
Gottschalk, A. *Histoire de l'alimentation et de la gastronomie*. 2 vols. Paris: Hippocrate, 1948.
Heiser, C. B. *Seed to Civilization: The Story of Food*. 2d ed. San Francisco: W. H. Freeman, 1981.
Tannahill, R. *Food in History*. New York: Stein and Day, 1973.

Chang, K. C., ed. *Food in Chinese Culture*. New Haven: Yale Univ. Press, 1977.
Darby, W. J., P. Ghalioungiu, and L. Grivetti. *Food: The Gift of Osiris*. 2 vols. New York: Academic Press, 1977. Food in ancient Egypt.
Drummond, J. C., and A. Wilbraham. *The Englishman's Food*. London: J. Cape, 1959.
Root, W., and R. de Rochemont. *Eating in America*. New York: Morrow, 1976.
Wheaton, B. K. *Savoring the Past*. Philadelphia: Univ. of Penn. Press, 1983. French culinary history, 1300–1789.

Histories of Science and Technology

Gillespie, C. C., ed. *Dictionary of Scientific Biography*. 16 vols. New York: Scribners, 1970–80.
Singer, C., E. J. Holmyard, A. R. Hall, and T. I. Williams, eds. *A History of Technology*. 7 vols. Oxford: Clarendon, 1954–78.

Histories of Words

Klein, E. *A Comprehensive Etymological Dictionary of the English Language*. 2 vols., 1966–67. Reprint. Amsterdam and New York: Elsevier, 1971. Unlike the *OED*, this dictionary does not give historical citations, but it does draw on more recent studies.
The Oxford English Dictionary. 12 vols., 1933. Reprint. Oxford: Clarendon, 1961, with later supplements.

Food Science

Charley, H. *Food Science*. 2d ed. New York: J. Wiley, 1982. Comprehensive and up-to-date.
Griswold, R. M. *The Experimental Study of Foods*. Boston: Houghton Mifflin, 1962.
Paul, P. C., and H. H. Palmer, eds. *Food Theory and Applications*. New York: J. Wiley, 1972. Older than Charley but often more detailed.
Vaill, G. E., J. A. Phillips, L. O. Rust, R. M. Griswold, and M. M. Justin. *Foods*. 7th ed. Boston: Houghton Mifflin, 1978.

General Works on Science and Technology

These encyclopedias include entries on utensil materials, methods of heat transfer, and on such industrially manufactured foods as chocolate and confectionery, fermented foods, natural and artificial sweeteners.
Kirk-Othmer Encyclopedia of Chemical Technology. 2d and 3d eds. New York: J. Wiley, 1963–71 and 1978.
McGraw-Hill Encyclopedia of Science and Technology. 15 vols. New York: McGraw-Hill, 1977.

INTRODUCTION

Plato. *Gorgias*. Translated by W. D. Woodhead. In E. Hamilton and H. Cairns, eds. *The Collected Dialogues of Plato*. Princeton: Princeton Univ. Press, 1961.
Spedding, J. *An Account of the Life and Times of Francis Bacon*. 2 vols. London: Trübner, 1878.

CHAPTER 1   MILK AND DAIRY PRODUCTS

Androuët, P. *The Complete Encyclopedia of French Cheese*. Translated by J. Githens. New York: Harper and Row, 1973.
Aristotle. *Generation of Animals*. Translated by A. L. Peck. Cambridge, Mass.: Harvard Univ. Press, 1953.
Dickson, P. *The Great American Ice Cream Book*. New York: Atheneum, 1973.
Fussell, G. E. *The English Dairy Farmer, 1500–1900*. London: F. Cass, 1966.
Ginzburg, C. *The Cheese and the Worms*. Translated by J. Tedeschi and A. Tedeschi. Baltimore: Johns Hopkins Univ. Press, 1980.
Grant, A. J., trans. *Early Lives of Charlemagne*. London: Chatto and Windus, 1922.
Herodotus. *History*. Translated by J. E. Powell. Oxford: Clarendon, 1949.
Polo, Marco. *Travels* (ca. 1300). Translated by W. Marsden. New York: Dutton, 1908.
Riepma, S. F. *The Story of Margarine*. Washington, D.C.: Public Affairs Press, 1970.

Arbuckle, W. S. *Ice Cream*. 2d ed. Westport, Conn: AVI, 1972.
Atherton, H. V., and J. A. Newlander. *Chemistry and Testing of Dairy Products*. 4th ed. Westport, Conn: AVI, 1977.
Davis, J. G. *Cheese*. 2 vols. London: J. and A. Churchill, 1965.

Fomon, S. J., ed. *Infant Nutrition*. 2d ed. Philadelphia: Saunders, 1974.

Kalab, M. Microstructure of dairy food. *Journal of Dairy Science* 62 (1979): 1352–64.

———. Scanning electron microscopy of dairy products: an overview. *Scanning Electron Microscopy* 3 (1979): 261–72.

———, and D. B. Emmons. Milk gel structure. *Milchwissenschaft* 33 (1978): 670–73.

Kretchmer, N. Lactose and lactase. *Scientific American*, October 1972, 71–78.

Patton, S. Milk. *Scientific American*, July 1969, 59–68.

Quinn, T. *Dairy Farm Management*. New York: Van Nostrand Reinhold, 1980.

Simoons, F. J. Traditional use and avoidance of foods of animal origin. *Bioscience* 28 (1978): 178–84.

Stephenson, L. S., and M. C. Latham. Lactose intolerance and milk consumption. *American Journal of Clinical Nutrition* 27 (1974): 296–303.

Webb, B. H., and A. H. Johnson, eds. *Fundamentals of Dairy Chemistry*. Westport, Conn: AVI, 1965.

CHAPTER 2   EGGS

American Egg Board. *Eggcyclopedia*. Park Ridge, Ill.: 1981.

Encyclopédie, ou dictionnaire raisonné des sciences, des arts, et des métiers. Paris, 1751–77. Reprint. 35 vols. Stuttgart-Bad Cannstatt: Frommann, 1966.

Smith, P., and C. Daniel. *The Chicken Book*. Boston: Little, Brown, 1975.

Alexander, A. E., and P. Johnson. *Colloid Science*. 2 vols. Oxford: Clarendon, 1949.

Carter, G. S. *Structure and Habit in Vertebrate Evolution*. Seattle: Univ. of Washington Press, 1967.

Chang, C. M., W. D. Powrie, and O. Fennema. Microstructure of egg yolk. *Journal of Food Science* 42 (1977): 1193–1200.

Feeney, R. E., and R. G. Allison. *Evolutionary Biochemistry of Proteins*. New York: J. Wiley, 1969.

Graham, D. E., and M. C. Phillips. The conformation of proteins at the air-water interface and their role in stabilizing foams. In R. J. Akers, ed. *Foams*, 237–55. London and New York: Academic Press, 1976.

Heath, J. L. Chemical and related osmotic changes in egg albumen during storage. *Poultry Science* 56 (1977): 822–28.

McGee, H. J., S. R. Long, and W. R. Briggs. Why whip egg whites in copper bowls? *Nature* 308 (1984): 667–68.

Romanoff, A. L., and A. J. Romanoff. *The Avian Egg*. New York: J. Wiley, 1949.

Stadelman, W. J., and O. J. Cotterill, eds. *Egg Science and Technology*. 2d ed. Westport, Conn.: AVI, 1977.

Taylor, T. G. How an eggshell is made. *Scientific American*, March 1970, 89–94.

Walker, J. Physics and chemistry of the lemon meringue pie. *Scientific American*, June 1981, 194–200.

Wilson, W. Q. Poultry production. *Scientific American*, July 1966, 56–64.

CHAPTER 3   MEAT

Aristotle. *Meteorologica*. Translated by H. D. P. Lee. Cambridge, Mass.: Harvard Univ. Press, 1952.

Digby, Kenelm. *The Closet of Sir Kenelm Digby, Knight, Opened* (1669). Edited by A. Macdonnell. London: P. L. Warner, 1910.

Harris, M. *Cannibals and Kings*. New York: Random House, 1977.

Moulton, F. R., ed. *Liebig and After Liebig*. Washington, D.C.: American Association for the Advancement of Science, 1942.

Reed, C. A., ed. *Origins of Agriculture*. The Hague: Mouton, 1977.

Soler, J. The semiotics of food in the Bible. Translated by E. Forster. In R. Forster and O. Ranum, eds. *Food and Drink in History*, 126–38. Baltimore: Johns Hopkins Univ. Press, 1979.

Zeuner, F. E. *A History of Domesticated Animals*. London: Hutchinson, 1963.

Bone, Q. Locomotor muscle. In W. S. Hoar and D.J. Randall, eds. *Fish Physiology*. Vol. 7, 361–424. New York: Academic Press, 1978.

Briskey, E. J., R. G. Cassens, and B. B. Marsh, eds. *Physiology and Biochemistry of Muscle as a Food*. Vol. 2. Madison: Univ. of Wisconsin Press, 1970.

Fox, J. B. The chemistry of meat pigment. *Journal of Agricultural and Food Chemistry* 14 (1966): 207–10.

Herz, K. O., and S. S. Chang. Meat flavor. *Advances in Food Research* 18 (1970): 1–83.

Laakonen, E. Factors affecting tenderness during heating of meat. *Advances in Food Research* 20 (1973): 257–323.

Lawrie, R. A. *Meat Science*. 3d ed. Oxford: Pergamon, 1979.

Love, R. M. *The Chemical Biology of Fishes*. London and New York: Academic Press, 1970.

Lutz, H., H. Weber, R. Billeter, and E. Jenny. Fast and slow myosin within single skeletal muscles of adult rabbits. *Nature* 281 (1979): 142–44.

Paul, P. C., and H. H. Palmer, eds. *Food Theory and Applications*. New York: J. Wiley, 1972.

Pearson, A. M., J. D. Love, and F. B. Shorland. "Warmed-over" flavor in meat, poultry, and fish. *Advances in Food Research* 23 (1977): 1–74.

Rubenstein, N. A., and A. M. Kelly. Development of muscle fiber specialization in the rat hindlimb. *Journal of Cell Biology* 90 (1981): 128–44.

Urbain, W. M. Food irradiation. *Advances in Food Research* 24 (1978): 155–227.

Weeds, A. Myosin: polymorphism and promiscuity. *Nature* 274 (1978): 417–18.

CHAPTER 4 FRUITS AND VEGETABLES, HERBS AND SPICES

Leopold, A. C., and R. Ardrey. Toxic substances in plants and the food habits of early man. *Science* 176 (1972): 512–13.

Lincoln, Mrs. D. A. *Mrs. Lincoln's Boston Cook Book*. Boston: Roberts Bros., 1883.

Szent-Györgyi, A. V. *On Oxidation, Fermentation, Vitamins, Health, and Disease*. Baltimore: Williams and Wilkins, 1939.

Theophrastus. *De causis plantarum*. Edited by F. Wimmer. Leipzig: 1854. Latin translation.

Borgstrom, G. *Principles of Food Science*. 2 vols. New York: Macmillan, 1968.

Brown, M. S. Frozen fruits and vegetables. *Advances in Food Research* 25 (1979): 181–236.

Cronquist, A. *The Evolution and Classification of Flowering Plants.* Boston: Houghton Mifflin, 1968.

Desor, J. A., O. Maller, and L. S. Greene. Preference for sweet in humans. In J. M. Weiffenbach, ed. *Taste and Development*, 161–72. Bethesda, Md.: National Institutes of Health, 1977.

Dodson, E. O., and P. Dodson. *Evolution.* 2d ed. New York: Van Nostrand Reinhold, 1976.

Esau, K. *Anatomy of Seed Plants.* 2d ed. New York: J. Wiley, 1977.

Graham, A. *Floristics and Paleofloristics of Asia and Eastern North America.* Amsterdam and New York: Elsevier, 1972.

Haard, N. F., and D. K. Salunkhe, eds. *Symposium: Postharvest Biology and Handling of Fruits and Vegetables.* Westport, Conn.: AVI, 1975.

Hulme, A. C., ed. *The Biochemistry of Fruits and Their Products.* 2 vols. London and New York: Academic Press, 1970–71.

Iwata, T., and B. B. Stowe. Probing a membrane matrix regulating hormone action. The kinetics of lipid-induced growth and ethylene production. *Plant Physiology* 51 (1973): 691–701.

Jadhav, S. J., and D. K. Salunkhe. Formation and control of chlorophyll and glycoalkaloids in tubers of *Solanum tuberosum*.... *Advances in Food Research* 21 (1975): 307–54.

Liener, I., ed. *Toxic Constituents of Plant Foodstuffs.* 2d ed. New York: Academic Press, 1980.

Merory, J. *Food Flavorings: Composition, Manufacture, Use.* Westport, Conn.: AVI, 1968.

Pijl, L. van der. *Principles of Dispersal in Higher Plants.* 2d ed. Berlin and New York: Springer-Verlag, 1973.

Rosenthal, G. A., and D. H. Janzen, eds. *Herbivores: Their Interaction With Secondary Plant Metabolites.* New York: Academic Press, 1979.

Salunkhe, D. K. *Storage, Processing, and Nutritional Quality of Fruits and Vegetables.* Cleveland: CRC Press, 1974.

Sondheimer, E., and J. B. Simeone, eds. *Chemical Ecology.* New York: Academic Press, 1970.

Steiner, J. E. Facial expressions of the neonate infant indicating the hedonics of food-related chemical stimuli. In J. M. Weiffenbach, ed. *Taste and Development*, 173–89. Bethesda, Md.: National Institutes of Health, 1977.

Swain, T. Biochemical evolution of plants. *Comprehensive Biochemistry* 29A (1974): 125–302.

Toxicants Occurring Naturally in Foods. 2d ed. Washington, D.C.: National Academy of Sciences, 1973.

Notes on Common Fruits, Vegetables, Herbs, and Spices

Condit, I. J. *The Fig.* Waltham, Mass.: Chronica Botanica, 1947.

Harrison, S. G., B. E. Nicholson, G. B. Masefield, and M. Wallis. *The Oxford Book of Food Plants.* Oxford: Oxford Univ. Press, 1969.

McPhee, J. *Oranges.* New York: Farrar, Straus, and Giroux, 1967.

Robinson, E. F. *The Early History of Coffee Houses in England.* London: Kegan Paul, 1893.

Rosengarten, F. *The Book of Spices.* New York: Jove, 1973.

Schapira, J., D. Schapira, and K. Schapira. *The Book of Coffee and Tea.* New York: St. Martin's Press, 1975.

Theophrastus. *Enquiry Into Plants.* 2 vols. Translated by A. Hort. New York: Putnam, 1916.

Allison, A. C., and K. G. McWhirter. Two unifactorial characters for which man is polymorphic. *Nature* 178 (1956): 748–49. The effects of asparagus and beets on urine.

Bartoshuk, L. M., C. Lee, and R. Scarpellino. Sweet taste of water induced by artichoke *(Cynara scolymus). Science* 178 (1972): 988–90.

Blakeslee, A. F. A dinner demonstration of threshold differences in taste and smell. *Science* 81 (1935): 504–07.

Brouk, B. *Plants Consumed by Man.* London and New York: Academic Press, 1975.

Claus, R., H. Ø. Hoppen, and H. Karg. The secret of truffles: a steroidal hormone? *Experientia* 37 (1981): 1178–79.

Dodt, E., A. P. Skouby, and Y. Zotterman. Effect of cholinergic substances on the discharges from thermal receptors. *Acta Physiologica Scandinavica* 28 (1953): 101–14. Menthol.

Gilman, A. G., L. S. Goodman, and A. Gilman, eds. *Goodman and Gilman's The Pharmacological Basis of Therapeutics.* 6th ed. New York: Macmillan, 1980. Caffeine as a drug.

Hammond, J. B. W. Changes in the composition of harvested mushrooms. *Phytochemistry* 18 (1979): 415–18.

Harler, C. R. *The Cultivation and Marketing of Tea.* 3d ed. Oxford: Oxford Univ. Press, 1964.

Hensel, H., and Y. Zotterman. Effect of menthol on the thermoreceptors. *Acta Physiologica Scandinavica* 24 (1952): 27–34.

Lison, M., S. H. Blondheim, and R. N. Melmed. A polymorphism of the ability to smell urinary metabolites of asparagus. *British Medical Journal* 281 (1980): 1676–78.

Mangalakumari, C. K., V. P. Sreedharan, and A. G. Mathew. Studies on blackening of pepper *(Piper nigrum,* Linn.) during dehydration. *Journal of Food Science* 48 (1983): 604–06.

Rick, C. M. The tomato. *Scientific American,* August 1978, 76–87.

Rozin, P., and D. Schiller. The nature and acquisition of a preference for chili peppers by humans. *Motivation and Emotion* 4 (1980): 77–101.

Simmonds, N. W., ed. *The Evolution of Crop Plants.* London: Longman, 1976.

Todd, P. H., Jr., M. G. Bensinger, and T. Biftu. Determination of pungency due to capsicum by gas-liquid chromatography. *Journal of Food Science* 42 (1977): 660–65.

Wickremasinghe, R. L. Tea. *Advances in Food Research* 24 (1978): 229–87.

CHAPTER 5   GRAINS, LEGUMES, AND NUTS

Augustine. *The City of God Against the Pagans* (ca. A.D. 415). Vol. 4. Translated by P. Levine. Cambridge, Mass: Harvard Univ. Press, 1966.

Carson, G. *Cornflake Crusade.* New York: Rinehart, 1957.

Champlain, S. de. *The Voyages, 1619.* Vol. 3 of *The Works of Samuel de Champlain.* Translated by H. H. Langton and W. F. Ganong. Toronto: The Champlain Society, 1929.

Katz, S. H., M. L. Hediger, and L. A. Valleroy. Traditional maize processing techniques in the New World. *Science* 184 (1974): 765–73.

Leitch, N. L. *C. W. Post*. Washington, D.C.: privately printed, 1963.

Mangelsdorf, P. C. *Corn: Its Origin, Evolution, and Improvement*. Cambridge, Mass: Harvard Univ. Press, 1974.

Menninger, E. A. *Edible Nuts of the World*. Stuart, Fla.: Horticulture Books, 1977.

Navarrete, Fr. D. *The Travels and Controversies* (1676). 2 vols. Translated by A. Churchill and J. Churchill (1704); edited by J. S. Cummins. Cambridge, U.K.: Hakluyt Society, 1962.

Roe, D. A. *A Plague of Corn*. Ithaca, N.Y.: Cornell Univ. Press, 1973.

Schultz, B. *The Wild Ricer's Guide*. Berkeley, Ca.: Appleseed, 1979.

Thornton, H. J. *History of the Quaker Oats Company*. Chicago: Univ. of Chicago Press, 1933.

Zohary, D., and M. Hopf. Domestication of pulses in the Old World. *Science* 182 (1973): 887–94.

Bressani, R., and N. S. Scrimshaw. Effect of lime treatment on in vitro availability of essential amino acids . . . in corn. *Journal of Agricultural and Food Chemistry* 6 (1958): 774–78.

Calloway, D. H., C. A. Hickey, and E. L. Murphy. Reduction of intestinal gas-forming properties of legumes. . . . *Journal of Food Science* 36 (1971): 251–55.

Dure, L. S. Seed formation. *Annual Review of Plant Physiology* 26 (1975): 259–78.

Fordham, J. R., C. E. Wells, and L. H. Chen. Sprouting of seeds and nutritional composition of seeds and sprouts. *Journal of Food Science* 40 (1975): 552–56.

Gould, M. F., and R. N. Greenshields. Distribution and changes in galactose-containing oligosaccharides in ripening and germinating bean seeds. *Nature* 202 (1964): 108–09.

Grist, D. H. *Rice*. 5th ed. London: Longman, 1975.

Houston, D. F. *Rice: Chemistry and Technology*. St. Paul, Minn.: American Association of Cereal Chemists, 1972.

Kent, N. L. *Technology of Cereals*. 2d ed. Oxford: Pergamon, 1975.

Leonard, W. H., and J. H. Martin. *Cereal Crops*. New York: Macmillan, 1963.

Matz, S. A. *Cereal Technology*. Westport, Conn.: AVI, 1970.

Pfeiffer, C. J. Gastroenterologic aspects of manned space flight. *Annals of the New York Academy of Sciences* 150 (1968): 40–48.

Pusztai, A. Metabolism of trypsin-inhibiting proteins in germinating seeds of kidney bean. *Planta* 107 (1972): 121–29.

Robutti, J. L., R. C. Hoseney, and C. E. Wassom. Modified opaque-2 corn endosperms. *Cereal Chemistry* 51 (1974): 173–80.

Smith, A. K., and S. J. Circle, eds. *Soybeans: Chemistry and Technology*. Westport, Conn.: AVI, 1978.

U.N. Food and Agriculture Organization. *FAO Production Yearbook*. Vol. 34 (1980). Rome: 1981.

Wang, H. L., E. W. Swain, C. W. Hesseltine, and H. D. Heath. Hydration of whole soybeans. . . . *Journal of Food Science* 44 (1979): 1510–13.

Woodruff, J. G. *Tree Nuts: Production, Processing, Products*. 2 vols. Westport, Conn.: AVI, 1967.

CHAPTER 6   BREAD, DOUGHS, AND BATTERS

Arberry, A. J., trans. A Baghdad cookery-book. *Islamic Culture* 13 (1939): 21–47; 189–214.

Archestratus. *Gastronomia*. Translated into Italian by D. Scina. Venice: 1842.

Hippocrates. *Regimen*. Translated by W. H. S. Jones. Vol. 4 of *The Works of Hippocrates*. New York: Putnam, 1931.

Jacob, H. E. *Six Thousand Years of Bread*. Translated by R. Winston and C. Winston. Garden City, N.Y.: Doubleday, Doran, 1944.

Leslie, E. *Miss Leslie's New Cookery Book*. Philadelphia: 1857.

Sheppard, R, and E. Newton. *The Story of Bread*. London: Routledge and Kegan Paul, 1957.

Spicer, A., ed. *Bread: Social, Nutritional, and Agricultural Aspects of Wheaten Bread*. London: Applied Science Publishers, 1975.

Storck, J., and W. D. Teague. *Flour for Man's Bread*. Minneapolis: Univ. of Minnesota Press, 1952.

Bernardin, J. E., and D. D. Kasarda. The microstructure of wheat protein fibrils. *Cereal Chemistry* 50 (1973): 735–45.

Chan, W. S., and R. T. Toledo. Dynamics of freezing and their effects on water-holding capacity of a gelatinized starch gel. *Journal of Food Science* 41 (1976): 301–03.

Charley, H. Effects of baking pan material on heat penetration during baking. . . . *Food Research* 15 (1950): 155–67.

Dexter, J. E., B. L. Dronzek, and R. R. Matsuo. Scanning electron microscopy of cooked spaghetti. *Cereal Chemistry* 55 (1978): 23–30.

Grosskreutz, J. C. A lipoprotein model of wheat gluten structure. *Cereal Chemistry* 38 (1961): 336–49.

Hoerr, C. W. Morphology of fats, oils, and shortenings. *Journal of the American Oil Chemists' Society* 37 (1960): 539–46.

Hoseney, R. C., and P. A. Seib. Structural differences in hard and soft wheat. *Bakers Digest* 47 (1973): 26–28.

Kulp, K., P. M. Ranum, P. C. Williams, and W. T. Yamazaki. Natural levels of nutrients in commercially milled wheat flours. *Cereal Chemistry* 57 (1980): 54–58.

Matz, S. *Bakery Technology and Engineering*. 2d ed. Westport, Conn.: AVI, 1972.

Pomeranz, Y., ed. *Wheat: Chemistry and Technology*. 2d ed. St. Paul, Minn.: American Association of Cereal Chemists, 1971.

Sugihara, T. F., L. Kline, and M. W. Miller. Microorganisms of the San Francisco sour dough bread process. *Applied Microbiology* 21 (1971): 456–65.

CHAPTER 7   SAUCES

Bugialli, G. *The Fine Art of Italian Cooking*. New York: Times Books, 1977.

Child, J., L. Bertholle, and S. Beck. *Mastering the Art of French Cooking*. Vol. 1. New York: Knopf, 1961.

David, E. *French Provincial Cooking*. Harmondsworth, U. K.: Penguin, 1970.

Kitchiner, W. *The Cook's Oracle*. London, 1817.

Sokolov, R. *The Saucier's Apprentice*. New York: Knopf, 1976.

Bancroft, W. D. *Applied Colloidal Chemistry*. New York: McGraw-Hill, 1921.

Becher, P. *Emulsions: Theory and Practice*. New York: Reinhold, 1957.

Chang, C. M., W. D. Powrie, and O. Fennema. Electron microscopy of mayonnaise. *Canadian Institute of Food Science and Technology Journal* 5 (1972): 134–37.

Corn Industry Research Foundation. *Corn Starch*. Washington, D.C.: 1964.

Corran, J. W. Some observations on a typical food emulsion. In *Emulsion Technology: Theoretical and Applied*. 2d ed. Brooklyn: Chemical Publishing Co., 1946.

Eglandsal, B. Heat-induced gelling in solutions of ovalbumin. *Journal of Food Science* 45 (1980): 570–73.

Hegg, P., H. Martens, and B. Löfqvist. Effects of pH and neutral salts on the formation and quality of thermal aggregates of ovalbumin. *Journal of the Science of Food and Agriculture* 30 (1979): 981–93.

Miller, B. S., R. I. Derby, and H. B. Trimbo. A pictorial explanation for the increase in viscosity of a heated wheat starch-water suspension. *Cereal Chemistry* 50 (1973): 271–80.

Paul, P. C., and H. H. Palmer, eds. *Food Theory and Applications*. New York: J. Wiley, 1972.

Perram, C. M., C. Nicolau, and J. W. Perram. Interparticle forces in multiphase colloid systems: the resurrection of coagulated *sauce béarnaise*. *Nature* 270 (1977): 572–73.

Popiel, W. J. *Introduction to Colloid Science*. Hicksville, N.Y.: Exposition Press, 1978.

Radley, J. A. *Starch and Its Derivatives*. 4th ed. London: Chapman and Hall, 1968.

Small, D. M., and M. Bernstein. Doctor in the kitchen: experiments on sauce béarnaise. *New England Journal of Medicine* 300 (1979): 801–02.

Walker, J. The amateur scientist. *Scientific American*, January 1981, 168–69.

———. The physics and chemistry of a failed sauce béarnaise. *Scientific American*, December 1979, 178–90.

Whistler, R. L., and E. F. Paschall, eds. *Starch: Chemistry and Technology*. 2 vols. New York: Academic Press, 1965–67.

Yeadon, D. A., L. A. Goldblatt, and A. M. Altschul. Lecithin in oil-in-water emulsions. *Journal of the American Oil Chemists' Society* 35 (1958): 435–38.

CHAPTER 8   SUGARS, CHOCOLATE, AND CONFECTIONERY

Aykroyd, W. R. *The Story of Sugar*. Chicago: Quadrangle, 1967.

Benzoni, G. *History of the New World* (1565). Translated by W. H. Smyth. London: Hakluyt Society, 1857.

Deerr, N. *The History of Sugar*. 2 vols. London: Chapman and Hall, 1949.

Diaz del Castillo, B. *The True History of the Conquest of New Spain* (1568). Vol. 1. Translated by A. P. Maudslay. London: Hakluyt Society, 1908.

Gage, T. *The English-American His Travail by Sea and Land* (1648). Edited by J. E. S. Thompson. Norman, Okla: Univ. of Oklahoma Press, 1958.

Hendrickson, R. *The Great American Chewing-Gum Book*. Radnor, Pa: Chilton, 1976.

Hentzner, P. *A Journey Into England* (1612). Translated by R. Bentley. Strawberry Hill, U.K.: 1757.

Humboldt, A. von. *A Personal Narrative of Travels to the Equinoctial Regions of the New Continent, 1799–1804.* Vol. 4. Translated by H. M. Williams. London: 1819.

Kramer, S. N. *History Begins at Sumer.* Reprint. Philadelphia: Univ. of Penn. Press, 1981.

Lévi-Strauss, C. *From Honey to Ashes.* Translated by J. Weightman and D. Weightman. New York: Harper and Row, 1973.

Nearing, H., and S. Nearing. *The Maple Sugar Book.* New York: Schocken, 1970.

Ransome, H. *The Sacred Bee.* London: Allen and Unwin, 1937.

Rinzler, C. A. *The Book of Chocolate.* New York: St. Martin's Press, 1977.

Rush, B. *Letters.* Edited by L. H. Butterfield. Princeton: Princeton Univ. Press, 1951.

Bidwell, R. G. S. *Plant Physiology.* New York: Macmillan, 1974. Contains an explanation of sap run.

Crane, E., ed. *Honey: A Comprehensive Survey.* London: Heinemann, 1975.

The Hive and the Honey Bee. Hamilton, Ill.: Dadant and Sons, 1975.

Liebowitz, M. R., and D. F. Klein. Hysteroid dysphoria. *Psychiatric Clinics of North America* 2 (1979): 555–75.

Long, S. R. The elusive fudge factor. *Yale Scientific,* Winter 1979, 6–9.

Minifie, B. W. *Chocolate, Cocoa, and Confectionery: Science and Technology.* Westport, Conn.: AVI, 1970.

Pancoast, H. M., and W. R. Junk. *Handbook of Sugars.* 2d ed. Westport, Conn.: AVI, 1980.

Paul, P. C., and H. H. Palmer, eds. *Food Theory and Applications.* New York: J. Wiley, 1972.

Sauter, J. J. Maple. In *McGraw-Hill Yearbook of Science and Technology.* New York: McGraw-Hill, 1974.

CHAPTER 9   WINE, BEER, AND DISTILLED LIQUORS

Adams, L. D. *The Wines of America.* Boston: Houghton Mifflin, 1973.

Aristotle. *Meteorologica.* Translated by H. D. P. Lee. Cambridge, Mass.: Harvard Univ. Press, 1952.

————. *Problemata.* Translated by E. S. Forster. Oxford: Clarendon, 1927.

Arnold, J. P., and F. Penman. *History of the Brewing Industry and Brewing Science in America.* Chicago: 1933.

Athenaeus, *Deipnosophists.* Vol. 1. Translated by C. B. Gulick. New York: Putnam, 1927.

Benzoni, G. *History of the New World* (1565). Translated by W. H. Smyth. London: Hakluyt Society, 1857.

Brander, M. *The Original Scotch.* New York: C. N. Potter, 1975.

Carr, J. *The Second Oldest Profession.* Englewood Cliffs, N.J.: Prentice-Hall, 1972. A history of moonshining.

Conant, J. B., ed. Pasteur's study of fermentation. In *Harvard Case Studies in Experimental Science.* Cambridge, Mass.: Harvard Univ. Press, 1952.

Corran, H. S. *A History of Brewing*. Newton Abbott, U.K.: David and Charles, 1975.

Grossman, H. J. *Grossman's Guide to Wines, Beers, and Spirits*. 6th ed. Revised by H. Lembeck. New York: Scribners, 1977.

Jefferson, T. *The Writings of Thomas Jefferson*. Vol. 15. Edited by A. A. Lipscomb and A. E. Bergh. Washington, D.C.: Thomas Jefferson Memorial Association, 1903.

Johnson, H. *World Atlas of Wine*. New York: Simon and Schuster, 1978.

Loubère, L. A. *The Red and the White: A History of Wines in France and Italy in the Nineteenth Century*. Albany: State Univ. of New York Press, 1978.

McGuire, E. B. *Irish Whiskey*. New York: Barnes and Noble, 1973.

Multhauf, R. The significance of distillation in Renaissance medical chemistry. *Bulletin of the History of Medicine* 30 (1956): 329–46.

Plato. *Laws*. Translated by A. E. Taylor. In E. Hamilton and H. Cairns, eds. *The Collected Dialogues of Plato*. Princeton: Princeton Univ. Press, 1961.

Prakash, O. *Food and Drinks in Ancient India*. Delhi: Munshi Ram Manohar Lal, 1961.

Schelenz, H. *Zur Geschichte der Pharmazeutisch-chemischen Destilliergeräte*. Hildesheim: Olms, 1961.

Seward, D. *Monks and Wine*. London: Mitchell Beazley, 1979.

Simon, A. L. *The History of Champagne*. London: Octopus, 1972.

Varro, M. T. *On Agriculture*. Translated by W. D. Hooper. Cambridge, Mass.: Harvard Univ. Press, 1934. Bound with Cato, *On Agriculture*.

Watney, J. *Mother's Ruin: A History of Gin*. London: P. Owen, 1976.

Amerine, M. A., ed. *Wine Production Technology in the United States*. Washington, D.C.: American Chemical Society, 1981.

Amerine, M. A., and M. A. Joslyn. *Table Wines: The Technology of Their Production*. 2d ed. Berkeley and Los Angeles: Univ. of Calif. Press, 1970.

Amerine, M. A., and E. B. Roessler. *Wines: Their Sensory Evaluation*. San Francisco: W. H. Freeman, 1976.

Briggs, D. E. *Barley*. London: Chapman and Hall, 1978.

Gastineau, C. F., W. J. Darby, and T. B. Turner, eds. *Fermented Food Beverages in Nutrition*. New York: Academic Press, 1979.

Hough, J. S., D. E. Briggs, and R. Stevens. *Malting and Brewing Science*. London: Chapman and Hall, 1971.

MacGregor, A. W., and R. R. Matsuo. Starch degradation in endosperm of barley and wheat kernels during initial stages of germination. *Cereal Chemistry* 59 (1982): 210–16.

Majchrowicz, E., ed. *Biochemical Pharmacology of Ethanol*. New York: Plenum, 1975.

Majchrowicz, E., and E. P. Noble, eds. *Biochemistry and Pharmacology of Ethanol*. 2 vols. New York: Plenum, 1979.

Roe, D. *Alcohol and the Diet*. Westport, Conn.: AVI, 1979.

CHAPTER 10   FOOD ADDITIVES

Two reliable guides to food additives:

Jacobson, Michael F. *Eater's Digest: The Consumer's Factbook of Food Additives*. Garden City, N.Y.: Doubleday, 1972.

Winter, Ruth. *A Consumer's Dictionary of Food Additives*. New York: Crown, 1978.

Beaconsfield, P., N. Borlaug, J. Huxley, H. Krebs, and R. Peters. The chemical age. *Journal of the American Pharmaceutical Association* 17 (1977): 369–73.
Binkerd, E. F., and O. E. Kolari. The history and use of nitrate and nitrite in the curing of meat. *Food and Cosmetics Toxicology* 13 (1975): 655–61.
Kermode, G. O. Food additives. *Scientific American*, March 1972, 15–21.
Rousseau, J.-J. *Emile: Or, On Education* (1762). Translated by Allan Bloom. New York: Basic, 1979.
Smith, R. J. Latest saccharin test kills FDA proposal. *Science* 208 (1980): 154–56.
Stumpf, S. E. Culture, value, and food safety. *Bioscience* 28 (1978): 186–90.
Sullivan, M. *America Finding Herself*. Vol. 2 of *Our Times: The United States, 1900–1925*. New York: Scribners, 1927.
Tannenbaum, S. R. Ins and outs of nitrites. *The Sciences*, January 1980, 7–9.
Ziegler, P. T. *The Meat We Eat*. Danville, Ill.: Interstate, 1944.

Ames, B. N. Dietary carcinogens and anticarcinogens. *Science* 221 (1983): 1256–64.
MacDonald, B., J. I. Gray, and N. L. Gibbins. Role of nitrite in cured meat flavor. *Journal of Food Science* 45 (1980): 893–97.
MacDougall, D. B., D. S. Mottram, and D. N. Rhodes. Contribution of nitrite and nitrate to the colour and flavour of cured meats. *Journal of the Science of Food and Agriculture* 26 (1975): 1743–54.
Potter, N. N. *Food Science*. Westport, Conn.: AVI, 1966.
Rayman, M. K., B. Aris, and A. Hurst. Nisin: a possible alternative or adjunct to nitrite in the preservation of meats. *Applied and Environmental Microbiology* 41 (1981): 375–80.
Roberts, T. A. The microbiological role of nitrate and nitrite. *Journal of the Science of Food and Agriculture* 26 (1975): 1755–60.
Smith, L. D. *Botulism*. Springfield, Ill.: C. C. Thomas, 1977.
Tompkin, R. B., L. N. Christiansen, and A. B. Schaparis. Enhancing nitrite inhibition of *Clostridium botulinum*. . . . *Applied and Environmental Microbiology* 35 (1978): 59–61.
Yarbrough, J. M., J. B. Rake, and R. G. Eagon. Bacterial inhibitory effects of nitrite. . . . *Applied and Environmental Microbiology* 39 (1980): 831–34.

CHAPTER 11   NUTRITION: AMERICAN FADS, INTRICATE FACTS

Two excellent books on nutrition:
Brody, Jane. *Jane Brody's Nutrition Book*. New York: Norton, 1981.
Mayer, Jean. *A Diet For Living*. New York: D. McKay, 1975.

A.M.A. Council on Foods and Nutrition. Zen macrobiotic diets. *Journal of the American Medical Association* 218 (1971): 397.
Bryce, A. *Modern Theories of Diet*. New York: Longman's, Green, 1912.

Dubos, R. Nutritional ambiguities. *Natural History,* July 1980, 14–21.
Gillie, R. B. Endemic goiter. *Scientific American,* June 1971, 92–101.
McClure, F. J. *Water Fluoridation.* Bethesda, Md.: U.S. National Institute of Dental Research, 1970.
Tocqueville, A. de. *Democracy in America.* Translated by H. Reeve, F. Bowen, and P. Bradley. New York: Knopf, 1945.

Burkitt, D. P., A. R. P. Walker, and N. S. Painter. Dietary fiber and disease. *Journal of the American Medical Association* 229 (1974): 1068–74.
Goodheart, R. S., and M. E. Shils, eds. *Modern Nutrition in Health and Disease.* 6th ed. Philadelphia: Lea and Febiger, 1980.
Nutrition Reviews. *Present Knowledge in Nutrition.* 4th ed. New York: Nutrition Foundation, 1976.
Robinson, C. H., and E. S. Weigley. *Fundamentals of Normal Nutrition.* 3d ed. New York: Macmillan, 1978.
Symposium on the role of dietary fiber in health. *American Journal of Clinical Nutrition* 31 (1978): S1–S291.

CHAPTER 12   DIGESTION AND SENSATION

Burton, R. *The Language of Smell.* London: Routledge and Kegan Paul, 1976.
Fletcher, H. *The New Glutton, or Epicure.* New York: Stokes, 1903.
Rozin, E., and P. Rozin. Culinary themes and variations. *Natural History,* February 1981, 6–9.

Cabanac, M. The physiological role of pleasure. *Science* 173 (1971): 1103–07.
Denton, D. A., and J. P. Coghlan, eds. *Olfaction and Taste.* Vol. 5. New York: Academic Press, 1975.
Goldstein, L., ed. *Introduction to Comparative Physiology.* New York: Holt, Rinehart and Winston, 1977.
Kare, M. R., M. J. Fregly, and R. A. Bernard, eds. *Biological and Behavioral Aspects of Salt Intake.* New York: Academic Press, 1980.
Kare, M. R., and O. Maller, eds. *The Chemical Senses and Nutrition.* New York: Academic Press, 1977.
Landau, B. R. *Essential Human Anatomy and Physiology.* 2d ed. Glenview, Ill.: Scott, Foresman, 1980.
Left- and right-handed odors. *Scientific American,* August 1971, 46–47.
Moncrieff, R. W. *The Chemical Senses.* 3d ed. London: L. Hill, 1967.
Nicolaidis, S. Early systemic responses to orogastric stimulation. *Annals of the New York Academy of Science* 157 (1969): 1176–1200.
Romer, A. S. *The Vertebrate Body.* 4th ed. Philadelphia: Saunders, 1970.
Rozin, P. Psychobiological and cultural determinants of food choice. In T. Silverstone, ed. *Appetite and Food Intake,* 285–311. Berlin: Abakon, 1976.
———. The use of characteristic flavorings in human culinary practice. In C. M. Apt, ed. *Flavor: Its Chemical, Behavioral, and Commercial Aspects,* 101–27. Boulder, Colo.: Westview Press, 1977.
Rozin, P., and A. E. Fallon. The acquisition of likes and dislikes for foods. In J. Solms and R. Hall, eds. *Criteria of Food Acceptance.* Zurich: Forster, 1980.

Schaumburg, H. H., R. Byck, R. Gerstl, and J. H. Mashman. Monosodium L-glutamate: its pharmacology and role in the Chinese restaurant syndrome. *Science* 163 (1969): 826–28.

Wilkins, L., and C. P. Richter. A great craving for salt by a child with corticoadrenal insufficiency. *Journal of the American Medical Association* 114 (1940): 866–68.

CHAPTER 13   THE FOUR BASIC FOOD MOLECULES

Davis, K. S. *Water, the Mirror of Science.* Garden City, N.Y.: Doubleday, 1961.

Dickerson, R. E., and I. Geis. *Chemistry, Matter, and the Universe.* Menlo Park, Cal.: Benjamin-Cummings, 1976.

Wilson, E. O., T. Eisner, W. R. Briggs, R. E. Dickerson, R. L. Metzenberg, R. D. O'Brien, M. Susman, and W. E. Boggs. *Life on Earth.* 2d ed. Sunderland, Mass.: Sinauer, 1978.

CHAPTER 14   COOKING METHODS AND UTENSIL MATERIALS

Berk, Z. *Braverman's Introduction to the Biochemistry of Foods.* Amsterdam and New York: Elsevier, 1976.

Goldblith, S. A. Basic principles of microwaves and recent developments. *Advances in Food Research* 15 (1966): 271–301.

Scientific American issue on "Materials," September 1967. See especially articles by A. H. Cottrell on metals, J. J. Gilman on ceramics, R. J. Charles on glasses, and J. Ziman on thermal properties.

Index

Principal references to topics are in **boldface**; references to illustrations are in italics. Foods are generally indexed under the noun rather than the modifier (e.g., "Broad beans" under "Beans, broad"; "Whipped cream" under "Cream, whipped").